Coastal Karst Landforms

Coastal Research Library

VOLUME 5

Series Editor:

Charles W. Finkl
Department of Geosciences
Florida Atlantic University
Boca Raton, FL 33431
USA

The aim of this book series is to disseminate information to the coastal research community. The Series covers all aspects of coastal research including but not limited to relevant aspects of geological sciences, biology (incl. ecology and coastal marine ecosystems), geomorphology (physical geography), climate, littoral oceanography, coastal hydraulics, environmental (resource) management, engineering, and remote sensing. Policy, coastal law, and relevant issues such as conflict resolution and risk management would also be covered by the Series. The scope of the Series is broad and with a unique crossdisciplinary nature. The Series would tend to focus on topics that are of current interest and which carry some import as opposed to traditional titles that are esoteric and non-controversial. Monographs as well as contributed volumes are welcomed.

For further volumes:
http://www.springer.com/series/8795

Michael J. Lace • John E. Mylroie
Editors

Coastal Karst Landforms

Springer

Editors
Michael J. Lace
Coastal Cave Survey
West Branch, IA, USA

John E. Mylroie
Department of Geosciences
Mississippi State University
Starkville, MS, USA

ISSN 2211-0577 ISSN 2211-0585 (electronic)
ISBN 978-94-007-5015-9 ISBN 978-94-007-5016-6 (eBook)
DOI 10.1007/978-94-007-5016-6
Springer Dordrecht Heidelberg New York London

Library of Congress Control Number: 2013939786

Printed on acid-free paper

Springer is part of Springer Science+Business Media (www.springer.com)

Foreword

Bermuda, Bahamas, Barbados .... many of the places described in this book might appear to be inspired by travel brochures. But here they serve as the prime testing grounds for one of the most vibrant and far-reaching fields in geomorphology: the study of coastal karst and caves.

Seacoasts are among Earth's most dynamic places. While sedimentary deposits are constructed along some of them, erosion and dissolution tend to dominate elsewhere. These opposing processes result in a variety of unique caves, karst, and solutional pores. In water-soluble rocks, coastal caves and karst include a warren of pockets, chambers, channels, and depressions. Water seeping from the surface has decorated many of them with crystalline deposits. Where mechanical erosion dominates, sea caves can be produced in any kind of rock.

These features are significant not only for revealing geomorphic processes, but also for interpretation of past climates, sea levels, and geography. For economic geologists they provide important clues to paleoenvironments and the distribution of porosity. Biologists value them as refugia for rare plants and animals. Archeologists and paleontologists recognize their ability to shelter and protect artifacts and organic remains. Some of these coastal features have become well-known tourist sites. The majority of seacoast caves and karst, as well as the rocks that contain them, are relatively young in geologic terms – typically less than a million years old. Thus they contain records of the intense fluctuations in climate and sea level that took place during that time.

It was not until recent decades that coastal caves and karst received substantial attention from the scientific community. They had previously been the domain of scuba divers who entered them mainly for sport, as well as a few geochemists who were intrigued by the interaction of fresh water and seawater. Eventually, in their quest for new frontiers, several karst scientists were drawn from their original domains of continental karst to the attractive islands of the Caribbean and other low-latitude regions. There they found wide-open and welcoming field areas that led in many new directions of exploration and science. Since then, they have expanded their research throughout the world, and they have developed new terminology to encompass their findings. Terms such as *eogenetic karst*, *flank margin cave*, and *carbonate island karst model* are now widespread in the technical literature.

The authors of this book include many of the pioneers in the field, as well as their students and associates. Their goal is mainly geomorphic and speleogenetic, although relevant aspects of biology, archeology, and management are also included. Today many researchers utilize coastal caves for various other purposes, most notably the analysis of cave deposits for interpretation of past and present climates. Much fine science has been done in these derivative fields, but there is often a tendency for them to bypass the geomorphic fundamentals – as though building a tower with no foundation. This book provides that foundation.

Professor of Hydrogeology, Emeritus
State University of New York, Oneonta, NY

Arthur N. Palmer

Preface

Coastlines may be the most abundant landform unit on the planet earth. They come in a variety of types, shapes, origins, and function. Where the coastlines are rocky, an abundance of sea caves can develop. When those rocks are soluble, a unique set of karst features can appear, both on and within the rock. Even where the coastline is a beach, if soluble rocks are present immediately inland, coastal processes will create distinctive caves and karst not found in the interiors of continents. The literature on coastal caves, until recently, either dealt with sea (or littoral) caves, or for carbonate coasts, applied models that were developed on old limestones deep within continental platforms. The subject is further confused as dynamic coastal processes constantly modify all cave types present, obscuring cave origin. Caves have long been known as repositories of information on past ecologies, climate, hydrology, and human history. Coastlines, as the intersection point between the marine and terrestrial realms, have a wealth of information to provide if only it could be preserved. Coastal caves, especially those produced by dissolution, can create a long-term preservational environment that can enlighten us about the past. In the present, the caves harbor unique ecologies, present complex land use problems, help control fresh-water resources, and provide underground landscapes of incredible beauty and fragility.

This book is designed as a gateway to the concepts and examples of coastal karst and pseudokarst development. Part I deals with our understanding of the modern fundamental concepts that control cave development in coastal areas. Part II presents a variety of case histories from around the world, from simple islands to complicated continental coasts. The focus is on dissolutional caves, and given the paucity of evaporate rocks in coastal areas, the dissolutional caves are formed mostly in carbonate rock. The majority of carbonate coasts in the world today are in the tropics and subtropics, and consist of young carbonates that are eogenetic, that is, have not yet undergone burial and associated diagenesis. Coupled with the unique hydrology and geochemistry of these coastal environments, this youthful age creates caves and karst of a unique form and character. In areas where the rocks are older and diagenetically mature, such as the 6,000 km of carbonate coast found in Croatia, interesting parallels with younger carbonate coasts can be drawn. A chapter on the sea caves of the west coast of the USA offers an interesting comparison to carbonate coasts.

We hope that this book stimulates further interest and research in these interrelated fields of the coastal sciences. Much of the work presented in this volume is the result of extensive fieldwork conducted by numerous karst researchers, graduate students and ongoing research projects from various disciplines and organizations. We extend our thanks to the many researchers and cave explorers who have and continue to tirelessly document cave and karst resources in so many parts of the world. Our collective understanding of coastal caves and karst continues to develop as does the appreciation for the integral roles these shoreline and paleoshoreline features play in modeling coastal processes, past, present and future. We extend our thanks to the many colleagues and friends who supported and contributed to the compilation of this work with invaluable insights tempered from a range of perspectives from many fields of research.

West Branch, IA and Mississippi State, MS

Michael J. Lace
and John E. Mylroie

Contents

Contributors

Robert Brinkmann Department of Geology, Environment, and Sustainability, Hofstra University, Hempstead, NY, USA

David E. Bunnell Angel's Camp, CA, USA

David A. Burney National Tropical Botanical Garden, Kalaheo, HI, USA

James G. Coke IV Kent, OH, USA

Beth Fratesi SDII Global Corporation, Tampa, FL, USA

Angel Ginés Departamento de Biologia, Universitat de les Illes Balears, Palma de Mallorca, Spain

Joaquín Ginés Departamento de Biologia, Universitat de les Illes Balears, Palma de Mallorca, Spain

Francesc Gràcia Departamento de Ciències de la Terra, Universitat de les Illes Balears, Palma de Mallorca, Spain

John J. Jenson Water and Environmental Research Institute of the Western Pacific, University of Guam, Mangilao, GU, USA

Patricia N. Kambesis Department of Geosciences, Mississippi State University, Mississippi State, MS, USA

Miklós Kázmér Department of Palaeontology, Eötvös University, Budapest, Hungary

Johanna L. Kovarik Department of Geology, University of South Florida, Tampa, FL, USA

Michael J. Lace Coastal Cave Survey, West Branch, IA, USA

Hans G. Machel Department of Earth and Atmospheric Sciences, University of Alberta, Edmonton, AB, Canada

Gregory J. Middleton Department of Environmental Studies, University of Tasmania, Sandy Bay, TAS, Australia

Joan R. Mylroie Department of Geosciences, Mississippi State University, Mississippi State, MS, USA

John E. Mylroie Department of Geosciences, Mississippi State University, Mississippi State, MS, USA

Athena M. Owen Department of Geosciences, Mississippi State University, Mississippi State, MS, USA

Jason S. Polk Department of Geography and Geology, Western Kentucky University, Bowling Green, KY, USA

Kevin W. Stafford Department of Geology, Stephen F. Austin State University, Nacogdoches, TX, USA

Danko Taboroši Water and Environmental Research Institute of the Western Pacific, University of Guam, Mangilao, Guam

Island Research & Education Initiative, Palikir, Pohnpei, FM, Micronesia

Abbreviations

AAR	amino acid racemization
ASL	above sea level
$CaCO_3$	Calcium Carbonate
CIKM	carbonate island karst model
DEM	digital elevation model
DOC	dissolved organic carbon
ENSO	El Nino Southern Oscillation
ka	thousands of years ago
LIDAR	light detection and ranging
Ma	million years ago
MIS	marine isotope state
MLLW	tidal datum referring to the "mean lower level water" height of sea level observed
MPA	marine protected area
MSL	mean sea level
NGO	non-governmental organization
RYBP	radiocarbon years before present
U/Th	Uranium Thorium disequilibrium dating
XRD	x-ray diffraction

Part I

Principles of Coastal Karst Development

Pseudokarst Caves in the Littoral Environment

1

John E. Mylroie and Joan R. Mylroie

Abstract

Rocky coastal regions can host caves produced by karst (dissolutional) processes, and caves produced by pseudokarst (non-dissolutional) processes. On limestone coasts, which are common world wide, both processes can be active and a complex interplay can result. Lava tubes, calcaerous tufa deposition, and reef growth all produce constructional caves, voids formed as the rock itself is formed. Only reef growth is an obligatory result of the marine coastal environment. Tafoni result from subaerial weathering of a variety of lithologies exposed on a cliff or steep slope, and can mimic other types of pseudokarst caves and karst caves. Talus and fissure caves result from failure of steep slopes and cliffs, themselves a result of coastal erosion which can quickly remove these pseudokarst cave types. Sea arches and sea caves are abundant on rocky coasts, as the interaction of wave dynamics and rock properties create a variety of erosional voids. Sea cave processes can overprint other cave types to produce a hybrid cave. Sea caves are likely the most common cave type in the world, but on limestone coasts, dissolutional mixing zone caves also form in great numbers, and are commonly overprinted to make abundant hybrid caves.

1.1 Introduction

Caves that are found in coastal cliffs and rocky shorelines today (Fig. 1.1) can have a variety of origins. These can be classified as two major categories, karst caves and pseudokarst caves. Karst consists of the landforms developing on, and the water flow system forming within, soluble rock material; pseudokarst is defined as karst-like features that form by processes other than dissolution (modified from Palmer 2007). Common karst-forming rocks are the carbonate rocks limestone, dolomite, and marble (Fig. 1.2). Less common, especially in coastal areas, are evaporite rocks such as gypsum and halite (Fig. 1.2), although they can be found in very arid coastlines. Evaporite rocks are highly soluble, and readily dissolve when in contact with seawater. These rocks are also mechanically weak,

J.E. Mylroie (✉) • J.R. Mylroie
Department of Geosciences, Mississippi State University, Mississippi State, MS 39762, USA
e-mail: mylroie@geoscience.msstate.edu

M.J. Lace and J.E. Mylroie (eds.), *Coastal Karst Landforms*, Coastal Research Library 5,
DOI 10.1007/978-94-007-5016-6_1, © Springer Science+Business Media Dordrecht 2013

Fig. 1.1 Breached cave chamber, in a coastal location, north shore Curaçao. The cave is developed in late Pleistocene limestone, a large chamber that is laterally open to the sea and waves surge out on to the limestone bench. The cave origin is thought to be dissolutional, with later marine invasion and modification to make it a hybrid cave

and do not endure the mechanical pounding that occurs in wave-swept coastal areas. Carbonate rocks, on the other hand, are both mechanically stronger, and chemically less soluble than evaporite rocks, and can structurally persist for long periods of time in the energy-rich coastal environment. Other rock types, such as quartzites, granites and sandstones, have been known to produce true karst under certain conditions, but karst caves developed by dissolution in these rocks in coastal zones are not yet recognized. Soluble rocks of sufficient mechanical strength can also form caves in coastal settings by non-dissolutional, or pseudokarst, processes.

Pseudokarst processes operating in coastal environments are an amalgamation of geological processes that characteristically operate in continental interiors, and processes unique to coastal settings. Because coastal landforms are the interface between the marine and terrestrial realm, there is a focus of these pseudokarst processes at that interface. Pseudokarst caves can form from a wide variety of weathering mechanisms. In coastal areas, these can be expressed as tafoni, talus caves, fissure caves, sea (littoral) caves, or caves shaped by a combination of these processes.

1.2 Constructional Caves

Other pseudokarst cave types, such as lava tubes (Fig. 1.3), can also occur in the coastal realm, but these caves are a product of rock deposition, as opposed to rock weathering. The presence of lava tubes in a coastal environment is a geologic accident, independent of coastal processes. However, since many islands around the world are volcanic in origin, lava tubes reaching the coast are not uncommon. Lava tubes are a pseudokarst cave type, but also fall into the category called *constructional caves*, in that they form as an outcome of the deposition of the rock that hosts them. Lave tubes are made as lava flows are forming, moving, and solidifying. The Lanzarote anchialine lava tube system in the Canary Islands is a classic example of a lava tube overprinted by marine processes (Wilkens et al. 2009).

Fig. 1.2 Cave development by dissolution in a variety of soluble rocks. (**a**) Devonian limestone cave exposed in a quarry in New York State. (**b**) Small caves in Miocene dolomitized chalk exposed in a sea cliff, Barbados. (**c**) Cave passage in marble in Glomdalen, northern Norway (note remnant winter snow on floor). (**d**) Cave passage in rock salt (halite), northeastern Spain

Fig. 1.3 Lava tubes in coastal settings in Hawaii. (**a**) Lava tube intersected by coastal retreat. (**b**) Lava tube in a coastal setting, entered by roof collapse (Photos by D. Bunnell)

Other examples of constructional caves exist; one example are large primary voids in reef limestones, created as the coral reef grows, branches, and is buried in its earlier stages. In ancient rocks that have been buried and undergone diagenetic maturation, such voids are usually infilled and closed, at least at the macroscopic scale. However, in very young limestones that have not undergone burial, such as uplifted Pleistocene coral reefs or submerged modern reefs, such large voids may allow human entry into actual caves.

Fig. 1.4 Constructional caves formed by carbonate tufa deposition. (**a**) Cartoon of how tufa cascades can enclose a void to make a cave (Adapted from Bögli 1980). (**b**) Tufa cascades forming tufa caves on a small sea cliff, Margaret River, Western Australia

Another example is where surface waters have excessive loads of dissolved $CaCO_3$; degassing of CO_2 may result in massive deposition of $CaCO_3$ to form calcareous tufa and travertine. These deposits, especially on steep slopes and cliffs (Fig. 1.4), may grow outward and then downward, so as to enclose open spaces to form tufa caves (Bögli 1980). The reef and tufa examples may be somewhat counter-intuitive, as the process of enclosing the void is done by manipulation of the chemistry of $CaCO_3$. But the process in these two cases is $CaCO_3$ deposition, not dissolution; the $CaCO_3$ entered the water as dissolved species at some location distant in time and space from the final depositional site. The caves were made by construction, at the depositional site, and not by *in situ* dissolution.

1.3 Hybrid Caves

Caves that are formed by karst processes, or by depositional processes such as lava tubes or tufa caves, may eventually end up in the coastal environment. In that case, the caves may be overprinted by coastal processes to form a cave type called a *hybrid cave* (Machel et al. 2012). A hybrid cave is a cave structure produced by one geologic process that has been overprinted by a second geologic process. The overprinting of a lava tube, or a karst cave, by wave action along a rocky coast is the most common example of cave hybridization. Figure 1.1 presents a possible hybrid cave; Fig. 1.4 shows tufa caves undergoing wave attack.

1.4 Sea Level

As noted earlier, the coast is the interface between the marine and terrestrial realms. This interface is not static over geologic time. The migration of this interface is called sea-level change. There are two main ways in which sea level changes. *Eustatic sea-level change* is a change that occurs essentially simultaneously, and by similar (but not necessarily exact) amount and rate, at all coasts of the world (Kelletat 1995). The major ways to cause a eustatic change is to change the amount of ocean water, change the volume of the ocean basin holding the water, or change the physical properties of the ocean water.

An example of changing the amount of water would be glaciation-deglaciation cycles, called *glacioeustasy*, where ice is stored on continents, resulting in sea level fall; when the earth warms, the ice melts and water returns to the sea, raising sea level (Fig. 1.5a). An example of changing

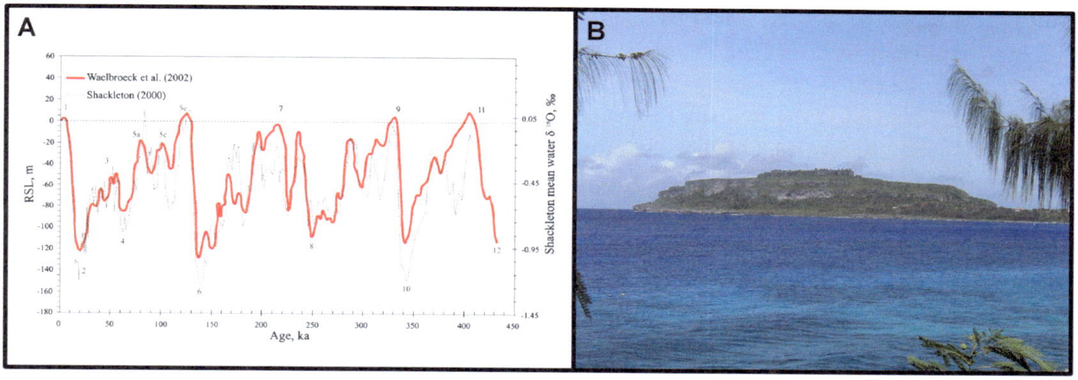

Fig. 1.5 Examples of sea level change. (**a**) The glacioeustatic sea-level curve for the last 450 ka, a summary of curves by Shackleton (2000) and Waelbroeck et al. (2002), amalgamated by Lascu (2005). Note how fast sea level change occurs, and how little time sea level has been spent at or near modern levels. (**b**) Tectonic uplift of Rota Island, Mariana Archipelago; the flat limestone terraces and steep cliffs represent periods of tectonic quiescence and uplift, respectively

the volume of the ocean basin would be sea-floor spreading rates, where rapid spreading leads to large, thermally inflated mid-oceanic ridges. This situation results in oceanic crust taking up more space on the ocean floor, and sea level rises. Slow sea-floor spreading results in colder, denser oceanic crust near the ridge axes, and a deeper and higher volume ocean basin, such that sea level falls. An example of changing the physical properties of seawater would be to change its temperature. For each degree Celsius rise in ocean temperature, sea level globally would rise a meter or two as a result of thermal expansion of the water.

The second sea-level change category is *local sea-level change*, a rise or fall of sea level restricted to a local area of coast. Such change is the result of the properties of the specific coastal area, as opposed to being an effect observed on coasts world wide. Tectonics is one of the more common means of local sea-level change, where the coast can be uplifted, or down warped, to give the appearance of a sea-level change (Fig. 1.5b). What has actually happened is sea level has remained constant and the land has moved vertically with respect to that constant datum. Coastal subsidence as the result of sediment compaction (as in the Mississippi River delta) is another way to produce local sea-level change on tectonically quiescent coasts. Isostasy, the vertical adjustment of the lithosphere to the addition or subtraction of a load, can also move tectonically quiescent coasts. Subsidence of the lithosphere due to ice loading, and lithosphere rebound following deglaciation, are significant factors in changing local sea level. For ice loading and removal, "local" can occur along thousands of kilometers of coast. Asthensosphere adjustment effects in the mantle can extend far away from the site of ice loading, called "far field effects", one of the reasons glacioeustasy effects are not exactly similar on all coasts world-wide. Persistent winds (*seiche*) or currents can also move sea level tens of centimeters, but usually for limited time periods of a few days. Even tide rise and fall represents a local sea-level change with a possible magnitude of meters and a duration of hours.

The causes of local and eustatic sea-level change are more varied and complex than the simple discussion presented here. The major sea-level changes that concern coastal caves are primarily glacioeustasy, acting globally, and tectonics and isostasy, acting locally. A couple of key points in this discussion: first, because of glacioeustasy, eustatic sea level has only been at its present position for about 3,000 years; second (Fig. 1.5a), ignoring seiche and tides, only tectonics at the local level consistently works faster than glacioeustasy (isostatic rebound from deglaciation is slower than glacioeustasy but still significant to coastal cave preservation by uplifting coastal caves away from wave attack; see Chap. 14). As a result, the many other mechanisms of sea-level change are mostly irrelevant to understanding coastal cave-forming processes today.

An additional point must also be emphasized: the coastlines we see today are the result of a short 3,000 year time window, in places modified by tectonics and isostasy. As illustrated in the following chapters, coastal caves and karst features evolve in conjunction with dynamic coastline evolution. Most contemporary coastlines are not static landforms; shore platforms, such as wave cut benches, are still advancing inland on rocky coasts, barrier islands are still migrating and modifying on depositional coasts (Pilkey 2003). Caves found in rocky coastal settings today have either been produced in the last 3,000 years by coastal processes, have persisted from an earlier sea-level position similar to modern, or are inherited from a previous non-coastal cave forming mechanism.

The migration of coastlines inland as sea level rises, and their migration seaward during sea-level's subsequent fall at a later time, mean that caves formed by past coastal processes may be found in inland settings today (i.e. paleoshorelines). Isostatic rebound following deglaciation has been especially efficient at removing caves formed at sea level to elevations above that datum. The understanding of cave-forming processes helps make caves an important indicator of past sea-level position. Lava tubes are poor indicators of past sea levels, whereas sea caves can serve as excellent indicators, as discussed in greater detail in the next chapter.

1.5 Minor Pseudokarst Cave Types

Tafoni are simple chambers and openings in the rock surface produced by weathering that causes rock grains to separate and fall away. They can form in almost any rock lithology (Fig. 1.6), and a wide variety of weathering mechanisms have been proposed for their development. They can develop on any steeply inclined or vertical rock surface in a variety of settings and a variety of weathering environments. Coastal locations seem to neither favor nor discourage tafoni development, except that coastal processes help create the cliffs and steep slopes deemed essential to tafoni development. A case history of tafoni in Quaternary eolianites, with a review of the subject, is presented in Chap. 8.

Talus caves are voids created when cliff failure produces a collection of blocks sufficiently large that people can enter and move around within the block interstices. They are very common in actively uplifting mountain areas, and glaciated areas, both environments where over-steepening of slopes and cliffs leads to rock failure and the collection of massive, large block material as a talus. Coastal areas are another place where over-steepening and basal erosion routinely cause cliff failure and the deposition of large blocks of rock (Fig. 1.7). In many cases, the wave energy that caused the cliff failure quickly acts to reduce and remove the collapsed block material, but in some areas the blocks can persist sufficiently long enough to create cave systems that can be mapped. But such caves are commonly transitory.

Fissure caves, sometimes called tectonic caves where internal earth forces are thought to have cracked the rock, are also common on over-steepened hillsides and cliffs. Coastal areas, as with talus caves, commonly have over-steepened slopes and cliffs which can begin to fail, opening fissures that may become partially or totally roofed with collapse material, creating linear cave systems which at times are complex and extensive (Fig. 1.8). Fissure caves are common in tectonic areas of continental interiors (Fig. 1.8d). Because the processes that create the fissures are on-going in coastal areas, cliff collapse is also on-going and fissure caves are constantly being destroyed and replaced by new fissure caves as the cliff retreats under the coastal erosion onslaught (e.g. Fig. 1.8b).

Fig. 1.6 Examples of tafoni in coastal settings. (**a**) In granite, Kangaroo Island, Australia (see Chap. 18). (**b**) In gneiss, Margaret River, Australia. (**c**) In Quaternary eolian calcarenites, San Salvador Island, Bahamas. (**d**) In Miocene limestones, Exmouth Gulf, Australia. Examples (**c**) and (**d**) are instructive as pseudokarst caves in a soluble rock type

1.6 Sea Caves

The most common cave found in coastal environments is the sea cave, or littoral cave. Fingal's Cave in Scotland, the Blue Grotto of Capri in the Mediterranean, Sea Lion Cave on the coast of Oregon, and Arcadia Cave on the coast of Maine are well-known examples of sea caves visited by tourists on a regular basis. Many organisms use sea caves as a refuge, particularly seals, sea lions and other marine mammals, as well as birds, which roost in the ceiling ledges above the reach of waves. From the viewpoint of cave exploration, sea caves are not a major category, primarily because they are short in length. In areas where other types of caves are rare, such as in southern California, sea caves offer the best cave exploration option (see Chap. 14). Occasionally, sea caves can have spacious chambers and over 400 m of passages. Exploration of sea caves can be very dangerous for people who are inexperienced with the littoral environment of strong waves, tides, and/or currents. Sea arches are a type of natural bridge formed by wave action on rocky coasts, and from overprinting of other cave types, both karst and pseudokarst, that occur in coastal environments.

Sea caves are pseudokarst caves. Sea caves can vary from arches (Fig. 1.9) and small voids only a few meters across to very large chambers up to 100 m deep and wide (Fig. 1.10). These caves, formed by wave action on rocky coastlines, can be found in almost any rock lithology on coastlines around the world, and in inland freshwater bodies as well. They may be the most

Fig. 1.7 Talus caves in coastal settings. (**a**) Talus blocks collecting on a wave-cut bench, Cala Figuera, Mallorca Island, Spain. (**b**) Talus cave within talus, north coast of Barbados. (**c**) North shore of South Island, Pohara, New Zealand. One of the blocks here has a fragment of a dissolution cave within it, from when the block was still in the cliff to the left of the image, an example of a hybrid cave. (**d**) Road seen in (**c**) going under a natural bridge formed by talus blocks (person in *black circle* for scale). All these pseudokarst talus caves are in soluble limestone

common coastal cave type in the world. Moore (1954, p. 71) presented an outline of the critical requirements for sea cave genesis: "*The prerequisite conditions of [sea] cave formation are: 1) the presence of a sea cliff which is in direct contact with the erosive forces of waves and currents; 2) the exposed face of the cliff must contain certain geologic structures, or textures, which will allow the establishment of differential erosion; 3) the rock of which the cliff is composed must be of sufficiently resistant nature so as to prevent rapid formation of a protective beach at its base.*" Unstated, but assumed is that the rock material has sufficient mechanical strength to support a void carved into it by wave action. Bunnell (1988), for Santa Cruz Island, California, demonstrated the importance of heterogeneity in the rock, such as joints and faults that allowed wave energy to exploit those pre-existing weaknesses to produce sea caves of a linear nature. Chapter 14 presents case histories involving sea caves of the west coast of the United States. The east coast south of New York City, and the Gulf Coast lack the rocky coastlines necessary for sea cave development, and few sea caves are found over those many thousands of kilometers of shoreline.

Waterstrat et al. (2010) examined sea caves formed in Quaternary eolianites of the Bahamas. These rocks are very uniform and isotropic, and being young and never buried, lack faults or

Fig. 1.8 Fissure caves (*white circles* around persons for scale). (**a**) Large open coastal fissure from cliff failure, Tinian Island, Mariana Archipelago. (**b**) Sequence of fissure development, Rota Island, Mariana Archipelago. *Black arrow* points to a widening fissure as the block to the right creeps seaward. *White arrow* points to an incipient fissure. (**c**) Fissure on Tinian Island, Mariana Archipelago, note blocks collecting in the fissure to roof it over. The young age of the limestone makes it susceptible to recrystallization when fluid first flowed down the incipient fissure. As the fissure widened, the more resistant fissure walls stand out in positive relief. (**d**) Gravity sliding of dolomite over weaker shales in the Bighorn Mountains of Wyoming creates a series of fissure caves. These pseudokarst fissure caves are all in a soluble rock

joints. Sea cave development in this case appears controlled by wave energy differences produced by wave interference, lagoon and reef physiography, tides, and wind direction and magnitude. Bioerosion may also play a part in these diagenetically immature carbonate rocks. So while differences in rock strength and character are important to sea cave formation, such differences are not the sole arbiter of sea cave placement or size.

Historic photographs of sea caves and arches show that they can undergo development and destruction at a fairly rapid pace. Given that sea level has only been at current elevations for ~3,000 years, it is clear that sea cave development can be relatively fast. Sea caves are commonly found on the inland side of wave-cut benches 50 m or more across (Fig. 1.11). As this bench must be only 3,000 years old or less, the age of the current sea-level highstand, one must assume that a series of sea caves has been created and destroyed as the sea cliff retreated through the late Holocene. At high latitudes, isostatic rebound following deglaciation has uplifted former wave-cut benches and associated sea caves

Fig. 1.9 Sea arches. (**a**) High sea arch, Tinian Island, Mariana Archipelago. (**b**) Broad sea arch, Curaçao. (**c**) Sea arch, Barbados. (**d**) Isolated sea arch, Rottnest Island, Australia. The arches are all in limestone

many meters above present datum (see Chap. 14 examples). The time window for development of these sea caves was necessarily restricted to much less than the 3,000 years of the current sea-level highstand. Despite the rapidity of sea cave formation and subsequent destruction, they can persist in their formational environment for a greater duration compared to talus caves and fissure caves, unless tectonic or isostatic uplift moves the various cave types from the destructive coastal regime.

1.7 Conclusions

In some cases, as noted earlier, caves formed by weathering processes other than wave action are coincidentally located in the intertidal coastal zone. These caves are rapidly overprinted to become hybrid caves. With sufficient overprinting, their origin may become obscured and they may be interpreted to have formed solely by wave processes. Of particular note on carbonate coasts are karst caves that develop in the freshwater lens, just inland from the coast, called *flank margin caves*. Their origin is tied to sea level (see Chap. 4), and their near-coastal location makes them vulnerable to exposure and overprinting by wave processes. Overprinted flank margin caves are the most common hybrid cave found in coastal areas.

Talus caves and fissure caves require cliff failure to develop. They exist in an active and ephemeral environment, and are short-lived. Tafoni require only an exposed cliff, which may be relatively stable. Tafoni are not tied to any specific sea-level position, and can develop at many elevations on a sea cliff simultaneously. Tafoni

Fig. 1.10 Sea caves. (**a**) Wave energy making a sea cave in Holocene eolian calcarenites, San Salvador, Bahamas. (**b**) Demonstration of wave energy on a limestone cliff 25 m high, Barbados. (**c**) Sea cave in Pleistocene subtidal deposits, San Salvador Island, Bahamas, with a fronting beach, contrary to Moore (1954). As has been consistently noted in earlier figures, these pseudokarst cave examples are all in soluble rock

Fig. 1.11 Implications of coastal wave-cut benches. (**a**) Wave cut bench at the Kaikoura Peninsula, South Island, New Zealand. People at *white arrow* too small to be seen, structure under the *black arrow* is 3 m high. (**b**) Expanded image of the scene in a; *white arrows* point to people barely visible. The entire bench must have been carved since sea level stabilized 3,000 years ago, therefore sea caves must have been repeatedly created and destroyed as the sea cliffs migrated landward

can appear by casual observation to be incorrectly identified as uplifted sea caves, or overprinted flank margin caves, and can create a false sense of past sea-level position (Walker et al. 2008).

Pseudokarst caves formed by coastal processes are necessarily ephemeral. The wave action that creates them immediately begins their subsequent removal. Tafoni, lava tubes, and tufa caves are accidental, and their presence in the coastal environment is somewhat decoupled from coastal processes, although one can argue that tafoni are tied to coastal cliff development. Some karst caves in coastal settings are also accidental, remnant fragments of earlier flow systems now decayed and abandoned. But other karst caves are directly tied to the coastal environment by the base level that sea level creates. Chap. 4 will investigate the unique karst processes that make dissolution caves in coastal environments.

References

Bögli A (1980) Karst hydrology and physical speleology. Springer, New York, 284 p

Bunnell D (1988) Sea caves of Santa Cruz island. McNally and Loftin, Santa Barbara, 123 p

Kelletat DH (1995) Atlas of coastal geomorphology and zonality, CERF special issue 13. Coastal Education & Research Foundation, Charlottesville, 286 p

Lascu I (2005) Speleogenesis of large flank margin caves of the Bahamas. M.Sc. Thesis, Department of Geosciences, Mississippi State University, 218 p. http://library.msstate.edu/etd/show.asp?etd=etd-05102005-132949

Machel HG, Kambesis PN, Lace MJ, Mylroie JR, Mylroie JE, Sumrall JB (2012) Overview of cave development on Barbados. In: Kindler P, Gamble DW (eds) Proceedings of the 15th symposium on the geology of the Bahamas and other carbonate regions, Gerace Research Centre, San Salvador, Bahamas, pp 96–106

Moore DG (1954) Origin and development of sea caves. Natl Speleological Soc Bull 16:71–76

Palmer AN (2007) Cave geology. Cave Books, Dayton, 454 p

Pilkey OH (2003) A celebration of the world's barrier islands. Columbia University Press, New York, 309 p

Shackleton NJ (2000) The 100,000-year ice-age cycle identified and found to lag temperature, carbon dioxide, and orbital eccentricity. Science 289:1897–1902

Waelbroeck C, Labeyrie L, Michel E, Duplessy JC, McManus JF, Lambeck K, Balbon E, Labracherie M (2002) Sea-level and deep water temperature changes derived from benthic foraminifera isotopic records. Quaternary Sci Rev 21:295–305

Walker LN, Mylroie JE, Walker AD, Mylroie JR (2008) The caves of Abaco Island Bahamas: keys to geologic timelines. J Cave Karst Stud 70(2):108–119

Waterstrat WJ, Mylroie JE, Owen AM, Mylroie JR (2010) Coastal caves in Bahamian eolian calcarenites: differentiating between sea caves and flank margin caves using quantitative morphology. J Cave Karst Stud 72:61–74

Wilkens H, Iliffe TM, Oromí P, Martínez A, Tysall TN, Konemann S (2009) The Corona lava tube, Lanzarote: geology, habitat diversity and biogeography. Mar Biodivers 39:155–167

Erosional and Depositional Textures and Structures in Coastal Karst Landscapes

2

Danko Taboroši and Miklós Kázmér

Abstract

Exposed surfaces of limestones on marine coastlines are characterized by a tremendous range of rock textures and structures. Many of them are features limited to coastal areas and are morphologically and genetically distinct from inland analogs. This distinction is due to idiosyncrasies of both coastal environments and coastal limestones. Processes operating in coastal settings are not limited to dissolution by fresh water and involve profound chemical and physical action of sea water and marine biota. In addition, these processes act upon rocks that are frequently younger and diagenetically less mature than inland limestones that have undergone deep burial and accompanying changes. The outcomes are distinct types of karren sculpturing, bioerosional markings, deposited and precipitated fabrics, bioconstructions, and compound structures that are unique to coastal karst. Many are limited to particular microenvironmental settings and certain elevations with respect to the sea level and can, therefore, be used as powerful paleoenvironmental and past sea level indicators.

2.1 Introduction

Limestones exposed along marine coastlines display a prodigious array of rock textures and structures. These include both erosional and depositional features, many of which are characteristic of coastal areas and distinct from analogs found in other karst environments. They are generally limited to the narrow coastal belt (from the always-submerged subtidal zone, through the intertidal zone, to the supratidal zone of wave splash and sea spray) and are shaped by a variety of marine weathering processes that occur in the

D. Taboroši (✉)
Water and Environmental Research Institute of the Western Pacific, University of Guam, Mangilao 96923, Guam

Island Research and Education Initiative, Palikir, Pohnpei, FM 96941, Micronesia
e-mail: taborosi@gmail.com

M. Kázmér (✉)
Department of Paleontology, Eötvös University, Pázmány sétány 1/c, H-1117 Budapest, Hungary
e-mail: mkazmer@gmail.com

M.J. Lace and J.E. Mylroie (eds.), *Coastal Karst Landforms*, Coastal Research Library 5,
DOI 10.1007/978-94-007-5016-6_2, © Springer Science+Business Media Dordrecht 2013

presence of sea water (Paskoff 2005) and marine biota (Spencer 1988).

Erosion of limestones and geomorphic evolution of exokarst (exposed karst surface) is usually equated with the process of solution of calcium carbonate. This is appropriate in inland areas, where the main agent that shapes rock surfaces is fresh water. However, most sea water at normal pressure is saturated with respect to calcium carbonate and is not expected to produce dissolutional features. Even so, limestone dissolution does occur in the coastal zone under a variety of conditions, many of which are insufficiently understood and involve input of fresh water or biologic agents. In addition to chemical and biologically-mediated dissolution (jointly known as corrosion, Guilcher 1953), coastal limestones are subject to a number of physical processes peculiar to the coastal zone: mechanical breakdown by the action of surf, salt weathering (haloclastism), wetting and drying, abrasion by wave-suspended sediment, and other forces that are absent or less intense in inland karst.

Also of great importance is the widespread and potent geomorphic action of marine biota, whose dwelling and feeding lifestyles involve effective reshaping of rock and account for a truly fundamental difference between the fates of karst rocks exposed in coastal and inland settings. Inundated or wetted by water, coastal rocks are superficially but in effect "alive" – coated with pervasive, persistent, and complex communities of organisms that engage in erosion, but also other, often opposing processes. In addition to powerful destructive effects of bioeroding organisms, some coastal biota protect the rock from erosion, baffle water currents and garner sediment, or precipitate their own calcium carbonate. This offsets and locally reverses the effects of erosion, resulting in a complex miniature landscape where the net result of removal and deposition of calcium carbonate may be different in any given spot. This is controlled by presence or absence of specific organisms and their own controlling factors: tidal and wave regime, coastal exposure and water energy fluctuations, shading and illumination patterns, and biologic interactions such as competition for space and predator-prey relationships. Therefore, unlike inland exokarst, where broad areas see comparable denudation rates and spatially consistent geomorphic reduction of the landscape, coastal exokarst surface can be imagined as a "battlefield" between destruction and construction (erosion and deposition), which, controlled by physicochemical and biologic microenvironmental variations, act in discrete locations, in various ways, at varying intensities, and at different scales; shift both spatially and temporally; and create a dynamic and multiplex overall pattern of rock textures and structures.

Furthermore, some baseline lithologic differences between coastal limestones and limestones that dominate inland karst settings should be considered. Many limestone coasts, especially those in the tropics and subtropics, are in young, diagenetically immature limestones that have never undergone deep burial and accompanying changes. Such rocks, described as eogenetic, tend to be coarse grained and retain much depositional heterogeneity that has not been smoothed and averaged by deep burial diagenesis. For the same reason, they also preserve high primary porosity. Extremely young units can include the calcium carbonate polymorph aragonite, which is more soluble than the traditional calcite found in older limestones. The combination of high textural variability, relatively high porosity, and differential solubility substantially adds to the complexity of exokarst forms that develop in coastal environments.

In addition to intrinsic scientific interest in such complex features and the sheer beauty of coastal karst landscapes, understanding their textures and small-scale (mm to cm) and medium-scale (dm to m) elements may have several important applications. They are increasingly utilized for their high potential in paleoenvironmental interpretation (e.g. Vescogni et al. 2008), tracking climate change (e.g. Silenzi et al. 2004), recognition of past catastrophic events (e.g. Benac et al. 2004), analysis of sea level fluctuations (e.g. Laborel and Laborel-Deguen 1994, 1995), studies of biologic community interactions (e.g. Jones 1989), and even research in planetary geology (Bourke and Viles 1997). The purpose of this chapter is to provide an overview of small-scale and medium-scale exokarst features produced in carbonate rock surfaces by various inorganic and biologic erosional and depositional processes that operate along marine coastlines. We will explore erosional and bioerosional textures and depositional and bioconstructional features; and consider how the compounding and overprinting of such basic forms on small- and medium-scales creates larger structures and engenders typical coastal karst landscapes.

2.2 General Characteristics

Small- and medium-scale erosional sculpturing of karst rocks has been of interest to geomorphologists for over a century. A collective term "lapiéz" was applied to various rills, flutes, channels, and cracks in limestone rocks already in the nineteenth century (Chaix 1895) and the first comprehensive studies and classifications ensued (Martel 1921 and Cvijić 1924). Most of this early research took place in mountainous regions of the Alps and classical karst of southern Europe and is well summarized by Ginés (2009). With the studies by Bögli (1951, 1960), and subsequent work by Trimmel (1965), Monroe (1970), Jennings (1971), and Sweeting (1972), the term "karren" came to stand for all dissolutional sculpturing in soluble rocks. Karren research became a vital element of geomorphologic, hydrologic, and paleoenvironmental studies in karst terrains and remains a dynamic field today (e.g. Ginés 2009; Veress 2010). Some of the first detailed descriptions of rock sculpturing as observed specifically in marine coastal settings were produced by Macfadyen (1930), Wentworth (1939, 1944), Emery (1946), Corbel (1952), Guilcher (1958) and others; and were paralleled with studies of biological zonation on rocky coasts (Stephenson and Stephenson 1949; Doty 1957; Southward 1958). Attempts to link the observed morphologies with specific processes (e.g. Guilcher 1953; Dalongeville 1977; Ley 1979) increasingly led to awareness that rock-shaping processes operating in coastal karst settings are unique and distinct from what is happening in other karstlands. An appreciation of idiosyncrasies in the forms observed, as well as processes operating along carbonate coasts had led to an understanding that small- and medium-scale rock sculpturing of coastal carbonates form unique "coastal karren" (or "marine karren" – Ley 1977 and "littoral karren" – Malis and Ford 1995) assemblages that are discrete from those of other karst settings ("normal" rainfall-solution karren, subsoil karren, cave karren, etc.). Just to what extent can the processes and morphologies that exist on karst coasts differ from those of inland karst settings was elucidated by the landmark paper of Folk et al. (1973). They described the extremely jagged and chaotic karren of Cayman Island coast and contrasted its highly irregular morphology with the orderly and linear classical karren features studied thus far. They named these features "phytokarst" to emphasize the role of cyanobacteria in their evolution. By that time, the studies of biological erosion were blossoming (e.g. Neumann 1966; Schneider 1976; Bromley 1978) and converged with research on coastal karst, as several influential studies calculated erosion rates of coastal karst surfaces and found that much of it was biological in nature (Hodgkin 1964; Trudgill 1976, 1987; Viles and Trudgill 1984; Spencer 1985a; Kelletat 1988). These advances were paralleled by increasing research on carbonate deposition along karst coastlines. Many geomorphic features constructed by living organisms were described for the first time (e.g. Bosence 1973; Focke 1978) and in due course became recognized as integral elements of coastal karst landscapes. Compound effects of entire biologic communities on erosional sculpturing and production of sediment along carbonate coasts were considered by Schneider and Torunski (1983) and biological construction of carbonate deposits was understood as concurrent and inseparable from biological destruction (Kelletat 1985). Ultimately, the concept of coastal "biokarst" (Viles 1984) solidified to recognize that carbonates along marine coasts are shaped by invertebrate action in addition to inorganic and microbially-mediated processes and that this includes both erosion and deposition. The term "halokarst" was also offered as an umbrella term because it emphasizes the role of salty marine water in the formation of coastal exokarst (Fairbridge 1982).

Eventually, investigations of coastal karren and the related erosional and depositional processes and features in contact with or in vicinity of seawater amalgamated with the concurrently growing body of data about coastal limestones in general. Work on formation of sea caves (Moore 1954), origin of marine notches (Higgins 1980), shaping of coastlines by groundwater (e.g. Back et al. 1986), dissolution of aquifer margin caves (Mylroie and Carew 1990), etc. made it clear that karst of coastal settings is a "world onto itself" – remarkable, complex, and highly distinct from karst topography elsewhere. Its small-scale and medium-scale components, particularly erosional textures, continue to be an object of fascination and subject of extensive research (see reviews by Lundberg 2009; De Waele and Furlani 2013).

2.2.1 Processes

The fact that various weathering and erosional processes work concurrently and often in synergy with each other is perhaps nowhere as clear as it is along coastlines. In coastal settings, chemical weathering can be quite aggressive because of mixing of fresh waters (meteoric, surface, groundwater) with seawater; physical destruction can be intense due to high energy released from the ocean; and biologic erosion at sea level is far more vigorous than in almost any other setting, subaerial or subaqueous. In most cases, a given rock surface exposed at the coast is subject to all three groups of processes operating together and under influence of each other. The texture and structure that ultimately develops at a given location depends on the particular balance of processes operating there and may only sometimes be linked to a specific mechanism that emerges as dominant. The remarkable diversity of rock appearances and sculpturing therefore derives from the fact that their overall form is controlled by the magnitude and relative importance of numerous chemical, physical, and biological erosion processes, which differ in space and time (Spencer 1985a).

Chemical dissolution by meteoric or surface water is the primary cause of surface sculpturing in most karst rocks. The specific mechanisms include direct impact of raindrops, laminar or turbulent flow in channels or sheets over rock surfaces, adherence of thin water films to rock surfaces, stagnation of water in pools, and so forth (Ginés and Ginés 1995). In coastal areas, effects of meteoric water alone are not different than in karst elsewhere, except that they may be less intense because of the general lack of acidifying soil and organic-rich sediment in immediate vicinity of shorelines. What makes rocks in coastal settings quite distinct, however, is that meteoric water there does not occur alone and dissolution involves both fresh and saline waters. Seawater is present in unlimited quantities, but is saturated with respect to calcium carbonate and unable to dissolve limestone under normal circumstances. Some researchers have postulated that seawater may become locally and briefly undersaturated due to respiration of intertidal organisms at night when there is no photosynthetic uptake of carbon dioxide (Holbye 1989). Trudgill (1976) suggested that in a calm tropical environment, nearshore waters can achieve undersaturation with respect to aragonite and high-Mg calcite even during daytime and that some 10 % of limestone erosion on raised reefs is due to dissolution by seawater. Miller and Mason (1994) have shown undersaturation in stagnant isolated seawater bodies on intertidal platform. Physical agitation by breaking surf (Fairbridge 1952) has also been suggested as a mechanism to achieve fleeting undersaturation. The first mechanism that was unequivocally demonstrated as effective in making seawater dissolutionally aggressive is the addition of fresh water (Bögli 1964). Higgins (1980) correlated the efficacy of erosion of limestone at sea level with discharge of fresh groundwater into the sea. Stoessell et al. (1989) have shown that dissolution of coastal limestone can take place in sea water mixed with very little fresh water. Working in Yucatán, they have observed that only 5 % addition of fresh water to ambient sea water can result in undersaturation with respect to aragonite, and a 10 % addition can produce undersaturation for calcite. Along protected shorelines, where coral reefs, mangrove swamps, sand bars, or other structures minimize circulation of coastal waters, discharge of fresh water from groundwater systems, surface drainage, or intense rainfall can produce a floating layer of dissolutionally aggressive brackish water on the ocean surface.

Physical impact of surf, hydraulic action of water, and pneumatic work by compressed air pushed into small crevices break and shape coastal rocks. However, they loosen and disintegrate bedrock and produce widened fractures and sea caves rather than small-scale erosional textures. For that reason, mechanical action of waves is usually not responsible for development of karren. Instead, it tends to destroy karren or prevent its formation by constantly renewing rock surfaces exposed to the ocean. In addition, sediment suspended in wave-agitated water acts as abrasive agent and polishes exposed surfaces to minimize their three-dimensional relief. In contrast, wave splash and sea spray drive incessant wetting and drying cycles that promote weathering of supratidal rocks. This may disintegrate some limestones, particularly eogenetic, chalky, or argillaceous units, or produce erosional sculpturing in more densely crystalline rocks. Evaporation of wave splash from pits and indentations in rocks leaves behind waters of various chemistries, which may corrode rock and become particularly aggressive when they mix with rain water. Crystallization of salt within pore spaces of coastal rocks results in salt weathering, much of which transcends the definition of karren and may, along with wetting and drying, create cavernous hollows known as tafoni (see Chap. 8). In polar areas, frost weathering may also play a role in formation of some karren and general erosion of coastal limestone (Cowell and Ford 1983).

Biologic action is extremely important and evident in almost all coastal karst settings. Much of it is reflected in bioerosion, the destruction and removal of lithic substrate by direct biologic action (Neumann 1966). The most common agents of bioerosion in coastal settings are cyanobacteria (Radtke et al. 1996; Schneider and Le Campion Alsumard 1999) and other microbes, fungi (Duane et al. 2003), algae (Kobluk and Risk 1977), lichens (Moses and Smith 1993), sponges (Holmes et al. 2009), polychaete worms (Hutchings and Peyrot-Clausade 2002), sipunculid worms (Williams and Margolis 1974), molluscs (Morton and Scott 1980), echinoderms (Mokady et al. 1996), and crustaceans (Ahr and Stanton 1973). Fish are important only in subtidal settings, particularly on coral reefs (Risk and McGeachy 1978). Living organisms erode rock by mechanical means (bioabrasion) or by chemical means (biocorrosion) (Tribollet and Golubic 2011). Many species rely on both when making their variously shaped excavations. Some of them penetrate rock in order to take advantage of endolithic (rock interior) habitat that is stable and buffered from outside stresses and relatively isolated from potential predators. Others wear it down as they graze on epilithic (rock surface) organisms or work to access prey that hides within the rock (Fórnos et al. 2006). Bioerosion is thus a collective process in which many organisms destroy rock by different means, in different ways, and for various reasons. Much of it is driven by predator-prey and competitive relationships between the bioeroders themselves and results in a composite and highly complex diminutive landscape in which traces of various organisms concur, modify, and overprint each other.

Of course, the coastal karst landscape is a highly complex and dynamic system in which there also exist chemical, physical, and biologic processes that act in direction opposite to the destruction of rock. Chemical precipitation and cementation occurs locally to produce oölitic sands (Newell et al. 1960), carbonate mud (Robbins and Blackwelder 1992), beachrock, and some less common deposits. Physical transport and deposition creates numerous seascapes, including beautiful sandy beaches on countless carbonate coasts. Biological processes are certainly not limited to bioerosion and encompass several important groups of processes that actively counter weathering and erosion. Many organisms precipitate calcium carbonate and some directly build standing structures – magnificent coral and algal reefs and other fascinating formations. This is bioconstruction, which is the production of sedimentary structures by living organisms (Spencer and Viles 2002; Naylor et al. 2002). It occurs in many parts of the ocean floor, but is particularly common in coastal areas, particularly the intertidal zone. It involves colonial and solitary skeleton-building and encrusting organisms: algae (especially coralline red algae), scleractinian corals and certain other cnidarians, bryozoa, polychaete worms, and some mollusks. In addition, there are biosedimentation, where shells and other biogenic hard parts lead to production of loose sediment; biocementation, where organisms foster lithification of sediment (e.g. Webb et al. 1999), and bioprotection, where organisms coat and isolate rock surfaces from erosive action (Carter and Viles 2005). Of course, any depositional structures, biogenic or not, are attacked by erosion even as they are being built, and persist only when and where local carbonate production or accumulation outweighs all erosion. As a result, some important features of coastal karst are depositional in nature, even though karst in general is by and large a dissolutional and erosional landscape.

2.2.2 Relief and Scale

Rock textures can be described in terms of differences in small- and medium-scale relief. Most apparent, and indeed defining, features are those of negative relief, where material has been eroded away. Positive relief features are merely residual forms that persisted where erosion did not occur as fast due to uneven contact with water, channeling, variations in flow, variations in lithology, etc. In coastal karst, however, the situation is more complicated. Residual forms regularly persist thanks to the countless microenvironmental factors that locally define net erosion, including being protected by epilithic organisms. Furthermore, there are positive relief forms that arise by local carbonate buildup. This is unusual in inland settings, where tufa and speleothems are the main exceptions, but is pervasive in coastal settings, where biologic calcium carbonate production is the norm. As previously discussed, the interplay of limestone destroying forces (physical, chemical, and biological weathering and erosion) and limestone creating forces (precipitation, deposition, biologic production of shells and skeletons, cementation) results in a very complex small- and medium-scale landscape of countless erosional and depositional facets. In a coastal setting, quite often, one cannot be properly considered without the other. This was best stated by Lundberg (2009), who conceptualized overall coastal karst relief at any location as negative (due to erosion), remnant (due to less or lack of erosion), and positive (due to buildup).

In general, negative relief is on cm or m scale. In exceptional circumstances, karren features may be over 10 m deep, both in classical karst (Cvijić 1924) and tropical (Salomon 2006) settings, but can always be qualified as small- or medium-scale when compared to most other karst landforms (10–100 m to km-scale dolines, poljes, caves, etc.) Expressing size and scale is relatively easy for most karren, because length, width, and depth of channels, flutes, ridges, pans, and other forms can be delimited and directly measured. In coastal karst settings, this is not always straightforward because many morphologies are chaotic. As we shall see later, the basic morphologic elements of many coastal limestone surfaces are irregular pits of various shapes and sizes. Though most appear to be within cm-scale range, a closer look should reveal that there is a continuum of comparable features from microscopic apertures to human-sized hollows. The composite landscape is often reminiscent of a cratered surface, in which smaller pits are contained within larger ones and basic morphology is repeated on a variety of scales superimposed on the same rock surface. Torunski (1979) has called this pseudofractal nature and Lundberg (2009) considered it a collection of basic elements ("building blocks") that are variously scaled and assembled into textures and structures ("modules").

2.2.3 Lithologic Controls

When dealing with features that develop on exposed carbonate rock surfaces, we must also consider the rocks' inherent properties. Lithology of host rock plays important role in the development of karren in coastal settings (Ley 1977) and elsewhere. General prerequisite is that the rocks are soluble (though some secondary textures on non-karst rocks have long been called karren, e.g. by Palmer 1927). True dissolutional sculpturing develops best and in greatest variety in massively bedded, fine grained, and homogeneous limestones (Ford and Williams 1989), whereas porous and mechanically weak rocks are unfavorable to karren development (Jennings 1985). For example, in Guam, Mio-Pliocene reef limestone exhibits deeper and more complex erosional sculpturing than coeval foraminiferal and argillaceous limestones; and Holocene reef limestone possesses a much rougher exposed surface than beachrock deposits of approximately the same age (Taboroši et al. 2004). Similarly, in the Bahamas, older eolianites show better developed erosional structures than younger eolianites (Mylroie and Mylroie 2009) and in the Balearic islands Miocene reef limestone exhibits more complex karren assemblages than Miocene muddy calcarenite (Gómez-Pujol and Fornós 2009). In western Ireland, karren development on Carboniferous limestone was quantitatively correlated with the purity of limestone (Burke 1994, cited in Drew 2009). This indicates that constitution and diagenetic maturity is an important factor in determining the type of texture that develops on a certain rock surface. This tendency is very significant in coastal settings. Many limestones are deposited under nearshore marine conditions and become karstified following relative sea level change, without intervening episodes of burial and diagenesis. Such eogenetic rocks are far more porous and physically and mineralogically heterogeneous than diagenetically mature units. Their lithologic characteristics exert strong controls over the type of dissolutional and erosional features that form on their exposed surfaces. Efficacy of bioerosion is enhanced because pre-existing voids facilitate rock excavation by invertebrates; and pore spaces, irregularly shaped grains, bedding planes, and fractures lead to easy microbial and water penetration. Pre-existing heterogeneities in eogenetic limestones thus cause dissolutional and bioerosional attack on them to be scattered and uneven, making them predisposed to development of chaotic textures rather than well-structured karren.

2.2.4 Rates of Development

Rates at which dissolutional textures develop in coastal karst tend to be different from those of inland karst. The tempo of sculpturing of limestone rocks exposed along coastlines depends on the type of rock, shaping mechanisms, climate, exposure to wave energy, composition of biologic community, and other factors. In general, however, the formation of secondary textures seen on exposed limestone surfaces must outpace the frequency and intensity of large-scale reshaping of the overall coastal landscape by major erosion events (Mylroie and Mylroie 2009). This is because large storms and ensuing coastal breakdown tend to erase secondary textures of exposed rocks by breaking them and exposing virgin surfaces.

In a seminal study, Trudgill (1976) measured erosion rates between 0.3 and 4 mm/year in young reef limestone in a tropical Indian Ocean atoll setting. Donn and Boardman (1988) measured rates between 1.8 and 2.6 mm/year in intertidal zone and 0.4 mm/year in supratidal zone in the Bahamas. Spencer (1985b) also worked in the tropics and recorded rates over 2.5 mm/year on exposed coasts and under 0.5 mm/year on reef protected coasts in the Cayman Islands. Perhaps surprisingly, the lowest net erosion, less than 0.2 mm/year, was found on coasts exposed to high wave action. This is because high water flux promotes growth of filter-feeding organisms whose bioconstruction activities offset effects of bioerosion. Erosion rates in temperate climates are comparatively lower. Maximum measured in intertidal limestones in Europe were ~1 mm/year in the northern Adriatic Sea (Torunski 1979) and just under 0.4 mm/year in Ireland (Trudgill et al. 1981). Comparable values were observed in New Zealand and Australia (Gill and Lang 1983; Stephenson and Kirk 1998). Summaries of published rates from all over the world are provided by Furlani et al. (2009) and De Waele and Furlani (2013). The former also provide new data from the Adriatic Sea. They found the denudation rate at sea level to be 0.1 mm per year. More specifically, that rate was characteristic of intertidal and splash zones, from 0.25 m below mean sea level to 0.75 m above mean sea level. In slightly deeper water, the rate drops by half to ~0.05 mm/year. Erosion rates are by far the lowest in inland areas away from marine influence. Inland areas in the same region and with comparable rainfall, but at elevations several hundred meters above sea level exhibit maximum erosion rates between 0.01 and 0.03 mm/year. A very valuable observation by Furlani et al. (2009) is that while maximum erosion rates in coastal and inland settings can differ by a factor of 10, minimum erosion rates measured at various coastal and inland sites are comparable. This corroborates observations that coastal erosional sculpturing appears random and extremely variable from site to site. Rock removal rates at any particular position in coastal areas clearly depend more on microenvironmental parameters such as wave exposure and biologic community structure rather than large-scale phenomena such as rainfall and soil cover that control the rates inland.

2.3 Karren and Erosional Textures

The variety of dissolutional textures that form in exposed limestone surfaces is vast. There are many dozens of types of basic karren, each with defined form, proportions, technical name, and a number of synonyms. Terms such as rillenkarren, rinnenkarren, meanderkarren, trittkarren, rundkarren, kluftkarren, etc. stand for various types of flutes, grooves, channels, scallops, tubes, clefts, and other common features of exposed soluble rock surfaces (see definitions in Bögli 1960 or Ginés et al. 2009). They are best seen in homogeneous and diagenetically mature limestones of inland settings, where regular patterns are the norm and a formal classification based on recurring morphologies and deduced genetic origins is firmly in place. In striking contrast, surfaces of limestones (especially diagenetically immature units) exposed along marine coasts (but also brackish and sometimes lacustrine shores and inland areas of very humid tropical regions) tend to exhibit disordered assemblages of pits, protuberances, and various irregular morphologies. As a result, classification of karren and related features is particularly difficult in coastal areas. An attempt must be based on morphology (see, for instance, Gómez-Pujol and Fornós 2009) rather than genetic origin, as the latter is evident only in some hydrodynamically or structurally-controlled karren. In most cases, the genetic mechanism is difficult to isolate from so many concurrently operating and interrelated processes. As previously suggested, many coastal rock textures are composite and polygenetic. Analogous features can be created by different processes and single features can be created by multiple processes. For example, a small pit in limestone surface can be a result of dissolution by pooled water, selective mineral loss, removal of a detached fossil, or any of a number of different processes. It can also be a compound result of several processes, for instance attack by microbes, dissolution, and salt weathering (Moses and Smith 1994). It could also be an excavation made by some marine invertebrate. Bioerosional markings are sometimes considered a kind of coastal karren features (Lundberg 2009), but perhaps should not be whenever it can be determined that the scars were made by particular organisms, especially through bioabrasion. We discuss such traces separately in this chapter and consider them distinct from karren, which should be limited to predominantly dissolutional features. Of course, both karren and bioerosional scars are important and integral parts of coastal karst landscapes and a clear distinction cannot always be made between them. The two concepts converge and overlap just like the actual features coexist and overprint in nature (case in point being eogenetic karren, discussed further on).

2.3.1 Hydrodynamically-Controlled Karren

Archetypal karren are shaped as rills, runnels, grooves, flutes, and other forms that are evidently hydrodynamically controlled. They are produced by solution of carbonates and other soluble rocks through laminar or turbulent flow of water. They are most common in steeply dipping surfaces of diagenetically mature limestones and occur in all karst settings, from the humid tropics (Sweeting 1972) to deserts (e.g. Sweeting and Lancaster 1982). Their representation in coastal areas varies from common to absent, with the main deciding factor being diagenetic maturity of rock. Because many coastal limestones are diagenetically immature and highly porous, they do not support surface flow even across very short distances and effectively preclude the formation of hydrodynamically-controlled karren forms. Instead, they are host to chaotic karren collectively known as eogenetic karren (see eogenetic karren and pinnacles further in text). Essentially, hydrodynamically-controlled karren and eogenetic karren are opposite end members of a continuum ranging from dissolution by focused flow to widely dispersed corrosion, and are often mutually exclusive. To the best of our knowledge, hydrodynamically-controlled karren are either lacking or exhibit only rudimentary forms in eogenetic limestones. In contrast, they can be common in diagenetically mature limestones and exist in many different types.

Rillenkarren are solution flutes, about a cm wide and few tens of cm long, shallow, round-bottomed troughs that occur packed side-by-side and separated by sharp-crested ridges (Ford and Williams 1989). They are among most typical karren in alpine karst and have been noted in some coastal settings as well. For example, Gómez-Pujol and Fornós (2009) describe mm-scale sinuous microrills and cm-scale rillenkarren from fine-grained Miocene calcarenites exposed to wave spray on the Balearic Islands' coast. Taboroši (this chapter) has observed cm-scale rillenkarren in Cretaceous-aged coastal (and inland) karst in the northern Adriatic. A closely-related form are wandkarren, which are straight solution runnels that form on vertical and subvertical cliffs and shaft walls by dissolution by rainwater flowing down steep slopes (Bögli 1960). We have observed them in coastal cliffs of Permian and Carboniferous limestones descending into waters of Krabi Bay, Thailand and islands of Ha Long Bay, Vietnam. In the vast majority of cases, however, coastal locations of hydrodynamically-controlled karren are merely incidental and forms observed are no different than those seen on the same rocks inland. The only hydrodynamically-controlled karren that can be understood as restricted to coastal settings are rill sets and runnels made by surf. Inclined rocks regularly splashed by waves may exhibit some rilling (Fig. 2.1a), but that is often partial and lopsided (Fig. 2.1b) and not straight and regular as true rillenkarren, presumably due to the relative diagenetic immaturity of exposed units. Solution pans and tide pools where wave splash accumulates at different elevations may be connected by decantation runnels, or may have outlet runnels that conduct overflow back to the ocean. Finally, fluting or channels incised by wave swash and backwash are often evident in beachrock deposits (Fig. 2.1c).

2.3.2 Structurally-Controlled Karren

Water moving through rock may exploit previously existing discontinuities and create karren types that represent dissolutionally-enlarged joints and bedding planes. Coastal areas are often sites of stress resulting from seismic activity, bank-margin failures, oversteepening of scarp edges, etc. and develop numerous joints. Joints provide routes for infiltrating meteoric water and gradually become widened. They exist at various scales. Small varieties are widened only superficially and taper off with depth. Solutionally widened joints (Fig. 2.1d) and fractures are considered a type of karren and are known as kluftkarren. They are typical feature of limestone pavements (Ford and Williams 1989), and may also be seen in coastal settings in very fractured rocks (Fig. 2.1e). On meter and larger scales they transition into large grikes (Fig. 2.1f), karst corridors, fracture caves (see Chap. 13) and various other fissures. This commonly occurs in coastal settings, where structural discontinuities in rock are mechanically widened by hydraulic and pneumatic action of wave-compressed seawater and air and eventually become large enough to be considered sea caves (see Chap. 1) or are enlarged by focused discharge of groundwater into the ocean (Fig. 2.1g). In limestones where bedding

Fig. 2.1 Hydrodynamically and structurally-controlled coastal karren features. (**a**) Vertical dissolutional rills in diagenetically mature Cretaceous limestone, Sveti Marko island, Croatia [ruler is 10 cm long]; (**b**) Rudimentary channels and ridges formed by returning wave splash in Miocene-Pliocene limestone, Yucatán, Mexico; (**c**) Parallel grooves incised in Holocene beachrock by wave swash and backwash, Guam; (**d**) Solutionally-enlarged, steeply-inclined joint in Pleistocene reef limestone, Guam [scale bar is 15 cm]; (**e**) A top-view of coastal platform cut-up by kluftkarren, in Cretaceous chalky limestone, near Tripoli, Lebanon; (**f**) Series of large, vertical grikes dissecting Cretaceous chalky limestone near Tripoli, Lebanon [outcrop face is 7 m tall]; (**g**) Coastal fracture in Pleistocene reef limestone in Guam, widened by dissolution through mixing of ambient seawater and a steady supply of fresh groundwater that is captured inland and discharged into the ocean; (**h**) Subhorizontal linear karren features, guided by bedding planes, in otherwise irregularly dissolutionally sculpted eogenetic coastal eolianite near Cueva del Indio, Puerto Rico; (**i**) Naturally fractured beachrock deposit with slabs separating along bedding planes and vertical cracks, Guam [scale bar is 15 cm]

planes are present, water may also preferentially move along and widen them by dissolution. In coastal areas, this is often observed in eolianites, where dissolutional widening of bedding planes produces horizontal linear karren (Fig. 2.1h) and in beachrock deposits where fractured slabs separate along pre-existing planes of weakness (Fig. 2.1i).

2.3.3 Eogenetic Karren and Pinnacles

The exposed surfaces of diagenetically immature coastal limestones tend to be dominated by chaotic and very rough karren consisting of multitudes of densely packed pits of various sizes, separated by irregular ridges and sharp tips (Taboroši et al. 2004). This elementary form can be seen in very young limestones where corrosion has not gone too far and ridges and points clearly stand out as residual relief among adjacent pits (Fig. 2.2a). On more corroded surfaces, pits can be deeper and reminiscent of tiny craters (Fig. 2.2b). Extremely corroded surfaces can be so intricate that they evoke three-dimensional fretwork (Fig. 2.2c). This type of dissolutional sculpturing exhibits little, if any, gravitational control and is well developed on rock surfaces of all orientations, including vertical and overhanging. This has led to an understanding that biological activity and haloclastism play major roles, making this one of the most variable and least understood karren types (White 1988). Indeed, extremely jagged forms with convoluted pitting, knife-edge ridges, very sharp points, and completely penetrating holes (Fig. 2.2d) have been named "phytokarst" by Folk et al. (1973), who recognized that biocorrosive microorganisms (see microborings, discussed later) have a decisive role in the shaping of this karren type. The microorganisms' macroinvertebrate predators, especially littorinid gastropods, also play a geomorphic role by colonizing, scraping, and propagating pits and other negative relief (Fig. 2.2e) in search of prey. Effects of haloclastism are most obvious in units comprised of cemented sand and indirectly exposed to much seawater, where karren tends to develop delicate lace-like forms reminiscent of tiny tafoni (Fig. 2.2f).

Alongside extraordinary irregularity, the most remarkable feature of this karren is the apparent continuum of form across a range of scales. Just like that of the karren itself, the larger-scale topography is defined by negative relief consisting of irregular depressions and residual positive relief consisting of jagged ridges and upward-pointing pinnacles. Surfaces of pinnacles and other larger-scale surfaces are highly fretted by smaller-scale eogenetic karren (Fig. 2.2g). Because the pits are of so many different dimensions (from mm-scale and smaller boreholes by microbes to human-sized holes) and leave behind points and ridges of proportional sizes, the end result can be a fractal-like pattern (Fig. 2.2h), where mm- and cm-scale pits, points, and ridges appear to be repeated on dm- and meter-scale topography. In addition to diagenetically immature limestones of tropical coasts, this type of karren occurs on similar rocks in inland areas in the humid tropics. Inland varieties are somewhat more rounded and subdued in relief and exhibit paler color: the extreme coastal forms are black or dark gray, while the inland forms tend to be light gray to tan (Fig. 2.2i). The intensity of color is thought to reflect the type and amount of organic coating by epilithic and shallow endolithic microorganisms, which are at least partly responsible for the observed morphology (Viles 1987; Jones 1989).

After initial descriptions of this karren from the Aldabra Atoll (Stoddart et al. 1971) and the Cayman Islands (Folk et al. 1973), analogous forms were reported from Balearic Islands (Ginés and Ginés 1995), Nauru (Jacobsen et al. 1997), Christmas Island (Grimes 2001), Lord Howe Island (Moses 2003), Niue (Terry and Nunn 2003), Morocco (Duane et al. 2003), Guam and the Mariana Islands (Taboroši et al. 2004), Puerto Rico (Chacón et al. 2006), Bahamas (Mylroie and Mylroie 2009), etc. Despite being very conspicuous on numerous coastlines around the world, this type of karren lacks a widely accepted name. Both the centimeter-scale karren and meter-scale pinnacles have been assigned many different terms over the past few decades. Some of the synonyms include *champignon surface* (Stoddart et al. 1971), *phytokarst* (Folk et al. 1973), *lacework morphology* (Bull and Laverty 1982), etc. The term *spitzkarren* (Trudgill 1979) was also suggested, despite little similarity with spitzkarren originally described by Bögli (1960) (compare Trudgill's Fig. 1 and Bögli's Fig. 15). In recent years, the term *eogenetic karren* (Taboroši et al. 2004) has gained some

Fig. 2.2 Eogenetic coastal karren features. (**a**) Eogenetic limestone surface with chaotic karren consisting of irregular pits and residual ridges and points, Tinian, Mariana Islands; (**b**) Eogenetic limestone surface corroded as if parts of rock have been scooped out, leaving behind countless tiny craters, Guam; (**c**) Extremely intricately shaped, fretwork-like karren on eogenetic limestone, Boracay, Philippines; (**d**) Particularly jagged "phytokarst" morphology, with its convoluted pits, knife-edge surfaces, and completely-penetrating holes, as is typical of eogenetic rocks on tropical coasts, Rock Islands, Palau; (**e**) Littorinid snails grazing on a highly-pitted karren surface and preferentially hiding within holes whose sizes appear to correspond to the animals' own shell sizes, near Cueva del Indio, Puerto Rico; (**f**) Calcareous sandstone surface on the protected landward side of a small island exposed to high surf; it receives much seawater input but no direct hits by waves and is shaped by salt weathering and grazing by numerous small littorinid snails, Isabela, Puerto Rico; (**g**) Meter-scale ridge and pinnacle morphologically comparable to cm-scale karren with which they are covered, Pleistocene eolian calcarenite, near Cueva del Indio, Puerto Rico; (**h**) A small outcrop (approximately 1 m across) of grotesquely corroded young reef limestone exposed in a mangrove swamp in Palau; (**i**) Landward change from dark-colored sharply-corroded "phytokarst" by the sea (background) to lighter-colored rocks with less-pronounced relief and some vegetation (foreground), Santo Domingo, Dominican Republic

acceptance (Mylroie and Mylroie 2009; Lundberg 2009). This is based on an understanding that this type of karren is polygenetic and arises from a number of processes (including biocorrosion, wetting and drying, salt weathering and hydration, and salt spray and rain water mixing) that are superimposed on the highly heterogeneous texture and high primary porosity of diagenetically immature limestone. Eogenetic lithology is deemed to be the crucial factor controlling the development of this type of karren by making the rocks predisposed to the development of multitudes of irregular hollows and residual forms that accompany them (Taboroši et al. 2004). Occasionally and somewhat unexpectedly, jagged karren with circular depressions separated from one another by sharp ridges get reported from old, diagenetically altered limestones (e.g. by Mark 2009), but the analogy with eogenetic karren remains to be examined. Many pitted and sharp textures reminiscent of eogenetic karren can be created in the intertidal zone of both tropical and non-tropical places as composite results of bioerosional activities by a variety of concurrent (and competing) or successive (and overprinting) taxa (Kázmér and Taboroši 2012).

2.3.4 Other Small Pits

In addition to the pervasive and closely-packed pitting that is the hallmark of eogenetic karren, there are many other examples of pit-like karren occurring in isolation or clusters. They belong to a wide variety of genetic origins and are common in coastal areas as well as in inland karst. Examples include small dissolutional pits that are commonly observed on floors of marine notches (Fig. 2.3a), beachrock deposits (Fig. 2.3b), and other bedrock at sea level. The last is especially true of very high latitudes, where cm- and dm-scale dissolutional pits (Fig. 2.3c) are the predominant secondary texture in coastal limestone. They have been given the name of bowl karren (Holbye 1989) and are typical coastal karren of arctic coasts, being well developed only in the wave swash zone and immediately adjacent areas above (Lundberg and Lauritzen 2002). Small pits of any origin can be further enlarged by dissolution to produce solution pans (see next section), or expanded by other processes, notably mechanical force (see potholes further in text) and bioerosional activity (see bioerosional textures further in text). Salt weathering produces clusters of tafoni pits that mimic karren on shore platforms (Matsukura and Matsuoka 1991).

2.3.5 Solution Pans

Solution pans are idiosyncratic features of coastal karst (Emery 1946). They are circular, elliptical, or irregular in plan view, and are clearly distinguished from pits and other topographically enclosed depressions by two key characteristics. Their hallmark is conspicuously flat or almost flat bottom (floor) that always has horizontal or nearly horizontal original orientation. In addition, the pans' walls are steepened by undercutting and may posses a corrosion notch at the base which creates an overhanging edge around the perimeter. Individual pans are typically on a decimeter or meter scale, with diameters several times greater than depth. Synonyms used for this form in karst literature include *solution basins*, *tinajitas*, and *kamenitzas* (Cucchi 2009). Solution pans are especially common in coastal karst and may be entirely absent from inland areas, such as is the case in the Mariana Islands (Taboroši et al. 2004). They are common in the wave spray zone of most karst coasts. In the tropics, they are ubiquitous in pinnacled terrain and fields of eogenetic karren reached by wave splash and spray (Fig. 2.3d), and are also recurrent in the topographically-less-dissected coastal rocks around the Mediterranean and in temperate regions (Fig. 2.3e).

Solution pans are initiated in depressions where organic detritus and evaporation residue accumulate on the floor and cause dissolution to be even and slow downward but focused and strong around the perimeter (Ford and Williams 1989). In coastal settings, they are regularly filled with wave splash and rainwater in

Fig. 2.3 Pits, pans, and other circular-plan coastal karren features. (**a**) A group of small coalescing pits in the coastal bench, approximately at high tide level, Bali, Indonesia [tape measure is extended 20 cm]; (**b**) Shallow smooth-floored pits formed in Holocene beachrock, Guam, exhibit yellow coloration because they provide regularly wet microenvironment for epilithic algae [scale bar is 15 cm]; (**c**) Bowl karren and smaller cm-scale pits dominate the coastal erosional ramp in marble, on Helgeland coast, northern Norway [ruler is 20 cm long] [photograph and © by Stein-Erik Lauritzen]; (**d**) Active solution pans with overhanging rims, formed in Pleistocene eolian deposits, Puerto Rico; (**e**) Active solution pans with characteristically flat floors and overhanging rims, Baix des Guix, Menorca, Balearic Islands [photograph and © by Ignacio Benvenuty Cabral]; (**f**) Solution pans with coral fossils (gray, indicated by black arrows) and paleosol pockets (brown, indicated by white arrows) protruding from otherwise flat floor (beige), coast in Santo Domingo, Dominican Republic; (**g**) A cup-shaped pothole with a well-rounded cobble acting as abrasion tool, in beachrock, Puerto Rico [ruler is 30 cm long]; (**h**) A pothole undercut by erosion of the marine notch, Boracay, Philippines; (**i**) Numerous holes eroded in Quaternary calcarenite and residual pinnacled relief, within wave swash zone in a very exposed location at Isabela, Puerto Rico [scale bar is approximately 30 cm long]

different proportions. Due to regular wetting and drying, chemistries of waters that accumulate in the pans are highly variable both temporally and spatially (Emery 1946, 1962) but, their starting point being seawater, are generally considered to be saturated with respect to $CaCO_3$ (Schneider 1976). Occasional freshwater pooling during rains or rainwater input into seawater pools can cause some dissolution, but not on such a grand scale to make solution pans so common on karst coasts, including in very arid regions. Extreme microenvironmental variability makes this habitat too stressful for anything but specialized microbes, notably cyanobacteria. It is thought that physiological activity of these autotrophs, specifically absence of photosynthetic CO_2 consumption during nighttime, causes the water to be temporarily undersaturated and boosts dissolutional potential of pooled water (Schneider 1976).

Over time, wetting and drying causes the pan floors to become cemented, which leads to selective widening of the pan and creation of the overhanging rim. Originally circular or oval-shaped pans amalgamate to create irregularly-shaped compound features that often reach several meters across. Occasionally, lithologic heterogeneities in limestone may be left protruding from the flat floors due to being slightly more resistant to dissolution (Fig. 2.3f). In places where evaporation rates exceed filling rates, the pans may contain deposits of sea salt. In places where filling rates are faster than evaporation rates, pans may develop outflow channels that conduct spillage water back to the ocean or a pan at a lower elevation. Pooling of water ceases only when lowered floor of a pan intercepts a joint, bedding plane, or another feature where water can easily escape. It is this lowering of pans' floors and overall deepening of depressions that contain them that is considered one of the key geomorphic factors producing the rough pinnacled terrain (Gómez-Pujol and Fornós 2010) typical of many coastal karsts, especially in eogenetic limestone. Pinnacles, ridges, and other residual relief is, to some extent, what is left behind as undissolved material between various solution pans.

It should be noted that solution pans are limited to supratidal areas where there is irregular input of seawater by wave splash and spray. Any similar depressions found in intertidal areas are either former solution pans colonized by marine algae and invertebrates or depressions created by bioerosive action and turned into tidal pools (De Waele et al. 2009). Sea urchins, in particular, can excavate rather flat-floored and laterally expanding basins reminiscent of solution pans (Kázmér and Taboroši 2012).

2.3.6 Potholes

Potholes are erosional features that are widely reported from fluvial channels incised in bedrock (e.g. Whipple et al. 2000). They are also found in marine coastal settings and were described early on (e.g. Wentworth 1944; Abbott and Pottratz 1969; Tschang 1966). They are roughly bowl-shaped depressions eroded by moving water (Richardson and Carling 2005). Consequently, they are neither karren nor karst, though marine potholes may be considered pseudokarst when they form in limestone. They are included in this chapter because they are rarely discussed in literature (see review by Dionne 1964) despite being very common on limestone coasts. They can also be confused with solution pans, which are true karst features. The main distinction between them is that potholes tend to have concave floors and lack any undercutting and corrosion notches (Fig. 2.3g). Marine potholes develop in zones of breaking waves where sand, gravel, pebbles, and cobbles are trapped within depressions and repeatedly swirled by vortices, acting as abrasion tools that physically excavate bedrock (Huggett 2007). They are most common right at or just above the sea level in rocks regularly scoured by waves (Fig. 2.3h). In eogenetic rocks, particularly raised coral and algal reefs or calcarenites, potholes can be so densely packed and rapidly deepened that they engender an intense relief of wave-washed sharp points and ridges (Fig. 2.3i). In places of especially rough surf, potholes can also develop in deeper water, down to wave-base depths.

2.3.7 Projections and Casts

Uneven dissolution and mechanical erosion can etch rock by accentuating pre-existing structural, lithological, and mineralogical heterogeneities. Less resistant minerals or portions will be preferentially attacked, leaving behind various new patterns. This is all very commonly observed in coastal limestones where certain portions of heterogeneous units resist dissolution better than others. For example, on a small scale, aragonite grains can be preferentially removed from calcite surroundings; on outcrop scale, interlaminated chalk and chert layers will develop uneven textures due to their differential resistance to erosion. Coral fossils within micrite matrix (Fig. 2.4a), chert nodules within chalk (Fig. 2.4b), secondary calcite veins within fractured limestone, and rhizoliths or pockets of paleosol (Fig. 2.4c) may all turn out as knobs, ridges, or other positive relief features that withstood erosion better than the surrounding surface. If the feature is preferentially removed due to being less resistant than the surrounding medium, or simply falls out as a result of erosion, it will leave behind a cast.

2.3.8 Smooth Surfaces

Though the "signature" textures of marine karst are harshly eroded, many coastal outcrops exhibit smooth and polished surfaces. This is typical on limestone boulders or beachrock deposits that are embedded in sand on high-energy beaches. Turbulent flow of water carries sand grains in suspension and acts as "sand-paper" upon immobile surfaces it breaks against (Fig. 2.4d), smoothening karren features in the process. Alternatively, regularly wetted surfaces may be covered with fast-growing turfs of green or brown algae, creating luxurious bioprotective carpets that isolate the rock from direct impact of splash and spray. In either case, the resultant smooth surfaces are very distinct from karren-rich areas that may be nearby but out of reach of polishing action of surf-suspended sand or buffering by algae. In addition, smooth surfaces are also encountered in places where there is no nearby source of sand (Fig. 2.4e). It is hypothesized that they are promulgated by presence of very thin and nearly stationary films of water isolating the rock from erosive agents (Trudgill 1985). Existence of such boundary layer of laminar flow is suggested as the cause of some smooth rock surfaces in areas regularly sprayed by waves (Gómez-Pujol and Fornós 2009).

Smooth surfaces can also be produced beyond the reach of water by "sand-papering" by wind-blown sand grains. The resultant features are called ventifacts (Cooke et al. 1993) and are most common in various lithologies in deserts. However, they also occur in relatively homogeneous and dense carbonate rocks in coastal settings where there is ample wind and sand supply to act as abrasive agent (Veress et al. 2006). Diagenetically mature fine-grained limestone boulders exposed in vicinity of beaches or coastal dunes are known to develop wind polished facets over time (Knight 2005). Finally, one unusual and relatively smooth texture may form in laminated argillaceous limestones. Marls that are recurrently wetted by waves and dried by the sun manifest spheroidal weathering through polygonal cracking and exfoliation of clay-rich layers (Fig. 2.4f).

2.3.9 Fractured Surfaces

Karren and other secondary textures will be absent in coastal areas where cliffs and scarps retreat too quickly. If this is so, fractured facets repeatedly replace incipient dissolutional sculpturing and reproduce unaffected rock surfaces. This may occur on large scale, resulting in relatively karren-free stretches of coastline; or on local scale, creating recently-stripped patches devoid of karren and appearing rather featureless in comparison with adjacent surfaces that were subaerially exposed for longer time (Mylroie and Mylroie 2009). With time, dissolutional textures develop again but persist only until a major storm or another disturbance breaks the rocks again

Fig. 2.4 Smooth and fractured surfaces and relict karren-like features. (**a**) Coral colony embedded in fine-grained matrix of Pleistocene limestone in Okinawa, Japan; (**b**) Chert nodule embedded in Cretaceous chalk near Beirut, Lebanon; (**c**) Closeup of a preserved lump of paleosol emerging from the floor of a solution pan shown in Fig. 2.3f; note the highly pockmarked surface and littorinid snails [lens cover is 52 mm in diameter]; (**d**) Limestone outcrop polished by sand-laden wave swash at Ras Al Jinz, Oman; (**e**) Smooth surfaces in Cretaceous limestone, just above mean sea level and regularly splashed by waves, Sveti Marko island, Croatia [6.5 cm of the ruler is showing]; (**f**) Unusual texture on the surface of Cretaceous marly chalk subjected to wetting, drying, cracking, and volume changes of clay content, near Tripoli, Lebanon [scale bar is 10 cm]; (**g**) Natural breakage, caused by storm waves, in very exposed Quaternary calcarenite, with freshly broken karren-free surface (top arrow) in color and texture contrast with longer-exposed karren-covered surface (bottom arrow) [ruler is 30 cm long]; (**h**) Small voids created by mixing zone dissolution within limestone and subsequently exposed by coastal erosion in Saipan, Mariana Islands; these features should not be confused with rock surface textures and structures [photograph and © by John E. Mylroie]; (**i**) Sponge-like porosity developed by very aggressive mixing zone dissolution in a coastal cave, Guam; it is reminiscent of very rough eogenetic karren

(Fig. 2.4g). In contrast, removed clasts generally do not renew dissolutional texturing because they are redistributed by waves and currents. In the process, they are subject to numerous impacts with adjacent particles. With time, sharp edges are broken off, and angular clasts gradually become rounded. Roundness of clasts increases with time of mechanical reworking and particle size and decreases with rock hardness (Allen 1985). Limestone is softer than most rocks and readily produces rounded and smooth pebbles and cobbles along karst coastlines. When clasts collide with sufficient force, they may fracture. In heterogeneous, eogenetic limestones, fracturing occurs along structural weaknesses and re-creates angular and irregular surfaces. In homogeneous, fine-grained, densely crystalline limestones, breakage may create percussion fractures and curvilinear facets (Bourke and Viles 1997).

2.3.10 Relict Morphologies

This discussion of small- and medium-scale coastal exokarst features excludes textures and structures that form below the land surface and by action of groundwater. Subsoil karren (Zseni 2009), such as smooth-walled and sinuous tubes and hollows that originally formed by slowly moving acidic water beneath a cover of soil and vegetation, can be exposed along modern coasts by erosional retreat and are often encountered in recently barren erosional surfaces on upper shore platforms and in the walls of marine notches. In general, subsoil karren co-occur with pockets of lithified soil. Once exposed to subaerial weathering, both are overprinted by various karren, particularly prolific small pits and sponge-like corrosion features that readily develop in carbonate-rich paleosol (Fig. 2.4c).

A somewhat similar, though more extremely dissolved, type of cavernous weathering (see Ford and Williams 1989) forms by highly aggressive dissolution within halocline zone, particularly in cenotes and flank margin caves (Mylroie and Carew 1990). It involves sponge-like porosity, with numerous completely penetrating holes and often rough edges (Fig. 2.4h). For lack of better term, this is known as *swiss-cheese morphology* (after Baceta et al. 2001, etc.), though *spongework morphology* and *boneyard morphology* are also used (e.g. Chap. 13). This type of sculpturing is initiated exclusively within an aquifer's phreatic zone in areas where freshwater-seawater mixing corrosion and microbially-mediated processes maximize dissolution. It may be exposed along the coast where it should not be considered karren but a remnant feature of voids opened by erosion and collapse. Extremely corroded varieties (Fig. 2.4i) may be reminiscent of coastal eogenetic karren and could be confused with it. Care should be taken to avoid this because the two bear entirely different paleoenvironmental connotations.

2.4 Bioerosional Textures

Bioerosional textures are karren-like traces made directly by living organisms that penetrate or otherwise damage rock surfaces. Substrates are attacked by mechanical means (Ansell and Nair 1969) – using teeth, shell, spines, and other organs, as well as by chemical means – via production of metabolic acids or excretion of ligands and enzymes. In terms of mechanics of rock destruction, organisms engage in boring, rasping, scraping, drilling, and other activities, which leaves behind a great variety of bioerosional markings, including some that are highly distinct. This makes them potentially valuable tools in paleoenvironmental interpretation when taxon and behavior of organism that made them can be identified. Bioerosional textures are classified here by general architecture of features, though more rigorous studies systematize them into ichnofossils that are assigned names based on behavior and identity of perpetrating organism (Ekdale et al. 1984).

2.4.1 Microborings

Coastal limestone surfaces are overrun by microorganisms living in several ecological guilds.

Epilithic microbes produce rock surface biofilms, chasmolithic and cryptoendolithic microbes inhabit existing cracks and pores within the substrate (Ginsburg 1953), and true endolithic microbes penetrate into the rock and produce bioerosional traces referred to as microborings. The term micropits is also used, especially for superficial markings as opposed to deeper holes, and though it tends to describe solutional microtopography, it largely overlaps with microscopic biogenic boreholes when gravitational control is not evident (Gómez-Pujol and Fornós 2009). Despite being ubiquitous, microborings (or micropits) are generally not visible to the naked eye due to their sub-millimeter size. Instead, they are studied in petrographic thin sections and by scanning electron microscope (SEM). There are countless species-specific morphologies, which often correspond to body outlines of producing organisms and can be recognized as belonging to a certain taxon (Glaub et al. 2007). Main groups of organisms that bore into limestone substrates of coastal and nearshore environments are cyanobacteria, algae, and fungi. They attack the substrate by chemical means via production of acids and chelating compounds or manipulation of photosynthetic and respiratory activities during daily cycles (Tribollet 2008) or by physical action of hyphae (Chen et al. 2002). They bore into rock in order to escape environmental stresses and predators. Photoautotrophic microbes remain ecologically limited to near-surface layers reached by light, as filaments retain connections with the surface (Trenhaile 1987). They can thoroughly pervade rock down to what is known as light compensation depth (Torunski 1979), below which respiration exceeds photosynthetic assimilation. This was seen in microscopic thin sections to be about 1 mm (Horwitz and Roberts 2010) and is not expected to exceed 1 cm (De Waele and Furlani 2013). In contrast, heterotrophic endoliths develop deeper-boring behavior because they do not have requirements imposed by photosynthesis. Schneider (1976) distinguished the two and considered shallow borers not true endoliths. He called them cariants – microorganisms that produce surface pitting and corrode the rock surface to give it a decayed (carious) aspect. They are instrumental in the formation of highly irregular, gravity-independent forms of karren. Deep microborings are the domain of heterotrophic endoliths, fungi in particular, which are known to drill into coastal limestone to depths of 0.5 m and produce stromatolite-like structures via trapping and binding of carbonate by calcified spores and filaments within the rock (Duane et al. 2003).

The products of combined activity of microborers, both cariants and true endoliths, can be observed along carbonate coastlines in many parts of the world. Microborers can colonize all carbonate substrates but locations and intensity of their activity in specific places is controlled primarily by pressure from predators, and spatial and temporal availability of water and light (Kleemann 2001). In the intertidal zone, cyanobacteria and algae bore into the substrate primarily to escape predation by invertebrates that feed on them (Schneider and Torunski 1983; Tribollet 2008). This provokes bioerosional attack by mollusks, echinoderms, and other algivorous organisms which acquire food by destroying the microbe-inhabited top layer of rock, already structurally weakened by microborings. In turn, the microborers are stimulated to penetrate deeper into the substrate. The combined action of microbes and their grazers results in extremely effective synergistic bioerosion concentrated in the intertidal zone and leads to production of large scale erosional morphologies, notably erosional benches and notches (Torunski 1979).

In the constantly wetted and dried zone of wave splash, microbes, algae, and fungi bore into the substrate in order to find a more stable microenvironment buffered from the stresses of high insolation and desiccation. Microbial infestation decreases rock strength and increases surfaces exposed to dissolution and other corrosive processes, helping produce the extremely corroded forms. SEM studies have shown that rock surfaces bored by cyanobacteria and algae can have up to 50 % void space, resulting in extremely fretted rock textures (Trudgill 2003). Eogenetic karren, in particular, can be considered, to some extent, a large-scale by-product of rock infestation by filamentous cyanobacteria (Folk et al. 1973; Jones 1989) and legion of other microbes (Duane et al. 2003). It is so perforated by microbes that profusion of tiny (under millimeter in diameter) circular holes, named alveoli (Moses 2003), can be seen by the naked eye. In some places, coastal karst may exhibit light-oriented rock textures referred to as photokarren (Simms 1990) and considered to be relatively pure by-products of photosynthetic microborers.

2.4.2 Dark Belts and Patches

Surfaces of rocks that are home to prolific microbial communities can often be identified by the naked eye because they appear as stained, usually darker, patches compared to otherwise pale limestone rock. This is familiar from fractured coastal limestones, where cracks in the rock develop a darker hue due to being preferentially colonized by microbes (which respond to microenvironmental variations in shade and water availability) and give a web-like appearance to the rock surface (Fig. 2.5a). On more uniform rock surfaces, microbial colonization can be more haphazard and creates random dark patches that coalesce over time. Darker patches are often accompanied by a difference in microrelief, with depressions forming in places where the top layer of rock was colonized and gradually destroyed by microbes (Fig. 2.5b). This results in surface lowering and augments color contrast with the surrounding rock. On the large scale, microbial colonization is readily apparent in microenvironments that are periodically, but not permanently, wetted by sea water, resulting in the darkest rocks being localized to upper intertidal and lower supratidal zones. Below it are yellowish-brownish colored rocks colonized by marine algae and grazed upon by invertebrates (Fig. 2.5c). In places with low wave energy, microenvironmental zonation is tightly controlled by tidal wetting regime and produces narrow and sharply-defined dark-colored belts parallel to the sea level in intertidal and wave splash zones. Laterally extensive belts of differently-colored rock at and just above the water level are a familiar feature of many temperate limestone coasts (Fig. 2.5d).

In addition to continuous belts, localized dark patches form in places of intense surf and tend to expand along topographic lows and fractures in rock where wave splash pools or flows. Despite ubiquity, these dark areas appear to have had no geomorphic term assigned to them. Local names used along the northern Adriatic coast are *mrkine* and variants (Lovrić et al. 2002). Comparable color-belt zonation can also be observed in the tropics (Fig. 2.5e). Folk et al. (1973) described their coastal "phytokarst" in the Cayman Islands as having black color, quite unlike the light color of rock surfaces inland. In the Bahamas, several color belts have been noted. Intertidal areas that are regularly wetted are known as the "yellow zone" and those less wet in the supratidal zone are "dark zone" or "black zone" (Mylroie J, 2012, personal communication). Further inland is the "light zone" in which wetting by sea water is uncommon (Mylroie and Mylroie 2009). The color contrasts derive from microenvironmentally determined differences in microbial assemblages (Fig. 2.5f), which, directly or indirectly also affect the rock texture. The wave-splashed "yellow zone" will exhibit bioerosional scars made by individual marine invertebrates, the sprayed "black zone" will be extremely corroded with eogenetic karren, and the dry "light zone" will have somewhat smoother, less pitted texture. In the humid tropics, the black color of cyanobacteria-rich coastal limestones is so intense (Fig. 2.5g) that it places broad swathes of coastline in stark contrast with inland rocks. Depending on local conditions, these "black zones" abruptly terminate at the limit of normal sea-spray wetting or lack distinct boundaries as they imperceptibly transition into cyanobacteria-poor "light zone" rock surfaces inland (Taboroši et al. 2004). The acquired microbial coloration of rocks is most apparent in places of abrupt microenvironmental change (Fig. 2.5h) or when juxtaposed with freshly broken surfaces (Fig. 2.5i).

Fig. 2.5 Texture and coloration pattern developed in coastal karst rocks by microbial action. (**a**) Fractured surface of Cretaceous limestone approximately 0.5 m above sea level shows that cracks are preferentially colonized by microbes and gradually develop a darker color and lower relief than the main body of the rock, Sveti Marko island, Croatia [scale bar is approximately 15 cm long]; (**b**) Corrosion of light-colored and smooth surface of Cretaceous limestone exposed within wave splash zone and the gradual development of a darker, pitted surface colonized by endolithic microbes, Sveti Marko island, Croatia [small water bottle cap for scale]; (**c**) Vertical zonation in rock color and texture observed at the same location as in the previous photograph: yellowish-brownish area at the base is regularly wetted by tides and exhibits rich algal coating and some rough bioerosional surface, dark area above is dry but wetted by enough spray to support colonies of rock-corroding microbes, and the light-colored smooth area at the top is the original rock surface largely unaffected by marine and biological erosion [scale bar is 15 cm and its position marks the sea level at the time the photograph was taken]; (**d**) A broader view of the same general location as the previous two photographs; (**e**) Horizontal belts showing different rock color and texture along the coast of Guam: intertidal yellowish-brownish belt with epilithic algae grazed upon by marine invertebrates, lower supratidal dark belt with endolithic cyanobacteria beyond the reach of most marine predators, and uppermost light-colored rocks whose surface is similar to those of inland rocks; (**f**) Differential coloration of the surface of coastal eolianite in the Bahamas, as produced by differences in microenvironmental parameters and ensuing microbial communities [photograph and © by John E. Mylroie]; (**g**) Originally light-colored coastal limestone that acquired nearly black color due to infestation by epilithic and endolithic microbes, Guam; (**h**) Supratidal Pleistocene reef limestone in Guam with pronounced color contrast between biofilm-poor grey surface that was relatively recently uncovered by natural removal of beach sand and the longer-exposed biofilm-rich brown surface [scale bar is 20 cm long]; (**i**) Color contrast between biofilm-covered dark-colored exposed surface of Pleistocene eolian *calcarenite* and its freshly-broken light-colored interior, coast near Cueva del Indio, Puerto Rico

2.4.3 Raspings and Scrapings

Mollusks, echinoids, crabs, and other animals feeding on rock surface biofilms, turf algae, and endolithic microbes engage in superficial forms of mechanical erosion. Chitons have extremely hard teeth capped by magnetite; and limpets and possibly other gastropods have radulas that contain silica and goethite (Stone et al. 2005). This imparts hardness far greater than calcite and aragonite substrates and allows the animals to easily remove uppermost layers of rock as they graze upon them. Rasp marks engraved into the substrate by chitons are prominent (Fig. 2.6a) and are usually shaped as meandering or straight paths of longitudinal grooves (Wisshak 2006). Ingested material is processed to remove nutrients and is then excreted as whitish fecal pellets, composed of as much as 96 % pulverized $CaCO_3$ (Rasmussen and Frankenberg 1990). Analysis of pellets of limpets has shown that they consume up to 5 g of substrate per year per individual, causing up to 0.5 mm of surface lowering per year (Andrews and Williams 2000). Littorinid snails (winkles) also "scrape a living" (Norton et al. 1990) by feeding on algal biofilms. They lack a mineralized radula (Spencer 1988) and do not seem to leave individual traces visible to the naked eye. However, they can remove carbonate grains loosened by other agents and are effective in destroying weakened rock (see Fig. 2f) fraught with microborings by cyanobacteria upon which they feed (De Waele and Furlani 2013). Echinoids have a highly specialized, pentaradiate chewing organ consisting of five united jaws and known as the Aristotle's lantern. As they graze upon a rock, they leave characteristic star-shaped pattern of grooves. The tip of each jaw bears a rapidly growing calcite tooth to balance for loss during rasping. In addition, echinoids rely on their tough spines to fasten themselves in their hiding places and may scrape surrounding rock in the process. Some crabs can also abrade limestone surfaces in search of food. For example, scratches made by grapsid crabs (Fig. 2.6b) are usually the only visible bioerosional markings in supratidal areas that cannot be reached by less motile invertebrates.

2.4.4 Homing Places

Many rasping and scraping bioeroders create their own resting sites in the rock. Chitons produce pronounced pits that accommodate own body size and represent an individual's long term residence (Fig. 2.6c). They make regular journeys away from these homing scars in order to graze on surrounding rocks, but return to the exact spot they previously inhabited. Limpets also live in self-made scars that correspond exactly to the size and shape of an individual's shell (Fig. 2.6d). Many echinoids, such as *Paracentrotus* in temperate regions and *Echinometra* in tropical regions excavate individual hiding boreholes (Fig. 2.6e) and expand them during their lifetime (Fig. 2.6f). Some sea urchins even engage in gardening of algal turf within their boreholes (Asgaard and Bromley 2008) so their homes may contain a hiding hole for protection and extended V-shaped (Fig. 2.6g) or winding galleries that host algal gardens on which the animal feeds (Fig. 2.6h). Deepening and coalescence of adjacent cavities may produce undercut ledges and widening of the gardens eventually creates meter-scale echinoid-made tide pools (Fig. 2.6i) that further amalgamate and become major intertidal features and unique habitat for other organisms (Schoppe and Werding 1996).

2.4.5 Borings

Many invertebrates bore into rock in order to live within it, mostly for protection from predators. They create boreholes that completely and permanently house the occupant. These structures in hard substrate are not to be confused with burrows in soft sediment. Typical endolithic organisms that create boreholes in coastal limestones are certain bivalves, barnacles, sipunculid and polychaete worms, echinoids, etc. Organisms can often be identified based on the type and shape of the borehole opening in the rock surface. In some cases, organisms (such as worms and crustaceans) penetrate living substrate (such as coral and coralline algae) and allow their entry holes to be sealed by the host's growth, leaving behind no

Fig. 2.6 Rasp marks, scrape marks, and home scars produced by invertebrates in intertidal limestone. (**a**) Texture formed in the floor of a marine notch exposed to long-term grazing by chitons, Railay, Thailand; (**b**) Fresh grazing marks left by grapsid crabs in the dark-colored epilithic biofilm on supratidal rocks, Guam; inset shows light-colored $CaCO_3$-rich crab fecal pellets from the same location, at the same scale; (**c**) Chiton homing scars, only one occupied, during low tide, Palau; (**d**) Limpets clinging to wave-splashed rock in intertidal zone, Guam; note the color and texture contrast between unoccupied homing scars and surrounding rock [flashlight is 12.5 cm long]; (**e**) Boring sea urchins in own cavities, Okinawa, Japan [tape measure extended 19 cm]; (**f**) Fossil cavities originally made by sea urchins and now seen in the roof of an uplifted marine notch, Railay, Thailand ; (**g**) Boring sea urchin hiding in a self-made borehole whose V-shape was created by regular grazing, Okinawa, Japan [image width is 16.5 cm across]; (**h**) Adjacent trough-like boreholes created by sea urchins that graze regularly in "algal gardens" that grow within, Okinawa, Japan [tape measure extended 20 cm]; (**i**) Large tidal pans formed by amalgamation of numerous sea urchin cavities; their perimeters are overhanging and colonized by sea urchins whose activity continues to expand the pans' size, Okinawa, Japan

surface expression. This results in fully enclosed boreholes known as embedment cavities.

Best known rock-boring bivalves belong to the *Lithophaga* genus, whose Greek name means "rock eater." They use chemical or physical means to create deep club-shaped cavities that accommodate the shell and increase in diameter with the growth of the organism (Wilson 2007).

The boreholes have openings to the rock surface in order to provide access to seawater that these filter-feeders require. Their characteristic dumbbell-shaped surface expression corresponds to the organism's inhalant and exhalant siphons (Fig. 2.7a). Infestation by *Lithophaga* bivalves enhances erosion of coastal rocks by making the often densely-riddled rock very susceptible to fracturing and breakdown. Original openings are quickly lost with erosion and lowering of the rock surface, which reduces boreholes (Fig. 2.7b) to shallow pits (Fig. 2.7c) and pockmarks (Fig. 2.7d) before obliterating them. Similar holes are produced by *Hiatella arctica* in coastal limestones of arctic regions (Brookes and Stevens 1985). In the tropics, *Tridacna* bivalves make larger, lenticular holes to whose bottoms they are permanently attached and which have wide amygdaloid openings through which the shell walls and animals' soft tissues are clearly visible (Fig. 2.7e). Rock-boring barnacles, notably *Lithotrya* genus, live a similarly sessile lifestyle. They create cylindrical or pouch-shaped holes with surface openings and remain attached to the bottoms of their holes. A diagnostic feature of their boreholes is an oval cross-section (Ahr and Stanton 1973). Boring sipunculid and polychaete worms produce boreholes that are generally thinner than those of bivalves and barnacles (Trudgill 2003). Deposit feeding worms may create U-shaped or winding domicile tunnels (Liu and Hsieh 2000). Their width corresponds to body size but longer length allows an individual to move throughout and collect detritus trapped within. The characteristic surface expression of worm boring are paired openings (Fig. 2.7f).

2.4.6 Complex Networks

Over time, intensive boring by bivalves, worms, and other invertebrates may reduce a host rock to a hollow mass full of holes. Truly sponge-like interconnected network of voids, however, is the signature pattern of none other than boring sponges. These sponges, notably *Cliona* spp., are known to penetrate calcareous substrates such as rock and shells and produce interconnected networks of voids whose overall morphology is reminiscent of sponge's own anatomy (Ekdale et al. 1984). From the outside, sponge borings appear as numerous mm-scale apertures in rock surfaces (Fig. 2.7g), which, if the surface is broken, reveal connections to complex internal networks of chambers (Fig. 2.7h). While the sponge is alive, brightly colored sponge tissue can be seen emerging from the openings or entirely coating the rock surface. Upon death, the lattice eventually collapses to carbonate "chips" which are redistributed as sediment (Wilkinson 1983).

2.4.7 Etchings

Some organisms attach themselves to a hard rock substrate but do not significantly penetrate into it. When removed, they may leave characteristic markings etched in the rock surface. Scars left by sessile epilithic organisms permanently attached to rock substrate are known as etchings (Ekdale et al. 1984). They are commonly produced by bryozoans, bivalves, brachiopods, and barnacles, and only in sites where an individual was affixed to the substrate. They are seen in coastal carbonate rocks only after the individual has died and been removed. The morphology of etchings varies with species, but they are usually difficult to notice due to overprinting by more significant bioeroders.

2.4.8 Drill Holes

Finally, an interesting type of bioerosional trace is produced by predatory gastropods that feed on bivalves. They produce circular drill holes commonly seen in sea shells on sandy beaches (Fig. 2.7i). Drill holes are not observed in rock substrates because their purpose is to provide access to the soft tissue of living prey. If the process is not completed, there may be a partial excavation (Wilson 2007). Predatory drilling is locally common on epilithic bivalves covering intertidal rocks (Sawyer and Zuschin 2010) and thus contributes to coastal erosion by disrupting the bioprotective layer. In addition to gastropods,

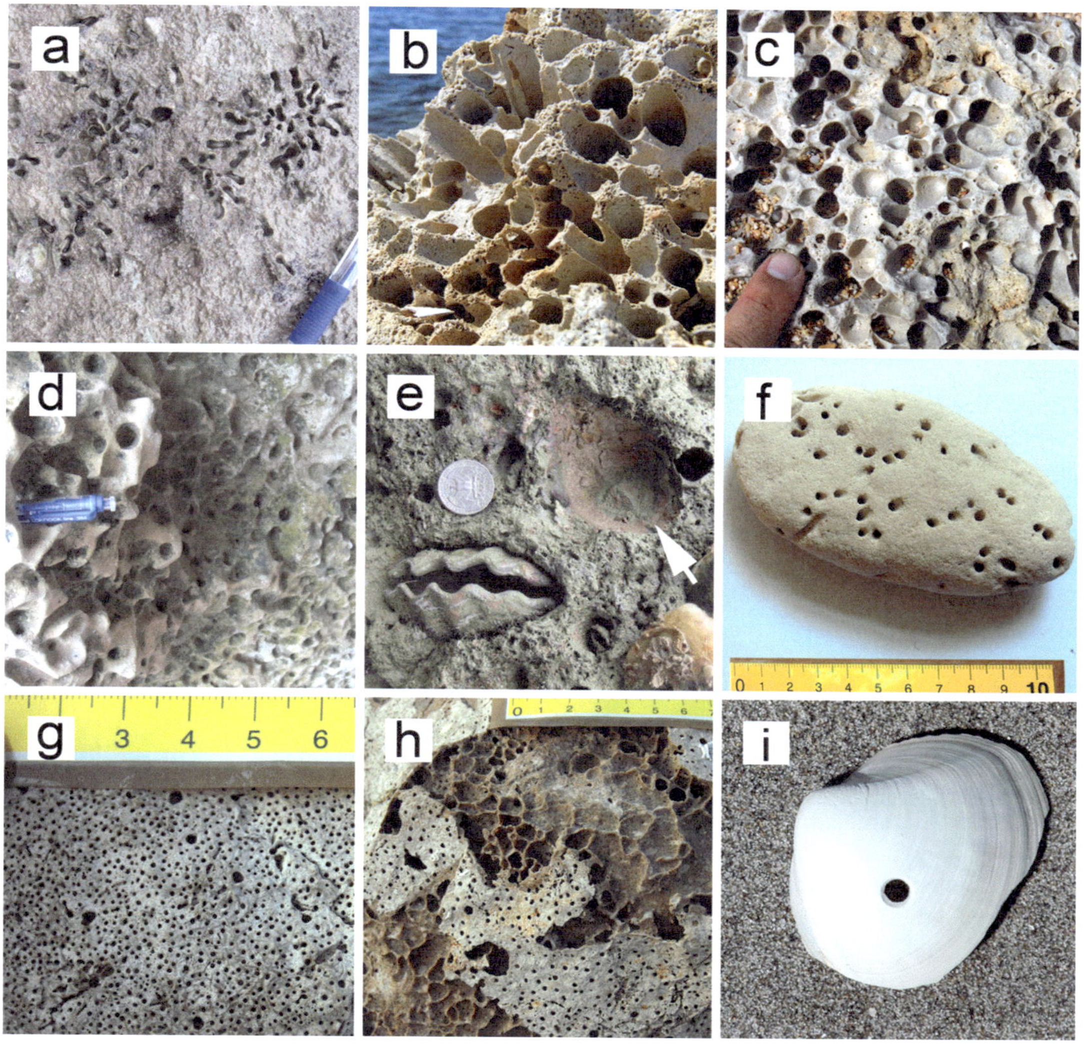

Fig. 2.7 Boreholes produced by invertebrates in intertidal limestone. (**a**) Cluster of actively-bored bivalve holes whose dumbbell shape is an expression of the organisms' double siphons, Railay, Thailand; (**b**) Limestone boulder heavily bored by bivalves (large holes) and sponges (small holes), Istria, Croatia [hole diameter is approximately 1 cm]; (**c**) Rock formerly colonized by boring bivalves whose holes have been partly eroded away, Ras Al Jinz, Oman [finger for scale]; (**d**) Texture consisting of innermost portions of former bivalve boreholes that have been partly destroyed and reshaped by erosional lowering of the rock surface and smoothening by waves, Permian limestone, Phang Nga, Thailand; (**e**) Recently died *Tridacna* bivalve *in situ* within its hole (lower left) and a vacant borehole of the same species where organism has been removed by erosion but the impression of the commissure of its shell valves remains (indicated by arrow), Palau; (**f**) Wave-rounded piece of coral exhibiting small boreholes with paired openings thought to have been made by worms, Socotra, Yemen; (**g**) Numerous small-diameter boreholes created and used by a boring endolithic sponge to interface with the outside environment, Kraljevica, Croatia; (**h**) Honeycomb-like galleries that used to host the main body of a boring sponge, as revealed by natural breakage of the surface rock layer, in the same location as the previous photograph; (**i**) Drill hole made by a predatory gastropod in a shell of a living bivalve, Hokkaido, Japan

spionid polychaetes also drill into bivalve shells and can interfere with health of intertidal mollusk beds (Wargo and Ford 1993).

2.5 Depositional Features

Like all interfaces of land and the ocean, karst coasts can be predominantly erosional or depositional depending on the local conditions. As discussed above, erosion can take numerous forms resulting from physical action of the ocean, chemical processes of karstification, and biological effects of biota. These suites of processes can be locally reversed to produce the opposite effect: accumulation of calcium carbonate. This is reflected in deposition of loose sediment, cementation and lithification of sediment, precipitation of calcium carbonate, and bioconstruction (discussed in the next section).

2.5.1 Loose Sediment

Karst coasts commonly include sites of clastic sediment accumulation. Sand in karst settings is composed of predominantly carbonate or of mixed carbonate-siliciclastic material. It may include land-derived limestone clasts, bioclastic material composed of whole and fragmented skeletal grains (Fig. 2.8a), locally precipitated ooids, peloids, as well as other particles, including non-carbonate components (see overview of carbonate beaches by Richmond 2002). Bioclastic material may account for 100 % of deposits on some coastlines, particularly on atolls and other low carbonate islands. Lower energy settings, such as deeper parts of reef lagoons, mangrove swamps, and some tidal flats, acquire deposits of calcareous mud (Fig. 2.8b). It derives from physical breakdown and biogenic micritization of larger particles, as well as biotic and abiotic precipitation of micrite directly (Reid et al. 1990). Higher energy beaches and shallows, e.g. on windward sides of carbonate islands, contain carbonate gravel and cobbles. Those are usually made of limestone clasts, pieces of coral and algal buildups (Fig. 2.8c), and whole or broken bivalve and gastropod shells. There are also very coarse limestone blocks which accumulate along rapidly eroding karst coastlines, typically as talus deposits at the bases of retreating cliffs and collapsing caves. This includes boulder-sized colluvium induced by rockfall (Fig. 2.8d).

2.5.2 Cemented Features

Beach sand (as well as mud and gravel, e.g. Scoffin 1970) can become cemented by locally precipitated cements and turned into lithified deposits known as beachrock. Beachrock is recurrent in, but not restricted to, coastal karst areas and usually involves poorly-sorted grains and cement of aragonite or high-Mg calcite (Scoffin and Stoddart 1983). It typically takes the form of consolidated, layered beds or slabs that gently dip toward the sea at the approximate angle of original beach slope (Fig. 2.8e). Beachrock develops through physicochemical precipitation of aragonite and calcite from seawater (Stoddart and Cann 1965), groundwater, sediment pore water (Hanor 1978), and microbial activity (Krumbein 1979) occurring at sea level and beneath beach sand. Beachrock is subject to corrosion as soon as it is exposed (Revelle and Emery 1957) and develops numerous karren and bioerosional markings. Somewhat related deposits are eolianites (Fig. 2.8f), which are also a common feature of karst coasts and develop from well-sorted, wind-transported sand particles that become cemented by calcite precipitated from downward percolating meteoric water in a vadose environment, above the sea level (Russell 1962).

2.5.3 Littoral Tufa

Calcareous tufa deposits are friable deposits of calcium carbonate, somewhat reminiscent of chalk and travertine. There exists a unique type of calcareous tufa that is found exclusively in coastal karst settings (Taboroši and Stafford 2004). The basic mechanism of tufa formation is similar to that which gives rise to speleothems (see Dreybrodt 1988), but is driven by water evaporation and biotic processes in open atmosphere instead of groundwater degassing

Fig. 2.8 Depositional and precipitated features along karst coasts. (**a**) Carbonate sand, composed of bioclastic material, produced by breakdown of mostly mollusk shells, and a small admixture of terrigenous siliciclastic grains, Ras al Jinz, Oman [scale bar is 3 cm]; (**b**) Calcareous mud and fine sand from mangrove-protected landward edge of a small cay, La Parguera, Puerto Rico; (**c**) Carbonate gravel composed of coral, algal, and molluskan fragments, from seaward edge of the same cay as in the previous photograph; (**d**) Talus of collapsed boulders produced by breakdown of a flank margin cave in Guanica, Puerto Rico; (**e**) Slabs of beachrock, occurring in several layers and matching the current dip of the beach slope, Guam; (**f**) Cross-bedded eolianite deposits of Tamala Limestone, Australia [photograph and © by S. K. Lowry]; (**g**) Marine-influenced tufa deposit, reminiscent of cave flowstone, covering the back wall of a raised marine notch in Tinian, Mariana Islands; note that tufa deposition occurred subsequent to bioerosional pockmarking of the rock surface; (**h**) Speleothem-like deposit of tufa in a raised marine notch at the same general location as the previous photograph; (**i**) Unusual deposit of tufa with coralloid surface and growing under conditions of regular and vigorous wave splash in an active marine notch, Tinian, Mariana Islands

and slow inorganic growth of sparry crystals inside caves. Because carbonate precipitation from saturated waters at the land surface is rapid, it results in poorly arranged and randomly oriented microcrystalline aggregates. Calcareous tufa typically forms in karst springs, streams, waterfalls, and some cliffs and cave entrances, but the special littoral subtype of tufa is commonly encountered in the roofs and walls of marine notches in the humid tropics (Fig. 2.8g). Its main distinctions from other types of tufa are speleothem-like (Fig. 2.8h) and especially coralloid morphologies (Fig. 2.8i), often almost purely aragonitic composition, and the presence of rich communities of halophilic microbes (that tolerate or necessitate the presence of seawater). Littoral tufa is often not recognized as an *in situ* actively forming notch deposit and has been neglected by researchers (but see study by Jones 2010). Its major significance is that diagenetic changes can transform it into deposits hardly distinguishable from old speleothems, complicating the field interpretation of raised marine notches vs. breached flank margin caves (Taboroši et al. 2006; Reece et al. 2006).

2.6 Bioconstructional Features

The vast majority of carbonate precipitation in the coastal zone is biologically driven. Plenty of marine taxa produce biomineralized skeletons which become loose sediment upon the organism's death. Some groups, however, are also capable of building free-standing calcium carbonate structures. Such bioconstructions occur in both shallow and deep waters, and include features as diverse as stromatolites, microbialites, rhodoliths (Basso and Tomaselli 1994), *Halimeda* bioherms (Orme and Riding 1995), sponge reefs (Conway et al. 2005), byssal mats (Frey 1973), and of course, algal and coral reefs (Hopley 2005a). In this discussion, we are concerned only with small- and medium-scale features that develop directly upon littoral rock and as integral parts of coastal karst settings. This typically involves organisms that live colonially or in aggregates, lead sessile lifestyles on benthic substrate in intertidal or shallow areas, and reproduce and grow relatively quickly. In the majority of cases, these are coralline red algae, bryozoans, serpulid and sabellarid worms, vermetid snails, and oysters. They produce encrusted rock surfaces and different types of unattached $CaCO_3$ concretions and buildups. In addition, mussels and some other bivalves produce organic filaments to attach themselves to substrate and cover subtidal and intertidal bedrock.

2.6.1 Encrustations

Original rock surfaces of many intertidal carbonates at all latitudes cannot be examined directly due to rife biogenic encrustation. Some rock encrusters can occupy large areas and impart new morphological, biological, and geological characteristics upon the environment they settled. Probably the most widespread encrusters along karst coastlines are coralline red algae, which colonize and thoroughly cover substrates by growing to fuse together into dense and rigid coatings (Fig. 2.9a). As they expand, they follow the relief of the substrate by folding over microtopography and overlapping to produce multiple and partly fossilized layers (Giaccone et al. 2009). The resultant covering may be monospecific (e.g, *Lithothamnion* in temperate areas, *Porolithon* in the tropics) under some circumstances (e.g. in ecologically narrow niches). More commonly, encrusting organisms interact with adjacent encrusters to develop a polyspecific patchwork and sometimes even cover and entomb other organisms. Coralline red algae are limited by their phototrophy to shallow settings and are widespread in high-energy environments. Some species are extremely prolific on very exposed coasts and are often the only sessile organisms sufficiently tough to withstand constant surf. They have very narrow range around mean sea level and are considered excellent zero-elevation indicators (Laborel et al. 1994). In low-energy conditions, algae are able to modify their building technique and switch from thick encrustations to various erect and branching structures (Fig. 2.9b). They can also grow as unattached concretions that freely roll with water movement along the bottom (rhodoliths, Fig. 2.9c) and in some cases form extensive subtidal deposits (*maërl beds*). Other common encrusters are foraminifera – amoeboid protists that may attach their calcareous shells (tests) to a wide variety of substrates (Fig. 2.9d),

Fig. 2.9 Bioconstructional and bioprotectional features on limestone coasts. (**a**) Broken piece of beachrock thoroughly covered by a calcareous coating of coralline red algae; it comes from a coastal freshwater spring in Guam, where monospecific encrustations by red algae with preference for brackish water give coastal rocks in the vicinity of groundwater discharge points unusual purple color; (**b**) Beach sediment comprised almost exclusively of broken pieces of branching coralline algae, Caroline Islands, Micronesia; (**c**) Free concretion produced by red algae without attachment to an immobile substrate, Socotra, Yemen; (**d**) Deep red encrustations of benthic foraminifera on a rock originally made by encrusting algae, Socotra, Yemen; (**e**) Piece of carbonate rock consisting almost exclusively of calcareous tubes of serpulid worms, Qurayyat, Oman; (**f**) Sabellarid bioherm in Natal, Rio Grande do Norte, Brazil [lens cover is 52 mm in diameter]; (**g**) Wave-splashed water cascading off an intertidal bioconstruction created predominantly by vermetid snails, near Beirut, Lebanon; (**h**) Oyster ledges attached to coastal rocks and boulders and photographed during very low tide, Socotra, Yemen; (**i**) Barnacle-covered rock on a mixed carbonate-siliciclastic sand beach in Qurayyat, Oman

bryozoans – highly diverse colonial animals whose mineralized skeleton crusts are especially common on high latitude coasts, serpulid worms – which secrete calcareous home tubes attached to and winding across the rock surface (Fig. 2.9e), sabellarid worms – whose parchment-like tubes are armored with cemented sand grains (Fig. 2.9f), and vermetid snails – whose hard shells are uncoiled and permanently cemented to the substrate. Most of these sessile animals are filter-feeders and their growth is boosted by steady flow of water (Fig. 2.9g). They proliferate in high-energy conditions at or near sea level, and may occur in relatively pure accumulations or in association with and as contributors to buildups by coralline algae.

In addition to coating rock surfaces, many encrusting organisms are capable of producing large and long-lasting structures that dominate the coastal zones of karst landscapes. The wave-battered outer margins of coral reefs and exposed rocky shores are sites of algal buildups that surpass mere encrustations and raise several tens of centimeters above the base surface. As colonies develop and organisms persistently attach their shells to bedrock and their antecedents' remains, they grow from surface coatings to irregularly shaped raised colonies. These buildups are often in the form of upward growing bands known as algal ridges or algal rims (Laborel 2005). Though they commonly grow along exposed margins of coral-algal reefs, they are independent of coral reefs proper. Though encrusting organisms are in many places limited to producing epilithic crusts (e.g., Azzopardi and Schembri 1997), they may also, with prolific growth, create standalone intertidal and shallow subtidal reefs (e.g. *boilers* of Bermuda, Ginsburg and Schroeder 1973) or, as is common along exposed subtropical and tropical rocky shores, substantial buildups directly upon littoral rock in the intertidal zone (see sea-level benches and platforms).

2.6.2 Bivalve and Barnacle Beds

In general, any sessile organisms that attach their carbonate skeletons to substrate and live at sufficient densities can thoroughly cover wide areas of bedrock. Thus, many solitary organisms, notably bivalves and barnacles, can conceal previously exposed rock surfaces. For example, many species of oysters have one valve permanently cemented to the substrate and warped to match underlying microtopography. After death, unattached valve may break off but the lower one remains as permanent rock coating, upon which new oysters can settle and grow. This is a true bioconstructional process and may result in large ledge-shaped accumulations (Fig. 2.9h) in the intertidal zone from temperate to tropical coasts. If preserved, they can be used as indicators of former sea levels.

Mussels, on the other hand, adhere to the substrate and to each other by using byssal threads. They form tight clusters that cover rock surfaces while individuals are alive, but are unfastened after death. Strictly speaking, this is not bioconstruction but bioprotection (Carter and Viles 2005) as the tightly packed shells attached to rock buffer it from erosion. In addition, they alter topography and microenvironment (Cocito 2004), and, while minimizing exposed rock surface, increase overall surface area and provide new substrate to be colonized and strengthened by encrusting organisms. Mussels are more common along cool shorelines and thrive near low-tide levels (Riding 2002). They have very high accretion rates locally and form thick beds. In the Adriatic, for example, mussels were found to build up in excess of 100 kg/m^2/year (Relini 2009).

Barnacles are crustaceans whose larvae settle on some marine substrate and begin sessile life within calcareous shells they secrete. They engage in numerous specialized lifestyles, including efficient bioerosion of live coral and limestone (Ahr and Stanton 1973) but are best known as rock surface dwellers in the intertidal zone (Fig. 2.9i). In places where they cover large patches of coastal rock substrate (Stephenson and Stephenson 1972), barnacles act as significant bioprotecting agents. Barnacle larvae are highly predisposed to settle in areas where adults of the same species are already present (Newman and Abbott 1980), resulting in tendency to carpet extensive stretches of exposed bedrock and minimize its exposure to bioeroding organisms. In addition to impeding bioerosion, barnacles are locally significant as sediment producers (carbonate sediments on the Florida shelf consist of up to 50 % barnacle shells; Milliman 1974).

2.7 Compound Structures

The various erosional, depositional, and constructional features discussed in this chapter can be understood as small-scale and medium-scale components that comprise larger coastal exokarst structures. Lundberg (2009) aptly called the former "building blocks" and the latter "modules", explaining that they often exist in a fractal-like continuum of scale. For example, bioerosion scars produced at the scale of individual organisms multiply to yield marine notches – prominent scars produced at the scale of rock outcrops. Simultaneously, as coastal outcrops are undercut, organisms encrust parts of eroded bedrock splashed by waves, but eventually proliferate to build up larger bioconstructions, including massive sea-level platforms and pools. Beyond the direct impact of waves, but within reach of sea spray, land surface is denuded and covered with karren features and solution pans, which escalate until they cover wide areas of shore platforms with karrenfelds – barren expanses of dissolutionally sculpted bedrock.

2.7.1 Marine Notches

The marine notch is an approximately semi-circular groove, up to several meters in diameter, that is cut horizontally into coastal cliffs and smaller outcrops near and parallel with the sea level (Fig. 2.10a). Its actual morphology ranges from inconspicuous nips to deeply incised and highly visible notches floored by extensive erosional benches. The formation of notches is attributed to biological erosion (Abensperg-Traun et al. 1990), mechanical erosion (Wziatek et al. 2011), and other factors (see Pirazzoli 1986). Rock textures within the notch give some idea of the dominant mechanism: rough, heavily scarred surfaces indicate intense bioerosion, smoother surfaces suggest abrasion (Fig. 2.10b). Dissolution is not expected in a marine environment but can influence notch formation in areas where there is much fresh groundwater discharge (Higgins 1980).

Marine notches are most pronounced in the tropics (De Waele and Furlani 2013) and become progressively smaller with distance north and south. They can be observed in many subtropical and some temperate (Trenhaile 1987) regions but are absent on high latitude coasts (Lundberg 2009), presumably due to minimal bioerosion. In addition, they are best developed in areas where low tidal range concentrates erosional processes in a narrow horizontal belt with limited vertical extent. Increased tidal range will cause the notch to be taller but less deeply incised (Lundberg 2009). Some areas exhibit a double notch (Focke 1978), with a horizontal raised lip dividing upper and lower parts. Sloping or low cliffs undercut by marine notches exhibit *visors* of overhanging bedrock that periodically collapse as notch incision progresses (Kogure and Matsukura 2010). Similarly, small islets may be entirely circumscribed by marine notches and form *mushroom rocks* (Fig. 2.10c), destined to eventually topple.

Cliffs in uplifted areas tend to exhibit fossil notches corresponding to previous sea-level stillstands. Marine notches are thus indicators of former sea levels (Kershaw and Guo 2001) though the precision at which they are useful is still discussed because their overall shape and the precise elevation of maximum penetration are under control of many factors. Even in relatively limited geographic areas, actively forming notches exhibit variable architecture due to the site-specific balance of local factors contributing to their formation and transform laterally along the coast as the conditions, particularly exposure to wave energy, change. Though it has been suggested that notches can form subtidally in moderately exposed places (Lundberg 2004) and supratidally in very exposed coasts subject to highly turbulent wave action (as in many Pacific Islands, and in the Mediterranean; Rust and Kershaw 2000), their vertical positions with respect to the intertidal zone may result from tectonic displacement. For example, the submerged notches in the northern Adriatic are being intensively studied to understand whether their subtidal position indicates a tectonic change in recent times (Antonioli et al. 2004).

Fig. 2.10 Notches and benches on tropical karst coasts. (**a**) Marine bioerosion notch, carved in Pleistocene Ryukyu Limestone, with extensive pitting and karren sculpturing on the floor, Misaki koen, Okinawa, Japan; (**b**) Marine abrasion notch in Paleogene chalk and marl deposits, near Sūr, Oman; (**c**) Small island circumscribed by a bioerosional notch, Koror, Palau; (**d**) Marine notch and a narrow erosional bench produced at its base, with some bioconstruction evident from the way edge of the bench floor projects outward and creates an overhang and water cascading effect, Santo Domingo, Dominican Republic; (**e**) Erosional bench transformed into a platform with a large pool whose water surface is close to a meter above the sea level and is held back by a well-developed bioconstructional rampart. Pinnacled eogenetic karrenfeld visible in the foreground, Tinian, Northern Mariana Islands. (**f**) Rimmed pools (*vasques*) on a coastal bench in Guam; each rim is a narrow ridge where filter-feeding vermetid snails thrive in the steady flow of water as it drains to lower-level pool

2.7.2 Sea-Level Benches and Platforms

As coastal cliffs and outcrops are undercut and progressively destroyed by erosion, an erosional bench is left behind at sea level by the receding coast. In tropical and subtropical regions, this bench extends seaward from and is continuous with the bases of marine notches. Its size ranges from narrow (<1 m wide) wave-cut benches with somewhat sloping and uneven floors carpeted by green and brown algae (Fig. 2.10d) to much wider planar platforms subject to vigorous bioconstructional activity due to prolific growth of encrusting algae and other organisms in wave-battered environment. They tend to coat the freshly-eroded surfaces and form thick, mostly algal, deposits called *trottoirs* (this term remains poorly defined, see Hopley 2005b). Growth of algae is most prolific in the seaward edges of the platforms where water flux from swash and backwash is most constant. They build up to produce bioconstructional ramparts that grow upward and seaward and act as dams that prevent wave-splashed water from quickly returning to the sea. In turn, that creates nearly-permanently filled pools atop of the platforms (Fig. 2.10e). The resultant platform-and-pool morphology is known by the French term *plate-form à vasques* (after the term *vasque* used for pools in these settings by Debrat 1974). The pools may develop additional sets of positive relief rims at places of most agitated water (where the growth of encrusters, particularly vermetids (Molinier 1955) is most stimulated). As the rims between basins grow to unequal heights, they partition the pools into stepwise series (Fig. 2.10f), in which higher pools spill over to lower ones. Their growth creates a cascading effect that further stimulates the growth of encrusters via positive feedback at spillover sites where water is more agitated than in the pools' inner parts. Concurrently, the pools become habitats in their own right and attract colonization by organisms that include bioeroders, notably sea urchins whose activity deepens and widens the pools, and appear to eventually attain a dynamic equilibrium between bioconstruction and erosion (Trenhaile 2003a).

These complex bioconstructional features form primarily at low latitudes and are particularly imposing in the tropics: Caribbean (e.g. Focke 1978; Jones and Hunter 1995), tropical Atlantic (e.g. Kempf and Laborel 1968), tropical Pacific (e.g. Emery 1962; Hadfield et al. 1972), and Madagascar (Battistini and Guilcher 1982). They are also found in the subtropics, such as South Africa (Miller and Mason 1994) and many parts of the Mediterranean (Dalongeville and Guilcher 1982), where, in absence of coral reefs, they tend to be the only major bioconstructions and are locally considered to be natural monuments (Bressan et al. 2009). In Southern Europe, these bioconstructions may develop on platforms without an associated notch profile (Gómez-Pujol and Fornós 2009) and even in areas without an eroded platform to act as support. They may be directly attached to cliffs as laterally continuous ledges at or just below the sea level. Such overhangs are known as *corniches* (Trenhaile 2003b) or *encorbellements* (Dalongeville 1995). In addition to encrusting algae, some of these structures can be dominated by other taxa, producing vermetid reefs (Safriel 1975), serpulid reefs (Bosence 1973; Glumac et al. 2004), and sabellarid reefs (Chen and Dai 2009). The majority, however, appear to be polyspecific and complex, containing algal, gastropod, polychaete, and other elements, and varying widely in appearance based on overall community structure, local substrate configuration, tidal range, exposure to waves and other factors.

2.7.3 Shore Platform Karrenfelds

Marine coasts are generally free of soil and normal vegetation up to the inland limits of regular reach of seawater. This is irrespective of the climate and geomorphic configuration of the coast and is true of subhorizontal shore platforms typical of tropical and subtropical regions, sloping erosional ramps of temperate and polar regions and vertical cliffs. All of them are characterized by diminishing contact with seawater from the shoreline landward (and upward): from permanently submerged subtidal zone, to periodically exposed intertidal zone, to supratidal areas that are intermittently wetted by wave splash and, further inland, occasionally reached by sea spray. Geomorphic reflection of this waning marine influence is a spectrum of different types of karren and bioerosional scars scattered over largely barren coastal limestone surface. Such scenery can be referred to as shore platform karrenfeld (or coastal karrenfeld), after more general German term meaning "karren field". The inland extent of shore platform karrenfeld depends primarily on the relative exposure/sheltering of the coast and penetration of sea spray inland. The more exposed an area is to surf, the wider the karrenfeld belt is expected to be, and vice versa. For example, coastal karren are observed up to 24 m inland at exposed sites and only 4–10 m inlands in sheltered sites on Jurassic limestone in Mallorca (Gómez-Pujol and Fornós 2009). The actual appearance of shore platform karrenfeld depends on practically all factors that influence coastal karst development and as such is subject to great complexity of features and their arrangements. Conceptual models are, therefore, multiple and usually based on the general climatic setting as one of the main overriding factors.

Along tropical and subtropical karst coasts, shore platform karrenfeld begins at the surface contact of the previously discussed flat bioconstructional platform with the eroded coastal bedrock on which it is superimposed. This is usually accompanied by a major change in slope, from horizontal algal-encrusted surface to oblique wall of the marine notch or upsloping erosional shore platform. Holes made by boring bivalves, boring barnacles, and sea urchins in submerged areas are replaced by marks of intense grazing by chitons and limpets, which relentlessly attack surfaces that are also subject to microbial and chemical corrosion. Rocks in this zone of wave swash are so disfigured that bioerosional markings and chaotic karren are overprinted and indistinguishable. Bedrock is reduced to isolated and soon-to-be-destroyed pyramid-shaped remnants (Fig. 2.11a). In the zone of splash, the remaining relief is more preserved and is often shaped as pinnacles that are closely packed and joined at their bases by ridges (Fig. 2.11b), containing between them irregular pools (Gómez-Pujol and Fornós 2009). The predominant grazers here are winkles, which congregate in shaded areas and pits as they feed on biofilms of epilithic and shallow-boring cyanobacteria. In the zone of sea spray, exposed rock surfaces are entirely covered by chaotic pits, ridges, and sharp points that typify karren of eogenetic limestones (which frequently dominate karst coasts of low latitudes). The terrain is barren and almost impassable due to razor-sharp eogenetic karren at various scales, up to several meters tall and sometimes grotesquely corroded pinnacles (Fig. 2.11c). Nestled among the pinnacles are many flat-bottomed solution pans. Away from the shore, both small-scale karren and meter-scale pinnacled relief are subdued, gradually giving way to smoother textures and flatter land surface. As the landscape transitions to inland karst, depressions are filled with soil and vegetated by progressively less salt-tolerant plants. In diagenetically more mature rocks, classical karren such as fluting and runnels readily appear. Because marine notches are continually deepened and overhanging rocks toppled, wave swash, splash, and spray zones slowly migrate inland, causing the interior areas to be progressively relieved of vegetative and soil cover and included in the eogenetic coastal karrenfeld.

In temperate and cooler regions, dissolutional sculpturing on all scales is much less pronounced than in warmer regions (as predicted by Guilcher (1953) who stated: "les formes de dissolution sont d'autant plus évoluées que les eaux sont plus chaudes"). To that it should be added that higher latitude limestones are generally more diagenetically mature than those in tropical and subtropical areas, which also minimizes their surface relief. Consequently, shore platform karrenfeld at mid-latitudes is not so strikingly different from karrenfelds further inland (Fig. 2.11d), at least when observed on a landscape-scale. Like overall relief, bioerosional and bioconstructional activity is also reduced compared to tropical latitudes. Tidal ranges may be greater and there is increased force of waves and mechanical erosion. This tends to preclude the development of benches and notches. Instead, the typical morphology manifests as erosional ramps (Lundberg 2009), which are sloping shore platforms dominated by mechanical erosion. Despite large-scale uniformity, surfaces of erosional ramps are host to various karren assemblages arranged in distinct zones. One of the best-studied coastal karrenfelds in temperate zones is in western Ireland (Fig. 2.11e), described

Fig. 2.11 Coastal karrenfelds. (**a**) Boundary between flat erosional bench with bioconstructional coating and outer rim and the seaward-most portions of coastal karrenfeld, with bedrock in swash zone reduced to isolated, pyramid-shaped dm-scale remnants, Santo Domingo, Dominican Republic; (**b**) Pinnacles and interposed pits that typify coastal karrenfeld of the wave splash zone in Pleistocene limestone of Guam, Mariana Islands; (**c**); Grotesquely corroded pinnacled terrain typical of a coastal karrenfeld in sea spray zone of tropical eogenetic limestones, Saipan, Northern Mariana Islands; (**d**) Coastal karrenfeld on an erosional ramp in the Mediterranean; horizontal erosional or bioconstructional bench is absent, but a prominent but fully submerged marine notch exists below the low tide mark, Sveti Marko island, northern Adriatic Sea; (**e**) Erosional ramp and tidepools in a temperate zone, with pronounced subtidal and intertidal bioerosional relief, but modest dissolutional topography in supratidal areas, Burren coast, County Clare, Ireland [photograph and © by Sandy Skipper]; (**f**) Erosional ramp in the Arctic, with visible bowl-like pits in the tidal belt and to a lesser extent the supratidal zone, carved in marble, Helgeland coast, northern Norway [photograph and © by Stein-Erik Lauritzen]

in detail by Lundberg (1977) and subsequent workers. The intertidal zone is dominated by bioerosional activity. Its lower and mostly submerged parts exhibit extremely hollowed out relief with jagged pinnacles and deep depressions made by boring sea urchins and other organisms. These macro-bioeroders prolifically produce pit-like scars and create extremely dissected surface that is so imprinted with adjacent gouge marks and residual points and ridges that it is reminiscent of, though not genetically related to, eogenetic karren sculpturing. This is evident from the fact that, unlike in eogenetic coastal karst, relief in the upper intertidal zone, beyond immediate reach of marine invertebrates, is much reduced and consists of shallow tide pools separated by rounded residual relief. Zone of wave splash has yet shallower basins and only slightly fretted rock surface. In the zone of spray, topography is rather subtle, with flat-floored pans separated by smooth rock surfaces colonized by salt-tolerant lichens, acting bioerosively or bioprotectively (Carter and Viles 2005). Further inland, the exokarst surface changes to limestone pavement and then to soil and plant cover. With further coastal erosion and bioerosion, especially by urchins laterally enlarging their pools to undermine and consume rocks higher up, all the zones migrate landward across the erosional ramp (Lundberg 1977).

Cold and periglacial karst coasts also exhibit overall erosional ramp profile and its shaping is clearly dominated by physical processes (Fig. 2.11f). Frost and salt weathering play important roles, and chemical and biological corrosion are inhibited by low temperature and ice-armoring of the coast during large part of the year. Consequently, dissolutional and bioerosional sculpturing on the exposed rocks is comparatively small (Malis and Ford 1995). Below the waterline, bioerosional scars are minimal, though boring bivalves do occur (e.g. Brookes and Stevens 1985). Immediately above the waterline, dissolutional texture is generally limited to numerous small pits (bowl karren of Holbye 1989). Beyond the zone of swash, the rock surface is generally smooth and karren-free. It is important to note here that coastal karstification at high latitudes is never complete because dissolutional and bioerosional processes, already slowed down by low temperatures and freezing, cannot catch up with rapid isostatic uplift experienced since the removal of ice cover after the last glacial maximum. Lundberg (2009) presents an illuminating diagram that portrays the polar karst coast as a sloped shore platform on which, in conveyer-belt fashion, glacially-polished smooth surfaces rise into the intertidal zone, experience rudimentary and then maximum pitting as the pass through the wave swash and splash zones, and deteriorate as they elevate further beyond marine influence.

2.8 Zonation and Landscape

The prospects of any particular small- and medium-scale feature or compound structure of coastal exokarst occurring in a particular location depend first and foremost on the baseline lithological, structural, tidal, climatic, and ecological context of the general geographic setting. Clearly, different types of features are limited or most common in particular geographic areas. In addition, their distribution is also closely controlled within a particular area by the local-scale physical, chemical, and biological gradients that permeate the immediate, often micro, environment. Consequently, the existence and mode of expression of different features of coastal exokarst turn up in a way that distinct geomorphic zones can be distinguished, as was evident from previous descriptions of shore platform karrenfelds. These zones can be traced to the overriding influence exerted by seawater: its tidal regime, wave energy, temperature, salinity, and other factors; they determine the type, timing, and chemistry of water-rock contact. Proximity of seawater boosts physical erosion and salt weathering, whereas its disengagement away from the coast boosts inorganic dissolution by normal karst processes. The same parameters affect biologic colonization of substrate and determine the type, timing, and intensity of biota-rock contact. Biologic gradients are coupled with hydrodynamic gradients and essentially reinforce them because both abundance and diversity of microbial biofilms and invertebrate bioeroders are proportional to the amount of water present at a given spot (Palmer et al. 2003). For that reason, the intensities of both biological erosion by invertebrates as well as direct and indirect corrosion by microbes generally diminish with distance away from the shore, though each of the countless taxa exhibits their own separate abundance vectors driven by the requirements of their own ecological niches. Other vectors that exist are the near immediate diminishing of bioconstruction, abrasion, and wave quarrying landward from regularly wetted parts; and the peaking of salt weathering, wetting and drying, and mixing corrosion in areas of occasional sea water input. The relative importance of various processes as a function of distance from the shoreline is beautifully illustrated by De Waele and Furlani (2013).

Because key abiotic parameters vary vertically (elevation above base level, tidal wetting and drying), horizontally transverse to the shore (distance away from waterline, wave wetting and drying, persistence of marine aerosols), and laterally along the coast (variations in exposure, wave energy, climate), they cause commensurate alteration in biologic community structure and all the ensuing biotic parameters. These environmental gradients control the presence and intensity of processes that drive geomorphic change and ultimately result in various coastal exokarst landforms being formed and arranged according to those gradients. This produces a zonation of ecological and geomorphic features that form in subtidal, intertidal, and supratidal zones; or experience water contact as inundation, swash, splash, or spray; or correspond to sheltered, somewhat exposed or very exposed coasts. Considering that all of this is superimposed on lithologic, structural, and other baseline characteristics of a particular area, the variety of overall landscapes is endless. However, as presented in this chapter, some common small and medium-scale textures and structures can be recognized as recurring components in all coastal karst settings. Their characteristics and configurations vary immensely and, together with tectonics, sea-level change, marine, groundwater, and subsurface karst processes (discussed elsewhere in this book) bring about the wider coastal karst landscapes in all their beauty and diversity.

References

Abbott AT, Pottratz SW (1969) Marine pothole erosion, Oahu, Hawaii. Pac Sci 23(3):276–290

Abensperg-Traun M, Wheaton GA, Eliot IG (1990) Bioerosion, notch formation and micromorphology in intertidal and supratidal zones of calcareous sandstone stack. J R Soc West Aust 73:47–56

Ahr WM, Stanton RJ Jr (1973) The sedimentologic and paleoecologic significance of *Lithotrya*, a rock-boring barnacle. J Sediment Petrol 43:20–23

Allen JRL (1985) Principles of physical sedimentology. George Allen & Unwin, London

Andrews C, Williams RBG (2000) Limpet erosion of chalk shore platforms in southeast England. Earth Surf Proc Land 25:1371–1382

Ansell AD, Nair NB (1969) A comparative study of bivalves which bore mainly by mechanical means. Am Zool 9:857–868

Antonioli F, Carulli GB, Furlani S, Auriemma R, Marocco R (2004) The enigma of submerged marine notches in northern Adriatic Sea. Quaternaria Nova 8:263–275

Asgaard U, Bromley RG (2008) Echinometrid sea urchins, their trophic styles and corresponding bioerosion. In: Wisshak M, Tapanila L (eds) Current developments in bioerosion. Springer, Berlin, pp 279–303

Azzopardi L, Schembri PJ (1997) Vermetid crusts from the Maltese Islands (Central Mediterranean). Mar Life 7(1–2):7–16

Baceta JI, Wright VP, Pujalte V (2001) Palaeo-mixing zone karst features from Palaeocene carbonates of north Spain: criteria for recognizing a potentially widespread but rarely documented diagenetic system. Sediment Geol 139(3–4):205–216

Back W, Hanshaw BB, Herman JS, Van Driel JN (1986) Differential dissolution of a Pleistocene reef in the ground-water mixing zone of coastal Yucatán, Mexico. Geology 14(2):137–140

Basso D, Tomaselli V (1994) Palaeoecological potentiality of rhodoliths: a Mediterranean case history. Bollettino della Societa Paleontologica Italiana Special volume 2:17–27

Battistini R, Guilcher A (1982) Les plates-formes littorales à vasques en roches calcaires: repartition dans le monde, mer Méditerranée non comprise. Karst Littoraux, Comite National Français de Géographie, Actes du Colloquium de Perpignan 1:1–11

Benac Č, Juračić M, Bakran-Petricioli T (2004) Submerged tidal notches in the Rijeka Bay NE Adriatic Sea: indicators of relative sea-level change and of recent tectonic movements. Mar Geol 212(1–4): 21–33

Bögli A (1951) Probleme der Karrenbildung. Geogr Helv 3:191–204

Bögli A (1960) Kalklösung und Karrenbildung. Zeitschrift für Geomorphologie Supplementband 2:4–21

Bögli A (1964) Mischungkorrosion – ein Beitrag zum Verstärkungsproblem. Erdkunde 18:83–92
Bosence DWJ (1973) Recent serpulid reefs, Connemara, Eire. Nature 242:40–41
Bourke MC, Viles H (1997) A photographic atlas of rock breakdown features in geomorphic environments. Planetary Science Institute, Tucson/Arizona, 88 p
Bressan G, Chemello R, Gravina MF, Gambi MC, Peirano A, Cocito S, Rosso A, Tursi A (2009) Other bioconcretions. In: Relini G (ed) Other types bioconstructions. Friuli Museum of Natural History. Udine, Italy, pp 90–114
Bromley RG (1978) Bioerosion of Bermuda reefs. Palaeogeogr Palaeoclimatol Palaeoecol 23(3–4):169–197
Brookes IA, Stevens RK (1985) Radiocarbon age of rock-boring, *Hiatella arctica* (Linné) and postglacial sea-level change at Cow Head, Newfoundland. Can J Earth Sci 22:136–140
Bull PA, Laverty M (1982) Observations on phytokarst. Zeitschrift für Geomorphologie 26:437–457
Burke MA (1994) A quantitative analysis of marine kamenitza on the Carboniferous limestone between Skerries and Loughshinny, Co. Doublin. Thesis submitted as part of B.A. degree, Geography Department, Trinity College, Dublin, 198 p
Carter NEA, Viles HA (2005) Bioprotection explored: the story of a little known earth surface process. Geomorphology 67(3–4):273–281
Chacón E, Berrendero E, Garcia-Pichel F (2006) Biogeological signatures of microboring cyanobacterial communities in marine carbonates from Cabo Rojo, Puerto Rico. Sediment Geol 185:215–228
Chaix E (1895) Contribution a l'etude des lapies: la topographie du desert de Plate. Le Globe 34: 67–108
Chen C, Dai C-F (2009) Subtidal sabellarid reefs in Hualien, eastern Taiwan. Coral Reefs 28(1):275
Chen J, Blume HP, Beyer I (2002) Weathering of rocks induced by lichen colonization, a review. Catena 38:121–146
Cocito S (2004) Bioconstruction and biodiversity: their mutual influence. Sci Mar 68(Suppl 1):137–144
Conway K, Barrie J, Krautter M (2005) Geomorphology of unique reefs on the western Canadian shelf: sponge reefs mapped by multibeam bathymetry. Geo-Mar Lett 2005:205–213
Cooke RU, Warren A, Goudie AS (1993) Desert geomorphology. UCL Press Limited, London
Corbel J (1952) Les lapiaz marins. Revue Géographique de Lyon 37:379–380
Cowell DW, Ford DC (1983) Karst hydrology of the Bruce Peninsula, Ontario, Canada. J Hydrol 61(1–3): 163–168
Cucchi F (2009) Kamenitzas. In: Ginés A, Knez M, Slabe T, Dreybrodt W (eds) Karst rock features – karren sculpturing. Carstologia 9. Založba ZRC/ZRC Publishing, Ljubljana, pp 139–150
Cvijić J (1924) The evolution of lapiés: a study in karst physiography. Geogr Rev 14:26–49
Dalongeville M (1977) Formes littorales de corrosion dans les roches carbonatées au Liban: Etude morphologique. Méditerranée 3:21–33
Dalongeville R (1995) Le rôle des organismes constructeurs dans la morphologie des littoraux de la mer Méditerranée: algues calcaires et vermetidés. Norois 165:73–88
Dalongeville M, Guilcher A (1982) Les plates-formes à vasques en Méditerranée, notamment leur extension vers le nord. Karst Littoraux, Comite National Français de Géographie, Actes du Colloquium de Perpignan 1:13–22
De Waele J, Furlani S (2013) Seawater and biokarst effects on coastal limestones. In: Schroder J, Frumkin A (eds) Treatise on geomorphology. Academic Press, San Diego, pp 341–350
De Waele J, Mucedda M, Montanaro L (2009) Morphology and origin of coastal karst landforms in Miocene and Quaternary carbonate rocks along the central-western coast of Sardinia (Italy). Geomorphology 106:26–34
Debrat JM (1974) Etude d'un karst calcaire littoral méditerranéen. Exemple du littoral de Nice à Menton. Méditerranée 17(17):63–85
Dionne J-C (1964) Notes Sur Les marmites Littorales. Revue de Geographie de Montreal 18(2):244–277
Donn TF, Boardman MR (1988) Bioerosion of rocky carbonate coastlines on Andros Island, Bahamas. J Coastal Res 4(3):381–394
Doty MS (1957) Rock intertidal surfaces. In: Hedgpeth J (ed) Treatise on marine ecology and paleoecology, vol 67. Memoirs of the Geological Society of America, Boulder, pp 535–585
Drew D (2009) Coastal and lacustrine karren in western Ireland. In: Ginés A, Knez M, Slabe T, Dreybrodt W (eds) Karst rock features – karren sculpturing. Carstologia 9. Založba ZRC/ZRC Publishing, Ljubljana, pp 503–512
Dreybrodt W (1988) Processes in karst systems: physics, chemistry, and geology. Springer, Berlin
Duane MJ, Al-Mishwat AT, Rafique M (2003) Weathering and biokarst development on marine terraces, northwest Morocco. Earth Surf Proc Land 28(13): 1439–1449
Ekdale AA, Bromley RG, Pemberton SG (1984) Ichnology: the use of trace fossils in sedimentology and stratigraphy. Society of Economic Paleontologists and Mineralogists, Tulsa, pp 108–128
Emery KO (1946) Marine solution basins. J Geol 54: 209–228
Emery KO (1962) Marine geology of Guam, U. S. Geological Survey professional paper 403-B. U. S. Geological Survey, Washington, D.C., pp B1–B76
Fairbridge RW (1952) Marine erosion. Seventh Pac Sci Congr Proc 3:347–359
Fairbridge RW (1982) Karst coast. In: Schwartz ML (ed) The encyclopedia of beaches and coastal environments. Hutchinson Ross Publishing Company, Stroudsburg, pp 500–502

Focke JW (1978) Limestone cliff morphology on Curaçao (Netherlands Antilles), with special attention to the origin of notches and vermetid/coralline algal surf benches ("corniches", "trottoirs"). Zeitschrift für Geomorphologie N.F. 22:329–349

Folk R, Roberts H, Moore C (1973) Black phytokarst from hell, Cayman Islands, British West Indies. Geol Soc Am Bull 84:2351–2360

Ford DC, Williams PW (1989) Karst geomorphology and hydrology. Unwin Hyman, Winchester, 320 p

Fórnos JJ, Pons GX, Gómez-Pujol L, Belaguer P (2006) The role of biological processes and rates of downwearing due to grazing organisms on Mallorcan carbonate coasts (western Mediterranean). Zeitschrift für Geomorphologie Supplementband 44:161–181

Frey RW (1973) Concepts in the study of biogenic sedimentary structures. J Sediment Res 43:6–19

Furlani S, Cucchi F, Forti P, Rossi A (2009) Comparison between coastal and inland karst limestone lowering rates in the northeastern Adriatic region (Italy and Croatia). Geomorphology 104:73–81

Giaccone T, Giaccone G, Basso D, Bressan G (2009) Algae. In: Relini G (ed) Marine bioconstructions. Friuli Museum of Natural History. Udine, Italy, pp 29–48

Gill ED, Lang JG (1983) Micro-erosion meter measurements of rock wear on the Otway coast of southeast Australia. Mar Geol 52:141–156

Ginés À (2009) Karrenfield landscapes and karren landforms. In: Ginés A, Knez M, Slabe T, Dreybrodt W (eds) Karst rock features – karren sculpturing. Carstologia 9. Založba ZRC/ZRC Publishing, Ljubljana, pp 13–24

Ginés À, Ginés J (1995) Les formes exocarstiques de l'illa de Mallorca. ENDINS 20. Mon Soc Hist Nat Balears 3:59–70

Ginés À, Knez M, Slabe T, Dreybrodt W (eds) (2009) Karst rock features – karren sculpturing. Carstologia 9. Založba ZRC/ZRC Publishing, Ljubljana

Ginsburg RN (1953) Intertidal erosion on the Florida keys. B Mar Sci 3:55–69

Ginsburg RN, Schroeder JH (1973) Growth and submarine fossilization of algal cup reefs, Bermuda. Sedimentology 20(4):575–614

Glaub I, Golubic S, Gektidis M, Radtke G, Vogel K (2007) Microborings and microbial endoliths: geological implications. In: Miller WC (ed) Trace fossils: concepts, problems, prospects. Elsevier, Amsterdam, pp 368–381

Glumac B, Berrios L, Greer L, Curran HA (2004) Holocene tufa-coated serpulid mounds from the Dominican Republic: depositional and diagenetic history, with comparison to modern serpulid aggregates from Baffin Bay, Texas. In: Proceedings of the 11th symposium on the geology of the Bahamas and other carbonate regions, Gerace Research Centre, San Salvador, pp 49–65

Gómez-Pujol L, Fornós JJ (2009) Coastal karren in the Balearic Islands. In: Ginés A, Knez M, Slabe T, Dreybrodt W (eds) Karst rock features – karren sculpturing. Carstologia 9. Založba ZRC/ZRC Publishing, Ljubljana, pp 487–502

Gómez-Pujol L, Fornós JJ (2010) Coastal karren features in temperate microtidal settings: spatial organization and temporal evolution. Studia Universitatis Babeş-Bolyai, Geologia 55(1):37–44

Grimes KG (2001) Karst features of Christmas Island (Indian Ocean). Helictite 37(2):41–58

Guilcher A (1953) Essai sur la zonation et la distribution des formes littorals de dissolution du calcaire. Ann Geogr 62:161–179

Guilcher A (1958) Coastal corrosion forms in limestones around the Bay of Biscay. Scot Geogr Mag 74(3): 137–149

Hadfield MG, Kay EA, Gillette MU, Lloyd MC (1972) The Vermetidae (mollusca gastropoda) of the Hawaiian Islands. Mar Biol 12:81–98

Hanor JS (1978) Precipitation of beachrock cements: mixing of marine and meteoric waters vs. CO_2-degassing. J Sediment Petrol 48:489–501

Higgins CG (1980) Nips, notches, and the solution of coastal limestone: an overview of the problem with examples from Greece. Estuar Coast Mar Sci 10(1): 15–30

Hodgkin EP (1964) Rate of erosion of intertidal limestone. Zeitschrift für Geomorphologie 8(4):385–392

Holbye U (1989) Bowl-karren in the littoral karst of Nord-Arnoy, Norway. Cave Sci 16:19–26

Holmes G, Ortiz J-C, Schönberg CHL (2009) Bioerosion rates of the sponge *Cliona orientalis* Thiele, 1900: spatial variation over short distances. Facies 55: 203–211

Hopley D (2005a) Coral reefs. In: Schwartz ML (ed) Encyclopedia of coastal science. Kluwer, Dordrecht, pp 343–349

Hopley D (2005b) Trottoirs. In: Schwartz ML (ed) Encyclopedia of coastal science. Kluwer, Dordrecht, p 1017

Horwitz MH, Roberts TM (2010) Geomorphic zoning and eogenetic karst on limestones within the supratidal environment: San Salvador, Bahamas. Studia Universitatis Babes-Bolyai, Geologia 55(1):17–27

Huggett RJ (2007) Fundamentals of geomorphology. Routledge, Abingdon

Hutchings PA, Peyrot-Clausade M (2002) The distribution and abundance of boring species of polychaetes and sipunculans in coral substrates in French Polynesia. J Exp Mar Biol Ecol 269:101–121

Jacobsen G, Hill PJ, Ghassemi F (1997) Geology and hydrogeology of Nauru Island. In: Vacher HL, Quinn T (eds) Geology and hydrology of carbonate Islands, vol 54, Developments in sedimentology. Elsevier Science BV, Amsterdam

Jennings JN (1971) Karst. Australian National University Press, Canberra, 252 p

Jennings JN (1985) Karst geomorphology. Basil Blackwell, Oxford, p 293

Jones B (1989) The role of microorganisms in phytokarst development on dolostones and limestones, Grand Cayman, British West Indies. Can J Earth Sci 26:2204–2213

Jones B (2010) Speleothems in a wave-cut notch, Cayman Brac, British West Indies: the integrated product of subaerial precipitation, dissolution, and microbes. Sediment Geol 232:15–34

Jones B, Hunter I (1995) Vermetid buildups from grand Cayman British West Indies. J Coastal Res 4:973–983

Kázmér M, Taboroši D (2012) Bioerosion on the small scale – examples from the tropical and subtropical littoral. Hantkeniana 7:37–94, incl. 96 figs and 1 table, Budapest, Hungary

Kelletat D (1985) Bio-destruktive und bio-konstruktive Formelemente en den spanischen Mittelmeerküsten. Geodynamik 6:1–20

Kelletat D (1988) Quantitative investigations on coastal bioerosion in higher latitudes: an example from northern Scotland. Geoökodynamik, Bensheim 9:41–51

Kempf M, Laborel J (1968) Formations de Vermets et d'Algues calcaires des côtes du Brésil. Rec Trav Stat Mar Endoume 43:9–23

Kershaw S, Guo L (2001) Marine notches in coastal cliffs: indicators of relative sea-level change, Perachora Peninsula, central Greece. Mar Geol 179(3–4): 213–228

Kleemann K (2001) Marine bioerosion. Lecture given at the University of Vienna, compiled by P Madl. Available on-line at: biophysics.sbg.ac.at/transcript/bioeros.pdf. Last Accessed 16 Mar 2012

Knight J (2005) Controls on the formation of coastal ventifacts. Geomorphology 64:243–253

Kobluk DR, Risk JM (1977) Rate and nature of infestation of a carbonate substratum by a boring algae. J Exp Mar Biol Ecol 27:107–115

Kogure T, Matsukura Y (2010) Critical notch depths for failure of coastal limestone cliffs: case study at Kuroshima Island, Okinawa, Japan. Earth Surf Proc Land 35(9):1044–1056

Krumbein WE (1979) Photolithotropic and chemoorganotrophic activity of bacteria and algae as related to beachrock formation and degradation (Gulf of Aqaba, Sinai). Geomicrobiology 1:139–203

Laborel J (2005) Algal rims. In: Schwartz ML (ed) Encyclopedia of coastal science. Kluwer, Dordrecht, pp 24–25

Laborel J, Laborel-Deguen F (1994) Biological indicators of relative sea-level variation and of co-seismic displacements in the Mediterranean area. J Coastal Res 10(2):395–415

Laborel J, Laborel-Deguen F (1995) Biological indicators of Holocene sea-level and climatic variations on rocky coasts of tropical and subtropical regions. Quatern Int 31:53–60

Laborel J, Morhange R, Lafont J, Campion L, Laborel-Deguen F, Sartoretto S (1994) Biological evidence of sea level rise during the last 4500 years on the rocky coasts of continental southwestern France and Corsica. Mar Geol 120:203–223

Ley RG (1977) The influence of lithology on marine karren. Abhandlung zur Karst und Höhlenkunde 15: 81–100

Ley RG (1979) The development of marine karren along the Bristol Channel Coastline. Zeitschrift für Geomorphologie Supplementband 32:75–89

Liu P-J, Hsieh H-L (2000) Burrow architecture of the spionid polychaete *Polydora villosa* in the corals *Montipora* and *Porites*. Zool Stud 39(1):47–54

Lovrić AŽ, Rac M, Milenković MH (2002) Diversity of old-Croatian names for seaweeds and maritime nature in the Adriatic Islands. Nat Croat 11(4):455–477

Lundberg J (1977) Karren of the littoral zone, Burren District, Co. Clare, Ireland. In: Proceedings of the 7th international speleological congress, Sheffield, pp 291–293

Lundberg J (2004) Coastal karst. In: Gunn J (ed) Encyclopedia of cave and karst science. Fitzroy Dearborn, New York/London, pp 231–233

Lundberg J (2009) Coastal karren. In: Ginés A, Knez M, Slabe T, Dreybrodt W (eds) Karst rock features – karren sculpturing. Carstologia 9. Založba ZRC/ZRC Publishing, Ljubljana, pp 249–264

Lundberg J, Lauritzen SE (2002) The search for an arctic coastal karren model in Norway and Spitzbergen. In: Hewitt K, Byrne ML, English M, Young G (eds) Landscapes in transition. Kluwer Academic Publishers, Dordrecht, pp 185–203

Macfadyen WA (1930) The undercutting of coral reef limestone on the coasts of some Islands in the Red Sea. Geogr J 75:27–34

Malis CP, Ford DC (1995) Littoral karren along the western shore of Newfoundland. Geol Soc Am, Abstracts with Programs 27(6):A9–A56

Mark H (2009) Karst landscapes in the Bay of Ha Long, Vietnam. Geographische Rundschau International Edition 5(1):48–51

Martel EA (1921) Nouveau traité des eaux souterraines. Doin, Paris, 838 p

Matsukura Y, Matsuoka N (1991) Rates of tafoni weathering on uplifted shore platforms in Nojima-zaki, Boso Peninsula, Japan. Earth Surf Proc Land 16:51–56

Miller WR, Mason TR (1994) Erosional features of coastal beachrock and eolianite outcrops in Natal and Zululand, South Africa. J Coastal Res 10:374–394

Milliman JD (1974) Marine carbonates. [Part 1, recent sedimentary carbonates]. Springer, New York, 375 p

Mokady O, Lazar B, Loya Y (1996) Echinoid bioerosion as a major structuring force of Red Sea coral reefs. Biol Bull 190:367–372

Molinier R (1955) Les plate-formes et corniches récifals de vermets (*Vermetus cristatus Biondi*) en Méditerranée occidentale. Comptes Rendus de l'Académie des Sciences 240:361–363

Monroe WH (1970) A glossary of karst terminology, Water supply paper, 1899-K. USGS, Washington, pp 1–26

Moore DG (1954) Origin and development of sea caves. Bull Natl Speleological Soc 16:71–76

Morton B, Scott PJB (1980) Morphological and functional specializations of the shell, musculature and pallial glands in the Lithophaginae (Mollusca: Bivalvia). J Zool Soc London 192:179–203

Moses CA (2003) Observations on coastal biokarst, Hells Gate, Lord Howe Island, Australia. Zeitschrift für Geomorphologie 47:83–100
Moses CA, Smith BJ (1993) A note on the role of Colema auriforma in solution basin development on a Carboniferous limestone substrate. Earth Surf Proc Land 18:363–368
Moses CA, Smith BJ (1994) Limestone weathering in the supratidal zone: an example from Mallorca. In: Robinson DA, Williams RBG (eds) Rock weathering and landform evolution. John Wiley and Sons, Chichester, pp 433–452
Mylroie JE, Carew JL (1990) The flank margin model for dissolution cave development in carbonate platforms. Earth Surf Proc Land 15:413–424
Mylroie JE, Mylroie JR (2009) Coastal eogenetic karren of San Salvador Island. In: Ginés A, Knez M, Slabe T, Dreybrodt W (eds) Karst rock features – karren sculpturing, Carstologia 9. Založba ZRC/ZRC Publishing, Ljubljana, pp 475–485
Naylor LA, Viles HA, Carter NEA (2002) Biogeomorphology revisited: looking towards the future. Geomorphology 47:3–14
Neumann AC (1966) Observations on coastal erosion in Bermuda and measurements of the boring rates of the sponge *Cliona lampa*. Limnol Oceanogr 11:92–108
Newell ND, Purdy EG, Imbrie J (1960) Bahamian oölitic sand. J Geol 68(5):481–497
Newman WA, Abbott DP (1980) Cirripedia. In: Morris RH, Abbott DP, Haderlie EC (eds) Intertidal invertebrates of California. Stanford University Press, Stanford, pp 504–535
Norton TA, Hawkins SJ, Manley NL, Williams GA, Watson DC (1990) Scraping a living: a review of littorinid grazing. Hydrobiologia 193(1): 117–138
Orme R, Riding R (1995) *Halimeda* segment reefs of the northern Great Barrier Reef. British Sedimentological Research Group. In: 1995 Annual meeting abstracts, Durham, p 64
Palmer HS (1927) Lapies in Hawaiian Basalts. Geogr Rev 17(4):627–631
Palmer M, Fórnos JJ, Balaguer P, Gómez-Pujol L, Pons GX, Villanueva G (2003) Spatial and seasonal variability of the macro-invertebrate community of a rocky coast in Mallorca (Balearic Islands): implications for bioerosion. Hydrobiologia 501:13–21
Paskoff RP (2005) Karst coasts. In: Schwartz ML (ed) Encyclopedia of coastal science. Kluwer, Dordrecht, pp 581–586
Pirazzoli PA (1986) Marine notches. In: Plassche OVd (ed) Sea-level research: a manual for the collection and evaluation of data. Geo Books, Zurich
Radtke G, Le Campion-Alsumard T, Golubic S (1996) Microbial assemblages of the bioerosional "notch" along tropical limestone coasts. Algol Stud 83: 469–482
Rasmussen KA, Frankenberg EW (1990) Intertidal bioerosion by the chiton *Acanthopleura granulata*: San Salvador, Bahamas. B Mar Sci 47/3:680–695
Reece M, Mylroie JE, Jenson JW (2006) Notches in carbonate cliffs and hillslopes: origin and implications. In: Davis RL, Gamble DW (eds) Proceedings of the 12th symposium on the geology of the Bahamas and other carbonate regions, Gerace Research Centre, San Salvador, Bahamas, pp 143–152
Reid RP, Macintyre IG, James NP (1990) Internal precipitation of microcrystalline carbonate: a fundamental problem for sedimentologists. Sediment Geol 68(3):163–170
Relini G (2009) Introduction. In: Relini G (ed) Marine bioconstructions. Friuli Museum of Natural History. Udine, Italy, pp 7–12
Revelle R, Emery KO (1957) Chemical erosion of beach rock and exposed reef rock, U. S. Geological Survey professional paper 260-T. U. S. Geological Survey, Washington, pp 699–706
Richardson K, Carling PA (2005) A typology of sculpted forms in open bedrock channels, Geological Society of America, Special Paper 392. Geological Society of America, Boulder, p 108
Richmond BM (2002) Overview of Pacific island carbonate beach systems. In: Robbins LL, Magoon OT, Ewing L (eds) Carbonate beaches 2000, Key Largo, Florida, American Society of Civil Engineers conference proceedings, 5–8 Dec 2000, pp 218–228
Riding R (2002) Structure and composition of organic reefs and carbonate mud mounds: concepts and categories. Earth Sci Rev 58:163–231
Risk MJ, McGeachy JK (1978) Aspects of bioerosion of modern Caribbean reefs. Rev Biol Trop 26:85–105
Robbins LL, Blackwelder PL (1992) Biochemical and ultrastructural evidence for the origin of whitings: a biologically induced calcium carbonate precipitation mechanism. Geology 20:464–468
Russell RJ (1962) Origin of beach rock. Zeitschrift für Geomorphologie 6:1–16
Rust D, Kershaw S (2000) Holocene tectonic uplift patterns in northeastern Sicily: evidence from marine notches in coastal outcrops. Mar Geol 167: 105–126
Safriel UN (1975) The role of vermetid gastropods in the formation of Mediterranean and Atlantic Reefs. Oecologia 20(1):85–101
Salomon J-N (2006) Les tsingy et leur genese. Spelunca 103:45–50
Sawyer JA, Zuschin M (2010) Intensities of drilling predation of molluscan assemblages along a transect through the northern Gulf of Trieste (Adriatic Sea). Palaeogeogr Palaeoclimatol Palaeoecol 285(3–4):152–173
Schneider J (1976) Biological and inorganic factors in the destruction of limestone coasts. Contrib Sedimentology 6:1–112
Schneider J, Le Campion Alsumard T (1999) Construction and destruction of carbonates by marine and freshwater cyanobacteria. Eur J Phycol 34:417–426
Schneider J, Torunski H (1983) Biokarst on limestone coasts, morphogenesis and sediment production. Mar Ecol 4:45–63

Schoppe S, Werding B (1996) The Boreholes of the Sea Urchin Genus Echinometra (Echinodermata: Echinoidea: Echinometridae) as a Microhabitat in Tropical South America. Mar Ecol 17(1–3):181–186
Scoffin TP (1970) A conglomeratic beachrock in Bimini, Bahamas. J Sediment Petrol 40:756–759
Scoffin TP, Stoddart DR (1983) Beachrock. In: Goudie AS, Pye K (eds) Chemical sediments and geomorphology: precipitates and residua in the near-surface environment. Academic, London, pp 401–425
Silenzi S, Antonioli F, Chemello R (2004) A new marker for sea surface temperature trend during the last centuries in temperate areas: Vermetid reef. Global Planet Change 40:105–114
Simms MJ (1990) Phytokarst and photokarren in Ireland. Cave Sci 17:131–133
Southward AJ (1958) The zonation of plants and animals on rocky sea shores. Biol Rev 33(2):137–177
Spencer T (1985a) Marine erosion rates and coastal morphology of reef limestones on Grand Cayman Island, West Indies. Coral Reefs 4:59–70
Spencer T (1985b) Weathering rates on a Caribbean reef limestone; results and implications. Mar Geol 69: 195–201
Spencer T (1988) Limestone coastal morphology: the biological contribution. Prog Phys Geog 12:66–101
Spencer T, Viles H (2002) Bioconstruction, bioerosion and disturbance on tropical coasts: coral reefs and rocky limestone shores. Geomorphology 48:23–50
Stephenson WJ, Kirk RM (1998) Rates and patterns of erosion on shore platforms, Kaikoura, South Island, New Zealand. Earth Surf Proc Land 23:1071–1085
Stephenson TA, Stephenson A (1949) The universal features between tide marks on rocky coasts. J Ecol 37:289–305
Stephenson TA, Stephenson A (1972) Life between tidemarks on rocky shores. Freeman, San Francisco, 425 p
Stoddart DR, Cann JR (1965) Nature and origin of beach rock. J Sediment Petrol 35:243–273
Stoddart DR, Taylor JD, Farrow GR, Fosberg FR (1971) The geomorphology of Aldabra. In: Westoll TS, Stoddart DR (eds) A discussion on the results of the Royal Society Expedition to Aldabra, 1967–1968. Philosophical Transactions of the Royal Society, London, pp 31–65
Stoessell RK, Ward WC, Ford BH, Schuffert JD (1989) Water chemistry and $CaCO_3$ dissolution in the saline part of an open-flow mixing zone, coastal Yucatán Peninsula, Mexico. Bull Geol Soc Am 101:159–169
Stone ED, Weiner S, Addadi L (2005) Morphology of goethite crystals in developing limpet teeth: assessing biological control over mineral formation. Cryst Growth Des 5(6):2131–2138
Sweeting MM (1972) Karst landforms. Macmillan/Columbia University Press, London/New York
Sweeting MM, Lancaster N (1982) Solutional and wind erosion forms on limestone in the Central Namib Desert. Zeitschrift für Geomorphologie 26:197–207
Taboroši D, Stafford K (2004) Littoral dripstone and flowstone – non-spelean carbonate secondary deposits. Int J Speleol 32:85–106
Taboroši D, Jenson JW, Mylroie JE (2004) Karren features in Island Karst: Guam, Mariana Islands. Zeitschrift für Geomorphologie 48:369–389
Taboroši D, Mylroie JE, Hirakawa K (2006) Stalactites on tropical cliffs: remnants of breached caves or subaerial tufa deposits? Zeitschrift für Geomorphologie 50: 117–139
Terry JP, Nunn PD (2003) Interpreting features of carbonate geomorphology on Niue Island, a raised coral atoll. Zeitschrift für Geomorphologie 131:43–57
Torunski H (1979) Biological erosion and significance for the morphogenesis of limestone coasts and for nearshore sedimentation (Northern Adriatic). Senckenbergiana Maritima 11:193–265
Trenhaile AS (1987) The geomorphology of rocky coasts. Clarendon, Oxford, 384 p
Trenhaile AS (2003a) Trottoir. In: Goudie A (ed) Encyclopedia of geomorphology. Routledge, London, p 1069
Trenhaile AS (2003b) Corniche. In: Goudie A (ed) Encyclopedia of geomorphology. Routledge, London, p 191
Tribollet A (2008) The boring microflora in modern coral reef ecosystems: a review of its roles. In: Wisshak M, Tapanila L (eds) Current developments in bioerosion. Springer, Berlin, pp 67–94
Tribollet A, Golubic S (2011) Reef bioerosion: agents and processes. In: Dubinsky Z, Stambler N (eds) Coral reefs: an ecosystem in transition, part 5. Springer, Dordrecht, pp 435–449
Trimmel H (1965) Speläologisches Fachwörterbuch. Landesverein für Höhlenkunde in Wien und Niederösterreich, Wien, 109 p
Trudgill ST (1976) The subaerial and subsoil erosion of limestones on Aldabra Atoll, Indian Ocean. Zeitschrift für Geomorphologie Supplementband 26: 201–210
Trudgill ST (1979) Spitzkarren on calcarenites, Aldabra Atoll, Indian Ocean. Zeitschrift für Geomorphologie Supplementband 32:67–74
Trudgill ST (1985) Limestone geomorphology. Longman, London/New York, 196 p
Trudgill ST (1987) Bioerosion on intertidal limestone, Co. Clare, Eire: Zonation, process, and form. Mar Geol 74:99–109
Trudgill ST (2003) Boring organism. In: Goudie A (ed) Encyclopedia of geomorphology. Routledge, London, pp 90–92
Trudgill ST, High CJ, Hanna KK (1981) Improvements to the micro-erosion meter (MEM). British Geomorphol. Research Group. Tech Bull 29:3–17
Tschang HL (1966) Marine potholes of Hong Kong. Chung Chi Journal 6:50–58
Veress M (2010) Karst environments: karren formation in high mountains. Springer Verlag, Dordrecht/Heidelberg, 230 p

Veress M, Szunyogh G, Tóth G, Zentai Z, Czöpek I (2006) The effect of the wind on karren formation on the Island of Diego De Almagro (Chile). Zeitschrift für Geomorphologie 50:425–445

Vescogni A, Bosellini FR, Reuter M, Bracher TC (2008) Vermetid reefs and their use as palaeobathymetric markers: new insights from the Late Miocene of the Mediterranean (Southern Italy, Crete). Palaeogeogr Palaeoclimatol Palaeoecol 267:89–101

Viles HA (1984) *Biokarst:* review and prospect. Prog Phys Geog 8:523–542

Viles HA (1987) Blue-green algae and terrestrial weathering on Aldabra atoll: an SEM and light microscope study. Earth Surf Proc Land 12:319–330

Viles HA, Trudgill ST (1984) Long term remeasurements of micro-erosion meter rates, Aldabra Atoll, Indian Ocean. Earth Surf Proc Land 9:89–94

Wargo RN, Ford SE (1993) The effects of shell infestation by *Polydora* sp. and infestation by *Haplosporidium nelsoni* (MSX) on the tissue condition of oysters, *Crassotrea virginica*. Estuaries 16:229–234

Webb GE, Jell JS, Baker JC (1999) Cryptic intertidal microbialites in beachrock, Heron Island, Great Barrier Reef: implications for the origin of microcrystalline beachrock cement. Sediment Geol 126(1–4): 317–334

Wentworth CK (1939) Marine bench-forming processes II, solution benching. J Geomorph 2:3–25

Wentworth CK (1944) Potholes, pits and pans, subaerial and marine. J Geol 52:117–130

Whipple K, Hancock G, Anderson R (2000) River incision into bedrock: mechanics and relative efficacy of plucking, abrasion, and cavitation. Geol Soc Am Bull 112:490–503

White WB (1988) Geomorphology and hydrology of karst terrains. Oxford Press, New York, 464 p

Wilkinson CR (1983) Role of sponges in coral reef structural processes. In: Barnes DJ (ed) Perspectives on coral reefs. Australian Institute of Marine Sciences and B. Clouston, Townsville/Canberra, pp 263–274

Williams JA, Margolis SV (1974) Sipunculid burrows in coral reefs: evidence for chemical and mechanical excavation. Pac Sci 28(4):357–359

Wilson MA (2007) Macroborings and the evolution of marine bioerosion. In: Miller WC (ed) Trace fossils: concepts, problems, prospects. Elsevier, Amsterdam, pp 356–367

Wisshak M (2006) High-latitude bioerosion: the Kosterfjord experiment, vol 109, Lecture notes in earth sciences. Springer, Berlin/New York, pp 1–122

Wziatek D, Vousdoukas MV, Terefenko P (2011) Wave-cut notches along the Algarve coast, S. Portugal: characteristics and formation mechanisms. J Coastal Res Special Issue 64:855–859 (Proceedings of the 11th international coastal symposium, Szczecin, Poland)

Zseni A (2009) Subsoil shaping. In: Ginés A, Knez M, Slabe T, Dreybrodt W (eds) Karst rock features–karren sculpturing, Carstologia 9. Založba ZRC/ZRC Publishing, Ljubljana, pp 103–121

Hydrology and Geochemistry of the Freshwater Lens in Coastal Karst

3

Beth Fratesi

Abstract

Interactions of freshwater and saltwater dominate the hydrogeology and hydrochemistry of coastal karst aquifers. The distribution of freshwater and saltwater in coastal aquifers is influenced by the nature and rates of recharge, size and boundary conditions of the system, distribution of hydraulic conductivities, and spatial and temporal variations in these parameters. The freshwater-saltwater mixing zone can be very thick and may intrude upward or inland due to pumping or decreased recharge. Dissolutional potential for karst development in coastal carbonates derives partially from carbonic acid from atmospheric and soil CO_2, and from mixing of freshwater and saltwater. Solubility of carbonate minerals in freshwater is affected by variables including P_{CO2}, temperature, pressure, and pH, a relationship complicated by the presence of other ions in natural waters. Freshwater/saltwater mixing induces several chemical effects that influence carbonate solubility, the strongest of which tend to cause undersaturation. Processes that complicate simple mixing dissolution include dissolution or outgassing of CO_2, dissolution or precipitation of carbonate minerals, reduction of sulfur compounds, and kinetics effects. The rate of coastal karst development depends on discharge rates along the coast, degree of mixing, kinetic effects along the lens margin, and stability of sea level.

3.1 Introduction

The hydrogeology and hydrochemistry of coastal areas is characterized by the interactions of freshwater hydrogeologic systems and seawater. In coastal aquifers, fresh groundwater flows towards the sea, floating over a body of slightly denser saline groundwater. Freshwater and saltwater are separated by a mixing zone, which is a location of enhanced chemical reactivity and potential dissolution of carbonate rocks (e.g. Back et al. 1979, 1986; Budd 1988; Hanshaw and Back 1980; Mylroie and Carew 1990; Plummer 1975; Randazzo and Bloom 1985; Smart et al. 1988; Stoessell 1992; Wicks and Herman 1995). While it is reasonable to say that the interaction

B. Fratesi (✉)
SDII Global Corporation, 4509 George Road, Tampa, FL 33634, USA
e-mail: bethfratesi@gmail.com

M.J. Lace and J.E. Mylroie (eds.), *Coastal Karst Landforms*, Coastal Research Library 5,
DOI 10.1007/978-94-007-5016-6_3, © Springer Science+Business Media Dordrecht 2013

of freshwater and saline water dominates the chemical processes within coastal karst systems, the end result of this interaction strongly depends upon the particulars of a coastal hydrogeological setting such as the flow dynamics, the specific chemical compositions of waters, and other sources of chemical reactivity within the system.

This chapter presents a theoretical overview of the major chemical processes that control karst development in coastal karst systems. Because the distribution of saline and fresh waters within the system is so chemically important, this chapter includes a brief overview of freshwater lens geometry, with discussion of human-induced saltwater intrusion. The basic equilibrium chemistry of limestone dissolution is covered, followed by a discussion of the chemistry of freshwater-saltwater mixing and a discussion of the rates of dissolution and porosity development.

3.2 Freshwater and Saltwater Configurations

In the phreatic zone of coastal aquifers containing both freshwater and saltwater, the freshwater generally occupies a lens- or wedge-shaped body overlying the saltwater portion of the aquifer. The source of freshwater is seaward flow of water derived either from rain falling directly onto the carbonate rocks at the land surface above the aquifer (*autogenic* recharge), or from water migrating from an inland recharge zone or an adjacent area with non-carbonate rocks (*allogenic* recharge) (Mylroie 1984). Autogenic recharge infiltrates downward through the vadose zone to the water table. Allogenic recharge enters limestone aquifers at discrete points as streams flowing off of nearby non-carbonate rocks (see for instance Upchurch 2002; Humphrey 1997).

The freshwater body tends to take on a lens shape whose upper surface is the water table and whose lower surface is the freshwater-saltwater boundary (Fig. 3.1). Often, the freshwater body abuts an aquitard or other geologic boundary, forming a wedge shape, such as in Barbados (Humphrey 1997), South Florida (Cooper 1964), and parts of Guam (Mink 1976; Mink and Vacher 1997). Freshwater lens geometry is a dominant factor in understanding and predicting hydrogeology and cave development in small carbonate islands: Vacher's (1997) general classification of carbonate islands and the Carbonate Island Karst Model (Mylroie et al. 2001) are both based on the geometric combinations of carbonate aquifers, aquitards, and noncarbonate basement rocks, all of which define the spaces that freshwater bodies might occupy on the islands.

A similar comparison has been made for larger-scale karst systems on continental margins, where the Yucatan Peninsula and the Florida Peninsula stand as contrasting examples of coastal karst development (Back and Hanshaw 1970). The Yucatan has no surface drainage system and a very thin freshwater body with low potentiometric gradient and low residence times. The Florida Peninsula has a much thicker freshwater system with higher potentiometric gradient, higher residence times, and deeper circulation. Back and Hanshaw (1970) attributed many of these physiographic and geochemical differences to the thick confining layers overlying much of the Florida Peninsula and the lack of similar confining layers in the Yucatan Peninsula.

In freshwater lenses and wedges, the transition between fresh water and seawater may comprise a sharp boundary or, more commonly, a broad transition zone (or "mixing zone") containing brackish water of gradually varying salinity. Flow in the freshwater portion of the wedge is generally towards the sea, with groundwater velocities increasing as the cross-sectional area of the lens or wedge decreases in the vicinity of the coastline. It should be noted, however, as Vacher (1988a, 1997) points out, that in most depictions of freshwater-saltwater systems, including the ones in this chapter, the vertical scale is greatly exaggerated for ease of illustration. This tends to overemphasize vertical flow in freshwater lenses and wedges; in reality, horizontal flow strongly dominates within freshwater wedges or lenses, except near the discharge zone (Fig. 3.1).

The geometry of a freshwater body is defined by the density contrast and head relations

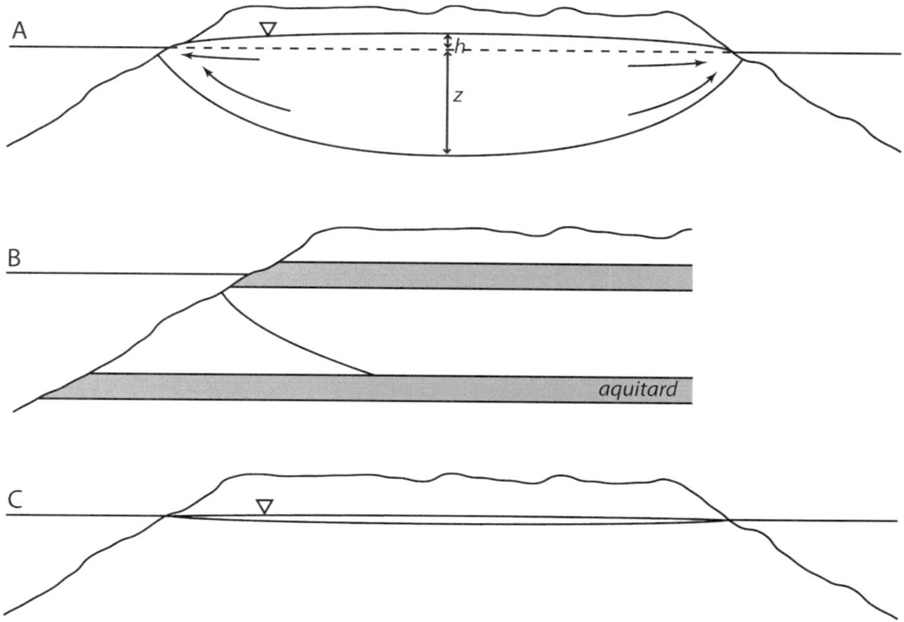

Fig. 3.1 Basic freshwater body configurations. (**a**) Freshwater lens. (**b**) Freshwater wedge in a confined aquifer. (**c**) Illustration of freshwater lens without vertical exaggeration

between seawater and overlying freshwater. This density contrast allows a thickness of freshwater to float over saltwater at any point in the freshwater lens. The depth to the freshwater-saltwater interface is proportional to the elevation of the water table or potentiometric surface at that point. The relationship is written

$$z = \frac{\rho_f}{\rho_s - \rho_f} h \qquad (3.1)$$

where z is the depth to the freshwater-saltwater interface, h is the elevation of the water table above mean sea level (freshwater head), ρ_f is the density of fresh water, and ρ_s is the density of salt water (Fig. 3.1). The ratio between z and h is 40:1 for freshwater and seawater (representing the 1.000 g/cm^3 vs 1.025 g/cm^3 density difference of typical fresh and marine waters, which is also 1 part in 40). This rule of thumb, referred to as the Ghyben-Herzberg principle (Badon Ghyben 1889; Herzberg 1901), is useful for studying freshwater bodies using idealized geometries, based on the assumptions that (1) the boundary between freshwater and saltwater is sharp, (2) flow within the freshwater lens is purely horizontal, and (3) water in the seawater portion of the aquifer is static (see for example Todd 1980; Fetter 1980; Vacher 1988a).

A wide array of analytical explorations of the geometry of freshwater lenses has been conducted, coupling analysis of the water table shape with some form of the Ghyben-Herzberg relationship to give general descriptions of the shape of the lens based on different sets of variables such as recharge, permeability, island width, and sea level (Fetter 1972; Glover 1964; Henry 1964; Hubbert 1940; Vacher 1978, 1988a, b; Van der Veer 1977). The Ghyben-Herzberg relationship yields results that compare relatively favorably to those produced by these more advanced analytical treatments except near the coastline, where flow has a more pronounced vertical component and the presence of an outflow face offsets the freshwater-saltwater interface somewhat (Vacher 1988a).

In general, the maximum thickness of the lens is proportional to island size and recharge, and inversely proportional to hydraulic conductivity. For a long, narrow "strip" island (similar to a

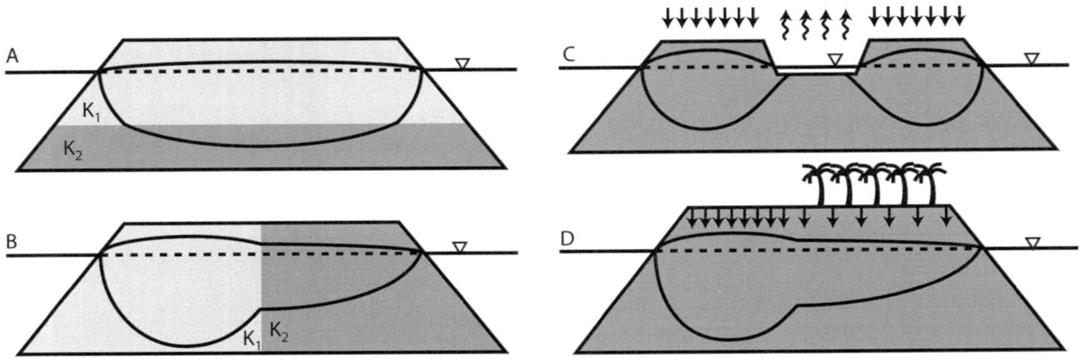

Fig. 3.2 Effects of spatial variations in aquifer/recharge variables on shape of freshwater lenses (After Vacher 1988a). (**a**) Freshwater lens configuration under the "Dual aquifer" model of aquifer permeability on atolls and other small carbonate islands. (**b**) Freshwater lens configuration with an "across-the-island" variation in permeability (Vacher 1978). (**c**) Freshwater lens fragmented by presence of an evaporative lake. (**d**) Freshwater lens impacted by variations in recharge to the aquifer

barrier island) with homogeneous hydraulic conductivity,

$$\frac{H}{L} = 3.2\left(\frac{R}{K}\right)^{\frac{1}{2}} \tag{3.2}$$

where *H* is the maximum thickness of the lens at the midline of the island, *L* is the width of the island, *R* is recharge, and *K* is hydraulic conductivity (Vacher 1988a). The equation indicates that the width of the land mass is theoretically the strongest direct influence on lens thickness and volume (see also Whitaker and Smart 1997). However, in karstic rocks, the hydraulic conductivity has the potential to vary by several orders of magnitude (Vacher and Mylroie 2002) and can therefore account for significant variability in lens geometry.

In limestone undergoing karstification, diagenesis tends to increase permeability over time by re-configuration of porosity by selective dissolution, conversion of aragonite to calcite, and recrystallization (Hanshaw and Back 1979; Vacher and Mylroie 2002). Freshwater lenses therefore tend to be thicker in younger, less permeable limestones than in older, more permeable limestones.

If recharge and hydraulic conductivity (the ratio *R*/*K*) vary spatially within a landmass, the shape of the freshwater lens is modified accordingly. For instance, the "dual-aquifer model" of atoll hydrogeology posits that many atoll islands comprise Holocene carbonate sediments overlying older, karstified Pleistocene limestones (Buddemeier and Holladay 1977; Oberdorfer et al. 1990; Wheatcraft and Buddemeier 1981), and that, if the permeability contrast is large enough, the freshwater lenses in these islands are often distorted or truncated at the contact between the two limestones (Fig. 3.2a). Similar scenarios exist in the Florida Keys (Vacher et al. 1992), the Bahamas (Whitaker and Smart 1997), Tarawa (Falkland and Woodroffe 1997) and the Marshall Islands (Peterson 1997), to name only a few. If carbonate sediments are deposited by lateral accretion onto existing units, as is the case in Bermuda (Vacher 1978), the contact between these units may be vertical, offsetting the thickest part of the lens towards the low-permeability unit and creating an asymmetric lens profile (Fig. 3.2b). Atoll islands also may exhibit a lateral gradation in permeability due to depositional facies, with lower permeabilities on the lagoon side of the island, which result in asymmetrical freshwater lenses (Anthony 1997). Wicks and Herman (1995), using numerical modeling, found that even small-size zones of high permeability (representing cavernous zones) impact the shape of a freshwater wedge, depending largely on their position relative to the coastline.

Freshwater lens geometry may also be modified by variations in recharge across the landmass, such as uneven rainfall distribution (Jones et al. 1997) and evapotranspiration caused by coconut trees and other deep-rooted vegetation (Falkland and Woodroffe 1997) (Fig. 3.2d). Evaporative lakes are present on many islands in relatively dry climates, such as the "Exuma-type islands" in southern Bahamas (Vacher and Wallis 1992; Wallis et al. 1991). The reduction in recharge due to evaporation can cause the freshwater lens to be thin or absent beneath the lake, changing the effective island width and substantially reducing the volume of the lens (Fig. 3.2c). In addition to geographic variations in recharge, temporal variations are also observed in many areas, most notably the wet and dry seasonal patterns in rainfall commonly seen in the tropics and subtropics (see for example Buddemeier and Oberdorfer 1997; Halley et al. 1997; Humphrey 1997).

Often, such broad-scale variations can be treated analytically using simplified geometries, and such models have been crucial for interpreting field data and building conceptual models. However, additional complexities such as irregular island geometries, temporal variations in recharge, tidal signals, and density-driven flow usually require the use of numerical modeling to fully characterize the configuration of the freshwater body (for example, Ayers and Vacher 1983; Essaid 1986; Hunt 1979; Ghassemi et al. 1993, 1996; Gingerich 1992; Langevin 2003; Contractor 1983; Oberdorfer et al. 1990; Underwood et al. 1992).

3.3 The Freshwater/Saltwater Mixing Zone

Although the theoretical treatments discussed to this point have treated the freshwater-saltwater boundary as a sharp interface, the mixing zone can be very thick, such that most of the meteoric water in the aquifer is mixed with seawater to form brackish water and the mixing zone extends almost to the water table. Near the coastline, or in very highly mixed lenses, seawater and brackish water may occupy the entire water column. Buddemeier and Oberdorfer (1997) present Enewatek Atoll as a prime example of a highly mixed lens and make a useful distinction between *freshwater inventory*, used to describe the amount of water in the lens that has a relatively low salinity (i.e., potable water), and *meteoric water*, referring to the amount of groundwater that is derived from rainfall, much of which may be mixed with seawater in the transition zone and have a relatively high salinity. If the transition zone is very thick, the freshwater inventory will constitute only a small percentage of the meteoric-water inventory.

At the transition zone, freshwater and saltwater flow parallel to one another along the freshwater/saltwater interface; however, mixing may be caused by dispersion due to spatial variations in hydraulic conductivity, as well as migration of the interface due to tidal oscillations, seasonal fluctuations in recharge, or pumping perturbations (Cooper 1964; Wentworth 1948). In locales with relatively homogeneous aquifers, the tidal signal propagates laterally inland from the coast, and attenuates with distance from the coast, causing a corresponding thinning of the mixing zone with distance inland (Vacher 1978; Whitaker 1992). The mixing zone is thicker in more transmissive karstified limestones; largely because of the efficient propagation of the tidal signal through these units (Cooper 1964; Whitaker and Smart 1997). In islands with dual aquifer systems, there is no lateral trend in mixing zone thickness, because the tidal signal is transmitted relatively efficiently through the older, more transmissive limestones and vertically through the younger limestones to the lens (Buddemeier and Oberdorfer 1997; Hunt and Peterson 1980; Wheatcraft and Buddemeier 1981). In these islands, the tidal efficiency (ratio of tidal amplitude in wells to that in ocean) increases with depth, and tidal lag (time difference between ocean and well tidal signals) decreases with depth (Underwood et al. 1992).

Many studies define the freshwater portion of the aquifer by drinking-water standards (less than 600 [WHO 1971] or 250 mg/L Cl^- [WHO 2010]). There is no natural boundary to the mixing zone; salinity theoretically varies with depth across this zone according to the error function (Bear and Todd 1960; Cooper 1959; Kohout 1960). However, salinity profiles often deviate from this pattern, exhibiting deflections at hydrogeologic or lithologic boundaries (Buddemeier and Oberdorfer 1997; Harris 1967).

Below the mixing zone, the groundwater in the deeper, salt-water portion of the aquifer is generally not static. In South Florida, Cooper (1959) and Kohout (1965) described flow of saltwater below the freshwater zone inland, counter to the seaward flow of fresh and brackish water. Cooper (1959) attributed this flow to the fact that saltwater is entrained into the mixing zone and out to sea, and saltwater flows in to replace it. Kohout (1965) cited geothermal heating and convection as a cause. Whitaker and Smart (1990) identified large-scale saltwater circulation in the Great Bahama Bank and attributed it to sea-level head differences and to density-driven flow of seawater concentrated by evaporation. Beddows et al. (2007) found widespread and persistent inflow of saltwater through cave systems in the Yucatan caused by gradual cooling of warmer water flowing inland from the coast.

Movement of saltwater through the deeper portions of the aquifer impacts the shallow flow systems. For instance, in the Yucatan Peninsula, the freshwater lens is 40 % thinner than predicted by Ghyben-Herzberg theory; Moore et al. (1992) attribute this to convectional flow of seawater. In south Florida, deep flow paths bring brackish water and saltwater into contact with gypsum and anhydrite beds, adding sulfate compounds to the geochemical processes occurring in the shallow system (Copeland et al. 2011).

3.4 Saltwater Intrusion and Upconing

Intrusion of seawater has been widely recognized in the United States and in other countries as a consequence of groundwater pumping in coastal areas since at least the early 1900s (Back and Freeze 1983). The shape of the freshwater body and the position of the freshwater-saltwater boundary are the result of a balance of the flux of freshwater towards the sea and the aquifer variables such as hydraulic conductivity discussed earlier in this chapter. Changing any of these variables substantially can change the configuration of the freshwater lens or wedge. Most commonly, freshwater flux is decreased by pumping. At the aquifer scale, the decreased recharge causes the freshwater body to shrink, allowing the freshwater-saltwater boundary to "intrude" upward or inland. Seawater intrusion may also be caused by effectively decreasing the size of the landmass by building seawater canals or other major land excavations (Bear et al. 1999; Langevin et al. 1998).

In the presence of an underlying aquitard or shallow basement, seawater can intrude laterally from the coastline (Fig. 3.3b). In areas where freshwater vertically overlies saltwater, drawdown of the water table or potentiometric surface by pumping wells or other means can cause the freshwater/seawater boundary to migrate vertically upward, in general accordance with the Ghyben-Herzberg principle (Fig. 3.3a). This phenomenon is called "upconing" in reference to the fact that the updrawn interface mirrors (with a 40–1 exaggeration) the drawdown cone around the well (Schmorak and Mercado 1969).

Seawater intrusion is a pervasive issue in populated coastal areas around the world but is of special concern on most carbonate islands, where the freshwater lens is often of limited volume, highly mixed, and directly overlying saline groundwater (e.g. Contractor 1983; Falkland and Woodroffe 1997; Peterson 1997; Whitaker and Smart 1997).

The Biscayne Aquifer in Florida is a well-documented example of human-induced seawater intrusion in a karst aquifer in the United States (Cooper 1964; Klein and Hull 1978; Kohout 1965; Parker et al. 1955; Sonenshein 1997). The Biscayne Aquifer is an unconfined, highly transmissive, karstic aquifer underlying much of the Everglades, as well as the metropolitan area of Miami. In addition to drawdowns from well fields in the area, a series of canals, built in the early twentieth century, drew down the water table in the Everglades and allowed seawater to intrude far inland (Leach et al. 1972). Close monitoring

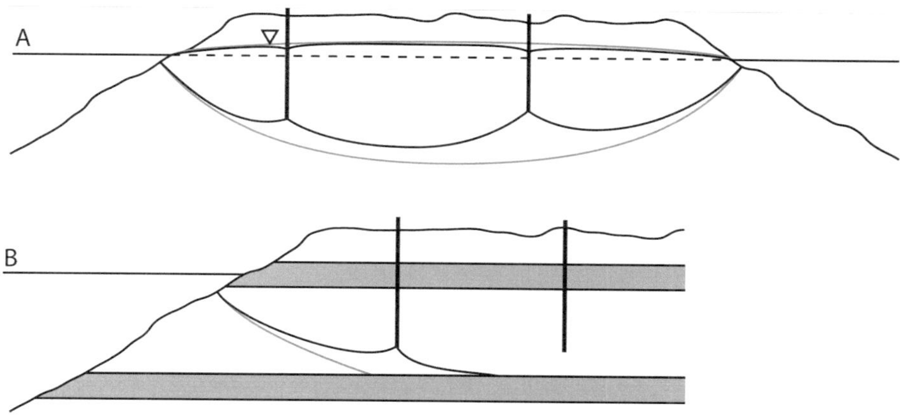

Fig. 3.3 Saltwater intrusion. (**a**) Upconing in a freshwater lens due to pumping. (**b**) Lateral migration of freshwater-saltwater boundary due in a confined freshwater wedge due to pumping

and restorative measures such as restoration of pumping and enhancement of recharge, followed by installation of salinity control structures on the canals, helped restore water levels in the area and have contributed to stabilization of the freshwater-saltwater interface (Sonenshein 1997).

Reclamation of well fields contaminated with seawater is very difficult. It is possible to restore freshwater recharge to the aquifer; however, it takes only a miniscule amount of seawater to make freshwater unpotable. Considering that virtually all of the residual seawater must be flushed from the aquifer to reclaim a saltwater-contaminated well field and the time it takes to do that, the only practical solution to seawater intrusion is prevention (Bear et al. 1999).

3.5 Limestone Dissolution

Several general discussions of the chemistry of limestone dissolution have been published. This brief overview is based on discussions found in Palmer (1991, 2007), Ford and Williams (1989), and Dreybrodt (2000).

There are several sources of dissolutional potential with respect to carbonate minerals. Extensive dissolution of calcite, aragonite, and dolomite is generally due to action by an acid, most commonly carbonic acid derived from atmospheric and soil CO_2. Under equilibrium conditions, the concentration of CO_2 in water whose surface is in contact with air is expressed as a pressure equal to that exerted by CO_2 on the surface of the water itself. The concentration ("partial pressure") of carbon dioxide in the atmosphere at sea level is 0.00038 atm (Zhang et al. 1997). In the pore spaces of soils, the CO_2 concentration can be orders of magnitude higher (Enoch and Dasberg 1971; Palmer 2007; Ford and Williams 1989). Meteoric waters gain much of their dissolutional potential percolating through the soil column.

Fresh water with dissolved carbon dioxide contains CO_2, H_2CO_3, HCO_3^-, and CO_3^{2-}. All of these species coexist within fresh waters under influence of the free atmosphere, in proportions relative to several interconnected variables, including temperature, pH, and partial pressure of CO_2 in the atmosphere. Limestone contributes Ca^+, as well as additional CO_3^{2-} and HCO_3^-. The species are related by the following processes:

In fresh water exposed to the atmosphere, CO_2 dissolves into the water:

$$CO_2(g) \leftrightarrow CO_2(aq) \qquad (3.3)$$

The dissolved CO_2combines with water to form carbonic acid:

$$CO_2 + H_2O \leftrightarrow H_2CO_3 \qquad (3.4)$$

The carbonic acid quickly dissociates, releasing a hydrogen ion and bicarbonate:

$$H_2CO_3 \leftrightarrow H^+ + HCO_3^- \quad (3.5)$$

Note that the reactions above do not depend on the presence of carbonate minerals; all species in Eqs. 3.3, 3.4 and 3.5 are derived from water and carbon dioxide. Carbonate minerals dissolve into water, releasing carbonate to the solution:

$$CaCO_3 \leftrightarrow Ca^{2+} + CO_3^{2-}; \quad (3.6)$$

$$CaMg(CO_3)_2 \leftrightarrow Ca^{2+} + Mg^{2+} + CO_3^{2-} \quad (3.7)$$

On its own, simple dissociation (such as happens with NaCl in water) does not cause much dissolution of carbonate minerals; however, in the presence of acid, the carbonate ion hydrates into bicarbonate,

$$H^+ + CO_3^{2-} \leftrightarrow HCO_3^-, \quad (3.8)$$

and allows further dissolution to occur.

To determine how far dissolution will proceed, we examine the degree of saturation of ions in a solution, comparing the effective concentration of ions in the solution of interest to their effective concentration in a saturated solution. The effective concentration is referred to as "activity" and represents the concentration of ions that are available for reaction.

The amount of a mineral able to dissolve in a solution is described by the *equilibrium constant* of the dissociation reaction (written here in terms of calcite dissolution):

$$K_{calcite} = (Ca^{2+})\left(CO_3^{2-}\right) \quad (3.9)$$

The two sides of Eq. 3.9 are equal for a saturated solution. In an undersaturated solution, the activities of the dissolved ions are lower than in the saturated solution:

$$K_{calcite} > (Ca^{2+})\left(CO_3^{2-}\right) \quad (3.10)$$

In a supersaturated solution, the activities are higher than in the saturated solution:

$$K_{calcite} < (Ca^{2+})\left(CO_3^{2-}\right) \quad (3.11)$$

This relation can be rearranged to provide a convenient expression of the degree of saturation of a solution called the *saturation index* (SI):

$$SI = \log \frac{(Ca^{2+})\left(CO_3^{2-}\right)}{K_{calcite}} \quad (3.12)$$

The logarithm is used to simplify comparisons of saturation states. For a saturated solution, $SI = 0$; for an undersaturated solution, SI is negative; for a supersaturated solution, SI is positive.

The *solubility* of a mineral is the concentration of that mineral in a saturated solution under given conditions. Solubility of calcite in freshwater is influenced by several variables, including P_{CO2}, temperature, pressure, and pH. An open system, in which P_{CO2} is held constant by transfer of CO_2 between an open water surface and the air, will result in more dissolutional potential than a closed system, wherein reactions run to completion with only a fixed starting amount of CO_2. Figure 3.4 illustrates the interrelationships between these variables. Atmospheric CO_2 directly influences the amount of CO_2 in solution, which in turn directly influences solubility of $CaCO_3$ through the calcite reaction series. An increase in temperature decreases the solubility of CO_2, thus decreasing the solubility of calcite as well. Additionally, and somewhat counterintuitively, $CaCO_3$ is more soluble in cold water than in warm water (even with CO_2 the same in both cases), an additive effect to the CO_2 solubility issue (White 1988; Palmer 2007). As expected, pH decreases with higher concentrations of CO_2 in solution.

The relationships between these variables are complicated by several chemical effects (Garrels and Christ 1965):

Ion pairs and ionic strength effects. Ions with opposite charges tend to pair together, forming complexes. The constituents tied up in complexes are unavailable for reaction, thus causing the

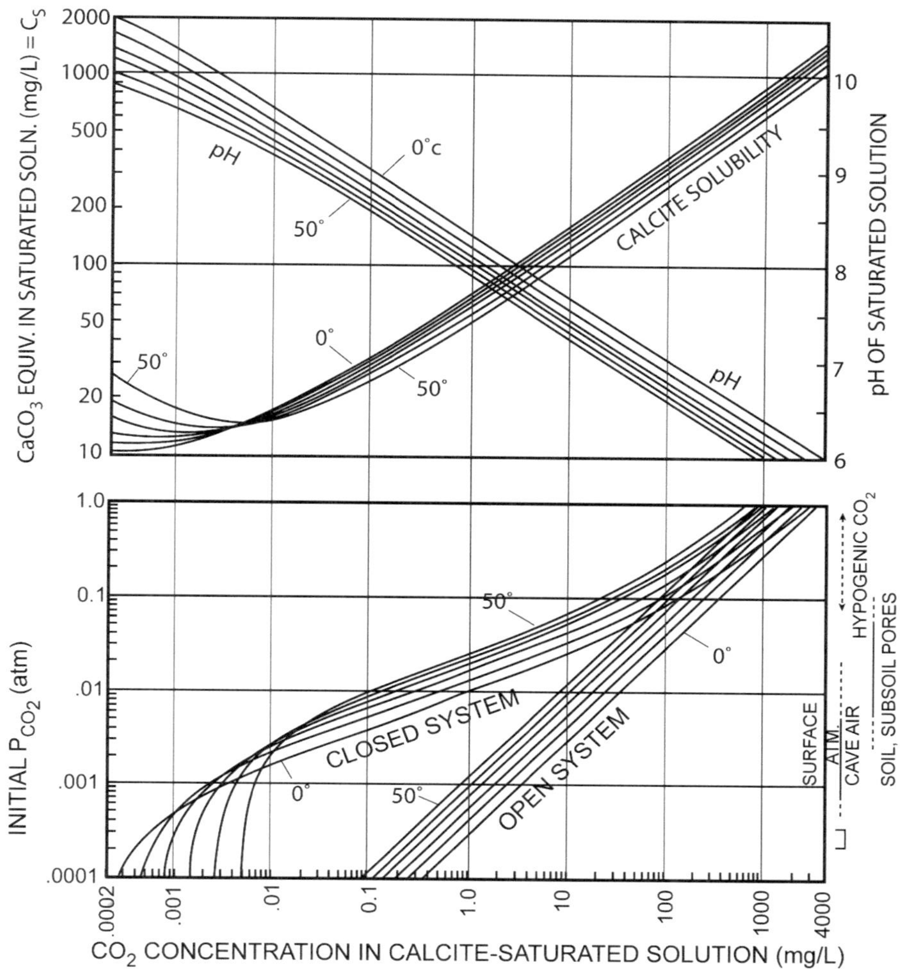

Fig. 3.4 Relationship between saturation concentration, dissolved calcite, initial P_{CO2}, equilibrium CO_2, and equilibrium pH (Reproduced with permission from Palmer 1991)

activity of the species to be less than its total mass concentration. The amount of mineral that can be dissolved therefore increases. This effect is important in solutions of high ionic strength (such as seawater). Large amounts of ions such as Na^+, K^+, and Cl^- can decrease the activity of ions such as Ca^{2+} and HCO_3^-, thus increasing the solubility of carbonate minerals.

Common ion effect. Calcite and dolomite release some of the same ions into solution when they dissolve, such that the reaction products for dolomite impact the calcite dissolution process and vice versa. The dissolutional potential of a solution is shared between calcite and dolomite, and the amount of dissolution of both minerals is therefore lower than if they were alone in the solution.

Other acids. The presence of acids such as hydrochloric acid and sulfuric acids provide additional dissolutional potential in limestone waters. These acids are often hypogenic. In Florida, for instance, deep circulation brings seawater into contact with gypsum and anhydrite deposits, resulting in high groundwater sulfate concentrations that impact the mixing-zone geochemistry in the area (Copeland et al. 2011; Rye et al. 1981).

Within the mixing zone itself, the density contrast traps organic matter and creates stratified microenvironments in which sulfate redox reactions generate acidity (Socki et al. 2002). Bacterially mediated sulfate reduction produces elemental sulfur or sulfide near the base of the mixing zone:

$$2CH_2O + SO_4^{2-} = 2HCO_3^- + S^0 + H_2 \quad (3.13)$$

$$2CH_2O + SO_4^{2-} = 2HCO_3^- + H_2S \quad (3.14)$$

Oxidation of the H_2S and S in the upper part of the mixing zone generates acidity:

$$FeS_2 + H_2O + \frac{7}{2}O_2 = Fe^{2+} + 2SO_4^{2-} + 2\,H^+ \quad (3.15)$$

$$H_2S + 2O_2 = SO_4^{2-} + 2H^+ \quad (3.16)$$

$$S^0 + \frac{3}{2}O_2 + H_2O = SO_4^{2-} + 2H^+ \quad (3.17)$$

Field studies from the Bahamas (Bottrell et al. 1991, 1993; Smart et al. 1988), Isla de Mona, Puerto Rico (Wicks and Troester 1998) and the Yucatan Peninsula (Perry et al. 2002; Stoessell 1992) have documented the contribution of sulfate redox reactions in systems open to a contribution of organic material form the surface (Wicks and Troester 1998).

Organics. Organic molecules can generate acidity, increasing dissolutional potential. Very large organic molecules can also form complexes with ions, isolating them from further reaction and thereby decreasing the activity of those ions (Upchurch et al. 1983). Calcium ions are particularly susceptible to organic complexing.

3.6 Limestone Dissolution in Coastal Areas

As described earlier in this chapter, chemical reactivity is enhanced in the areas where groundwaters of different chemistry mix. In unconfined coastal aquifers, mixing occurs not only within the freshwater-saltwater transition zone, but also at the water table, where phreatic water mixes with vadose water.

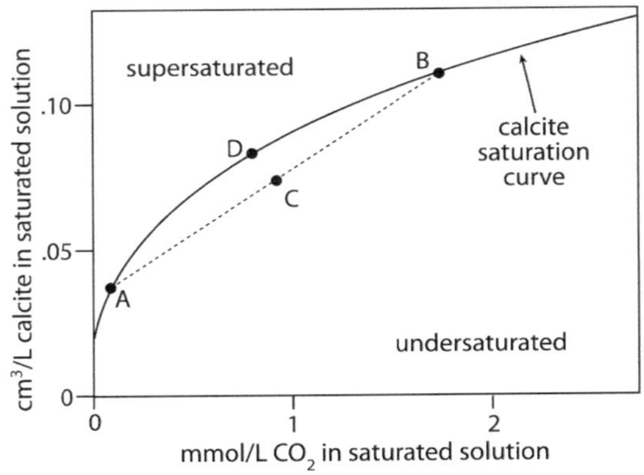

Fig. 3.5 The P_{CO2} effect. *Dotted line* is the mixing line showing the composition of a mixture of solutions *A* and *B* (After Palmer 1991)

When two limestone waters mix, depending on the chemistry of the end-member solutions and the proportions at which they are mixed, they produce a resulting solution that has a different set of characteristics with respect to the variables that affect the solubility of carbonate minerals. Because of the complexity of the interactions of these variables, the resulting solution may be undersaturated or supersaturated, and it may be difficult to predict which it will be (e.g. Bogli 1964; Plummer 1975; Wigley and Plummer 1976; Runnels 1969).

Several of the chemical interactions that influence carbonate solubility cause ionic activities to vary non-linearly with the mass concentrations of the ions in solution (Hanshaw and Back 1979; Wigley and Plummer 1976). When saturated groundwaters mix together, these chemical interactions yield a number of "mixing effects" that tend to cause the resulting mixture to be either undersaturated or supersaturated.

P_{CO2} Effect. The P_{CO2} effect (Bogli 1964, 1980) is perhaps the most dominant of the mixing effects and in most cases causes two saturated solutions to become undersaturated when they mix. The relationship between P_{CO2} and the solubility of calcite is nonlinear, exhibiting a convex upward slope (Fig. 3.5). In Fig. 3.5, the slope of the curve is steeper at lower P_{CO2} values and lessens as P_{CO2} increases. The relationship reflects the complex redistribution of carbonate species during mixing (Wigley and Plummer 1976); without these interactions, the relationship would be linear. When two calcite-saturated solutions are mixed, the composition of the resulting mixture is a weighted average of the compositions of the two end-member solutions. On a plot of calcite concentration and P_{CO2}, the mixture lies somewhere on the line connecting the two end members (Fig. 3.5, Palmer 1991). The new solution is below the saturation line, and is thus undersaturated. This effect is most pronounced where the difference in P_{CO2} between the two end members is high. The P_{CO2} effect has been singled out as a potential mechanism for cave development (Back et al. 1986; Mylroie and Carew 1990; Palmer et al. 1977).

Ionic strength effect. The addition of salts from saltwater to fresh groundwater causes limestone and dolomite to become more soluble. The ionic strength effect is only moderate in mixing of freshwaters but may be considerable in freshwater-saltwater mixing zones due to the high ionic strength of seawater.

Algebraic effect. The algebraic effect is easily demonstrated by derivation and describes a phenomenon wherein two saturated solutions tend to be supersaturated when they mix. As described above, the relative concentrations of species vary linearly during simple physical mixing. However, the saturation index depends upon the product of the ion activities (Eq. 3.12), which varies nonlinearly in these circumstances. As outlined in Wigley and Plummer (1976) it can be shown that

$$SI_c = \log\left[\frac{(1+b^2)}{4b}\right], \quad (3.18)$$

where b is the ratio of Ca^{2+} activities in the two end-member solutions. The part in brackets has a minimum value of 1 when the Ca^{2+} activities are equal (b of 1) and increases as b goes to zero or infinity. The algebraic effect is therefore strongest when two end-member solutions have greatly different activities of a particular ion of interest. This effect, however, can be overshadowed by other effects such as the ionic strength effect.

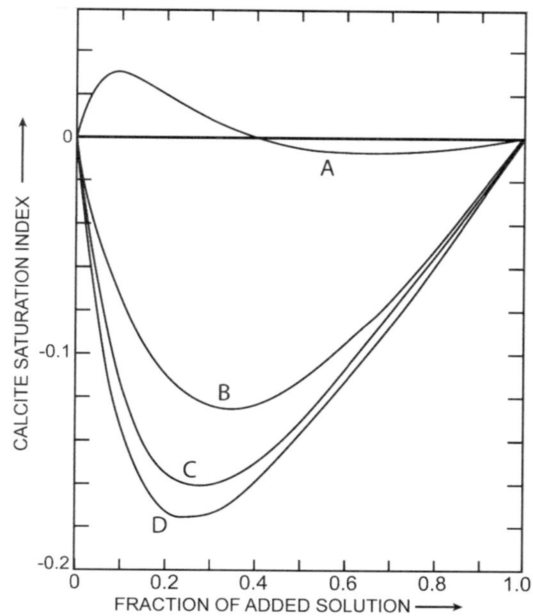

Fig. 3.6 Mixing effects on calcite saturation index. Each curve represents mixing of a pair of solutions. *Curve A* shows the influence of the P_{CO2} effect and algebraic effect in a pair of solutions for which the two effects are relatively balanced. *Curve B* shows the combination of P_{CO2} and temperature effects. *Curve C* shows the P_{CO2} effect alone. *Curve D* shows the P_{CO2} effect and the ionic strength effect (Reproduced with permission from Wigley and Plummer 1976)

Temperature. Mixing of waters of different temperatures produces a small undersaturation effect. The relationship of solubility to temperature traces a convex downward curve. The composition of a mixture lies somewhere along the line between the two end members, so that the concentration of any mixture of two saturated solutions lies above the saturation curve.

Figure 3.6 shows the effects of mixing on the saturation index for different sets of hypothetical waters, as calculated for demonstration purposes in Wigley and Plummer (1976). Note that the end-member solutions in these scenarios were carefully chosen to demonstrate various degrees of impact for the different mixing effects and do not represent the most commonly encountered scenarios. For mixing of freshwater and seawater, the P_{CO2} and ionic strength effects will often dominate the resulting mixture.

The chemistry of coastal karst waters may be dominated by processes other than mixing dissolution, especially when mixing effects cause only a minor degree of undersaturation (Sanford 1987). In these cases, the deviation from these relationships gives insight into the processes operating in a particular system. Some important processes that occur concurrently with mixing are (1) dissolution or outgassing of CO_2; (2) precipitation or dissolution of carbonate minerals; (3) reduction of sulfur compounds; and (4) kinetics effects.

Several field studies have found that the chemistry of groundwaters can be explained using some combination of these processes. For instance, Plummer et al. (1976) found that Bermuda groundwater chemistry is controlled by elevated CO_2 contents in soil; dissolution of aragonite; and mixing. The findings of Plummer et al. (1976) were inconsistent with simple mixing dissolution and indicate that the chemistry of Bermuda groundwaters is dominated by dissolution and precipitation under nonequilibrium conditions, influenced by dissolution and degassing of CO_2. In some Bermuda waters, outgassing of CO_2 overcomes the mixing dissolution effect and causes supersaturation with respect to aragonite or calcite.

Back et al. (1979, 1986) similarly concluded that the chemistry of waters in the Yucatan Peninsula is influenced not only by mixing, but also by the dissolution and precipitation of carbonate minerals and CO_2 dissolution and outgassing. Carbon dioxide from coastal mangrove swamps contributes to undersaturation, and outgassing causes saturation in open waters. Mixing dissolution therefore occurs mostly in underground mixing zones. Stoessell et al. (1993) and Socki et al. (2002) found that sulfur redox reactions appear to factor strongly in the dissolution process within the cenotes in the Yucatan.

Wicks and Troester (1998) found that groundwaters on Isla de Mona, Puerto Rico, were less saturated than predicted by theoretical mixing calculations. Within a water-filled cave passage, they found that chemistry was controlled by mixing and precipitation of aragonite; in an unconfined aquifer system, they found that the controlling processes were mixing, outgassing of CO_2, and reduction of sulfate.

In Barbados, sulfate reduction contributes to undersaturation with respect to calcite and aragonite (Stoessell 1992), and groundwaters are generally supersaturated with respect to dolomite (Humphrey 1997).

Jankowski and Jacobson (1991) found the groundwaters at Nauru Island in the South Pacific dominated by mixing, dissolution and precipitation of carbonate minerals, and dissolution and degassing of CO_2. Herman et al. (1985) found a similar result in Mallorca, Spain.

Fanning et al. (1981) found Ca enrichments and Mg depletions in waters from geothermal springs on the continental shelf of West Florida, interpreted to be from dolomitization and gypsum dissolution caused by interaction with geothermal fluids.

3.7 Rates of Porosity Development

Sanford and Konikow (1989) used a coupled reaction-transport model to study the rates of porosity development due to mixing dissolution. They found that porosity increases at rates proportional to three factors: the CO_2 content of the freshwater end member, the fluid flux through the system, and the rate of sea level change. Rezaei et al. (2005) and Sainz Garcia et al. (2011) have extended this work and reproduced Sanford and Konikow's model using reactive transport modeling and found that the highest dissolution rates occur in areas where mixing is greatest.

The CO_2 content of the freshwater end member strongly impacts the dissolutional potential of the mixed solution. In areas with typical groundwater velocities, kinetics effects in mixing dissolution are theoretically small (Sanford and Konikow 1989). Along the margin of the freshwater lens or wedge, groundwater velocities are fast enough that reaction rates become the limiting factor; kinetics effects thus become important in these areas (Sanford and Konikow 1989). Plummer et al. (1976) concluded that the rates of dissolution of aragonite and calcite are comparable to Bermuda groundwater residence times and that the kinetics effects are thus important.

In areas where kinetics effects are negligible, the rate of porosity increase varies linearly with specific discharge, which controls how fast the aggressive water is brought in and dissolved constituents taken away. In coastal mixing environments, this implies that the coastline-to-area ratio of the land mass may be an important predictor of the rate of porosity development. All other things being equal, a large land mass discharging large amounts of water through a coastline mixing zone will undergo mixing dissolution much faster than a small circular island, where the available recharge is likely much smaller in proportion to the amount of coastline. It is partially for this reason that it is recognized that mixing dissolution on small carbonate islands is probably enhanced by decay of organics (Mylroie and Carew 1995).

The position of the mixing zone is tied to sea level. If the coastline is migrating as a consequence of sea level change, the mixing zone may have limited potential for dissolution in any one area before it migrates away from that position (Sanford and Konikow 1989). The formation of macroporosity (i.e. caves) is thus much more likely at sea level stillstands. This is consistent with the model of flank margin cave development by Mylroie and Carew (1990). Flank margin caves are consistently found corresponding with relative elevations of known high- and low-stands around the world (e.g. Carew and Mylroie 1987; Frank et al. 1998; Jenson et al. 2006; Kelley et al. 2006; Mylroie et al. 1995, 2001, 2005; Mylroie and Carew 1995). Based on the duration of known highstands in the Bahamas, these caves are known to have formed in small freshwater lenses within a time span of 15,000 years or less (Carew and Mylroie 1995; Mylroie et al. 1991).

Hanshaw and Back (1980) found similarly high rates of dissolution when they calculated rates of limestone denudation in the Yucatan Peninsula and found that a block of limestone 1 km^2 by 3 m thick could have dissolved in 8,700 years. Specifically, they calculated that the incision of Xel Ha lagoon could have occurred as quickly as 3,000 years.

References

Anthony SS (1997) Hydrogeology of selected islands of the Federated States of Micronesia. In: Vacher HL, Quinn T (eds) Geology and hydrogeology of carbonate islands, Developments in sedimentology 54. Elsevier, Amsterdam, pp 693–706

Ayers JF, Vacher HL (1983) A numerical model describing unsteady flow in a fresh water lens. Water Resour Bull 19:785–792

Back W, Freeze RA (eds) (1983) Chemical hydrogeology. In: Benchmark papers in geology, vol 73(416). Hutchinson Ross Publication Co, Stroudsburg

Back W, Hanshaw BB (1970) Comparison of chemical hydrogeology of the carbonate peninsulas of Florida and Yucatan. J Hydrol 10:330–368

Back W, Hanshaw BB, Pyle TE, Plummer LN, Weidie AE (1979) Geochemical significance of groundwater discharge to the formation of Caleta Zel Ha, Quintana Roo, Mexico. Water Resour Res 15(6):1521–1535

Back W, Hanshaw BB, Herman JS, Van Driel JN (1986) Differential dissolution of a Pleistocene reef in the ground-water mixing zone of coastal Yucatan, Mexico. Geology 14:137–140

Badon Ghyben W (1889) Nota in verband met de voorgenomen putboring nabij Amsterdam. Tijdschrift van het Koninklijk Instituut van Ingenieurs, The Hague, pp 8–22

Bear J, Todd DK (1960) The transition zone between fresh and salt waters in coastal aquifers. Water Resources Center Contribution no. 19. University of California, Hydraulic Laboratory, Berkeley, 156 p

Bear J, Cheng AHD, Sorek S, Ouazar D, Herrera I (1999) Seawater intrusion in coastal aquifers – concepts, methods, and practices. Kluwer Academic Publishers, Dordrecht, p 625

Beddows PA, Smart PL, Whitaker FF, Smith SL (2007) Decoupled fresh–saline groundwater circulation of a coastal carbonate aquifer: spatial patterns of temperature and specific electrical conductivity. J Hydrol 346:18–32

Bogli A (1964) Mischungskorrosion: ein Beitrag zum Verkastungsproblem. Erkunde 18:83–92

Bogli A (1980) Karst hydrology and physical speleology. Springer, Berlin/Heidelberg/New York, 284 p

Bottrell SH, Smart PL, Whitaker F, Raiswell R (1991) Geochemistry and isotope systematics of sulphur in the mixing zone of Bahamian blue holes. Appl Geochem 6:99–103

Bottrell SH, Carew JL, Mylroie JE (1993) Bacterial sulphate reduction in flank margin environments: evidence from sulphur isotopes. In: White WB (ed) Proceedings of the 6th symposium on the geology of the Bahamas, Port Charlotte, Florida, Bahamian Field Station, pp 17–21

Budd DA (1988) Aragonite to calcite transformation during freshwater diagenesis of carbonates: insights from pore-water chemistry. Geol Soc of Am Bull 100: 1260–1270

Buddemeier RW, Holladay G (1977) Atoll hydrology; Island ground-water characteristics and their relationship to diagenesis. In: Proceedings of the 3rd international coral reef symposium, vol 2. University of Miami, Miami, pp 167–174
Buddemeier RW, Oberdorfer JA (1997) Hydrogeology of Enewetak Atoll. In: Vacher HL, Quinn TM (eds) Geology and hydrogeology of carbonate islands. Elsevier, Amsterdam, p 948
Carew JL, Mylroie JE (1987) Submerged evidence of Pleistocene low sea levels on San Salvador, Bahamas: symposia for undersea research. Natl Oc Atmos Adm 2(2):167–175
Carew JL, Mylroie JE (1995) Quaternary tectonic stability of the Bahamian archipelago: evidence from fossil coral reefs and flank margin caves. Quaternary Sci Rev 14:145–153
Contractor DN (1983) Numerical modeling of saltwater intrusion in the Northern Guam Lens. Water Resour Bull 19:745–751
Cooper HH Jr (1959) A hypothesis concerning the dynamic balance of fresh water and salt water in a coastal aquifer. J Geophys Res 64(4):461–467
Cooper HH Jr (1964) A hypothesis concerning the dynamic balance of fresh water and salt water in a coastal aquifer. In: Cooper HH Jr, Kohout FA, Henry HR, Glover RE (eds) Sea water in coastal aquifers, U.S. Geological Survey water-supply paper 1613-C. U.S. G.P.O, Washington, D.C., pp C1–C12
Copeland R, Doran NA, White AJ, Upchurch SB (2011) Regional and statewide trends in Florida's spring and well groundwater quality (1991–2003), Bulletin no. 69 (revised). Florida Geological Survey, Tallahassee, 393 p
Dreybrodt W (2000) Chemistry of speleogenetic processes. In: Klimchouk AB, Ford DC, Palmer AN, Dreybrodt W (eds) Speleogenesis: evolution of karst aquifers. National Speleological Society, Huntsville, p 527
Enoch H, Dasberg S (1971) The occurrence of high CO_2 concentrations in soil air. Geoderma 6:17–21
Essaid HI (1986) A comparison of the coupled fresh water-salt water flow and the Ghyben-Herzberg sharp interface approaches to modeling of transient behavior in coastal aquifer systems. J Hydrol 86:169–193
Falkland AC, Woodroffe CD (1997) Geology and hydrogeology of Tarawa and Christmas Island, Kiribati. In: Vacher HL, Quinn T (eds) Geology and hydrogeology of carbonate islands, Developments in sedimentology 54. Elsevier, Amsterdam, pp 381–431
Fanning KA, Byrne RH, Breland JA, Betzer PR (1981) Geothermal springs of the West Florida continental shelf: evidence for dolomitization and radionuclide enrichment. Earth Planet Sc Lett 52:345–354
Fetter CW Jr (1972) Position of the saline water interface beneath oceanic islands. Water Resour Res 8: 1307–1314
Fetter CW Jr (1980) Applied hydrogeology. Merrill Publishing Co., Columbus, 488 p
Ford D, Williams P (1989) Karst geomorphology and hydrology. Unwin Hyman Publishing, London/Boston, p 601
Frank EF, Mylroie JE, Troester J, Alexander EC Jr, Carew JL (1998) Karst development and speleologenesis, Isla de Mona Puerto Rico. J Cave & Karst Stud 60(2): 73–83
Garrels RM, Christ CL (1965) Solution, minerals, and equilibria. Harper & Row, New York, 450 p
Ghassemi F, Chen TH, Jakeman AJ, Jacobson G (1993) Two and three-dimensional simulation of seawater intrusion: performances of the "SUTRA" and "HST3D" models. AGSO J Aust Geol & Geop 14:219–226
Ghassemi F, Jakeman AJ, Jacobson G, Howard KWF (1996) Simulation of seawater intrusion with 2D and 3D models: Nauru Island case study. Hydrogeol J 4: 4–22
Gingerich SB (1992) Numerical simulation of the freshwater lens on Roi-Namur Island, Kwajalein Atoll, Republic of the Marshall Islands. M.S. thesis, Department of Geology & Geophysics, University of Hawaii, Honolulu
Glover RE (1964) The pattern of fresh-water flow in a coastal aquifer. In: Cooper HH Jr, Kohout FA, Henry HR, Glover RE (eds) Sea water in coastal aquifers, U.S. Geological Survey water-supply paper 1613-C. U.S. G.P.O, Washington, D.C., pp C32–C35
Halley RB, Vacher HL, Shinn EA (1997) Geology and hydrogeology of the Florida Keys. In: Vacher HL, Quinn T (eds) Geology and hydrogeology of carbonate islands, Developments in sedimentology 54. Amsterdam, Elsevier, pp 217–248
Hanshaw BB, Back W (1979) Major geochemical processes in the evolution of carbonate-aquifer systems. In: Back W, Stephenson DA (eds) Contemporary hydrogeology – the George Burke Maxey memorial volume. Journal of Hydrology 43. Elsevier, Amsterdam, pp 287–312
Hanshaw BB, Back W (1980) Chemical mass-wasting of the northern Yucatan Peninsula by groundwater dissolution. Geology 8:222–224
Harris WH (1967) Stratification of fresh and salt water on Barrier Island as a result of differences in sediment permeability. Water Resour Res 3:89–97
Henry HR (1964) Interfaces between salt water and fresh water in coastal aquifers. In: Cooper HH Jr, Kohout FA, Henry HR, Glover RE (eds) Sea water in coastal aquifers, U.S. Geological Survey water-supply paper 1613-C. U.S. G.P.O, Washington, D.C., pp C35–C70
Herman JS, Back W, Pomar L (1985) Geochemistry of groundwater in the mixing zone along the east coast of Mallorca, Spain. Karst water resources. In: Proceedings of the Ankara – Antalya symposium, IAHS publication no. 161, July 1985. IAHS Press, Wallingford, pp 467–479
Herzberg A (1901) Die Wasserversorgung einiger Nordseebder. Journal fur Gasbeluchtung und Wasserversorgung. 44: 815–819, 45: 842–844
Hubbert MK (1940) The theory of groundwater motion. J Geol 48:785–944

Humphrey JD (1997) Geology and hydrogeology of Barbados. In: Vacher HL, Quinn T (eds) Geology and hydrogeology of carbonate islands, Developments in sedimentology 54. Elsevier, Amsterdam, pp 381–431

Hunt B (1979) An analysis of the groundwater resources of Tongatapu Island, Kingdom of Tonga. J Hydrol 49:185–196

Hunt CD, Peterson FL (1980) Groundwater resources of Kwajalein Island, Marshall Islands. Water Resources Research Center technical report 126, University of Hawaii Press, Honolulu, 91 p

Jankowski J, Jacobson G (1991) Hydrochemistry of a groundwater-seawater mixing zone, Nauru Island, central Pacific Ocean. BMR J Aust Geol Geophys 12: 51–64

Jenson JW, Keel TM, Mylroie JR, Mylroie JE, Stafford KW, Taborosi D, Wexel C (2006) Karst of the Mariana Islands: the interaction of tectonics, glacioeustasy and fresh-water/sea-water mixing in island carbonates, Geological Society of America special paper 404. Boulder, Colorado, pp 129–138

Jones B, Mg KC, Hunter IG (1997) Geology and hydrogeology of the Cayman islands. In: Vacher HL, Quinn T (eds) Geology and hydrogeology of carbonate islands, Developments in sedimentology 54. Elsevier, Amsterdam, pp 299–324

Kelley K, Mylroie, JE, Mylroie JR, Moore C, Moore PJ, Collins L, Ersek L, Lascu I, Roth M, Passion R, Shaw C (2006) Eolianites and karst development in the Mayan Riviera, Mexico. In: Davis RL, Gamble DW (eds) Proceedings of the 12th symposium on the geology of the Bahamas and other carbonate regions, Gerace Research Centre, San Salvador, pp 88–99

Klein H, Hull JE (1978) Biscayne aquifer, southeast Florida. U.S. Geological Survey Water Resources Investigations, Tallahassee, pp 78–107, 52 p

Kohout FA (1960) Cyclic flow of saltwater in the Biscayne aquifer of southeastern Florida. J Geophys Res 65(7):2133–2141

Kohout FA (1965) A hypothesis concerning cyclic flow of salt water related to geothermal heating in the Floridan aquifer. T New York Acad Sci 28:249–271

Langevin CD (2003) Simulation of submarine ground water discharge to a marine estuary: Biscayne Bay, Florida. Ground Water 41(6):758–771

Langevin CD, Stewart MT, Beaudoin CM (1998) Effects of dredge and fill canals on freshwater resources of small oceanic islands. An example from Big Pine Key, Florida. Ground Water 36(3):503–513

Leach SD, Klein H, Hampton ER (1972) Hydrologic effects of water control and management of southeastern Florida, Report of investigations no. 60. Florida Bureau of Geology, Tallahassee, 115 p

Mink JF (1976) Groundwater resources of Guam: occurrence and development. WERI technical report 1, University of Guam

Mink JF, Vacher HL (1997) Hydrogeology of Northern Guam. In: Vacher HL, Quinn T (eds) Geology and hydrogeology of carbonate islands, Developments in sedimentology 54. Elsevier, Amsterdam, pp 743–762

Moore YH, Stoessell RK, Easley DH (1992) Freshwater/sea-water relationship within a ground-water flow system, northeastern coast of the Yucatan Peninsula. Ground Water 30:343–350

Mylroie JE (1984) Hydrologic classification of caves and karst. In: LaFleur RG (ed) Groundwater as a geomorphic agent. Allen and Unwin, Boston, pp 157–172

Mylroie JE, Carew JL (1990) The flank margin model for dissolution cave development in carbonate platforms. Earth Surf Proc Land 15:413–424

Mylroie JE, Carew JL (1995) Karst development on carbonate islands. In: Budd DA, Saller AH, Harris PA (eds) Unconformities in Carbonate Strata – their recognition and the significance of associated porosity, AAPG Memoir 63. American Association of Petroleum Geologists, Tulsa, Oklahoma, pp 55–76

Mylroie JE, Carew JL, Sealey NE, Mylroie JR (1991) Cave development on New Providence Island and Long Island Bahamas. Cave Sci 18:139–151

Mylroie JE, Carew JL, Vacher HL (1995) Karst development in the Bahamas and Bermuda. In: Curran HA, White WB (eds) Terrestrial and shallow marine geology of the Bahamas and Bermuda, Special paper 300. Geological Society of America. Boulder, Colorado, pp 251–267

Mylroie JE, Jenson JW, Taborosi D, Jocson JMU, Vann DT, Wexel C (2001) Karst features of Guam in terms of a general model of carbonate island karst. J Cave & Karst Stud 63(1):9–22

Mylroie JR, Mylroie JE, Jenson JW, MacCracken RS (2005) Fresh-water lens anisotropy and flank margin cave development, Fais Island, FSM. In: Geological Society of America Abstracts with Programs, vol 37, no 7, 2005 Salt Lake City annual meeting, Salt Lake City, 16–19 Oct 2005, p 436

Oberdorfer JA, Hogan PJ, Buddemeier RW (1990) Atoll island hydrogeology: flow and freshwater occurrence in a tidally dominated system. J Hydrol 120:327–340

Palmer AN (1991) The origin and morphology of limestone caves. Geol Soc Am Bull 103:1–21

Palmer AN (2007) Cave geology. Cave Books, Trenton, 454 p

Palmer AN, Palmer MV, Queen JM (1977) Geology and origin of the caves of Bermuda. In: Proceedings of the 7th international speleological congress, Sheffield, pp 336–339

Parker GG, Ferguson GE, Love SK et al (1955) Water resources of southeastern Florida with special reference to the geology and ground water of the Miami area, U.S. Geological Survey water-supply paper 1255. U.S. G.P.O, Washington, D.C., 965 p

Perry E, Velazquez-Oliman G, Marin L (2002) The hydrogeochemistry of the karst aquifer system of the northern Yucatan Peninsula, Mexico. Int Geol Rev 44(5):191–221

Peterson FL (1997) Hydrogeology of the Marshall Islands. In: Vacher HL, Quinn T (eds) Geology and hydrogeology of carbonate islands, Developments in sedimentology 54. Elsevier, Amsterdam, pp 381–431

Plummer LN (1975) Mixing of sea water with calcium carbonate ground water. In: Whitten EHT (ed) Quantitative studies in geological sciences, Memoir 142. Geological Society of America. Boulder, Colorado, pp 219–236

Plummer LN, Vacher HL, Mackenzie FT, Bricker OP, Land LS (1976) Hydrogeochemistry of Bermuda: a case history of ground-water diagenesis of biocalcarenites. Geol Soc Am Bull 87:1301–1316

Randazzo AF, Bloom JI (1985) Mineralogical changes along the freshwater/saltwater interface of a modern aquifer. Sediment Geol 43:219–239

Rezaei M, Sanz E, Raeisi E, Ayora C, Vázquez-Suñé E, Carrera J (2005) Reactive transport modeling of calcite dissolution in the fresh–salt water mixing zone. J Hydrol 311(1–4):282–298

Runnels DD (1969) Diagenesis, chemical sediments, and the mixing of natural waters. J Sediment Petrol 39:1188–1201

Rye RO, Back W, Hanshaw BB, Rightmire CT, Pearson FJ Jr (1981) The origin and composition of dissolved sulfide in groundwater. Geochim Cosmochim Acta 45(10):1941–1950

Sainz Garcia AM, Molinero Huguet JJ, Saaltink MW (2011) Numerical modeling of coastal aquifer karst processes by means of coupled simulations of density-driven flow and reactive solute transport phenomena. Carbonate Evaporite 26:19–27

Sanford WE (1987) Assessing the potential for calcite dissolution in coastal salt-water mixing zones. Ph.D. thesis, Pennsylvania State University, University Park, 109 p

Sanford WE, Konikow LF (1989) Porosity development in coastal carbonate aquifers. Geology 17:249–252

Schmorak S, Mercado A (1969) Upconing of fresh water-sea water interface below pumping wells, field study. Water Resour Res 6:1290–1311

Smart PL, Dawans JM, Whitaker F (1988) Carbonate dissolution in a modern mixing zone. Nature 335: 811–813

Socki RA, Perry EC, Romanek CS (2002) Stable isotope systematics of two cenotes from the northern Yucatan Peninsula, Mexico. Limnol Oceanogr 47:1808–1818

Sonenshein RS (1997) Delineation and extent of saltwater intrusion in the Biscayne aquifer, Eastern Dade County, Florida, 1995, Water-resources investigations report 96–4285. U.S. Geological Survey. U.S. G.P.O, Washington, D.C

Stoessell RK (1992) Effects of sulfate reduction on CaCO3 dissolution and precipitation in mixing-zone fluids. J Sediment Petrol 62:873–880

Stoessell RK, Moore YH, Coke JG (1993) The occurrence and effect of sulfate reduction and sulfide oxidation on coastal limestone dissolution in Yucatan cenotes. Ground Water 31(6):566–575

Todd DK (1980) Groundwater hydrology. John Wiley & Sons, New York, 525 p

Underwood MR, Peterson FL, Vos CI (1992) Groundwater lens dynamics of Atoll Islands. Water Resour Res 28:2889–2902

Upchurch SB (2002) Hydrogeochemistry of a karst escarpment. In: Martin JB, Wicks CW, Sasowsky ID (eds) Hydrogeology and biology of post-Paleozoic carbonate aquifers, Special publication 7. Karst Waters Institute, Charles Town, pp 73–75

Upchurch SB, Strom RN, Williams MJ (1983) Preservation of dolomite in coastal peats of the ten thousand Islands area, Florida. In: Raymond R Jr, Andrejko MJ (eds) Proceedings of workshop on mineral matter in peat: its occurrence, form, and distribution, Los Alamos National Laboratory, Los Alamos, 26–30 Sept 1983

Vacher HL (1978) Hydrogeology of Bermuda – significance of an across-the-island variation in permeability. J Hydrol 39:207–226

Vacher HL (1988a) Dupuit-Ghyben-Herzberg analysis of strip-island lenses. Geol Soc Am Bull 100:580–591

Vacher HL (1988b) Ground water in barrier islands: theoretical analysis and evaluation of the unequal sea level problem. J Coastal Res 4(1):139–148

Vacher HL (1997) Introduction: varieties of carbonate islands and historical perspective. In: Vacher HL, Quinn T (eds) Geology and hydrogeology of carbonate islands, Developments in sedimentology 54. Elsevier, Amsterdam, pp 1–34

Vacher HL, Mylroie JE (2002) Eogenetic karst from the perspective of an equivalent porous medium. Carbonate Evaporite 17(2):182–196

Vacher HL, Wallis TN (1992) Comparative hydrogeology of freshwater lenses of Bermuda and Great Exuma Island, Bahamas. Ground Water 30:15–20

Vacher HL, Wightman MJ, Stewart MT (1992) Hydrology of meteoric diagenesis: effect of Pleistocene stratigraphy on freshwater lenses of Big Pine Key, Florida. In: Fletcher CH, III, Wehmiller JF (eds) Quaternary coasts of the United States: marine and lacustrine system, Special publication 48. SEPM (Society of Sedimentary Geologists). Tulsa, pp 213–219

Van der Veer P (1977) Analytical solution for steady interface flow in a coastal aquifer involving a phreatic surface with precipitation. J Hydrol 34:1–11

Wallis TN, Vacher HL, Stewart MT (1991) Hydrogeology of the freshwater lens beneath a Holocene strandplain, Great Exuma, Bahamas. J Hydrol 125:93–100

Wentworth CK (1948) Growth of the Ghyben-Herzberg transition zone under a rinsing hypothesis. Am Geophys Union Trans 1:97–98

Wheatcraft SW, Buddemeier RW (1981) Atoll island hydrology. Ground Water 19:311–320

Whitaker FF (1992) Hydrology, geochemistry, and diagenesis of modern carbonate platforms in the Bahamas. PhD thesis, University of Bristol

Whitaker FF, Smart PL (1990) Active circulation of saline groundwaters in carbonate platforms: evidence from the Great Bahama Bank. Geology 18:200–203

Whitaker FF, Smart PL (1997) Hydrogeology of the Bahamian archipelago. In: Vacher HL, Quinn T (eds) Geology and hydrogeology of carbonate islands, Developments in sedimentology 54. Elsevier, Amsterdam, pp 381–431

White WB (1988) Geomorphology and hydrology of karst terrains. Oxford University Press, New York, 464 p

Wicks CM, Herman JS (1995) The effect of zones of high-porosity and permeability on the configuration of the saline fresh-water mixing zone. Ground Water 33(1–2):733–740

Wicks CM, Troester JW (1998) Groundwater Geochemistry of Isla de Mona, Puerto Rico. J Cave & Karst Stud 60:107–114

Wigley TML, Plummer LN (1976) Mixing of carbonate waters. Geochim Cosmochim Acta 40:989–995

World Health Organization (WHO) (1971) European standards for potable water, 2nd edn. WHO, Geneva

World Health Organization (WHO) (2010) Guidelines for drinking water quality, vol 1, 3rd ed. World Health Organization, Geneva

Zhang N, Tans PP, Thoning KW (1997) A high precision manometric system for absolute calibrations of CO_2 in dry air. J Geophys Res 102(D5):5885–5894

Coastal Karst Development in Carbonate Rocks

4

John E. Mylroie

Abstract

Coastal karst cave development globally is biased towards the tropics and subtropics, where carbonate deposition is ongoing, and therefore carbonate coasts are common. The Carbonate Island Karst Model (CIKM) delineates the unique conditions that separate coastal karst from traditional karst areas of continental interiors. In these warm-water carbonate islands, diagenetically immature, or eogenetic carbonate rocks are host to a fresh-water lens that creates flank margin caves in a diffuse flow environment. Diagenetically mature, or telogenetic carbonates, can also host flank margin caves. Flank margin caves can form rapidly, as carbonate sediment is deposited, to produce syndepositional caves called banana holes. Flank margin caves can survive as open voids for millions of years, and as infilled diagnostic features for tens of millions of years. Vadose fast flow routes called pit caves form as a result of surface micritization to provide point recharge to the fresh-water lens. The presence of non-carbonate rocks can perch vadose flow, creating stream caves that terminate in the fresh-water lens. When sea level falls to create large exposed carbonate platforms, phreatic conduit flow develops to carry recharge to the platform periphery. Collapse of these conduits, as well as bank margin fracture, account for the majority of blue holes in carbonate platforms. Closed depressions in eogenetic carbonate islands are commonly constructional, relicts from variable carbonate deposition. The most common sinkhole type is the cave-collapse sinkhole. Morphometric analysis of flank margin caves supports cave origin as the amalgamation of individual chambers, provides evidence of denudation rates, and can differentiate flank margin caves from some pseudokarst cave types.

4.1 Introduction

An overview of cave types found in rocky coastlines is presented in Chap. 1. The unique morphology of surficial karst exposed on the coast is

J.E. Mylroie (✉)
Department of Geosciences, Mississippi State University, Mississippi State, MS 39762, USA
e-mail: mylroie@geosci.msstate.edu

M.J. Lace and J.E. Mylroie (eds.), *Coastal Karst Landforms*, Coastal Research Library 5,
DOI 10.1007/978-94-007-5016-6_4, © Springer Science+Business Media Dordrecht 2013

described in detail in Chap. 2. The hydrological and geochemical processes at work inside the fresh-water lens of carbonate islands and coasts is discussed in Chap. 3. This chapter will focus on karst cave development by dissolution in coastal carbonates.

Carbonate rocks, as presented here, represent three common mineral species: aragonite, calcite, and dolomite. Aragonite and calcite are both calcium carbonate, or $CaCO_3$, but with a different crystal structure, and these minerals form limestone. Aragonite is slightly more soluble than calcite in surface conditions. Most modern calcium carbonate is deposited as aragonite, so it is abundant in very young limestones worldwide. After deposition, aragonite inverts to calcite within tens to hundreds of thousands of years, depending on the nature of the environment and fluids in contact with the limestones, so most older limestones are made of calcite. Dolomite is a calcium and magnesium mixture, with a formal composition of $CaMg(CO_3)_2$, and can occur in a wide variety of forms (Machel 2004). Most dolomite in coastal settings is initially deposited as calcium carbonate, and is subsequently altered to dolomite by the process of dolomitization. Dolomite is theoretically as soluble as calcite, but in the real world it dissolves much more slowly. Calcite enriched in magnesium (18–20 % Mg), but well below the level of dolomite, is called high-Mg calcite and it is a common skeletal material in certain marine organisms, such as echinoderms and red algae. As with aragonite, and unlike dolomite, high-Mg calcite tends to dissolve readily or alter quickly to regular calcite. The carbonate metamorphic rock marble can be made from either limestone or dolomite. Evaporites, for the reasons given in Chap. 1, are extremely rare in coastal settings and will not be discussed here.

Karst caves in coastal settings are found in very young carbonate rocks composed mostly of aragonite and high-Mg calcite, as well as in older carbonates composed mostly of calcite. Dolomite hosts karst caves as well, but in coastal regions there is uncertainty regarding the timing of cave formation as regards the dolomitization of the limestones. Three options can be presented: (1) the caves develop in limestone, after which the limestone is dolomitized but the caves remain; (2) the limestone is dolomitized, and then the caves form; and (3) cave formation and dolomitization occur concurrently with cave dissolution (perhaps as a linked process, i.e. "syngenetic"). Caves have been found in dolomite on numerous islands (e.g. Barbados, Curacao, Isla de Mona), and they seem comparable in form and morphology as those found in limestones.

As noted above, as Cenozoic limestones get older, their original aragonite and high-Mg calcite mineralogy changes to a low-Mg calcite mineralogy, generally within a few hundred thousand years at the most. This alteration can occur even when the limestones involved have not been buried, or subjected to heat and pressure. As geologic time proceeds, limestones are commonly buried, and as a result begin to undergo a change in character called *diagenesis*. This process is gradual, but over the length of geologic time, it can significantly alter the nature of a limestone unit. The stages of maturation of a limestone were described by Choquette and Pray (1970) as subdivisions of the postdepositional evolution of carbonate porosity into three *time-porosity stages* conforming to the rock cycle. They defined "the time of early burial as *eogenetic*, the time of deeper burial as *mesogenetic*, and the late stage associated with erosion of long-buried carbonates as *telogenetic*" (Choquette and Pray 1970, p. 215). Vacher and Mylroie (2002, p. 183) defined eogenetic karst as: "the land surface evolving on, and the pore system developing in, rocks undergoing eogenetic, meteoric diagenesis." They further defined *telogenetic karst* as: "the karst developed on and within ancient rocks that are exposed after the porosity reduction of burial diagenesis." Islands can exhibit both types of karst depending on their geologic history, for example the Bahamas as an eogenetic karst example (Chap. 7), and coastal New Zealand as a telogenetic karst example (Chap. 17). The stages of limestone diagenetic maturation and the subsequent impact on fluid flow are presented in Fig. 4.1.

The vast majority of carbonate coasts and islands are in the tropics and subtropics, where limestone deposition is currently ongoing. As a

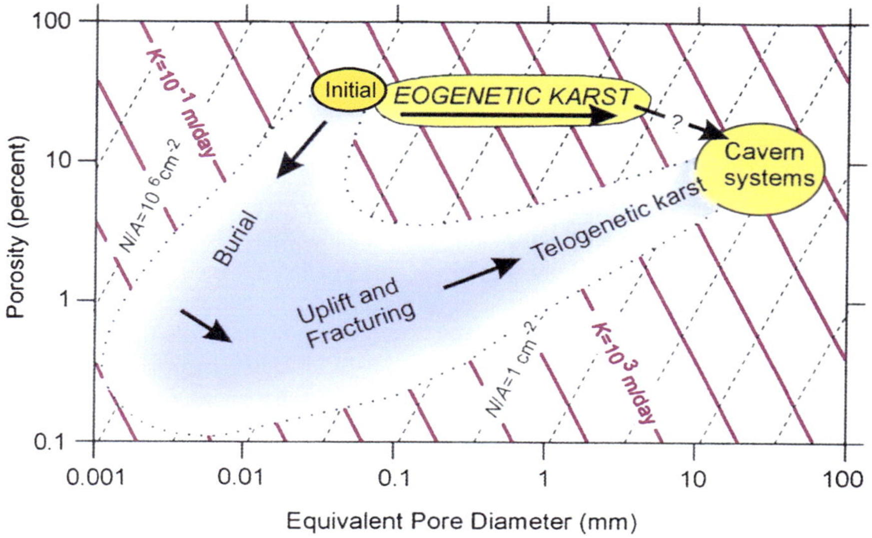

Fig. 4.1 The evolution of hydraulic conductivity in carbonate rocks (as modeled by straight tubes), showing the reduction in porosity and permeability with deep burial. Note that eogenetic rocks take a "short cut" to cavernous porosity (From Vacher and Mylroie 2002)

result, the limestones are considered eogenetic. They retain much of their original depositional porosity, and their allochems (organic and inorganically precipitated rock particles) are readily identifiable. As noted in Chap. 2, this eogenetic character of the rocks has a major influence on the nature of coastal karren. There is also a major impact on the development of karst caves. Vacher and Mylroie (2002) noted two important aspects of island karst development. First, one must differentiate between "karst on islands", and "island karst". The former describes karst features found on islands, almost always in the interior, where the karst landscapes resemble what is found in the interior of continents. Such karst is unaffected by the island setting, and lacks a signature associated with coastal conditions. Many features that the casual observer associates as island karst, such as mogotes in Puerto Rico or cockpit karst in Jamaica, are not restricted to the island setting at all. Those features are the result of the climate of the area, base level history, and the nature of the limestone and are identical to what would be found in a nearby continental setting, such as the interior of Belize. Island karst, on the other hand, refers to karst features that are linked to the coastal environment. These karst features are influenced by sea-level change and fresh water mixing with sea water, conditions not present in island or continental interiors.

Second, islands can be considered a special case of a carbonate coastline, which can be found along the edges of continents as well. South Florida and Yucatan would be examples of eogenetic carbonate coasts on continents (Chaps. 15 and 16), whereas the coast of the Adriatic Sea would constitute a telogenetic continental example (Chap. 17). Carbonate islands, especially small islands, offer limits in space that constrain models used to explain the karst features they contain. In tropical settings, these carbonate islands are commonly made up of eogenetic limestones, and in some cases such as the Bahamas, the limestones are only Middle Pleistocene to Holocene in age. This young age of the limestones provides limits on the time available to produce the observed karst features. Island water budgets are isolated and also limited, adding water budget constraints. The reason that much of the coastal cave and karst literature is focused on islands is because of the constraints offered by that setting. The Bahamas provide almost the ideal example (Chap. 7). The islands are all very young, with eogenetic limestones, and island size ranges from very small to very large. The small islands exist on both very small and on very large platforms, allowing island size versus platform size to be evaluated. They are in a quiescent tectonic setting, such that sea-level fluctuation is solely the result of glacioeustasy and long-term isostatic subsidence. Much of the early work on island karst was done in the Bahamas, as the simplicity of that location, and the time and spatial constraints in place, allowed easy recognition of the factors controlling the karst landforms. As other island groups were studied, many formed within more complex tectonic settings, the basic foundation factors could be determined despite the complexity of the site-specific situation. The result was the development of an island karst classification scheme.

4.2 The Carbonate Island Karst Model (CIKM) and Coastal Cave Development

The carbonate island karst model, or CIKM, is an attempt to characterize the levels of complexity found within carbonate islands, and by extension, continental carbonate coasts. The parameters that control the development of island karst were initially outlined by Mylroie and Vacher (1999) and codified as the Carbonate Island Karst Model, or CIKM, by that name after the initial study of Guam (Mylroie and Jenson 2000; Mylroie et al. 2001). As with any theoretical construct, the CIKM has been modified and evolved over the years as new data have emerged from many coastal karst areas (e.g. Mylroie and Mylroie 2007). The principles of the CIKM include:

1. Mixing of fresh and salt water at the boundaries of the fresh-water lens results in a localized area of preferential porosity and permeability development. Collection of organics at these boundaries may also enhance dissolution. The maximum dissolution occurs at the lens margin, where the water table and halocline mixing zones, and their associated organics, are superimposed, and where lens flow velocities are fastest.
2. Sea level change, which also moves the position of the fresh-water lens.
   (A) Glacioeustacy has moved sea level, and thus the fresh-water lens position, up and down more than 100 m throughout the Quaternary (note, for this chapter the glacioeustatic sea-level events will be identified by their Marine Isotope State, or "MIS"; see Fig. 1.5, Chap. 1).
   (B) Local tectonic movement, sediment compaction, and isostatic adjustment can cause overprinting of dissolutional and diagenetic features developed during different glacioeustatic events.
3. The karst is commonly eogenetic in that it has developed on rocks that are young and have never been buried below the zone of meteoric diagenesis; however telogenetic rocks do occur on carbonate islands and coasts.
4. Carbonate islands can be divided into four categories based on carbonate and non-carbonate relationships, and sea level (Fig. 4.2).
   (A) Simple Carbonate Island—Only carbonate rocks are present within the range of the fresh-water lens position (Fig. 4.2a). Meteoric catchment is entirely autogenic and flow within the fresh-water lens is controlled entirely by properties of the carbonate rock. The Bahamas are examples of simple carbonate islands.
   (B) Carbonate-Cover Island—Only carbonate rocks are exposed at the surface and the catchment is entirely autogenic (Fig. 4.2b). Non-carbonate rocks exist under the carbonate rocks and may partition and influence flow within the lens, including conduit flow at the carbonate and non-carbonate contact. Bermuda (at a glacioeustatic sea-level lowstand) is an example of a carbonate-cover island.
   (C) Composite Island—Both carbonate and non-carbonate rocks are exposed at the surface (Fig. 4.2c), allowing for allogenic and autogenic catchment. The lens is partitioned and conduit cave systems can develop at the contact of the carbonate

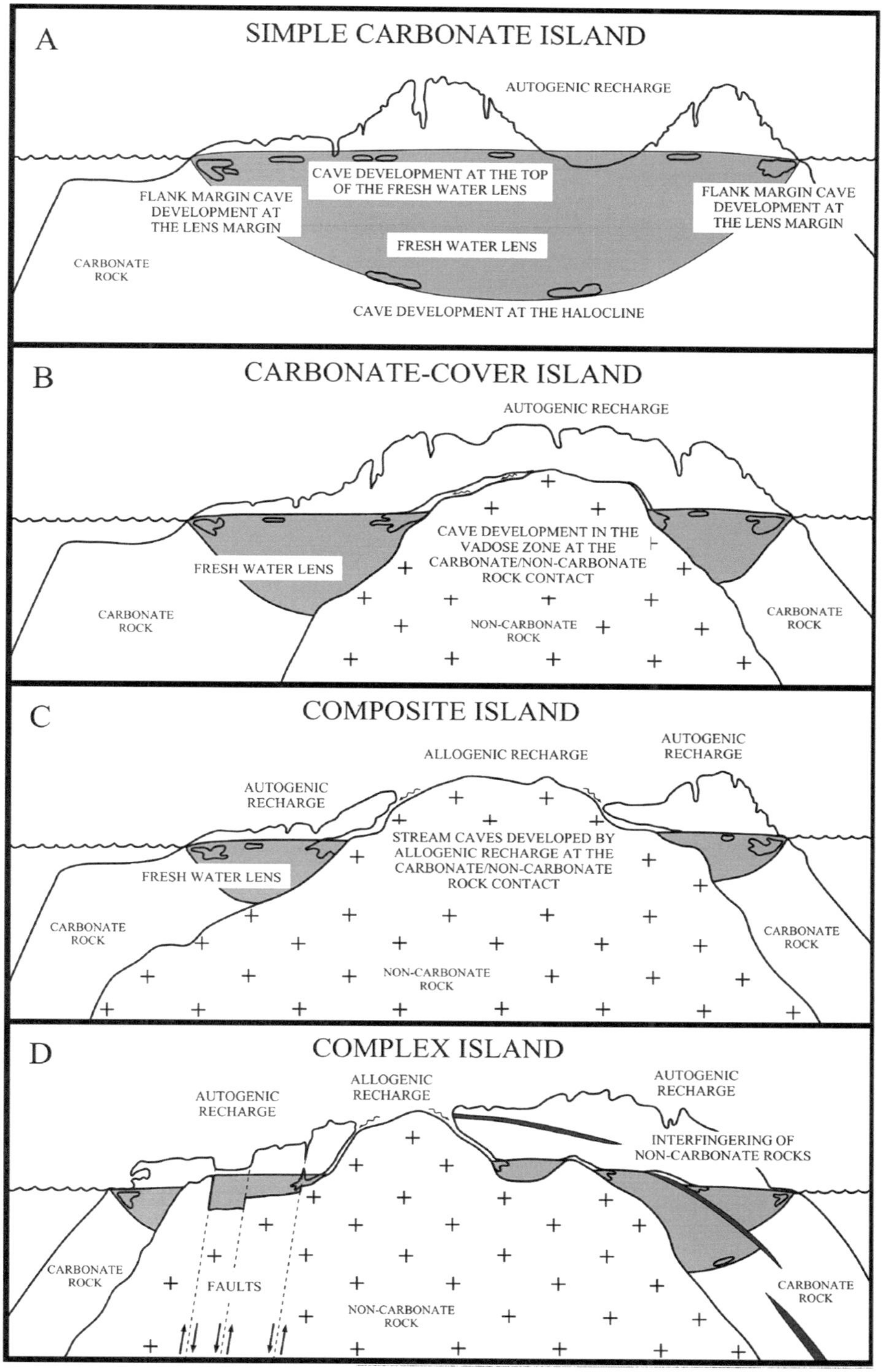

Fig. 4.2 The classification of island type within the CIKM, based on carbonate and non-carbonate rock interactions. Gradational conditions, or the subdivisions of portions of islands with respect to island type, are possible. (**a**) Simple Carbonate Island. (**b**) Carbonate Cover Island. (**c**) Composite Island. (**d**) Complex Island. See text for details

and non-carbonate rocks. Barbados and Guam are examples of composite islands.

(D) Complex Island—Carbonate and non-carbonate rocks are complexly interrelated by depositional relationships and/or faulting (Fig. 4.2d). Perching, isolation, and confining of the freshwater lens is possible. Saipan is an example of a complex island.

Autogenic recharge in Fig. 4.2 refers to collection of meteoric water directly on to the carbonate rock surface. As a result, the water is chemically modified in terms of its dissolutional potential as it descends by diffuse flow through the vadose zone to the top of the fresh-water lens. Allogenic recharge is collection of meteoric water on a non-carbonate surface, where the flow can concentrate into streams that flow laterally and then contact the carbonate rocks. The water enters the carbonate rocks as discrete point inputs that still retain most of their dissolutional potential. The difference between autogenic and allogenic catchment is both a geochemical and flow configuration difference.

The rigid classification of islands into one of these four categories is commonly not possible. Instead, regions or portions of islands can be classified according to the CIKM, which may assist in explaining the variety of cave and karst development that can occur on carbonate islands. Even the use of the term "carbonate island" needs care. An island can be primarily a non-carbonate rock, as in the Hawaiian Islands, and still contain isolated carbonate outcrops with caves and karst, for example the eolian calcarenites at Kahuku Point on the north end of Oahu, to which the CIKM can be applied. Chapter 12 presents another example of a volcanic island with minor carbonate outcrop.

While the allogenic catchment observed on composite and complex islands would appear to be karst on islands as opposed to island karst, in many islands the stream caves that developed from this allogenic flow discharge to the fresh-water lens, and not to the shoreline. This flow dynamic makes the allogenic catchment flow quite different from traditional stream caves found in continental interiors, or in the interior areas of large islands like Cuba or Jamaica.

4.3 Cave Development on Carbonate Islands

As Fig. 4.2 shows, a wide variety of dissolutional voids can develop on carbonate islands. To begin this discussion, the simplistic Bahama Islands will be examined. Chapter 7 is an overall perspective on cave development and distribution in the Bahamas, the discussion here will focus on the mechanisms involved.

4.3.1 Flank Margin Caves

In simple carbonate islands such as the Bahamas, extensive dry caves can be found that developed during a past glacioeustatic sea-level highstand. They are enterable today because while the present is a state of a high sea-level elevation, current sea level is not as high as earlier in the Quaternary (e.g. last interglacial or MIS 5e ~124–115 ka, Thompson et al. 2011). Within the fresh-water lens (see Chap. 3 and figures therein, or refer to Fig. 4.2), dissolution is favored at specific sites. The top of the lens is where descending vadose water mixes with the phreatic water of the lens. These waters are both commonly saturated with $CaCO_3$, but at different initial conditions. As explained in Chap. 3, this mixing of waters results in enhanced dissolution of $CaCO_3$. Similar mixing dissolution occurs at the base of the fresh-water lens, where fresh water and marine water mix (called a *halocline* if the boundary is sharp, a *mixing zone* if the boundary is gradational in salinity). Of these two mixing zones, the fresh water mixing with marine water at the bottom of the lens appears more important, perhaps because the top of the lens is vulnerable to CO_2 degassing, which may counter-act the mixing dissolution environment there.

Both the top and bottom of the fresh-water lens are density boundaries. Organics that collect at these two boundaries can oxidize to produce excess CO_2 to drive dissolution. If organic loading is significant, the system may become anoxic, allowing a variety of unusual biogeochemical environments that can promote $CaCO_3$ dissolution. In blue holes, the large surface openings accept major amounts of organic debris, including macroscopic vegetative remains as well as particulate matter, which collect at the halocline. In those cases, anoxia at the halocline produces H_2S which drives dissolution (Bottrell et al. 1991). For regular vadose flow in carbonate aquifers, large organic material is filtered out, and only dissolved organic carbon (DOC) and small particulate matter can reach the fresh-water lens, although vadose fast-flow routes called pit caves can transport large organic particles downward at specific sites. In continental settings, it has been shown that DOC and fine particulate organics can percolate downward through the vadose zone and promote dissolution within the aquifer (Wood and Petritus 1984). In the Bahamas it has been shown that organics can migrate to the fresh-water lens, and through biomediation of sulfur enhance dissolution in flank margin caves (Bottrell et al. 1993).

The fresh-water lens integrates all the catchment it receives and discharges it to the ocean. The lens also thins as it approaches the coast. These two factors result in the fastest flow velocities in the lens being found at the distal margin of the lens. Chemical reactions that occur here will have the advantage in that reactants are brought in, and products removed, faster than elsewhere in the lens. The lens margin also brings the top and bottom of the lens into the same area, such that the mixing impact of each environment is added together, as is the geochemical activity of any included organics at those boundaries. As a result, the greatest amount of $CaCO_3$ dissolution should occur at the lens margin. Decades of field work in the Bahamas and in other localities (Mylroie and Mylroie 2007) has shown that the largest caves found on small, simple carbonate islands are found in this location. These caves are named *flank margin caves* because of the location in what would have been the margin of the lens, under the flank of the enclosing land mass. The term was developed in the late 1980s (Mylroie 1988; Mylroie and Carew 1990) based on work in the Bahamas, and has since been expanded to a variety of island types around the world, including Isla de Mona (Frank et al. 1998), the Mariana Islands (Jenson et al. 2006), Puerto Rico (Lace 2008), Barbados (Machel et al. 2011), the Adriatic coast (Otonicar et al. 2010), Australia (Mylroie and Mylroie 2009a), and New Zealand (Mylroie et al. 2008b). The dissolutional mechanism is extremely powerful, and cuts across a variety of primary and secondary features in caves, as shown in Fig. 4.3.

Flank margin caves develop as dissolutional voids within a diffuse flow regime. They form without entrances, and human access can only be later gained after other erosional processes, such as hillslope retreat or collapse, open them to the outside environment. They have a consistent passage pattern in part controlled by their size. Their size is in turn controlled by how long the fresh-water lens was stable at a single position. Small flank margin caves are commonly singles chambers with a few short side passages (Fig. 4.4). Larger caves are a collection of chambers, and the largest caves are amalgamations of chamber collections. The caves fit the *ramiform* and *sponge-work* classifications of A. Palmer (1991), but with specific asymmetries related to their position at the lens margin. First, the chambers nearest the lens margin are usually the largest (Fig. 4.4a). To the rear of these chambers are smaller passages that head inward to what was once the lens interior. As the original chamber enlarged, marine water invaded the chamber and the mixing front moved to the back, inland wall of the chamber as a series of individual fresh-water discharge points and hence mixing locations. These passages grew headward into the lens. The passages may have cross-connected to create a maze-like set of passages, but they all end in blank bedrock walls (Fig. 4.4b). These passage terminations are the position of the mixing front when sea level fell and the caves became senescent. If the lens was stable for a sufficient time, a second set of chambers may develop, parallel to the original set of chambers and to the lens margin, but interior to the initial row of chambers. Figure 4.5a shows this idea conceptually, Fig. 4.5b, c shows how it actually appears in a very large flank margin cave. Because the geochemically active portion of the lens margin does not penetrate deep into the island interior, flank margin caves can

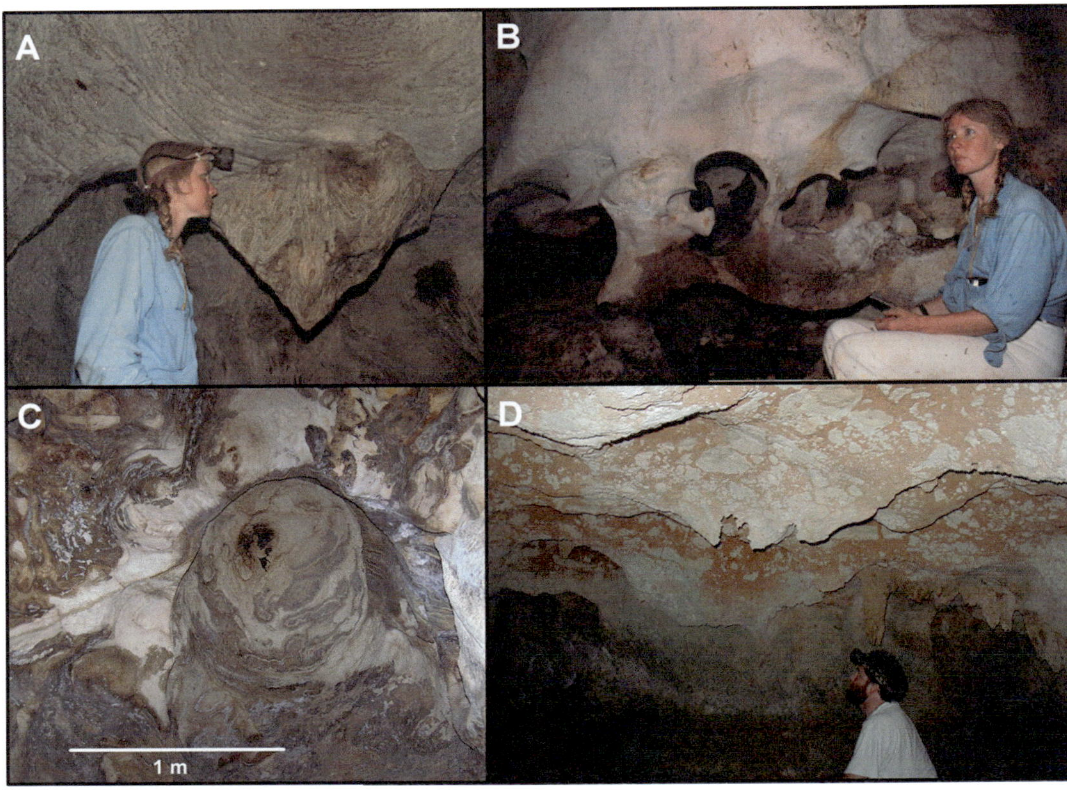

Fig. 4.3 Field examples of the power of dissolution in the fresh-water lens margin. (**a**) Smooth surface cut across both bedrock and flowstone, Hunts Cave, New Providence Island, Bahamas (indicating a phreatic-vadose-phreatic series of events). (**b**) Complex dissolutional structure, Harry Oakes Cave, New Providence, Bahamas. (**c**) Dissolutional surface cut smoothly through *Acropora Palmata* fossils and matrix, Suicide Cow Cave, Barbados. (**d**) Terra Rosa paleosol cut smoothly by dissolution, Red Roof Cave, Long Island, Bahamas

continue to grow in size only by amalgamating in a shore-parallel manner with other collections of chambers forming in the same way. Therefore the initially globular caves elongate along the lens margin by these connections (Fig. 4.5).

Flank margin caves contain a suite of phreatic dissolutional forms which all are consistent with slow flow. The features associated with turbulent stream flow, such as wall scallops or sediment bars and deposits, are absent in flank margin caves. The cave interiors are characterized by cuspate, curvilinear walls, isolated bedrock pillars, globular rooms, maze-like areas, and dead-end passages (Fig. 4.6). Another common feature are bell holes, vertical cylindrical holes in ceilings that can be meters high and up to a meter wide (Fig. 4.7). Their origin has been ascribed to three main options: condensation corrosion, bat activity, and convective phreatic flow. The debate has been significant, and the entire topic is discussed in Birmingham et al. (2010). We support the convective phreatic flow hypothesis and consider bell holes diagnostic of a diffuse flow regime that would not shear vertical convection cells in phreatic water.

Bermuda is presented in Fig. 4.2 as a carbonate cover island; however Bermuda has few flank margin caves in an environment that should have supported such cave development. Vacher and Mylroie (1991) and Mylroie et al. (1995a) explain this absence of flank margin caves as the result of the excess denudation the Bermuda eolianites have undergone. Comparing Bermuda to San Salvador, which are about the same size, are

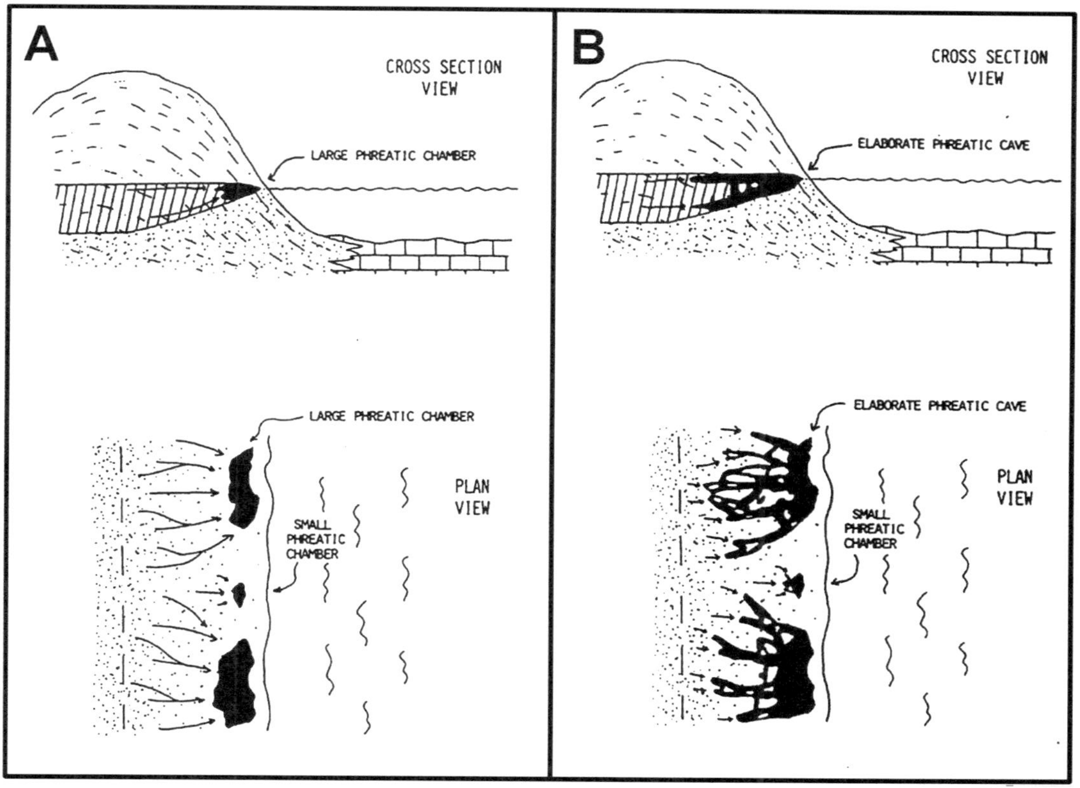

Fig. 4.4 Flank margin cave development. (**a**) Initial conditions, with chambers developing at the lens margin. (**b**) The fresh water lens retreats as cave porosity increases, moving the active mixing front to the back wall of the cave creating dead-end passages and maze-like interconnections (From Mylroie and Carew 1990)

dominated by Quaternary eolianites, and contain numerous interior water bodies, Mylroie et al. (1995a) argued that San Salvador Island is in a negative water budget environment, such that the interior water bodies are saline or hypersaline, and not capable of dissolving the lake coastlines. The negative water budget also retarded dissolutional denudation of the eolianites ridges, and the flank margin caves within them survive. On Bermuda, in contrast, there is a strongly positive water budget. The interior depressions are occupied by fresh water, unless they have been breached to the sea. During earlier sea-level highstands, more of the interior depressions were occupied by fresh water, which enlarged the lake coastlines by dissolution until those interior coastlines breached to the sea and were flooded by marine water. The eolian ridges underwent a higher degree of denudation, and existing flank margin caves were effectively removed. The large caves for which Bermuda is famous either formed as conduit caves when the platform was exposed by glacioeustasy, as in the Bahamas, or by progradational collapse of stream caves perched on the volcanic edifice that supports Bermuda. Other climatic implications for fresh-water lens configuration are explored in Mylroie and Carew 1995.

There are some caves in the Bahamas that fit the flank margin model almost perfectly in all aspects except for the elevation of where they are found, which is 10 m or more above current sea level. Osprey cave on Crooked Island, and Port Howe Sea Cave on Cat Island are examples of caves above 10 m. The higher passages in Hatchet Bay Cave, Eleuthera are also above 10 m elevation. But the crowning example is St. Francis Grotto, at 55 m elevation on Mount Alvernia,

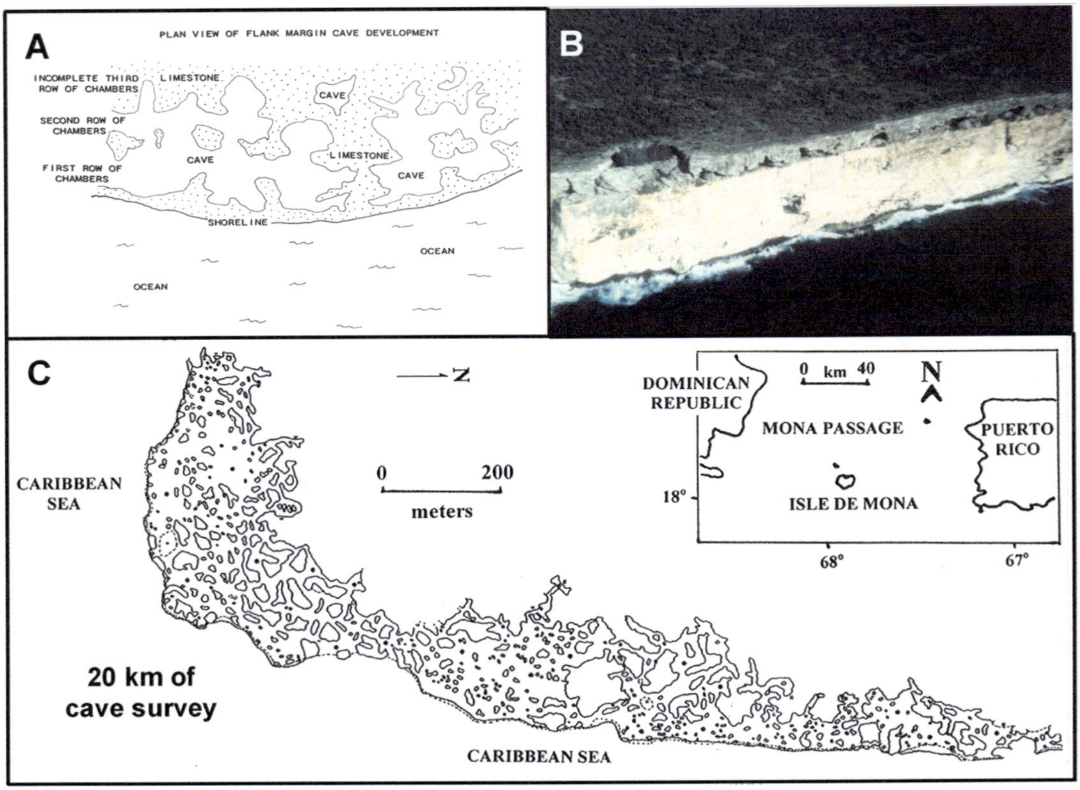

Fig. 4.5 The landward migration of the mixing front noted in Fig. 4.4b can occur as a step function. (**a**) Diagrammatic representation of shore-parallel cave chambers, with progressively less chamber linkage inland. (**b**) Coast of Isla de Mona, Puerto Rico, showing numerous cave entrances to Sistema Faro exposed by sea-cliff retreat. (**c**) Map of Sistema Faro (Cartography by M. Ohms, Isla de Mona Project), showing shore-parallel rows of passages and chambers. Note that the cave wraps around the easternmost margin of the island, and despite almost 20 km of mapped passage, extends inland less than 300 m

Cat Island (the highest point in the Bahamas at 63 m). The caves found at the +10 m elevation could perhaps be explained by accepting one of the higher values argued for the MIS 11 sea-level highstand about 440 ka (e.g. Hearty et al. 1999). But St Francis Grotto on Mount Alvernia is itself higher than any other topographic elevation in the rest of Bahamian Archipelago. Mylroie et al. (2006) attempted to explain the cave by the banana hole model of Harris et al. 1995, speculating that a thick terra rossa paleosol had perched a small water table, and the cave resulted from vadose and phreatic fresh-water mixing. As noted below in the banana hole section, the Harris et al. 1995 model may no longer apply. Cursory field work on upper Mount Alvernia did not locate the proposed perching paleosol either. The St Francis Grotto has classic phreatic dissolutional features, low, ovoid and globular rooms, and all the features described for flank margin caves. It has no pit cave features or morphologies. While most of the flank margin caves in the Bahamas found above +6 m elevation can be explained by proposing a sea level a few meters above the +6 m of MIS 5e, it would require melting almost all the ice on the planet to get sea level to +55 m elevation on Mount Alvernia in a tectonically stable environment like the Bahamas. These higher elevation caves remain a mystery, and perhaps a major thorn in the side of the flank margin model, which otherwise seems to explain phreatic coastal caves in the Bahamas so well.

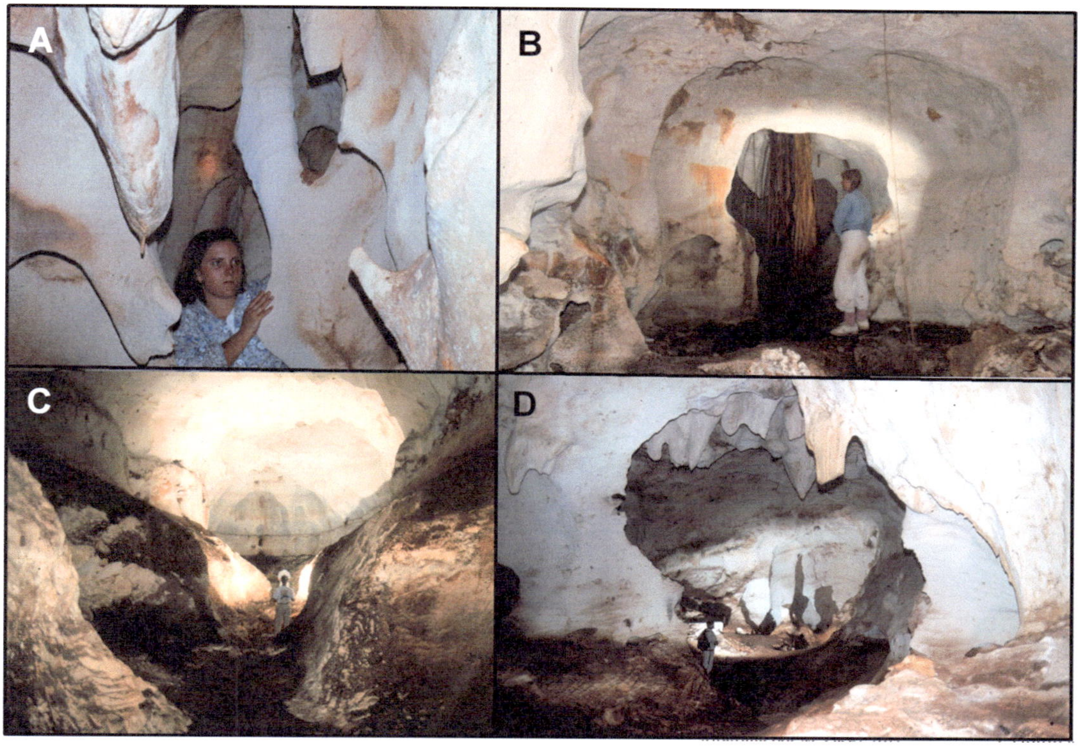

Fig. 4.6 Flank margin cave phreatic morphologies from the Bahamas. (**a**) Dissolutional sculpture cutting dipping eolianite beds, Lighthouse Cave, San Salvador. (**b**) Cuspate passage formed by linking globular chambers, Harry Oakes Cave, New Providence. (**c**) Large phreatic tube ending in a blank bedrock wall, Salt Pond Cave, Long Island. The blank wall indicates the position of the mixing front when sea level fell and the cave was abandoned by the fresh-water lens. (**d**) Large chamber, Hamilton's Cave, Long Island. The cave formed during the last interglacial (MIS 5e), a time span of 9,000 years

4.3.2 Telogenetic Flank Margin Caves

The eogenetic limestones common to many tropical islands and coasts allow significant matrix flow, especially as touching vug permeability develops (Vacher and Mylroie 2002). This eogenetic character has been presented within the CIKM as the reason for the mazy, ramiform and spongework shape of flank margin caves, and their origin as intersecting chambers and chamber collections within a coastal diffuse groundwater flow field. However, what happens when telogenetic rocks, with their almost non-existent matrix porosity, are placed in coastal situations? Shortly after the first presentation of the flank margin model in 1988 (Mylroie 1988), Proctor (1988) applied the model to Devonian limestones of Devon, southwest England, and indicated joints and bedding planes dominated the morphology of the resultant caves. Work in two disparate coastal areas with telogenetic carbonates, the Adriatic islands of Croatia and New Zealand (Chap. 17), showed that flank margin caves with passage characteristics of eogenetic limestones could be found in areas where the telogenetic rock offered a multitude of flow pathways, and in that manner imitated the flow pathways in eogenetic carbonates.

In Croatia, dense, hard Upper Cretaceous carbonates are locally mantled by a breccia interpreted to be a paleotalus. Otoničar et al. (2010) discovered flank margin caves preferentially developed in such paleotalus deposits in coastal settings, whereas adjacent Upper Cretaceous limestones had no caves (Chap. 17).

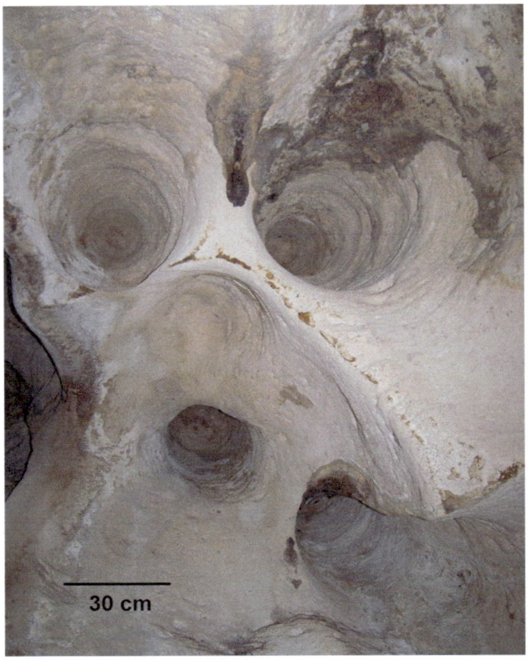

Fig. 4.7 Bell holes in the ceiling of Jumbey Hole, Acklins Island, Bahamas, looking straight up. Note the cylindrical shape, and the large height to width ratio

They hypothesized that the many flow pathways offered by the breccia allowed mixing to occur over a volume, as opposed to just along joints, fractures and bedding planes, and the resultant caves are typical in plan and morphology as those found in eogenetic island settings. The clasts in the paleotalus are Upper Cretaceous and telogenetic, but their re-deposition as a paleotalus created an appreciable diffuse flow field condition.

In New Zealand, Mylroie et al. (2008b) examined coastal carbonate outcrops along both North Island and South Island. Because of the extensive recent tectonism in New Zealand, Oligocene to Pliocene limestones found in the coastal setting have undergone deep burial and have been returned to the surface as dense, hard telogenetic limestones, despite their relatively young geologic age. In the Kaikoura area of northeastern South Island, intense deformation of the Paleocene Amuri Limestone has resulted in a highly fractured rock with cracks spaced ever few tens of centimeters (see Chap. 17). As a result, flank margin caves have formed with a morphology and pattern similar to that observed in eogenetic carbonates. On the west coast of South Island, near Punakaiki, the Oligocene Waitakere Limestone has a very dense joint spacing (Chap. 17); as a result, a series of small flank margin caves has formed along each joint. These voids are aligned horizontally, in accordance with the past fresh-water lens position, despite the dip of the beds being 20° to the southeast. In this case, the caves are not joined as in the first two telogenetic examples just presented, but the void abundance along a horizontal datum indicates a flank margin setting for cave origin.

The major difference between eogenetic and telogenetic limestones in coastal settings is the degree to which water flow can be spread out over a volume, and the degree to which it is restricted to fracture flow paths formed by joints, bedding planes, and faults. As seen in Devon, England, and Punakaiki South Island New Zealand, the lack of matrix flow creates cave voids spaced out along the fracture pattern, albeit with a horizontal datum control provided by the margin of the fresh-water lens at the time of void development. However, when the teleogenetic rock can be altered so as to create a closely spaced systems of openings (Croatian paleotalus) or fractures (Kaikoura fractured rock), then the flow mimics the diffuse flow regime of eogenetic rocks, and the flank margin caves that formed appear similar to those found in eogenetic carbonates.

On Mallorca, the largest of the Balearic islands in the coastal waters of Spain, Gines and Gines in Chap. 11 demonstrate the importance of facies control on cave pattern. In the reef facies, Vallegornera Cave is a series of globular chambers, but in the tight lagoonal mud facies, joint control is the only effective flow pathway and the cave pattern responds accordingly. In this case, eogenetic carbonate rocks, in the tight lagoonal facies, are imitating the lack of porous media flow routinely observed in telogenetic rocks.

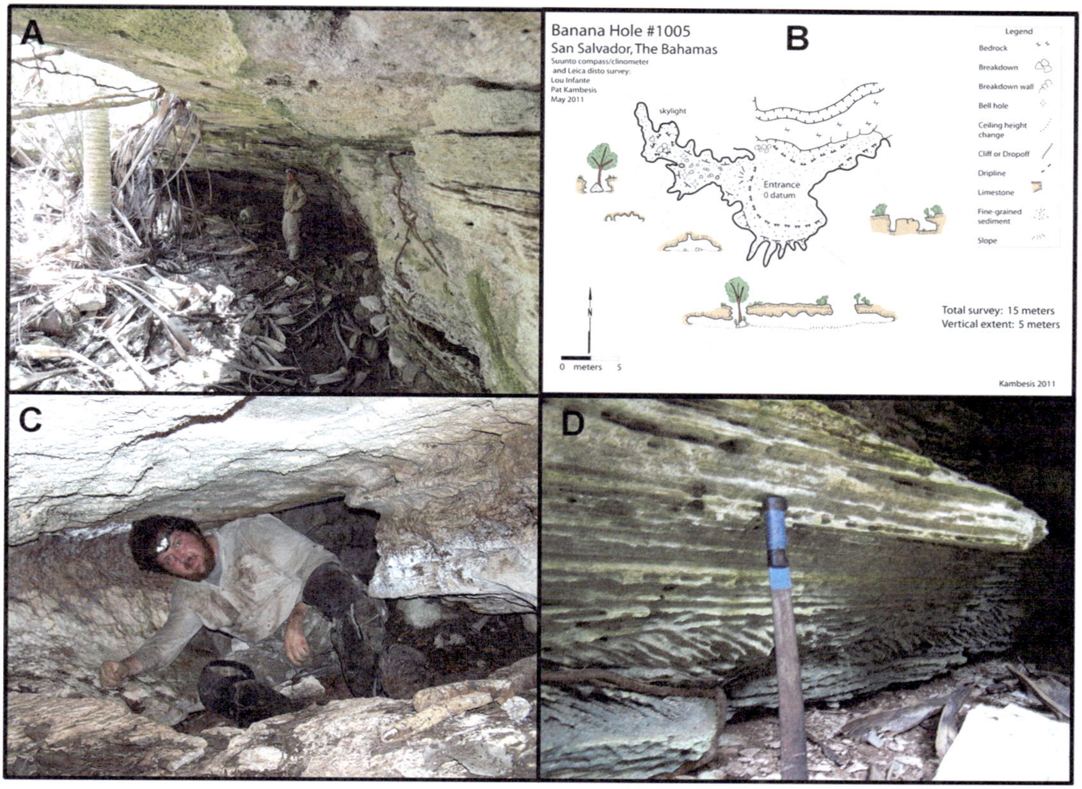

Fig. 4.8 Banana holes, San Salvador Island, Bahamas. (**a**) Large open banana hole. (**b**) Map of a typical banana hole, showing open and roofed sections. (**c**) Roofed section of a banana hole, with phreatic wall morphologies. (**d**) Wall rock of a banana hole, showing herring bone cross bedding, indicative of a subtidal depositional environment. The rock grades upward through beach to back-beach dune facies

4.3.3 Banana Holes

The Bahamas also contain another phreatic dissolution cave called the *banana hole*. The name is derived from native terminology for circular to oval bedrock openings in the land surface, meters to tens of meters across and a few meters deep (Fig. 4.8), that collect water and soil and are therefore excellent environments for growing specialty crops such as bananas (Harris et al. 1995). The surface openings commonly lead to an overhung chamber, the walls of which have phreatic dissolutional sculpturing. They are found in lowland plains in the Bahamas in large numbers. In many cases, there is no overhang and the walls are vertical to the surface. In other cases, a very small opening leads into an almost entirely intact cave. These features are usually a single oval to circular chamber, but connections to adjacent chambers, some which have completely intact ceilings, do occur. For many years the model proposed to explain banana holes (Harris et al. 1995) assumed vadose and phreatic fresh-water mixing at the top of the lens as the drive mechanism for dissolution. Given that most banana holes were far away from the current shoreline, the flank margin model using the discharging lens margin as the dissolutional mechanism was deemed inappropriate. The great abundance of banana holes in the Bahamas, and their relative scarcity in other carbonate settings, such as Isla de Mona or the Marianas Islands was attributed to the higher elevation of those islands, such that small voids at depth lacked the accommodation space to allow collapse to prograde tens of meters to the surface above. In the Bahamas, however, the banana holes are found in plains only 6–10 m above sea level, and the cave roofs are only a meter or two thick. These voids express quite easily by collapse. Because banana holes with intact roofs, partial roofs, and no roofs are all found in large numbers, their surface expression by collapse of a thin roof appears to be the appropriate mechanism.

Recently, several large-scale banana hole mapping projects in the Bahamas (Mylroie et al. 2008a; Infante et al. 2011) made an important observation: the banana holes were always located in subtidal facies (above modern sea level) that transitioned upward to beach, back beach and back-beach dune facies. In the Bahamas, subtidal facies above modern sea level are part of the Cockburn Town Member of the Grotto Beach Formation (Carew and Mylroie 1995, 1997), which correlates to the last interglacial sea-level highstand, MIS 5e approximately 124–115 ka (Thompson et al. 2011), which was ~6 m above modern sea level. These deposits were prograding strand plains during MIS 5e. The caves in these deposits must be younger than the rocks, but the only sea-level highstand that could have placed the fresh-water lens at the proper elevation (+6 m) was MIS 5e. In other words, the caves are *syndepositional*, forming in the rock immediately upon the rock's deposition (Mylroie and Mylroie 2009b). The banana holes are now seen as small, immature flank margin caves (Infante et al. 2011). They developed in the back-beach dune system shortly after the deposition of those carbonates during MIS 5e. As the strand plain continued prograding seaward (as a result of excess sediment supply), the fresh-water lens followed the advancing land, occupying the strand plain facies from subtidal units upward through the back-beach dune facies. As strand plain progradation continued, the initial flank margin caves were abandoned by the lens margin, and new ones began to form at the new lens margin position. With each stand plain progradational event, the sequence was repeated. Because the lens margin was migrating, void development was terminated at a given location quickly, and these immature flank margin cave chambers did not have time to become large, or to pervasively intersect their neighbors. This new hypothesis explains the occurrence of banana holes in the fossil strand plains, their distribution in what is today an inland setting, their development with thin roofs, and their small size (relative to regular flank margin caves), and their great abundance. This new interpretation avoids the controversy about the true dissolutional potential of the top of the fresh-water lens.

4.3.4 Pit Caves

Returning again to the Bahamas to exploit the time and space constraints available there, the discussion turns to pit caves. On many of the eolianite ridges of the Bahamas are vertical pits and shafts that extend from the surface downward as much as 10 m (Fig. 4.9). Many are above the position of past glacioeustatic sea-level highstands, which means they cannot be the result of dissolution in a past fresh-water lens. The caves contain vertical grooving and other indications of descending vadose flow. They may have a stair-step configuration. They rarely intersect flank margin caves at lower elevations. These caves are morphologically similar to classic vadose shafts that develop in the telogenetic karst areas of continental interiors (e.g. White 1988). Fieldwork (Harris et al. 1995) has shown that they can occur in large numbers in relatively small areas. This observation was initially thought to indicate that they could be a paleoclimatic indicator, and that their abundance was due to a higher rainfall in the past. However, Harris et al. (1995) studied these pit caves during major rainstorms, and were able to determine that they efficiently captured surface flow, but that they were also in competition with each other. The high abundance reflected shifting capture points over time, and abandonment of some pit caves, as opposed to concurrent pit cave development due to high precipitation. Pit cave complexes are found on other islands, for example, Isla de Mona (Chap. 9).

Pit caves are vadose fast flow routes that bypass the diffuse flow network and deliver water to depth in the carbonate outcrop quickly. On Guam,

Fig. 4.9 Pit caves on San Salvador Island, Bahamas. (**a**) Typical pit cave morphology as a deep but constricted vadose pathway. (**b**) Map of Triple Shaft Cave, showing pit competition for recharge gathered on the epikarst

it has been estimated that 40 % of the recharge to the fresh-water lens is by these vadose fast-flow routes (Jocson et al. 2002); the other 60 % can take months to reach the lens by diffuse flow. These observations help explain why some wells respond quickly to meteoric events, while other wells do not. It depends on the proximity of the well to a vadose fast flow route, which can mound the top of the lens and affect a nearby well, but the mound tails off before reaching more distant wells.

Why do pit caves develop on young eogenetic carbonates, where diffuse flow rates are high? The Bahamian study of Harris et al. (1995) suggests that cementation and micritization of the exposed limestone surface by repeated wetting and drying events (similar to caliche formation) results in a calcrete crust that can perch some of the meteoric water which then exploits any weak points in that crust to create a pit cave. Even in the absence of a calcrete crust, precipitation events can apply water to the land surface faster than infiltration can accommodate the flow. Antecedent events can limit infiltration even during moderate precipitation events. In hindsight, overland flow should be expected, and flow would focus downward to any spot where some characteristic of the rock accepted greater flow amounts. The result is a pit cave. Pit caves tend to end in sandy floors. The degree to which the floor of the pit cave is the downward limit of dissolution, or merely a washed in debris pile is unknown. Pit floors have been excavated for a meter or two in a few locations, but a true bedrock floor was not reached. In Majors Cave, San Salvador Island, a pit cave on the ridge top was forced downward through small passages into the rear of the main chamber of the cave, one of the few places in the Bahamas where such a vertical connection can be made.

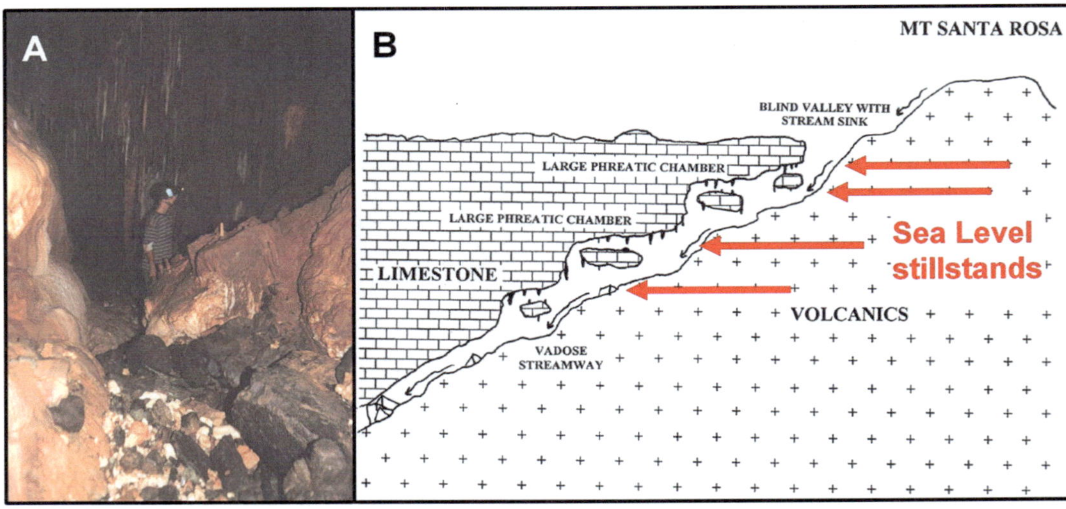

Fig. 4.10 Allogenic stream cave development, Guam, Marianas Archipelago. (**a**) Image of a cave passage developed at the limestone–basalt contact, Awesome Cave. (**b**) Cartoon cross section of Awesome Cave, showing stacked phreatic chambers undercut by vadose streamway. Each chamber represents a past fresh-water lens position

4.3.5 Allogenic Caves

Carbonate islands and coasts can form typical stream caves as found in continental interiors. Such caves are "karst on islands", as previously explained. However, in many cases, the development of a stream cave can occur in a manner consistent with "island karst". As shown in Fig. 4.2, when non-carbonate rock perches vadose flow, stream caves form. They can arise purely from autogenic catchment, as in carbonate-cover islands, or from allogenic catchment, as in composite or complex islands. In all cases, however, the opportunity may exist for the downstream end of these caves to end at the fresh-water lens, and create a vadose/phreatic mixing situation.

Because of glacioeustasy, during sea-level lowstands, such vadose stream caves can form downward to the new lens position. When an interglacial occurs, sea level and the fresh-water lens will rise, flooding the formally vadose passages. If the non-carbonate layer never reaches the coastline above any past lens position, then the discharge from the stream cave will always be into the lens. In some respects, the downstream end of the stream cave can be said to be hydraulically dammed, as it empties into a porous media aquifer (especially for eogenetic carbonates). As result, back flooding would be expected during high discharge events, as observed in Coles Cave on Barbados (Machel et al. 2011; Chap. 10), and large chambers should form in the adjacent lens as a result of water mixing. High organic loading would be expected in a vadose stream cave gathering allogenic recharge as in composite and complex island settings.

For tectonically uplifted islands, even at a glacioeustatic sea-level highstand, as is the case today, uplift should have brought the cave chambers associated with the vadose stream caves and fresh-water lens boundary up into the vadose zone. This situation is observed in Awesome Cave on Guam, Mariana Islands (Fig. 4.10), where a series of stacked, globular phreatic chambers extend away from the stream cave at past lens positions (Jenson et al. 2006; Chap. 13). On Saipan, Mariana Islands, the juxtaposition of normal faults and low permeability volcaniclastics has created hydrologic compartments where confined conditions occur (Chap. 13). The fresh water drained the compartment by flowing upwards and over the confining barriers. Uplift later drained the compartment from below, and an entirely phreatic cave can be observed today, with a 30 m high phreatic lift tube that drained a converging series of passages within the compartment below.

4.3.6 Closed Depressions

On continents, one of the most distinctive karst landforms is the closed depression, drained internally by karst hydrologic flow. These depressions range from small sinkholes or dolines up to depressions kilometers across known as poljes. On eogenetic carbonate islands, closed depressions also exist in a variety of sizes and shapes, however their origin is commonly different. Because of sea level change, the existing island topography may be in part constructional, meaning that closed depression may be the result of uneven deposition of carbonates, and not by subsequent removal through dissolution by internal karst flow. In eogenetic carbonates, the limestone may be porous and permeable enough to drain karst depressions, or such drainage may be facilitated by pit caves, sinking streams, and other traditional karst pathways. Constructional closed depressions are known from continental settings, most commonly in glaciated areas where rock basins are scoured out in limestones, or where valleys are dammed by glacial sediments and water flow escapes through limestones in the valley wall (e.g. as in New York, Mylroie and Mylroie 2004).

Two common constructional basins in eogenetic carbonate islands are raised lagoons, and swales between eolianite ridges. In these cases, the closed depression is maintained by karst processes, but was generated by differential limestone deposition. Large closed depressions, especially in very young limestone islands such as the Bahamas, are constructional. There hasn't been enough time between the deposition of the limestone, its exposure to the subaerial environment, and the present to allow the large amounts of carbonate dissolution necessary to create a large depression. The same time constraints also limit the development of true dissolution sinkholes or dolines. Combined with the porous and permeable nature of eogenetic limestones, surface flow does not result in the classic bowl or conical sinkhole shape. The development of a hard calcrete crust promotes the development of steep-sided pit caves instead. On continents, cover-collapse sinkholes, produce a sinkhole by subsidence of insoluble overburden downward through a small dissolution pipe or dissolutionally widened crack. Such action rarely occurs in young eogenetic limestones as there is little insoluble cover to be piped downward. As islands mature over time, and capture insolubles as dust fall, ash fall, or from adjacent non-carbonate terrain, the ability to form cover-collapse sinkholes improves.

The least common sinkhole in continental settings is the cave collapse sinkhole, produced by failure of the dissolutional void down inside the bedrock. On young carbonate islands, however, cave collapse sinkholes are the most common sinkhole type. In the Bahamas, where banana holes have formed in great abundance with very thin roofs, their expression as collapse features is extremely common. Flank margin caves, because they form on the flank of the landmass enclosing the fresh-water lens, are also prone to collapse, and many flank margin caves have numerous collapse entrances. Because the position of banana holes and flank margin caves are very predictable, the risk posed by their abundance can be minimized by knowing where unexpressed collapse-prone voids are likely to exist (See Chap. 6 for additional discussion on modeling void development).

4.3.7 Blue Holes

Blue holes are intriguing explorational and scientific sites (Fig. 4.11). In island karst, the term "blue hole" has a fairly specific meaning, but for karst in general, the meaning is more general. Essentially, the term is descriptive, referring to a water-filled hole that has a deep bluish cast or color, commonly associated with the Bahama Islands, but also used elsewhere, such as Belize (Dill 1977). The term has also been used to describe artesian springs in karst settings unrelated to islands (Zans 1951; Sweeting 1973), such

Fig. 4.11 Blue hole images. (**a**) Dean's Blue Hole, Long Island, Bahamas, an ocean hole with a connection to the sea. The blue hole is 200 m deep. (**b**) Watling's Blue Hole, San Salvador Island, Bahamas, an inland blue hole isolated from the open ocean

as Turner's Blue Hole in Kentucky, a typical continental interior karst spring (Mylroie and Mylroie 1991). In addition, features of similar morphology, that contain water, such as the cenotes of the Yucatan Peninsula, Mexico, are interpreted by some authors to be a form of blue hole (Ford and Williams 1989). The term blue hole proliferated in the last decades of the twentieth century because of more scientific research on karst environments in the Bahamas.

The Bahamas have been the focus of blue hole studies. Shaw (1993) reports the first blue hole reference in the literature to be by Catesby in 1725, who stated that local inhabitants in the Bahamas used the word "pit" to describe deep holes filled with water that fluctuated with the tide. The term blue hole appeared on the British Admiralty charts of the Bahamas in 1843 and 1844, to describe water-filled depressions and holes differentiated from the surroundings by a deep blue color (Shaw 1993). Nelson (1853) reported on fluctuating water levels in deep holes. Shaw (1993) also reports that Northrup in 1890 indicated the native Bahamians used the term "ocean hole", and that those holes demonstrating inflow and outflow were "boiling holes". Agassiz (1893) used the term blue hole in conjunction with the term ocean hole, to describe deep holes in the floor of Bahamian lagoons and banks. He gave no other definition or description. Shattuck and Miller (1905) used the terms ocean hole and blue hole to describe deep holes found on the banks whose sides flare-out beneath the opening and which usually have a constant circulation of water. The water in some of the blue holes they described responded in harmony with tidal fluctuations. Doran (1913) referred to blue hole as a local name given to ocean holes. Stoddart (1962) describes a blue hole as a deep depression in the floor of the lagoon which is perfectly round and reputedly bottomless. Stoddart preferred to use the term blue hole rather than the term ocean hole because of the deep blue color associated with the features. While obvious features of the Bahamas, blue holes did not receive much early scientific study (Sealey 1991). In-depth discussions of blue holes eventually became routine in textbooks about Bahamian geology and geomorphology (e.g. Sealey 1994).

Blue holes were popularized by a series of articles by cave diver George Benjamin (1970) in National Geographic magazine. The discovery of large underwater cave systems inside blue holes resulted in a wave of cave diver exploration, followed by scientific inquiry, much of it detailed in popular books on the subject (e.g. R Palmer 1985, 1989). A review of this period can be found in Mylroie et al. (1995b), who proposed a series of definitions to stabilize the varied use of the term "blue hole" in the literature. Their blue hole definition is: "subsurface voids that are developed in carbonate banks and islands; are open to the earth's surface; contain tidally-influenced waters of fresh, marine, or mixed chemistry; extend

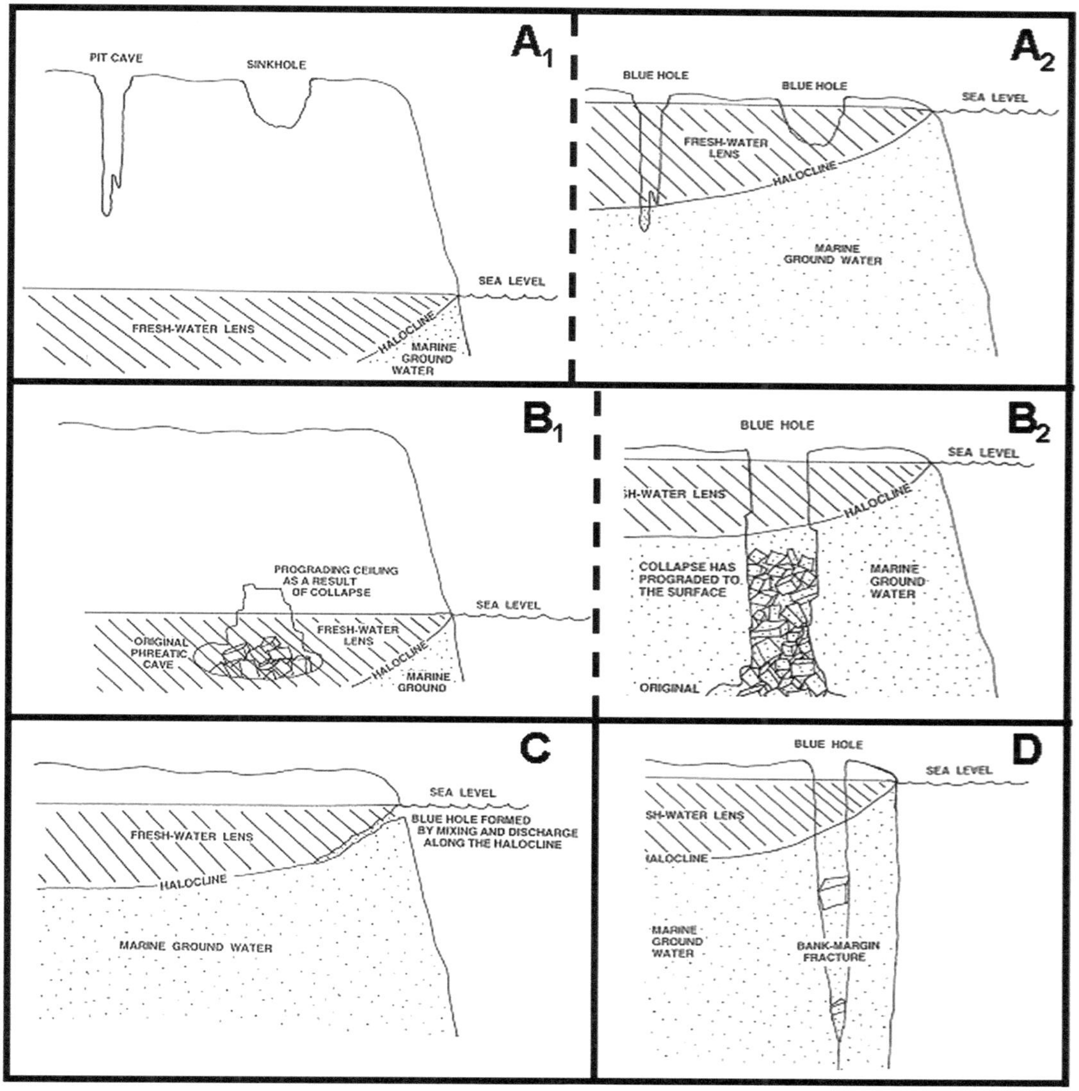

Fig. 4.12 Blue hole origin. (**a**) Vadose features such a pit caves and sinkholes can become small blue holes if sea level floods them. (**b**) Most blue holes result from progradational collapse of deep voids, perhaps originating as large conduits. (**c**) Downward trending passage following the halocline; the trend is misleading, a result of vertical exaggeration in making the figure. (**d**) Blue hole development by bank margin failure

below sea level for a majority of their depth; and may provide access to submerged cave passages." (Mylroie et al. 1995b, p. 225). Blue holes were further characterized by their surface connectivity to the open ocean: "An ocean hole opens directly into a lagoon or the ocean, is usually tidally influenced, and contains marine water. An inland blue hole opens directly onto the land surface, or into an isolated pond or lake, may be tidally influenced, and may contain a variety of water chemistries from fresh to marine." (Mylroie et al. 1995b, p. 230–231). See Fig. 4.11 for examples.

The origin of blue holes is believed to be polygenetic, as presented in Fig. 4.12 (Mylroie et al. 1995b). Because pit caves can form independently of sea level, requiring only the presence of a vadose zone, it is likely that pit caves have formed at elevations below modern sea level. With sea-level rise to modern position, these pit caves would be flooded and would appear as blue holes (Fig. 4.12a). However, their size, both in diameter and in depth, would be somewhat limited based on the size of pit caves available for observation above modern sea level today, and this mechanism cannot be used to explain the vast number of blue holes that are larger in diameter and depth than known pit caves.

It was also proposed (R Palmer and Williams 1984) that blue holes could develop by following the mixing zone or halocline downward as the fresh-water lens thickened inland (Fig. 4.12c). This interpretation was based on the observation that in descending underwater caves in the Bahamas, divers could follow the halocline to significant depth. Therefore, it was reasoned, mixing dissolution along the descending halocline had created the passage that curved upward to become a blue hole. This proposed explanation has two flaws. First, the steep incline of the fresh-water lens, as shown in Fig. 4.12c, is not real. In textbooks and in scientific papers, the fresh-water lens is commonly presented with significant vertical exaggeration. An island 10 km across might have a lens only 10 m thick, an aspect ratio of 1,000. Dissolution along the halocline would not result in a steeply descending cave. Second, the observation of the halocline within the descending caves is real, but cause and effect are reversed. The cave represents an easy flow path for the lens, and so captures the lens flow. The halocline is steeply descending because of the pre-existing cave; the halocline did not cause the cave in this case. Some other mechanism must have been at work to make the cave.

One important mechanism of blue hole formation is bank margin failure (Fig. 4.12d). The Bahama Banks have very steep margins, which are gravitationally unstable and prone to vertical failure (e.g. Mullins and Hine 1989). Quaternary glacioeustatic sea-level change, which subaerially exposes the upper 100 m+ of the bank wall, might promote gravitational failure as a result of the loss of buoyant support. Bank-parallel vertical fissures are produced which can lead to water both at low and high sea-level positions. Small graben-like structures form (Carew et al. 1998) as material subsides parallel to the facture walls. In the Bahamas, blue holes form as a series of openings along the fracture trace; see R Palmer (1989) for a full review. Because of the length of these fractures, they can form a large number of blue holes, but these blue holes are restricted to the bank margin area. The fractures can trace from lagoons to dry land, producing both ocean holes and inland blue holes. Many of these blue holes can be linked up underwater along the fracture trace. These fractures also act as interceptors of the flow within the fresh-water lens towards the coast, creating unique hydrological conditions (Smart and Whitaker 1997).

The most common blue hole is the progradational collapse structure, sometimes referred to as an Aston collapse (R Palmer and Williams 1984). As presented in Fig. 4.12b, these blue holes originate as deep-seated voids large enough to accommodate the collapse material as ceiling failure progrades its way upward. If enough collapse occurs, and the collapse debris can be accommodated at depth, then the collapse will reach the surface. If, as shown in Fig. 4.12b, the accommodating void has active geochemical processes at work, the collapse material can be dissolved and carried away, maintaining accommodation space as the collapse progrades upward. As noted earlier, there is abundant evidence of a well-developed fossil conduit flow system at depth in the Bahamas. The conduits associated with the maximum glacioeustatic sea-level depression of the Quaternary would be at the 100–125 m depth, where the time to reverse glacioeustatic sea-level fall to a sea-level rise would provide the time window for large conduit systems to develop. As with any large conduit system, collapses would occur at irregular sites along the conduit where site-specific conditions created roof instability (for a review of cavern breakdown processes, see White 1988). These collapse blue holes can also intersect former conduits at a variety of elevations. Current research (e.g. Larson and Mylroie 2012) is now modeling blue hole abundances on the Bahama Banks to determine if patterns will appear that reflect conduit traces in the subsurface.

In the Bahamas, voids have been located by exploratory drilling at a variety of depths. Meyerhoff and Hatten (1974) report voids at depths of 21–4,082 m. The deepest of these voids was able to accept 2,430 m of broken drill pipe, indicating an extremely large chamber. Such a large chamber, if at a shallower depth, could easily accept sufficient collapse material to allow collapse to prograde all the way to the surface. Deans Blue Hole, on Long Island, Bahamas, is over 200 m deep; arguably the deepest blue hole in the world (Fig. 4.11a). The deepest 15 m of that blue hole is an extremely large chamber (Wilson 1994). The vertical migration of these voids by collapse allows blue holes to integrate conduit cave systems at a variety of depths. As a consequence, blue holes are considered one of the most diverse habitats for anchialine ecosystems (Mylroie and Mylroie 2011).

As noted for flank margin caves, biogeochemistry has been implicated in the development of the dissolutional environment for void development. Blue holes can receive significant organic loading, especially inland blue holes that can accept woody vegetative material. The speleogenetic impact of organic decay in blue holes was first reported by Bottrell et al. (1991). Subsequent arguments have been made that make biogeochemistry the primary cause of blue hole formation (e.g. Schwabe and Carew 2006). It is important to remember that major organic loading of a blue hole cannot occur until it expresses itself at the earth's surface. While DOC and particulate organics may play a role in dissolutional production and modification of voids at depth, large scale organic activity must await opening to the surface.

Blue hole diving, as with any cave diving, can be extremely hazardous. The inflow and outflow of water from ocean holes in response to tidal cycles is particularly dangerous. Many blue holes contain elaborate vadose speleothems (stalactites, stalagmites, flowstone, etc.) deposited during glacioeustatic sea-level lowstands when the caves were in the vadose zone and air-filled. In addition to minerals, abundant fossil and archeological remains are also present (R Palmer 1989). In Sawmill Sink blue hole on Abaco, fossil material includes a bat fossilized in flowstone at 30 m depth, a former owl roost with thousands of small vertebrate bones at 25 m depth, and incredible preservation of crocodiles and tortoises on a talus cone within the anoxic mixing zone at shallower depth (Steadman et al. 2007).

4.4 Modeling the Chronology of Coastal Cave Development

The development of banana holes as syndepositional caves implies a very rapid rate of cave development. Supporting evidence is found in flank margin caves reported in the Holocene Merizo Limestone of Guam, where the Pacific mid-Holocene sea-level highstand first emplaced the rocks above modern sea level. Tectonic uplift made the outcrop subaerial such that a fresh-water lens could form and create the caves (Miklavič et al. 2012). As sea level dropped to its current position 2 m below the Mid-Holocene highstand, the caves were drained and became accessible. Submarine work on San Salvador Island, Bahamas (Mylroie and Mylroie 2007) located flank margin caves at specific depths of 105–125 m. The survey began at a depth of 60 m, because at shallower depths Holocene coral growth masked the original platform wall. In the depth from 60 m down to 105 m, however, no caves were found. The interpretation is that during the major glacioeustatic sea-level excursions of the Quaternary, sea level, and the fresh-water lens associated with it, was not at a vertical position long enough to leave a macroscopic dissolution signature in the platform wall. When the sea-level excursion bottomed out, the time necessary to stop and begin the sea-level return upward was long enough that dissolution could occur. The depths of 105–125 m match well with the four major troughs located on the Mid to Late Quaternary sea-level curve, Chap. 1, Fig. 1.5. The flank margin caves found subaerially today in the Bahamas are also the result of a sea-level turn around, in this case the peak turn around as opposed to the trough turn around found by submarine.

As demonstrated by the banana holes, the time window to form a flank margin cave is indeed small, but how long do such caves persist in the rock record? Isla de Mona, Puerto Rico has some of the largest known flank margin caves in the world. The island has been tectonically uplifted, and many of the caves exist in the coastal cliffs at elevations up to 70 m above the sea (Fig. 4.5). This position has saved them from over-printing by later glacioeustatic sea-level events. The key to their great size is related to their age. Paleomagenetic analysis of a sequence of sediment, flowstone and stalagmites in Cueva de Aleman demonstrated the caves to be at least two million years old (Panuska et al. 1998). These data indicate that the caves can persist for millions of years. The age of the caves also places them far enough back in time that glacioeustasy had not yet become the high amplitude, short wavelength observed from the Mid to Late Quaternary. Sea level, and the fresh-water lens, were in a fixed, stable position for a long period of time, allowing the flank margin cave dissolutional environment to act continuously and produce the large, extensive rows of chambers seen there (e.g. Fig. 4.5b, c).

In northern Spain, work by Baceta et al. (2007) has described flank margin caves preserved as infilled voids in Danian (lowest Paleocene, ~66–63 million years ago) limestones of the Urbasa-Andia plateau. The caves are in carbonate facies that correlate with the edge of the platform as regression occurred and the platform became subaerial, allowing a fresh-water lens to form. These observations indicate that flank margin cave signatures can survive deep burial and well into the geologic past. Flank margin caves form rapidly, yet persist through time. That makes the caves a high resolution yet durable sea-level indicator; forming in a few thousand years but surviving for millions of years and further illustrate the interrelated processes of cave development within the context of evolving coastal landforms.

4.5 Coastal Landform Configuration and Cave Development

In the Bahamas are many flank margin caves of appreciable size (Fig. 4.6). As noted earlier, this dissolution has been quite rapid, given the volume of the cave chambers and the relatively short glacioeustatic sea-level highstands necessary to elevate the fresh-water lens and produce them. The Bahamian flank margin caves are developed mostly in eolian calcarenite ridges (fossilized carbonate sand dunes). These eolianite ridges come in all sizes from a few meters high to over 60 m high. They can be kilometers long. During MIS 5e, they were long, linear islands with a fresh-water lens; a "strip island" in the sense of Vacher (1988). Somewhat surprising is the development of flank margin caves during MIS 5e in what would have been very small islands or cays. In some cases, the eolianite paleo-island would have been only a 100 m long and 30 m wide, or less, and only 10–15 m high, yet the ridges contain flank margin caves with linear dimensions in the meters to tens of meters (Walker et al. 2008). Even very small fresh-water lenses are able to exploit the flank margin condition to produce caves.

The development of flank margin caves is in part controlled by the configuration of the shoreline. Shoreline shape controls how the distal portion of the fresh-water lens discharges to the sea. The control of cave development by the lens margin position causes the flank margin caves to faithfully follow the lens margin as it curves in conjunction with the shape of the island coastline (Fig. 4.5c). In the eolianite ridges of the Bahamas, each ridge terminates at the two linear ends of the ridge. These ridges have the appearance of a doubly-plunging anticline, and field workers casually call the ends of the ridges "noses" in a similar manner. Flank margin caves found in the vicinity of the ridge terminations have a unique appearance, and have been informally called "nose caves" (Fig. 4.13). At the ridge nose the active portions of the lens from each side of the island begin to overlap. If the ridge is

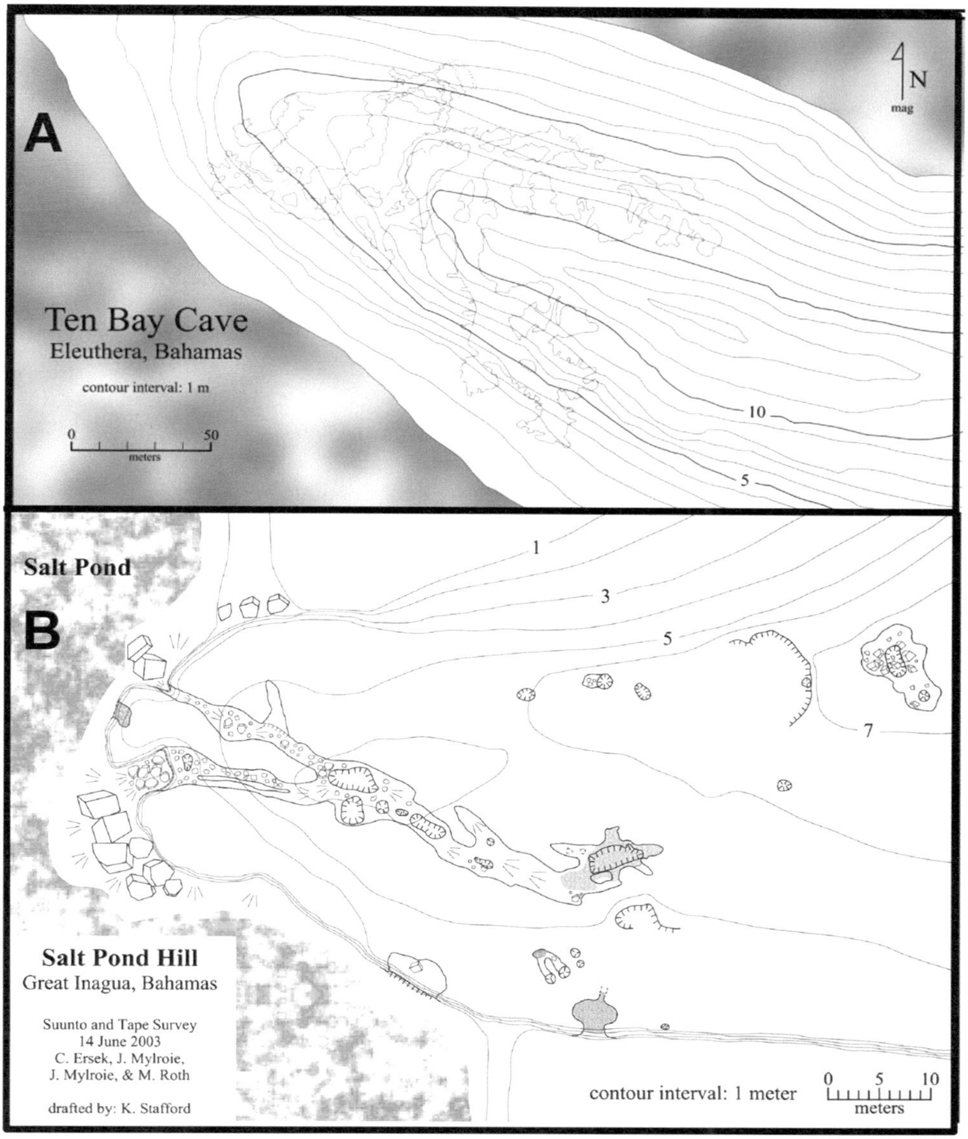

Fig. 4.13 Flank margin cave development at the end of Bahamian eolianite ridges, so-called "nose caves". (**a**) Ten Bay Cave, Eleuthera Island, wraps around the end of the dune and because the dune is wide enough the lens margin and resulting cave passages are separated. (**b**) Salt Pond Cave, Great Inagua Island, in a small dune ridge which overlapped the lens margin, creating a single passage down the dune axis

small, the overlap results in a single linear cave along the terminal axis of the ridge (Fig. 4.13b). If the ridge is large, the caves wrap around the nose of the ridge, cross-linking where the ridge is narrow, but as the ridge widens away from the nose, the cave separates along each flank of the ridge (Fig. 4.13a). Nose caves are another demonstration of the control the lens margin has on cave development, and the relative lack of dissolutional potential within the body of the fresh-water lens, where flow is slow and mixing does not occur.

Stream caves and linear conduits are not known in the Bahamas in the subaerial position today. However, cave divers report long, linear phreatic tubes, thousands of meters in length and meters wide and high at depths of 20–30 m in the Great Bahamas Bank (e.g. R. Palmer 1984). The openings to these caves were not visible on the platform edge during the submarine study as Holocene corals had obscured them. It has been argued (Mylroie and Vacher 1999) that the development of these cave passages is the result of sea-level position. Under current sea-level conditions, the islands on the Bahama Banks are relatively small, and highly linear (Chap. 7). This condition would be even more exaggerated during MIS 5e when more of the Bahamas Banks were flooded and most of the dry flank margin caves observed today developed. If sea level were only 20 m lower than today, however, almost the entire Bahama Banks would be subaerially exposed. The many islands of various sizes in existence today would combine to a few very large islands; small islands would exist only on the isolated small platforms (such as San Salvador Island). As the banks became exposed, their area would greatly increase with respect to their perimeter, as the former increases by the square but the latter does not. To discharge the collected meteoric water from the fresh-water lens by diffuse flow would become progressively less efficient, and at some area to perimeter ratio, conduit flow would be supportable. Cave divers have not located such conduit systems on the small platforms, as sea level fall does not increase their area to perimeter ratio by a significant degree. On large banks, flank margin caves could still develop, as water budget constraints would keep conduits widely spaced, and in the coastal locations between conduits, diffuse flow to create a lens margin would still exist. In the Akumal area of Quintana Roo in Mexico, flank margin caves from MIS 5e are found in the coastal eolianites, even as 100 km long conduit flow systems discharge below them from the interior of the Yucatan Peninsula (Kelley et al. 2006). The Yucatan, although it is part of a continental setting, is a possible analogue for behavior of the larger Bahamian banks when sea level was low enough to expose the bank tops (see Chap. 16).

4.6 Modeling the Interplay of Geomorphic Controls and Karst Processes in Littoral Settings

On Barbados (Machel et al. 2011), Kangaroo Island, Australia (Mylroie and Mylroie 2009a), Mallorca (Mylroie et al. 2012) and Curaçao, coastlines and paleocoastlines with deep inlets have flank margin caves developed within those inlet walls. Two models have been proposed to explain this cave distribution (Mylroie et al. 2012). One considers that expressions of flank margin cave development may exist several hundred meters inland from the active coast. Such a distance of cave development can be seen in Sistema Faro, Isla de Mona (Fig. 4.5c). If a field of cave voids was present, then as an inlet or incision was made into the coastline, by marine or fluvial processes, some of these caves would be exposed. The second model requires only a minor inland range of flank margin cave development. It suggests that after a valley or inlet has been incised into a coastline, a rise in sea level will result in marine invasion of the inlet, creating the flank margin speleogenetic environment in the walls of that coastal incision. After sea level falls, the caves are then exposed by slope retreat as happens on the open coast. The two models express a "chicken versus egg" situation; which came first, the caves or the inlet?

On Kangaroo Island, the inlets created a preservational bias situation in the eolian calcarenites (Mylroie and Mylroie 2009a; Chap. 18). The open coast fronts on the stormy, high wave amplitude Southern Ocean. As a result, the eolianites that could host flank margin caves have been stripped back in Holocene time, and many caves likely removed. Within the inlets, wave energy is low or non-existent as the inlets shoal to dry land, and numerous flank margin caves are preserved in the inlet walls. On Barbados (Chap. 10), work is currently ongoing to assess the inlet models in the development of flank margin caves found in the walls of steep-sided canyons called *gullies*. In the Mariana Islands, flank margin caves are commonly found ringing embayments into the coastline. Again, a chicken versus egg situation exists; are the flank margin caves there because the embayment places the lens margin along the embayment coast, or did flank margin cave development weaken the coast such that an embayment easily formed (Stafford et al. 2004)? Coastal configuration control by discharging ground water has been described for the Yucatan coast (see Chap. 16), creating inlets called *caletas* (Back et al. 1984).

Flank margin caves contain a record of past coastline configurations and conditions. For banana holes, they demonstrate a prograding strand plain. For nose caves, they show that marine water surrounded the ridge. In areas where the paleo-coastline is not well known, flank margin caves can clearly indicate where the coastline had to be for the caves to develop the configuration they present.

4.7 Flank Margin Cave Morphometrics

Flank margin caves, as noted earlier, develop without entrances. After abandonment by the fresh-water lens, usually due to sea level fall, the caves are breached by surface erosion. This breaching is most commonly from the side, but collapses or intersection by a pit cave (see below) can occur. Initially, slope retreat creates a tangential intersection of the cave. As time passes, and slope retreat continues, the cave will progressively disintegrate. Flank margin caves in all stages of erosional removal can be found (Fig. 4.14). The degree of removal is important, as it is a measure of denudation rate. On Tinian Island, Stafford et al. (2005) measured cave entrance widths and maximum cave interior

Fig. 4.14 Surface denudation intersects flank margin caves. (**a**) Caves Point West Cave, New Providence Island, Bahamas, displaying tangential surface intersection barely opening the cave chamber. (**b**) Harry Oakes Cave, New Providence Island, Bahamas, where slope retreat has removed almost all of the cave's outer wall. (**c**) Cave on Curaçao, where roof collapse has intersected the cave from above. (**d**) Faaligosey Cave, Fais Island, Federated States of Micronesia, where wave action has removed the outer wall of a flank margin cave

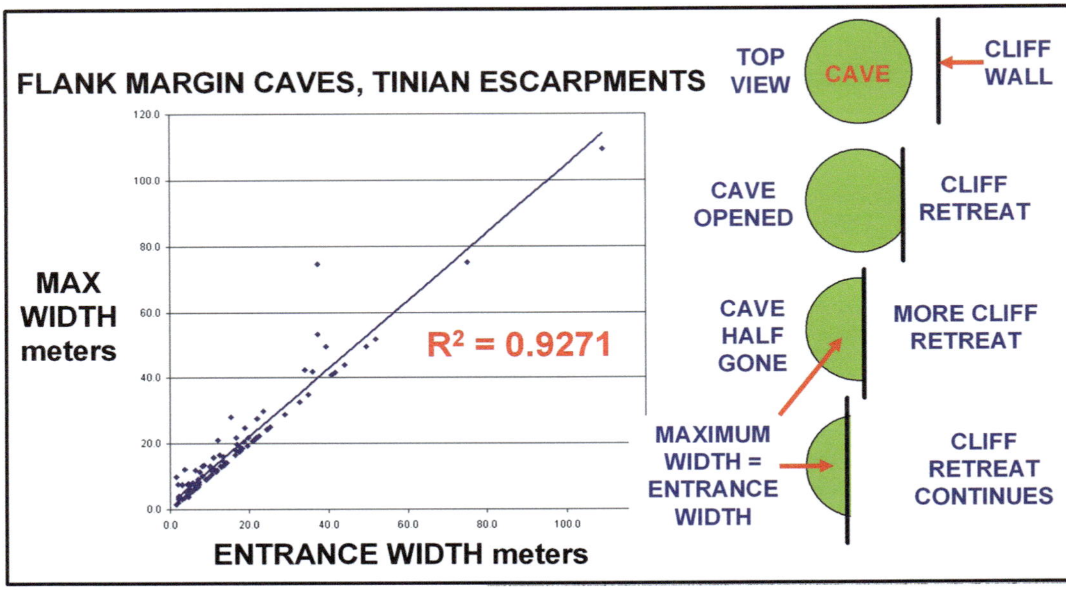

Fig. 4.15 Plot of maximum cave width versus entrance width for flank margin caves exposed in the cliffs on Tinian, Marianas Archipelago. Most caves fall on the 45° line, indicating the caves are 50 % or more eroded away, as shown diagrammatically at the far right of the figure

widths of flank margin caves found on the uplifted cliffs of the island. When these data were plotted (Fig. 4.15), they showed that the vast majority of the caves had an entrance width that equaled their maximum width. In other words, simplifying the cave chambers as simple flattened spheres, this equality between maximum width and entrance width required the cave to be half, or more than half, eroded away. Note that a few caves fall in the upper left of the graph, these are caves that are only partially breached, and they are few in number. It is not possible to have an entrance width greater than the maximum width, so no caves fall in the lower right part of the graph.

A series of other morphometric approaches have been used with flank margin caves. The simplest approach is a rank order plot of cave sizes. Cave size is itself somewhat argumentative, and for flank margin caves the convention is to use the areal footprint of the cave as seen on a high-quality cave map (see Mylroie 2007 for a full discussion of the issue). When first done for a suite of Bahamian flank margin caves (Roth et al. 2006), the resultant plot produced three straight line segments, each of which corresponded to a specific size range of caves (Fig. 4.16). When growth of flank margin caves in the distal margin of the lens was modeled on a computer (Fig. 4.17), the resulting rank order plot matched the empirical data set almost exactly (Labourdette et al. 2007). The computer model also displayed a fourth line segment at the very small end, representing caves so small they were not mapped. This lack of small cave data was an explorational bias of the survey teams, which did not bother to map very small caves. These size rank data support the growth model for flank margin caves as a series of small initial dissolution voids that enlarge and through time eventually intersect one another. As time proceeds, collections of chambers intersect. Each of these intersection events results in a jump in cave size, producing the individual straight line segments seen on the plot.

Another morphometric is the area to perimeter ratio. This approach takes advantage of the maze-like complexity of flank margin caves (Roth 2004), which produces a plot quite different from that created by linear stream caves (Fig. 4.18). Stream caves plot similar to an

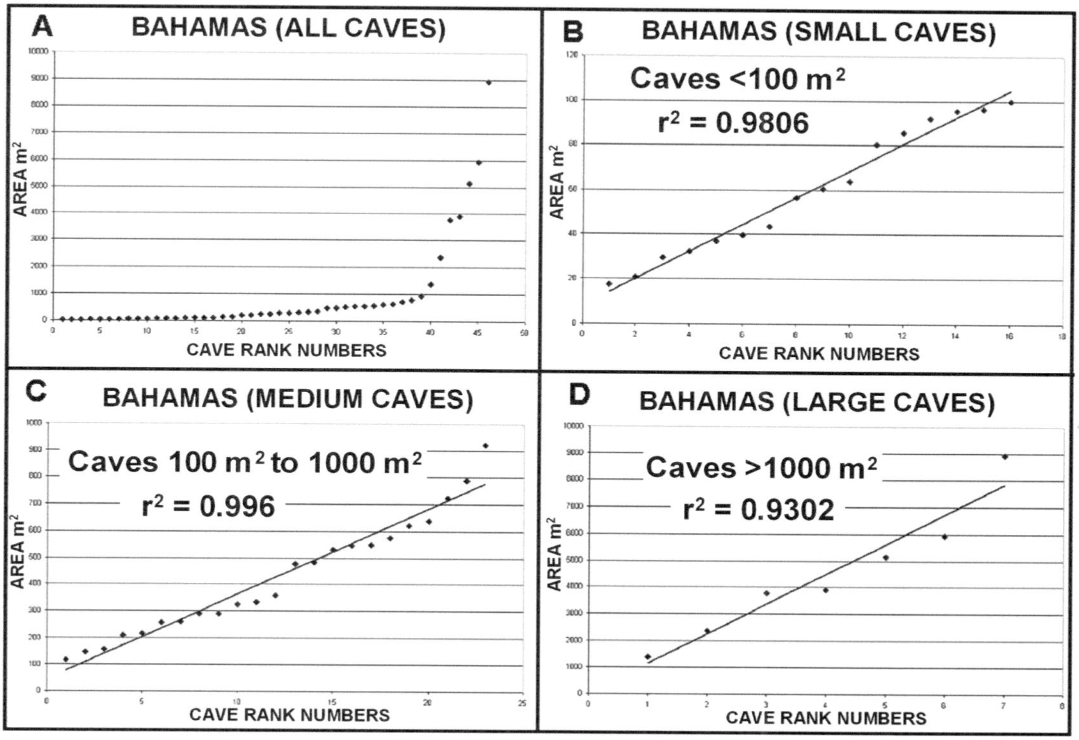

Fig. 4.16 Rank order plot of Bahamian flank margin caves, using areal footprint as a measure of cave size (From Roth et al. 2006). (**a**) All caves ranked, showing three segments of different slope. (**b**) Plot of caves under 100 m^2 area. (**c**) Plot of caves between 100 m^2 and 1,000 m^2 area. (**d**) Plot of caves over 1,000 m^2

extended rectangle with an aspect ratio of 1–100. Flank margin caves plot more as a 1–10 aspect ratio rectangle, a reflection of their globular, ramiform nature. However, as flank margin caves grow larger, they can only do so by connecting laterally along the lens margin; inward growth is limited. As a result, as Fig. 4.18 shows, the largest flank margin caves develop a linearity as a result of a switch from growth in all planar directions to a linear growth parallel to the lens margin. As with the area rank order plot, this pattern reinforces the model of flank margin cave growth as a function of chamber intersections.

The area versus perimeter (A vs P) approach was used by Lace (2008) to differentiate sea caves from flank margin caves in Puerto Rico. Waterstrat et al. (2010) did the same for San Salvador Island, Bahamas. The Bahamian approach utilized sea caves developed in Holocene eolianites as a control for sea cave pattern, as these eolianites could not host a flank margin cave from a Pleistocene sea-level event. Flank margin caves in Pleistocene eolianites, entered solely by roof collapse or pit caves were used as a control for the flank margin cave pattern, as wave energy could not carve a sea cave from above. The paper presented plots that indicated the A vs P ratio could separate sea caves from flank margin caves, but the statistical approach later proved incorrect (Curl and Mixon 2011), and when the correct approach was used, the statistical differentiation did not work (Waterstrat et al. 2011). The original argument that tafoni could be distinguished from sea caves and flank margin caves by their A vs P ratio did prove to be correct (see Chap. 8 for a discussion of tafoni).

The differentiation of sea caves from flank margin caves is not especially useful in sea-level studies, as the caves both form at sea level. However, differentiating a sea cave from a flank

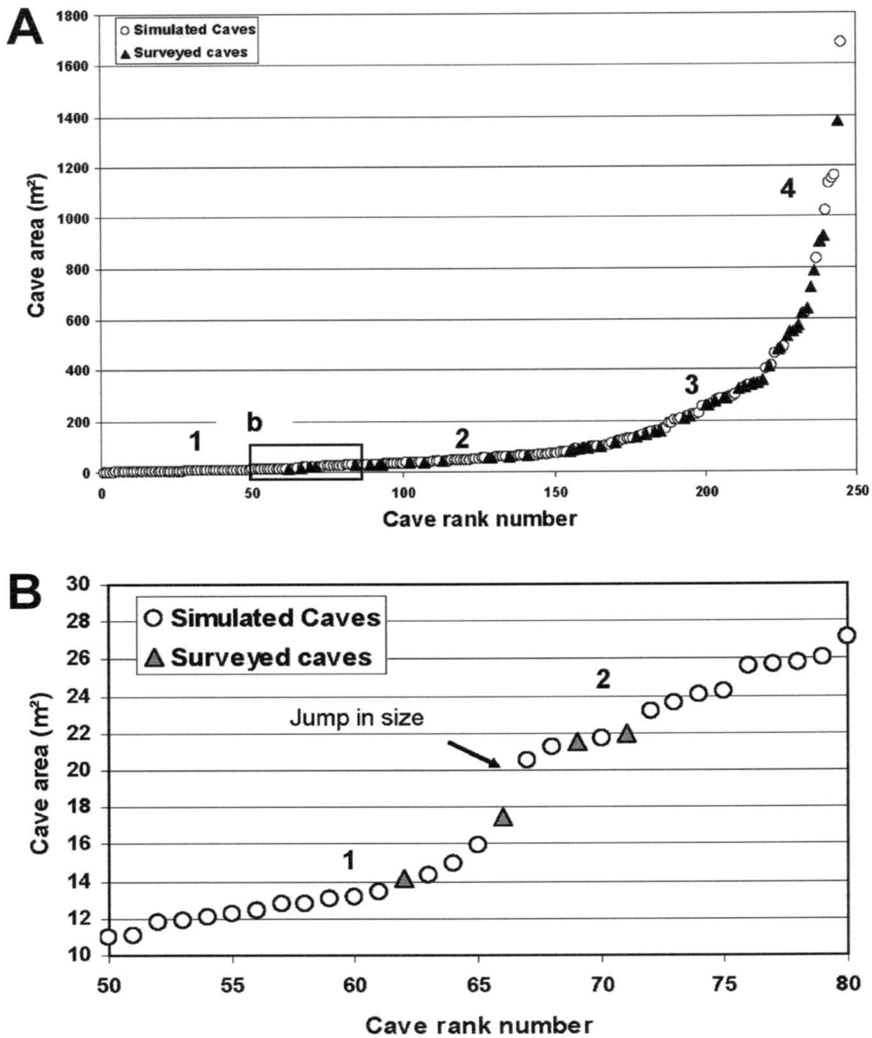

Fig. 4.17 A computer model was used to generate flank margin caves by the chamber amalgamation method. (**a**) The plot fits the empirical data of Roth et al. (2006) almost exactly, and (**b**) revealed a smaller subset of cave size, under 20 m^2, which cave mappers had ignored as too small (From Labourdette et al. 2007)

margin cave is critical for calculation of denudation rates based on degree of cave destruction (Waterstrat et al. 2010). Sea caves form from the outside inward, and are always open to the surface. Flank margin caves, on the other hand, form on the inside and are only enterable once they have been breached. The degree of denudation that could entirely remove a sea cave would only be enough to begin the opening of a flank margin cave (Fig. 4.19). A fossil sea cave incorrectly identified as a dissolutional cave would overestimate the amount of denudation; a dissolutional cave wrongly identified as a sea cave would underestimate the denudation.

Denudation has also been measured by measuring limestone pedestal heights where the limestone is protected by a cap rock. A common feature in glaciated areas, where a non-carbonate glacial erratic can protect the limestone beneath (Ford and Williams 2007), such features, called *karrentisch* (Miklavič et al. 2012), have been found on Guam in the Mariana Islands. In this setting, the cap rock is the Plio-Pleistocene Mariana Limestone, which has fallen from a cliff over

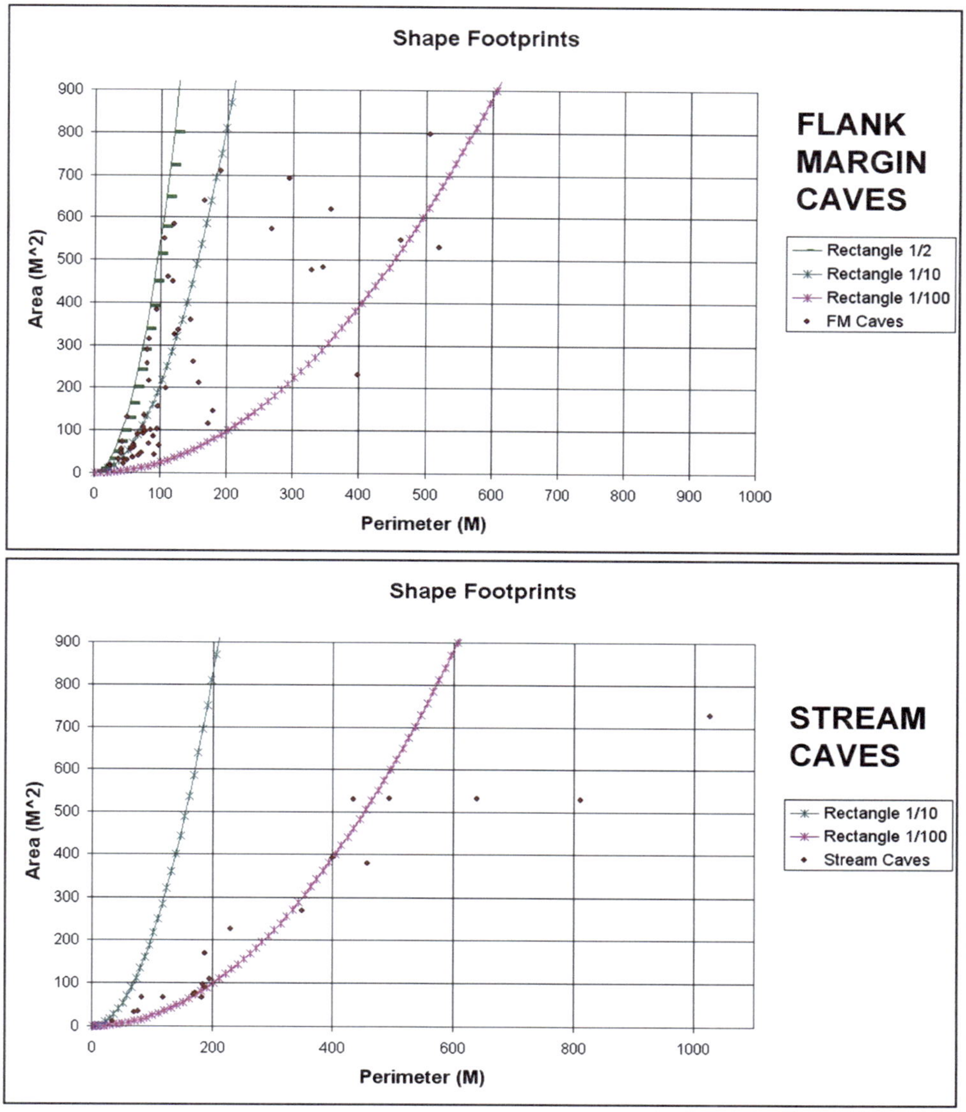

Fig. 4.18 Area vs Perimeter plots for flank margin caves and stream caves show that each cave type plots into different fields. The fields converge at large cave sizes, as flank margin caves can become large only by coast-parallel growth (see Fig. 4.5), which imposes a linearity on the globular cave pattern (From Roth et al. 2006)

100 m high on to a plain of Tarague Limestone, formed during the last interglacial (MIS 5e) sea-level highstand. Pedestals as much as 5 m high exist. Because the Tarague Limestone is so young, a minimum value for the denudation rate (assuming cap rock emplacement just after sub-aerial exposure about 114 ka) can be calculated.

4.8 Summary

Carbonate islands and coasts contain a variety of cave and karst features not found in the vastly more studied continental interiors of the world. The differences result from three major

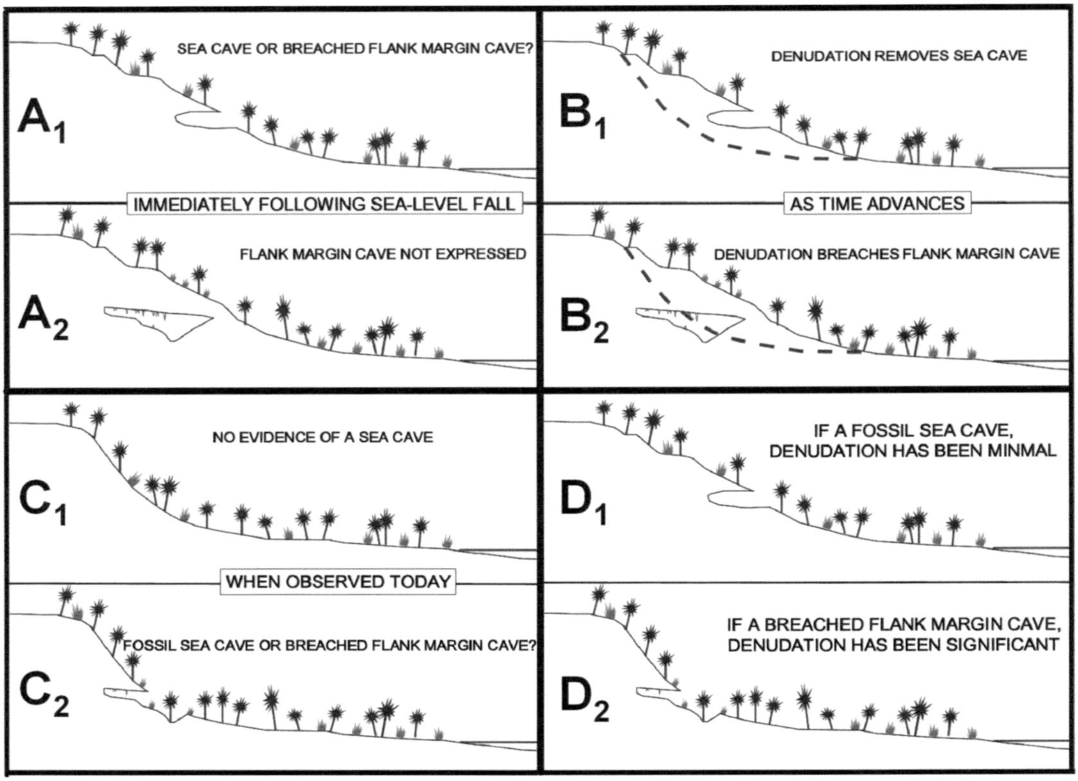

Fig. 4.19 Cartoon demonstrating the importance of differentiating relict flank margin caves from relict sea caves when assessing landscape denudation (From Waterstrat et al. 2010). (**a**) Initial condition. (**b**) Following denudation. (**c**) Denudation consequences. (**d**) Subsequent interpretations

influences: fresh water mixing with salt water, sea level and hence base level change, and commonly the young, eogenetic nature of the carbonate rocks and the continual reshaping of coastal landforms by littoral processes. The Carbonate Island Karst Model attempts to take into account the relationship of sea level, carbonate rocks, and non-carbonate rocks to explain the variety of caves and karst features found on islands, by creating four categories of island types, which can be discrete or form a continuum. The most common dissolutional cave in the world may well be the flank margin cave, as it has been found in great abundance on carbonate islands and continental coasts worldwide. Because sea-level change results in abandonment of old flank margin caves, and the production of new ones, the numbers of such caves are extremely high.

The study of island karst has many useful and practical applications. As shown by Labourdette et al. (2007), preservation of coastal caves in the rock record is of extreme interest to the hydrocarbon community. Carbonate island water resources are better managed when the CIKM can be applied to aquifer studies (e.g. Jocson et al. 2002; Mylroie et al. 2008c). The recreational caving community plays an important role in mapping of these caves (Mylroie 2007). The historical and archeological aspects of these caves are important (Chap. 5), as is the paleontology (Steadman et al. 2007). The complex relationship between carbonate deposition, sea level, and dissolution is an important contributor to our understanding of past Quaternary climate change, and by extrapolation, possible future consequences for society (Mylroie 2008).

References

Agassiz A (1893) A reconnaissance of the Bahamas and of elevated reefs of Cuba, in the steam yacht "Wild Duck". Mus Comp Zool Bull 25:41–43

Baceta JI, Wright PV, Beavington-Penney SJ, Pujalte V (2007) Palaeohydrological control of paleokarst macro-porosity genesis during a major sea-level lowstand. Danian of the Urbasa-Andia plateau, Navarra, North Spain. Sediment Geol 199:141–169

Back B, Hanshaw BB, Van Dreil JN (1984) Role of groundwater in shaping the eastern coastline of the Yucatan Peninsula, Mexico. In: LaFleur R (ed) Groundwater as a geomorphic agent. Allen and Unwin, Boston, pp 281–293

Benjamin GJ (1970) Diving into blue holes of the Bahamas. Natl Geogr Mag 138:347–363

Birmingham AN, Lace MJ, Mylroie JR, Mylroie JE (2010) Bell hole origin: constraints on developmental mechanisms, Crooked Island, Bahamas. In: Martin JB, Siewers FD (eds) Proceedings of the 14th symposium on the geology of the Bahamas and other carbonate regions, pp 18–30. Reprinted in Speleogenesis and evolution of Karst Aquifers, 2010, Issue 10. http://www.speleogenesis.info/content

Bottrell SH, Smart PL, Whitaker F (1991) Geochemistry and isotope systematics of sulphur in the mixing zone of Bahamian blue holes. Appl Geochem 6:99–103

Bottrell SH, Carew JL, Mylroie JE (1993) Bacterial sulphate reduction in flank margin environments: evidence from sulphur isotopes. In: White B (ed) Proceedings of the 6th symposium on the geology of the Bahamas. Bahamian Field Station, Port Charlotte, pp 17–21

Carew JL, Mylroie JE (1995) A stratigraphic and depositional model for the Bahama Islands. In: Curran HA, White B (eds) Terrestrial and shallow marine geology of the Bahamas and Bermuda, Special paper 300. Geological Society of America, Boulder, pp 5–31

Carew JL, Mylroie JE (1997) Geology of the Bahamas. In: Vacher HL, Quinn TM (eds) Geology and hydrogeology of carbonate islands. Elsevier Science Publishers, Amsterdam/New York, pp 91–139

Carew JL, Mylroie JE, Schwabe SJ (1998) The geology of South Andros Island, Bahamas: a reconnaissance report for the ninth symposium on the geology of the Bahamas field trip. Bahamian Field Station, San Salvador Island, Bahamas

Choquette PW, Pray LC (1970) Geologic nomenclature and classification of porosity in sedimentary carbonates. Am Assoc Petr Geol Bull 54:207–250

Curl R (2011) Comment on Coastal caves in Bahamian eolian calcarenites: differentiating between sea caves and flank margin caves using quantitative morphology'. J Cave Karst Stud 73:202. doi:10.4311/jcks2011es0195

Dill RF (1977) The blue holes – geologically significant submerged sink holes and caves off British Honduras and Andros, Bahama Islands. In: Proceedings of the third coral reef symposium 2, University of Miami, Florida, pp 237–242

Doran E (1913) Landforms of the southeast Bahamas. Department of Geography, the University of Texas, Austin, Texas, pp 10–12

Ford D, Williams P (1989) Karst geomorphology and hydrology. Chapman and Hall, New York

Ford DC, Williams PW (2007) Karst hydrogeology and geomorphology. John Wiley & Sons Ltd., West Sussex

Frank EF, Mylroie JE, Troester J, Alexander EC, Carew JL (1998) Karst development and speleogenesis, Isla de Mona, Puerto Rico. J Cave Karst Stud 60:73–83

Harris JG, Mylroie JE, Carew JL (1995) Banana holes: unique karst features of the Bahamas. Carbonate Evaporite 10:215–224

Hearty PJ, Kindler P, Cheng H, Edwards RL (1999) A +20m middle Pleistocene sea-level highstand (Bermuda and Bahamas) due to partial collapse of Antarctic ice. Geology 27:375–378

Infante LR, Mylroie JE, Kambesis PN, Leist JW (2011) Carbonate dissolution in a dynamic fresh-water lens: lens margin migration and the production of banana holes as syngenetic caves, San Salvador Island, Bahamas. Geol Soc Am, Abstracts with programs 43:341

Jenson JW, Keel TM, Mylroie JR, Mylroie JE, Stafford KW, Taborosi D, Wexel C (2006) Karst of the Mariana Islands, Geological Society of America special paper 404, The interaction of tectonics, glacioeustasy and fresh-water/sea-water mixing in island carbonates, Boulder, pp 129–138

Jocson JMU, Jenson JW, Contractor DN (2002) Recharge and aquifer responses: Northern Guam Lens Aquifer, Guam, Mariana Islands. J Hydrol 260:231–254

Kelley K, Mylroie JE, Mylroie JR, Moore C, Moore PJ, Collins L, Ersek L, Lascu I, Roth M, Passion R, Shaw C (2006) Eolianites and karst development in the Mayan Riviera, Mexico. In: Davis RL, Gamble DW (eds) Proceedings of the 12th symposium on the geology of the Bahamas and other carbonate regions, Gerace Research Centre, San Salvador, Bahamas, pp 88–99. Reprinted in Speleogenesis and Evolution of Karst Aquifers, 2011, Issue 11. http://www.speleogenesis.info/content

Labourdette R, Lascu I, Mylroie JE, Roth M (2007) Process-like modeling of flank margin caves: from genesis to burial evolution. J Sediment Res 77: 965–979

Lace MJ (2008) Coastal cave development in Puerto Rico. J Coastal Res 24:508–518

Larson EB, Mylroie JE (2012) Infilled blue holes on the Bahama Banks as point sources for whiting origin. Geol Soc Am, Abstracts and programs 44(4):25

Machel HG (2004) Concepts and models of dolomitization: a critical reappraisal. In: Braithwaite CJR, Rizzi G, Darke G (eds) The geometry and petrogenesis of dolomite hydrocarbon reservoirs, vol 235, Special publications. Geological Society, London, pp 7–63

Machel HG, Sumrall JB, Kambesis PN, Mylroie JR, Mylroie JE, Lace MJ (2011) Multiple cave types in Miocene and Pleistocene carbonates of Barbados, West Indies. In: Engel AS, Engel SA, Moore PJ, Duchene H (eds) Carbonate geochemistry: reactions and processes in aquifers and reservoirs, vol 16, Special publication. Karst Waters Institute, Leesburg/Virginia, pp 43–46

Meyerhoff AA, Hatten CW (1974) Bahamas salient of North America: tectonic frame-work, stratigraphy, and petroleum potential. Am Assoc Petrol Geologist Bull 58:1201–1239

Miklavič B, Mylroie JE, Jenson JW, Randall RH, Banner JL, Partin JW (2012) Evidence of the sea-level change since MIS 5e on Guam, tropical west Pacific. National Science Foundation workshop. Palma de Mallorca, Mallorca

Mixon B (2011) Comment on Coastal caves in Bahamian eolian calcarenites: differentiating between sea caves and flank margin caves using quantitative morphology. J Cave Karst Stud 73:202. doi:10.4311/jcks2011es0195

Mullins HT, Hine AC (1989) Scalloped bank margins: beginning of the end for carbonate platforms? Geology 17:30–39

Mylroie JE (1988) Caves and karst features of sandy point, San Salvador Island, Bahamas. In: Mylroie JE (ed) Field guide to the karst geology of San Salvador Island, Bahamas. Department of Geology and Geography, Mississippi State University and CCFL Bahamian Field Station, Ft, Lauderdale, Florida, pp 83–96

Mylroie JE (2007) Cave surveys, cave size, and flank margin caves. Compass Tape 17:8–16

Mylroie JE (2008) Late quaternary sea level position: Bahamian carbonate deposition and dissolution cycles. Quatern Int 183:61–75. doi:10.1016/j.quaint.2007.06.030

Mylroie JE, Carew JL (1990) The flank margin model for dissolution cave development in carbonate platforms. Earth Surf Proc Land 15:413–424

Mylroie JE, Carew JL (1995) Chapter 3, Karst development on carbonate islands. In: Budd DA, Harris PM, Saller A (eds) Unconformities and porosity in carbonate strata, Memoir 63. American Association of Petroleum Geologists, Tulsa, pp 55–76

Mylroie JE, Jenson J (2000) Guam and the carbonate island karst model. In: Onac BP, Tamas T (eds) Karst studies and problems: 2000 and beyond. Proceedings of the joint meeting of the friends of karst, theoretical and applied karstology, and IGCP 448. Cluj-Napoca, Romania, pp 82–86

Mylroie JE, Mylroie JR (1991) Meander cutoff caves and self piracy: the consequences of meander incision into soluble rocks. Natl Speleological Soc Bull 52:33–44

Mylroie JE, Mylroie JR (2004) Glaciated karst: how the Helderberg Plateau revised the geologic perception. Northeast Geol Environ Sci 26:82–92

Mylroie JE, Mylroie JR (2007) Development of the carbonate island karst model. J Cave Karst Stud 69:59–75

Mylroie JE, Mylroie JR (2009a) Caves as geologic indicators, Kangaroo Island, Australia. J Cave Karst Stud 71:32–47

Mylroie JE, Mylroie JR (2009b) Flank margin cave development as syndepositional caves: examples from the Bahamas. In: White WB (ed) Proceedings of the 15th international congress of speleology, vol 2. National Speleological Society, Huntsville, Alabama, pp 533–539

Mylroie JE, Mylroie JR (2011) Void development on carbonate coasts: creation of anchialine habitats. Hydrobiologia 677:15–32. doi:10.1007/s10750-010-0542-y

Mylroie JE, Vacher HL (1999) A conceptual view of carbonate island karst. In: Palmer AN, Palmer MV, Sasowsky ID (eds) Karst modeling. Karst Waters Institute special publication #5, Charles Town, pp 48–57

Mylroie JE, Carew JL, Vacher HL (1995a) Karst development in the Bahamas and Bermuda. In: Curran HA, White B (eds) Terrestrial and shallow marine geology of the Bahamas and Bermuda, Special paper 300. Geological Society of America, Boulder, pp 251–267

Mylroie JE, Carew JL, Moore AI (1995b) Blue holes: definition and genesis. Carbonate Evaporite 10: 225–233

Mylroie JE, Jenson JW, Taborosi D, Jocson JMU, Vann DT, Wexel C (2001) Karst features of Guam in terms of a general model of carbonate island karst. J Cave Karst Stud 63:9–22

Mylroie JE, Carew JL, Curran HA, Freile D, Sealey NE, Voegeli VJ (2006) Geology of Cat Island, Bahamas: a field trip guide. Gerace Research Centre, San Salvador Island

Mylroie JE, Mylroie JR, Owen AM, Waterstrat MJ (2008a) Cave and karst inventory of the primeval forest, New Providence Island, Bahamas: unexpected discoveries. In: Freile D, Park L (eds) Proceedings of the 13th symposium on the geology of the Bahamas and other carbonate regions. Gerace Research Centre, San Salvador, Bahamas, pp 107–118

Mylroie JE, Mylroie JR, Nelson CN (2008b) Flank margin cave development in telogenetic limestones of New Zealand. Acta Carsologica 37:15–40

Mylroie JE, Mylroie JR, Jenson JW, MacCracken RS (2008c) Fresh-water lens anisotropy and flank margin cave development Fais Island, FSM. In: Freile D, Park L (eds) Proceedings of the 13th symposium on the geology of the Bahamas and other carbonate regions, Gerace Research Centre, San Salvador, Bahamas, pp 135–139. Reprinted in speleogenesis and evolution of karst aquifers, 2011, Issue 11. http://www.speleogenesis.info/content/

Mylroie JE, Owen AM, Sumrall JB, Kambesis PK, Lace MJ, Larson EB, Mylroie JR (2012) Quantitative description and coastal mapping of flank margin caves on Mallorca Island, Spain: Studia Universitatis Babeşs-Bolyai, Geology, Special Issue 2012, pp 42–44

Nelson RJ (1853) On the geology of the Bahamas and coral reef formation generally. Q J Geol Soc Lond 9:200–221

Otonicar B, Buzijak N, Mylroie JE, Mylroie JR (2010) Flank margin cave development in carbonate talus breccia facies: an example from Cres Island, Croatia. Acta Carsologica 39:79–91

Palmer R (1985) The blue holes of the Bahamas. Jonathan Cape Ltd., London

Palmer R (1989) Deep into blue holes. Cambridge University Press, Cambridge

Palmer AN (1991) Origin and morphology of limestone caves. Geolog Soc Am Bull 103:1–25

Palmer RJ, Williams D (1984) Cave development under Andros Island. Cave Sci 11:50–52

Panuska BC, Mylroie JE, Armentrout D, McFarlane D (1998) Magnetostratigraphy of Cueva del Aleman, Isla de Mona, Puerto Rico and the species duration of Audobon's Shearwater. J Cave Karst Stud 60:96–100

Proctor CJ (1988) Sea-level related caves on Berry head, South Devon. Cave Sci 15:39–50

Roth MJ (2004) Inventory and geometric analysis of flank margin caves of the Bahamas. Thesis, Mississippi State University, http://sun.library.msstate.edu/ETD-db/theses/available/etd-07062004-164930/

Roth MJ, Mylroie JE, Mylroie JR, Ersek V, Ersek CC, Carew JL (2006) Flank margin cave inventory of the Bahamas. In: Davis RL, Gamble DW (eds) Proceedings of the 12th symposium on the geology of the Bahamas and other carbonate regions. Gerace Research Centre, San Salvador, Bahamas, pp 153–161

Schwabe SJ, Carew JE (2006) Blue holes: an inappropriate moniker for water-filled caves in the Bahamas. In: Davis RL, Gamble DW (eds) Proceedings of the 12th symposium on the geology of the Bahamas and other carbonate regions. Gerace Research Centre, San Salvador, Bahamas, pp 179–187

Sealey NE (1991) Early views on the geology of the Bahamas: 1837–1931. In: Bain RJ (ed) Proceedings of the fifth symposium on the geology of the Bahamas. Bahamian Field Station, San Salvador Island, Bahamas, pp 203–207

Sealey NE (1994) Bahamian landscapes, 2nd edn. Media Publishing, Nassau, Bahamas

Shattuck B, Miller B (1905) Physiography and geology of the Bahamas. In: Shattuck B (ed) The Bahama Islands. The Macmillan Co, New York

Shaw TR (1993) The history of caves studies in Trinidad, Jamaica, the Bahamas, and some other Caribbean Islands. Acta Carsologica 22:15–76

Smart PL, Whitaker F (1997) Hydrology of the Bahamian archipelago. In: Vacher HL, Quinn TM (eds) Geology and hydrogeology of carbonate islands. Elsevier Science Publishers, Amsterdam/New York, pp 91–139

Stafford KW, Mylroie JE, Mylroie JR, Jenson JW, Taborosi D (2004) Coves and pocket beaches: littoral erosion and flank margin caves. Geological Society of America abstracts and programs 36:284

Stafford KW, Mylroie JE, Taboroši D, Jenson JW, Mylroie JR (2005) Karst development on Tinian, commonwealth of the Northern Mariana Islands: controls on dissolution in relation to the carbonate island karst model. J Cave Karst Stud 67:14–27

Steadman DW, Franz LR, Morgan GS, Albury NA, Kakuk B, Broad K, Franz SE, Tinker KL, Pateman MP, Lott TA, Jarzen DM, Dilcher DL (2007) Exceptionally well preserved late quaternary plant and vertebrate fossils from a blue hole on Abaco, the Bahamas. Proc Natl Acad Sci USA 104:19897–19902

Stoddard DR (1962) Three Caribbean atolls: Turneffe Islands, lighthouse reef, and Glover's reef, British Honduras. The Pacific Science Board, no. 82, Washington, D.C., p 53

Sweeting MM (1973) Karst landforms. Columbia University Press, New York

Thompson WG, Curran HA, Wilson MA, White B (2011) Sea-level oscillations during the last interglacial highstand recorded by Bahamian corals. Nat Geosci 4:684–687

Vacher HL (1988) Dupuit-Ghyben-Herzberg analysis of strip-Island lenses. Geol Soc Am Bull 100:223–232

Vacher HL, Mylroie JE (1991) Geomorphic evolution of topographic lows in Bermudian and Bahamian Islands: effect of climate. In: Bain RJ (ed) Proceedings of the fifth symposium on the geology of the Bahamas. Bahamian Field Station, Port Charlotte, pp 221–234

Vacher HL, Mylroie JE (2002) Eogenetic karst from the perspective of an equivalent porous medium. Carbonate Evaporite 17:182–196

Walker LN, Mylroie JE, Walker AD, Mylroie JR (2008) The caves of Abaco Island, Bahamas: keys to geologic timelines. J Cave Karst Stud 70:108–119

Waterstrat WJ, Mylroie JE, Owen AM, Mylroie JR (2010) Coastal caves in Bahamian eolian calcarenites: differentiating between sea caves and flank margin caves using quantitative morphology. J Cave Karst Stud 72:61–74

Waterstrat W, Mylroie JE, Owen AM, Mylroie JR (2011) Reply to comments on: Coastal caves in Bahamian eolian calcarenites: differentiating between sea caves and flank margin caves using quantitative morphology. J Cave Karst Stud 73:203. doi:10.4311/2011ES0216

White WB (1988) Geomorphology and hydrology of karst terrains. Oxford University Press, New York

Wilson WL (1994) Morphology and hydrology of the deepest known cave in the Bahamas: Dean's Blue Hole, Long Island. In: Boardman MR (ed) Seventh symposium of the geology of the Bahamas. Bahamian Field Station, San Salvador Island, Bahamas, p 21

Wood WW, Petritus MJ (1984) Origin and distribution of carbon dioxide in the unsaturated zone of the southern high plains. Water Resour Res 20:1193–1208

Zans V (1951) On karst hydrology of Jamaica. Assoc Int Hydrol Scientfique Brux 2:267–279

The Biological and Archaeological Significance of Coastal Caves and Karst Features

5

Michael J. Lace and John E. Mylroie

Abstract

By virtue of their very structure, coastal caves can serve as protective environments, preserving a wide range of materials in otherwise dramatically evolving shorelines. Caves formed within littoral settings have emerged as critical environments supporting past and present floral and faunal endemic species. The biodiversity inherent in littoral ecosystems offers invaluable clues to modeling past climates and critical indicators of future change. Coastal karst landforms also harbor vestiges of anthropogenic uses spanning human history. These complex remnants have established various types of coastal cave structures as important repositories of preserved archaeological, historical and cultural materials. These repositories support a range of multidisciplinary studies such as geoarchaeology and cultural anthropology while continually reshaping our collective view of human history in the context of shoreline and paleoshoreline settings. The complex evolution of coastal communities in these settings is reflected in the long term uses of coastal landscapes and associated marine and terrestrial resources by archaic through modern indigenous cultures. Archaeologically significant coastal cave sites have been identified in an amazingly diverse set of climates, from continental coastal settlements to temperate equatorial island archipelagos to some of the most inhospitable and intemperate shorelines known – many of which are associated with the repeated migrations and complex settlement patterns of indigenous peoples over time from one coastal region to another.

5.1 Biodiversity in Coastal Cave and Karst Settings

Caves offer complex structural morphologies with unique microclimates that form habitats capable of supporting a broad range of troglobytic

M.J. Lace (✉)
Coastal Cave Survey, West Branch, IA 52358, USA
e-mail: michael-lace@uiowa.edu

J.E. Mylroie
Department of Geosciences, Mississippi State University, Mississippi State, MS 39762, USA
e-mail: mylroie@geosci.msstate.edu

M.J. Lace and J.E. Mylroie (eds.), *Coastal Karst Landforms*, Coastal Research Library 5,
DOI 10.1007/978-94-007-5016-6_5, © Springer Science+Business Media Dordrecht 2013

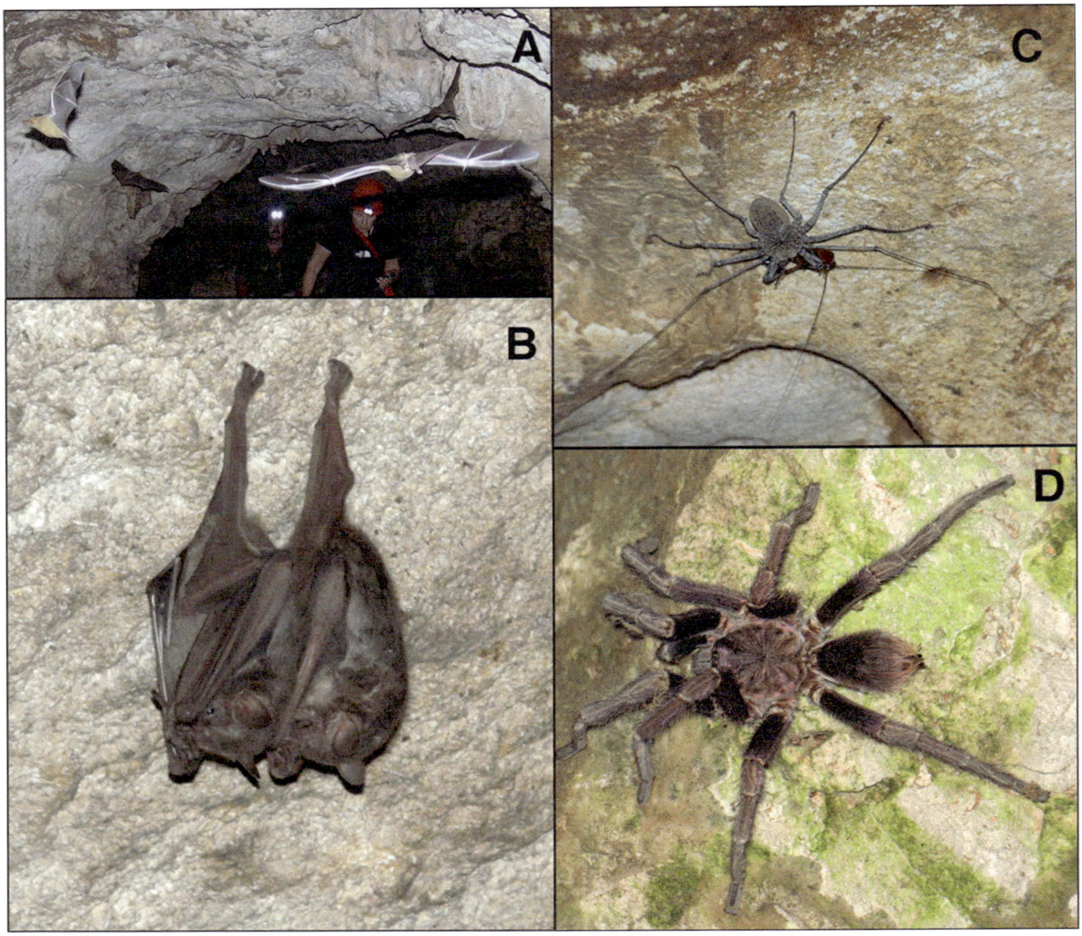

Fig. 5.1 (**a**) Coastal cave bat roost on Curaçao, Netherlands Antilles. (**b**) Roosting bats, Barbados (Photo by M. Ohms) (**c**) Amblipigyd (pseudoscorpion) devouring a cave cockroach, Curacao. (**d**) Tarantula (12 cm length) in a deep cave setting, Grotte-Marie Jeanne, Republique D'Haiti (Photo by J. Goodbar)

(cave-adapted), troglophylic (opportunistic cave-associated) and trogloxenic (refuge seeking) species associated with these environments (Iliffe et al. 1983; Martin et al. 2010; Kano and Kase 2008; Northrup and Lavoie 2001; Mylroie and Mylroie 2011; Culver and Pipan 2009; Romero 2009; Culver et al. 2009). Bats, for examples, utilize caves in coastal and inland karst settings as simple shelters, nurseries or hibernacula (Fig. 5.1b) (Hutson et al. 2001). In tropical climates, for example, coastal cave structure determines complex temperature gradients resulting in specific microclimates suitable for species-specific roosting behavior (Gamble et al. 2000; Rodríguez-Durán 2009). Like many other examples of cave macrofauna and microfauna, bats represent integral biological components not only of fragile cave ecosystems but in some cases also of related surface environments (Fleming and Raley 2009), for example, as pollinators of fauna semi-arid and tropical ecosystems such as Curaçao where the island ecology is mutually dependent on small bat communities. An estimated bat population of 2000 composed of seven species, many of which utilize coastal cave structures as their primary roosts (Fig. 5.1a), are critical for cactus pollination and limited seed dispersal. Within such an interdependent context, one can correlate the overall health of the bat population with overall cactus densities (Petit and Pors 1996).

As in karst and non-karst settings, bats and other cave fauna (Fig. 5.1c) serve critical roles in insect population control with wide ranging implications for agriculture and public health by limiting the spread of diseases transmitted by insect vectors in some areas, such a mosquito-borne malaria or dengue. Other species of cave fauna also play a role in forming complex and unusual cave ecosystems. Figure 5.1d, for example, illustrates one of several tarantula species that dominate the deep cave environment of Grotte Marie-Jeanne – a 4.7 km long, multi-level cave (102 m vertical extent) formed within Eocene-aged limestone in an uplifted coastal ridge on the southern peninsula of Haiti.

5.1.1 Anchialine Environments Within Coastal Caves

The development of an *anchialine* habitat requires two critical actions: (1) formation of a water body of mixed salinity; and (2) a rock or sediment structure which holds and preserves that water body over time. Anchialine habitats can occur in a variety of surface water configurations, but anchialine caves contain dark zones and are isolated from the land surface, creating a unique environment. While anchialine caves can form in the interior of continents due to some set of unusual hydrological and climatic circumstances, such caves are ubiquitous in coastal areas where a fresh-water lens meets marine water within carbonate rock to produce dissolutional voids, or where non-karst processes have made a coastal void (Mylroie and Mylroie 2011).

Anchialine caves in coastal locations develop by two mechanisms: pseudokarst processes that form talus caves, sea caves, tafoni, fissure caves and lava tubes (Chap. 1); and by karst processes using bedrock dissolution that form stream caves, flank margin caves, and blue holes (as discussed in Chap. 4). Pseuodokarst caves are relatively unimportant as regards to the overall degree of anchialine cave habitat development, with the occasional lava tube being a notable exception. Karst caves formed by dissolution provide the most extensive, variable, and long-term environments for anchialine habitats. These caves depend on fresh water and marine water mixing, sea level change, rock maturity, and interaction with adjacent non-carbonate rocks for their initiation and development. The Quaternary period is characterized by glacioeustasy, which has moved all coastal anchialine cave environments, whether karst or pseudokarst, consistently through a vertical range of over 100 m. The anchialine environments observed today did not develop at their current elevations until about 4,000 years ago when sea level reached its current elevation.

Blue holes are polygenetic structures, resulting from several mechanisms (Mylroie et al. 1995) but the most common are bank-margin fractures and upward progradational collapses from deep dissolutional voids. As a result, blue holes provide vertical connections between various independent levels of horizontal cave development produced by earlier sea-level positions. Blue holes have been overprinted by the cumulative sea-level excursions of the Quaternary; each sea-level rise and fall adds complexity to the habitats within blue holes and the cave systems they intersect and unite. Blue holes reach depths below the deepest glacioeustatic sea-level lowstand (up to 200 m, as at Deans Blue Hole, Long Island, Bahamas as reported by Wilson 1994), and thereby provide refugia for anchialine species when the complex of cave passages above are left dry by Quaternary sea-level fall. Blue holes continue to reveal newly described species (Daenekas et al. 2009) and represent some of the most significant anchialine cave environments in the world, providing clues to anchialine cave species colonization and speciation events.

Carbonate coasts from around the world contain abundant flank margin caves, as already discussed for the Bahamas, Puerto Rican islands, the Mariana Islands (Chaps. 7, 9 and 13, respectively) and New Zealand (Mylroie et al. 2008) and Australia (Mylroie and Mylroie 2009) (see also Chaps. 17 and 18). Submarine exploration has discovered these caves below modern sea level (Mylroie and Mylroie 2007). Flank margin caves can be found in great numbers over limited sections of carbonate coastline, as a result, caves of this type are probably the most abundant in the world. Flank margin caves form as isolated dissolution chambers which enlarge gradually, until adjacent chambers intersect, and ultimately collections of chambers intersect; each of these intersections creates a sudden jump in cave size. Flow in these caves is laminar or transitional (i.e. non-turbulent) because the caves are in a diffuse flow field where water enters and leaves the caves as porous media flow. Development by this mechanism indicates that interstitial colonizers would have a significant time advantage over larger organisms that require a macroscopic opening (cave entrance) to colonize the cave's anchialine environment.

The lava tube is the most significant type of pseudokarst cave in coastal areas. Lava tubes are common in volcanic areas across the planet, especially where basaltic lavas are produced. Plate tectonics and hot spots in the world's oceans produce many volcanic islands, creating conditions where lava flows, and their associated lava tubes will intersect the coastal environment. Lava tube creation is independent of coastal erosion dynamics, unlike other pseudokarst caves. Lava tube configuration is usually at right angles to the shoreline, such that they can persist for large amounts time despite coastal retreat by wave erosion, which can remove other cave types. This survivability allows lava tubes to host long-term anchialine cave habitats. The Lanzarote anchialine lava tube system in the Canary Islands is an excellent example of coastal cave development and persistence (Kornicker and Iliffe 1995; Wilkens et al. 2009).

A qualitative chart has been constructed (Fig. 5.2) that speculatively ranks coastal caves in terms of apparent abundance, and expected species diversity (Mylroie and Mylroie 2011). Blue holes are predicted to have the largest number of species and the greatest number of habitats for those species. Stream caves are considered to be less abundant than flank margin caves, but might be expected to have more diversity as they transit the fresh water to anchialine and marine conditions. Lava tubes

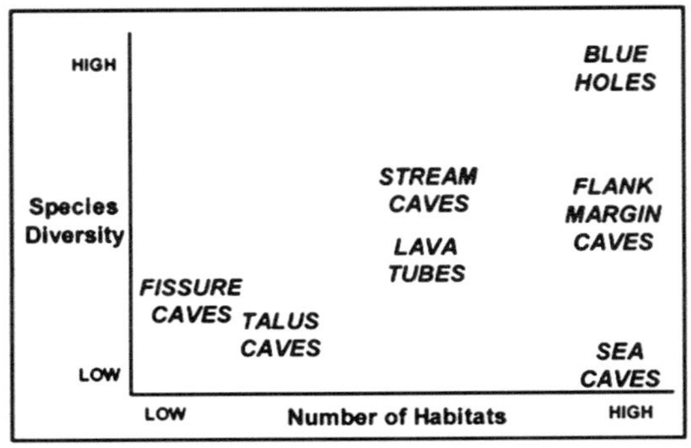

Fig. 5.2 Qualitative plot showing predicted number of habitats, and anchialine species diversity, based on geologic and hydrologic models of various expressions of coastal cave development. Blue holes, while fewer in number than sea caves or flank margin caves, are typically complex and contain a wide variety of interconnected habitats and therefore a predicted high species diversity (From Mylroie and Mylroie 2011)

are probably the most diverse of the pseudokarst caves; however sea caves are the most abundant of the pseudokarst caves but would be expected to be the least diverse. All pseudokarst caves are considered to have less species abundance than all dissolutional caves. Talus caves and fissure caves have been argued to have low biodiversity and low relative abundance, an outcome of their simple form, their young age, and their low survival rate as a result of multiple glacial sea-level cycles. More data need to be collected to determine if the various coastal cave types are significantly different as regards to the number and diversity of anchialine habitats.

5.1.2 The Role of Cave Sediments and Fossil Assemblages as Indicators of Paleoclimate, Paleobiodiversity and Coastal Karst Landform Genesis

Just as coastal caves can function as containment structures for fragile modern ecosystems, they can also harbor complex evidence of past biological systems or "paleobiodiversity". Fossil zooarchaeologic/paleobotanical remnants can serve as indicators of past periods of shifting biodiversity, such as faunal extinctions and/or anthropogenic utilization of marine or terrestrial species in coastal settings (Miracle 2005). In addition to the remains of more complex organisms, coastal cave sediments in particular have proven to be important matrices harboring microscopic archival indicators of paleoclimate (see Chap. 15) and paleodiversity (Sasowsky and Mylroie 2004; Woodward and Goldberg 2001). Preserved reef structures within coastal caves provide another opportunity to model cave development in the context of coastal landform evolution. Fossil coral densities and related analyses can be viewed in 3-D from the inside, offering remnant indicators of past coral reef environments and in turn serving as critical tools for defining coastal cave speleogenesis within complex littoral reef systems.

As discussed in Chap. 4, in addition to natural processes of coastal modification, anthropogenic activities have resulted in significant alterations of coastal karst in many areas. However, anthropogenic influences have not been limited solely to obscuring coastal speleogenesis patterns as long-term, progressive human uses of coastal areas have also profoundly modified the biodiversity associated with these coastal ecosystems and in turn reshaped the very evolution of many coastal geomorphologies (Fitzpatrick and Keegan 2007). The historical and modern effects of human influences on coastal caves and karst and coastal management strategies are outlined in Chap. 6 and discussed throughout this volume in the context of specific case studies in Part II. The remainder of this chapter focuses on the pivotal role of coastal caves in preserving past records of anthropogenic coastal landform uses.

5.2 Coastal Cave Geoarchaeology

Contrary to their historical treatment as marginal landscapes within the sweep of human evolution, coastal landscapes have emerged as critical components in reconstructing prehistoric migration and settlement patterns on a global scale (Baily 2004; Benjamin et al. 2011). While the topics of geoarchaeology and anthropology in coastal settings similarly represent expansive fields of study including many lines of research and related perspectives, the following examples will hopefully illustrate the complexity, significance and the broader scope of these coastal research efforts to date. Several excellent practical field guides to terrestrial, coastal and marine archaeologic fieldwork are available in the literature (Baily and Parkington 1988; Balme and Paterson 2006; Ford 2011) as well as numerous associated journals cited within the following references. This chapter therefore focuses on the significance of coastal cave settings within the context of such field research methods.

Interdisciplinary approaches have seen considerable success in integrating a functional understanding of geologic processes and anthropogenic land use patterns to reveal trends in cultural migrations and adaptive strategies evident in a variety of emergent and inundated coastal shorelines (Fitzpatrick 2004; Gusick and Faught 2011; Mason 1993). The archaeology of coastal karst naturally is not confined to terrestrial settings as maritime archaeologic studies continue to reveal cultural clues from submerged karstic and non-karst landforms, correlating them with terrestrial landscape data (Ford 2011). Efforts have included the analysis and reconstruction of coastal settlements and modeling coastal population migrations in an expansive array of island archipelagos and continental shorelines from a wide geographic distribution, displaying diverse coastal geomorphologies (Bicho et al. 2011; Busch 1994). From the Caribbean basin, the Polynesian and Japanese islands (Berman and Gnivecki 1995; Rouse 1986) to offshore submerged paleoshorelines of the Florida coast to the U.S. Channel Islands (as discussed in Chap. 14), the Mediterranean (Collina-Girard 1989; Shackleton et al. 1984), the South African Cape area (Hendey and Volman 1986) and in more remote shorelines, such as the Pantagonian (Maire et al. 2009) and Alaskan coastlines (Johnson and Stright 1991).

Cross-disciplinary approaches have also been effectively used to model cultural interaction between coastal and inland communities by integrating GIS and archaeological data with comparative karst geomorphologies (Reid 2008; Yuwono 2009). In geophysical terms, the classification of cave sites as coastal versus inland based on their location relative to contemporary shoreline can be useful when modeling cave development as it relates to specific geologic processes linked to past sea levels within the context of coastal landform evolution. In geoarchaeological terms, however, such a classification becomes more complex and potentially problematic, particularly on island platforms of limited surface area where, due to site proximity to coastal areas, cultural land uses such as exploitation of marine resources can be proposed to be ubiquitous (Keegan et al. 2008). Thus, in terms of anthropogenic uses, all cave sites in small island settings can be regarded as "coastal". Integrating a functional understanding of the origins and processes associated with a geophysical setting with the cultural and archaeological implications of sites within a given landform, be it littoral or interior, potentially provide a more durable conceptual framework for such models (Butzer 2008).

5.2.1 Archaeological and Cultural Resources Associated with Coastal Caves and Karst

Similar to their role as repositories of biological materials by providing a protective matrix for faunal and floral remains, cave sediments also serve the same function for archaeological materials in both inland and coastal contexts, ideally preserving invaluable chronological records of historic and contemporary human uses (Goldberg and Sherwood 2006; Hassan 1978). These physical manifestations of past anthropogenic cave uses take many forms as assemblages of vertebrate remains (both human and faunal), paleobotanical and inorganic traces, lithics, shell materials, ceramics, and rock art (Fig. 5.6).

Cultural materials are keys to understanding past anthropogenic exploitation of resources, historic and contemporary utilitarian and ritual/ceremonial uses and specific trade/industries in the evolution of coastal cultures. The practice of human burials in caves, for example, appears to be a widespread archaeological and historical trend manifested in numerous coastal areas (Anton and Steadman 2003; Fitzpatrick and Nelson 2008; Schaffer et al. 2012). Preserved human remains in coastal caves have offered glimpses into the evolution of social structure, the impact of changes in environmental conditions and paleopathology – defining the occurrence of specific human diseases within past human settlements and their role in shaping cultural progression.

Modeling past cultural cave uses from physical remnants has seen some success yet many past studies relied on limited lines of material evidence or an assemblage of materials often in the absence of a detailed correlation with their physical site context (Bonsall and Tolan-Smith 1997; Prufer 2005), i.e. including both cave structure and surface features of the surrounding landform. As Stone (2005: p. 210) concluded: "*in terms of artifact distribution at a given site, caves represent an ideal laboratory in which spatial analysis can illuminate specific ritual behaviors.*" For example, the distribution of rock art forms at a given site and within the broader context of its enclosing coastal landform can be used to potentially illuminate the complexities of past anthropogenic cave uses (Pastoors and Weniger 2011). The following sections delve into the cultural significance of rock art in coastal settings.

5.2.2 Rock Art in Coastal Caves

Examples of cave rock art, or images created on the canvas of rock surfaces within caves, have been documented worldwide in an ever-expanding inventory of karst settings. These graphic expressions of past cultural landscapes are recorded across a wide range of continental and island coastal landforms: in sea caves, flank margin caves, talus caves, fissures and fluvial structures and exposed surface sites. Coastal examples include caves formed within uplifted Pleistocene reef terraces of Curaçao and the eroded diorite boulders of Aruba (Hummelinck

Fig. 5.3 Coastal cave archaeologic site example (**a**) Coastal landform of Rum Cay, Bahamas, formed within a transgressive eolian dune ridge. (**b**, **c**) Hartford Cave, Rum Cay. (**d**) Vandalism within petroglyph panels in Hartford Cave

1991) and many other shorelines in the Caribbean Basin (Hayward et al. 2009; Lace 2012; Stone 1995; Winters 2009) (Figs. 5.3, 5.4 and 5.5). While other more ephemeral cultural remnants have frequently been damaged or looted from archaeologic cave sites, rock art images, though equally fragile, have in many cases endured and often require distinct preservation strategies and assessment protocols (Dorn et al. 2008; Lee 1991; Sanz 2008).

Cave rock art occurs in many forms: manifested as variations of *petroglyphs* – images carved, pecked or scratched into the surfaces of cave walls or speleothems which were often subsequently modified with sand and water to further shape the incisions or embellished (i.e. "painted") with a variety of patinas. *Pictographs* are images applied to a rock surface with a similar array of materials including plant-based pigments, mud, bat guano, charcoal or a mixture of such materials. *Mud glyphs* are often considered yet another distinct form of rock art often appearing as sculpture composed of mud or sediments while *geoglyphs* are forms associated with surface sites and typically composed of arrangements of stone (Whitley 2001). Figures 5.4 and 5.5 offers a broad representative glimpse of the range of rock art forms documented and the techniques used to generate them in coastal caves and surface sites from a variety of settings. Though often enigmatic in their meaning, they offer tantalizing clues to past anthropogenic uses of cave environments. While an accepted consensus of a uniform classification system for rock art forms has yet to be realized, several attempts in regional settings have been attempted to date (Roe 2009). In general, rock art imagery can be classified as anthropomorphic, zoomorphic or geometric (abstract) and further partitioned into stylistic scales including a range of simple to elaborate (complex)

Fig. 5.4 Coastal surfaces and rock art application. (**a**, **b**) Beach rock deposits on the northern coast of the Puerto Rican mainland and associated petroglyph. (**c**) Anthropomorphic petroglyphs carved in coastal basalt, Hawaii

variations. However, while seriation of ceramic styles has proven effective in archaeological research, the effectiveness of similar chronological categorization of rock art motifs as a function of cultural development remains to be proven.

Traditional methods of recording rock art included scale drawings yet these representations alone were inadequate to define both the cultural expressions and geospatial context. Photographic documentation has long been a useful tool in recording both the spatial distribution and subtle detail of cultural assemblages and has evolved to incorporate more modern techniques such as infrared and digital 3D imagery (Chippendale and Tacon 1998; Whitley 2001). Rudimentary cave maps are still employed as an additional tool but many modern studies now incorporate more detailed representative maps of cave sites with higher geophysical information density to complement other lines of evidence (Figs. 5.6a and 5.7).

Archaeologic dating of rock art sites has often been a product of incorporating observed rock art forms with accompanying cultural materials (such as ceramics, shell material and/or lithics) in what is defined as a cultural assemblage. Dating

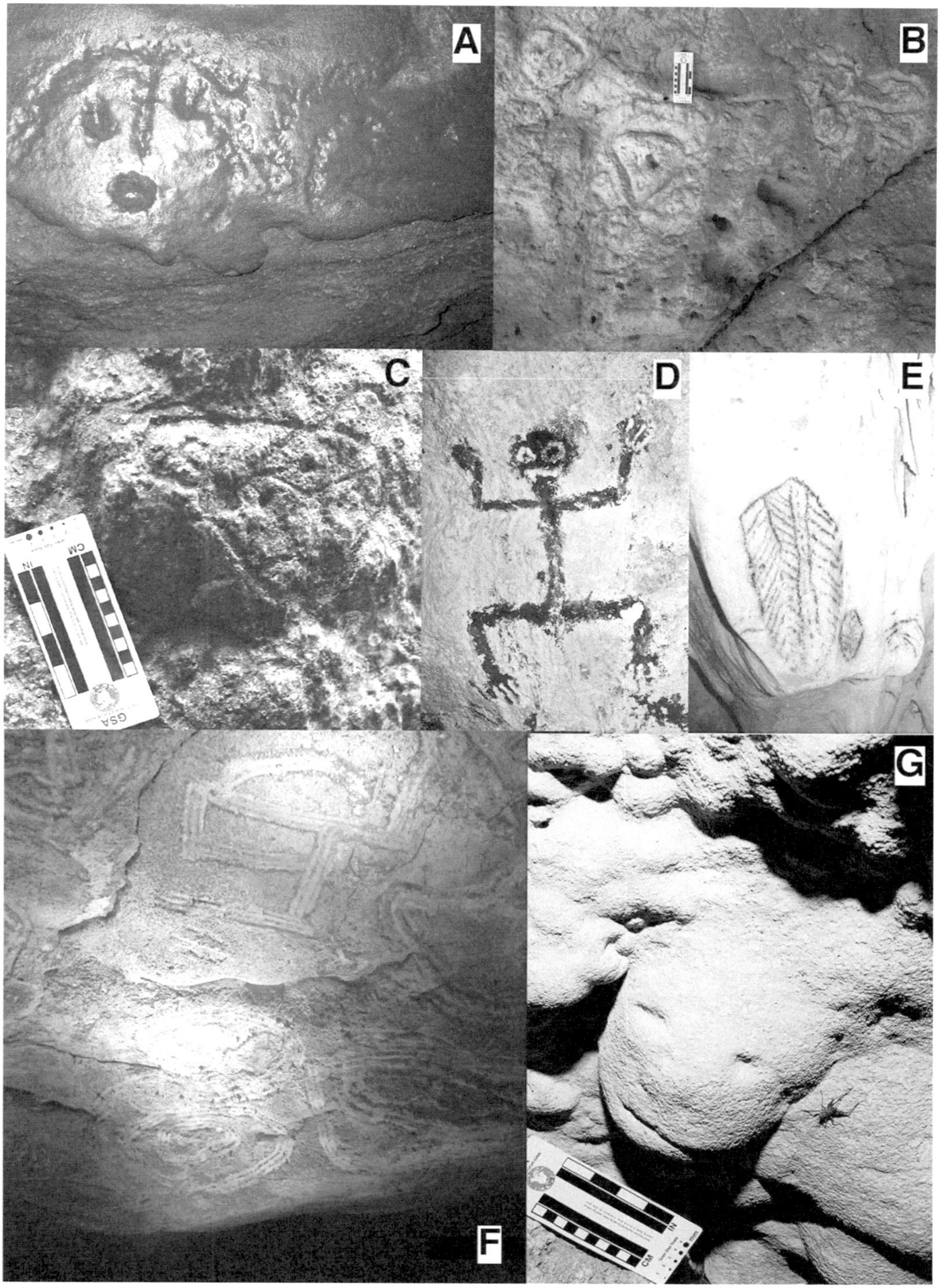

Fig. 5.5 (**a**) Coastal cave petroglyph (Barbuda). (**b**) Petroglyphs, Hato caves (Curaçao, Netherlands Antilles). (**c**) Petroglyph, cave on the north coast of Puerto Rico. (**d**) Mud Glyph in Cueva Balcones (Isla de Mona, Puerto Rico). (**e**) Pictograph in Cueva Ramos (Sancti Spritus, Cuba). (**f**) Images traced into soft corrosion residues (i.e. finger fluting), Isla de Mona, PR. (**g**) Petroglyph carved in relief, Springhead Cave, Barbados (note adjacent cave cricket – *right*)

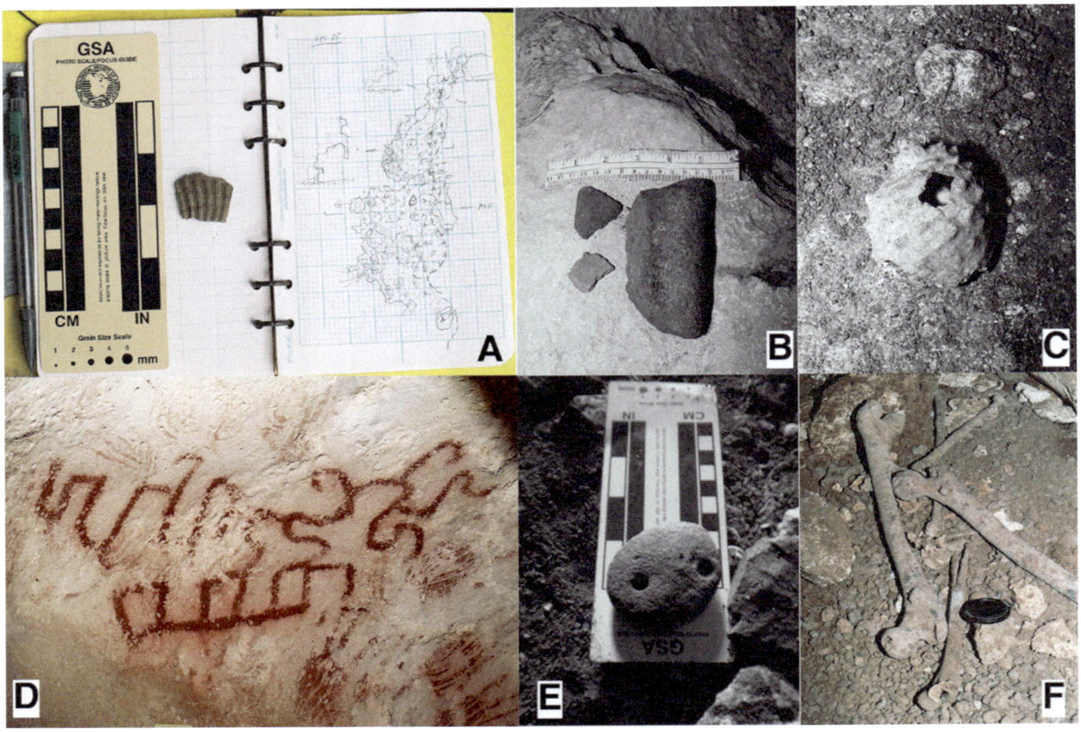

Fig. 5.6 (**a**) Documentation of archaeologic materials (left-potsherd; right-cave site survey), Curaçao. (**b**) Lithics and ceramics, Haiti. (**c**) Characteristic pre-contact utilization of marine resources, Golden Grove Cave, Barbados. (**d**) Handprints and geometric patterns in red clay on a cave ceiling, Aruba. (**e**) Carved lithic image ("cemi") and (**f**) Human skeletal remains in an unidentified coastal caves, Southern Peninsula of Haiti (note 35 mm lens cap for scale)

the age of rock art images themselves has historically been problematic due to destructive sampling of limited material suitable for available methods. Modern advances however have succeeded in dating trace quantities of resins in pigments or other organic materials associated with rock art application processes and secondary surface coatings though such techniques can prove problematic depending on the nature and condition of material and the dating techniques applied (Armitage et al. 2001; Pyatt et al. 2005; Pessis et al. 1999; Whitley 2001).

In contrast to the previous section defining certain cave types as offering more suitable environments for subterranean ecosystems, no clear pattern has emerged to identify one type of coastal cave structure or coastal surface over another in determining preferential archaeological or historical uses. Not surprisingly, use of such structures may be partially opportunistic but in many cases this is certainly not the sole determinant. Flank margin caves figure prominently in this distribution and likely due to their overall density in many coastal settings, as discussed in Chap. 4. However, in some areas, anthropogenic exploitation of specific natural resources within cave structures appears to be more of a determining factor in littoral settings. These can be manifested as economic exploitation of valued minerals or sediments at a given time to simple subsistence exploitation of endemic cave fauna or extraction of fresh water collected in some caves for either practical or ritualistic applications, as discussed below. Within a modern context, exploitation and modification of coastal karst may be also influenced by economic development, often on industrial scales as discussed in the next chapter (Chap. 6).

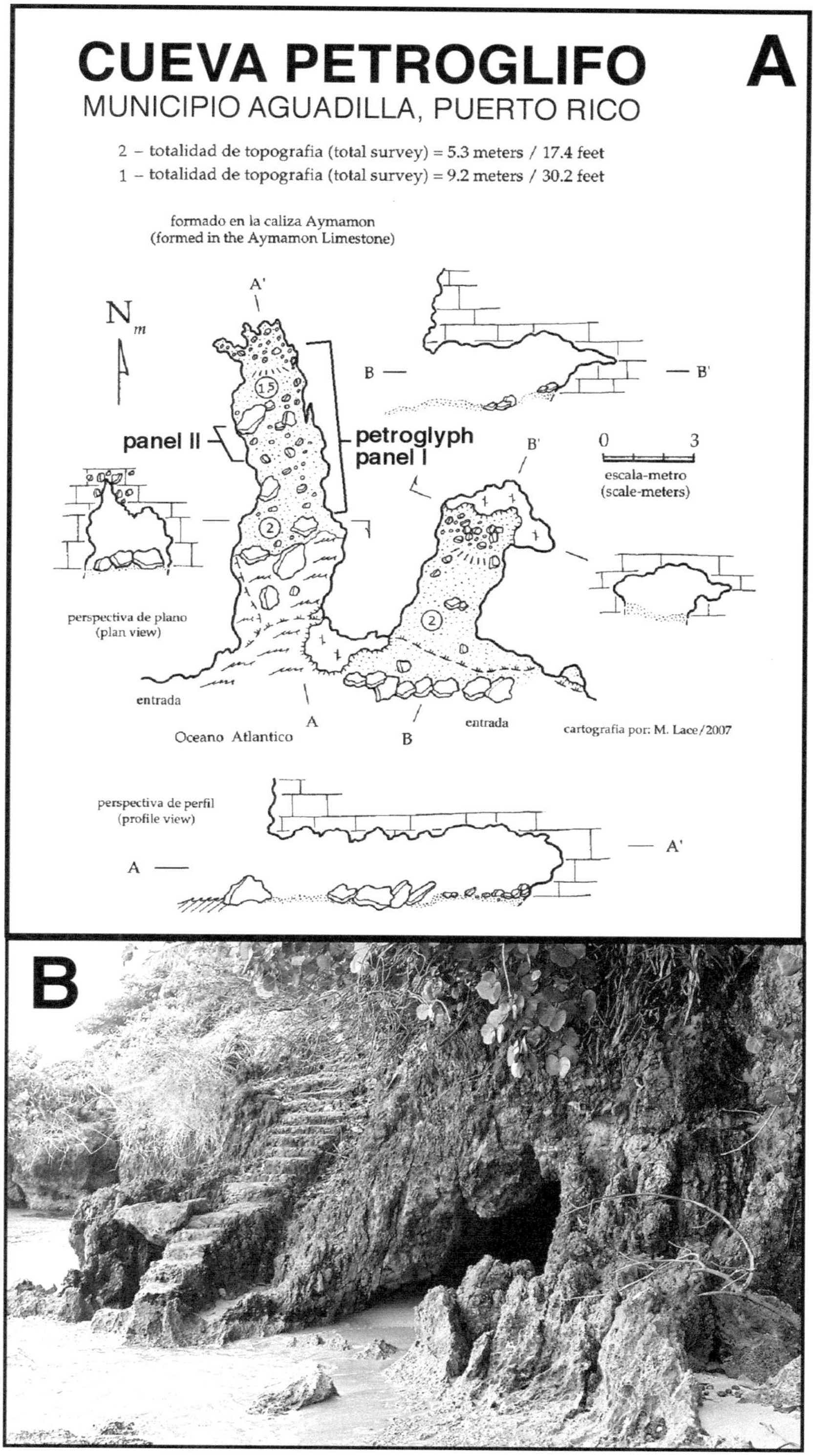

Fig. 5.7 Cartographic and photodocumentation of a coastal cave rock art site, Puerto Rico. (**a**) Map of cave structure. (**b**) Anthropogenic modification of landform surface

5.2.3 The Association of Cave Rock Art with Specific Human Activities

The opportunistic occurrence of cave sites with suitable surfaces selected for rock art placement would obviously play a role whether it be in coastal or interior island settings. For example, beachrock exposures or eolianitic dune surfaces adjacent to coastal settlements, the coastal basalts of the Hawaiian Islands (Fig. 5.4), ubiquitous volcanic surfaces in the Lesser Antilles (Hayward et al. 2009), to durable cave surfaces of sedimentary rock (Fig. 5.5a–g), in-cave flowstone (e.g. speleothems), and soft corrosion residues on cave walls and ceilings (Fig. 5.5f). The latter medium results from a structural cave feature, specifically the ubiquitous corrosional surfaces associated with this eogenetic karst setting and concomitant lithology of coastal caves on Isla de Mona, Puerto Rico (Chap. 9). Though not unique to the caves of Isla de Mona, these pliable corrosion surfaces provided a means of expression distinct from those available on the Puerto Rican mainland where such surfaces are less abundant and represents an interesting example of cave structure as a partial determinant in rock art expression and distribution (Lace 2012).

Additional environmental factors may have also influenced rock art placement, content and/or techniques and thus reflect varied and specific human uses of coastal cave sites. In tropical karst settings, for example, many of the dark zone areas harboring dense arrays of rock art are also associated with seasonal water catchments. Many small island platforms harbor a limited fresh water supply based on their specific hydrology, including the structure of the freshwater lens, and geomorphology (Chap. 3), thus the presence of water sources in what is now and/or has been considered in the past to be semi-arid environments may have conferred additional practical and/or ceremonial significance in site selection and rock art placement (Fig. 5.8).

As in other study areas within the Caribbean basin, such as the Yucatan peninsula, where access to adequate potable water sources was often limited, uses of cave water could have included both practical and ritual applications (Mercer 1896; Heyden 2005). Similarly, in the coastal caves of northern Quintanna Roo (Yucatan), research by the Yalahua Archaeological Cave Survey noted that the physical setting of this portion of the peninsula was distinct, where fresh water sources were comparatively more easily accessible than other areas. This permitted separation of ritual from utilitarian uses of cave water where special status was conferred to specific cave physiography. As Rissolo (Rissolo 2005, p. 365 and 362) described, "*in some caves, we encountered evidence of regular maintenance, and rock art marked the presence of many pools.*" And also noted, " … *the natural layout of caves has directed or channeled human interaction with cave space. This pattern is then reinforced by cultural modifications to cave environments.*" A scenario similar to that recorded on Isla de Mona where cave structure and seasonal water catchments in more remote areas may have included both ritual and practical uses by the native Taino (pre and post-contact) and indigenous cultures that preceded them. This pattern of anthropogenic modification of cave environments was then repeated in the nineteenth century when vast quantities of guano sediments were commercially extracted from caves on the island on a wide scale, resulting in significant alteration of these environments and again associated with historic graffiti of the period.

Preservation and management of cultural and biological resources within coastal karst settings presents unique challenges as discussed in greater detail in the next chapter. Both setting specific and broader regional management initiatives have been launched with varying success (Fig. 5.9), yet all require a clear understanding of specific coastal geomorphologies and patterns of speleogenesis in conjunction with in-depth field research that remains markedly incomplete in many coastal areas.

Fig. 5.8 (**a**, **b**) Rock art associated with natural water catchments in "dark zone" cave environments, Isla de Mona, Puerto Rico

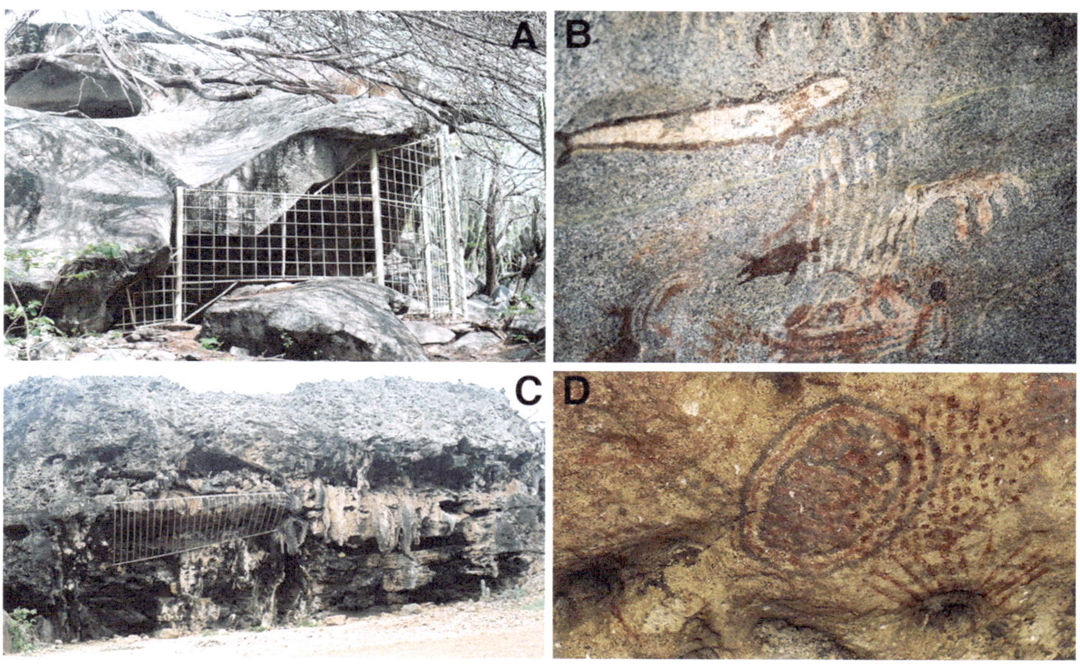

Fig. 5.9 (**a**, **b**) Protected rock art site within eroded diorite boulder (tafone), Arikok National Park, Aruba, Netherlands Antilles. (**c**, **d**) Rock art panel in cave remnant formed within uplifted carbonate reef terrace, Bonaire, Netherlands Antilles

References

Anton SS, Steadman DW (2003) Mortuary patterns in burial caves on Mangaia, Cook Islands. Int J Osteoarchaeol 13:132–146

Armitage RA, Brady JE, Cobb A, Southon JR, Rowe MW (2001) Mass spectrometric radiocarbon dates from three rock paintings of known age. Am Antiq 66(3):471–480

Baily G (2004) World history from the margins: the role of coastlines in human evolution. J Interdiscipl Stud Hist Archaeol 1(1):39–50

Baily G, Parkington J (1988) The archaeology of prehistoric coastlines. Cambridge University Press, Cambridge, UK

Balme J, Paterson A (2006) Archaeology in practice: a student guide to archaeological analyses. Blackwell Publishing, Malden, 438 p

Benjamin J, Bonsall C, Pickard C, Fischer A (eds) (2011) Submerged prehistory. Oxbow Books, Oxford, 352 p

Berman MJ, Gnivecki PL (1995) The colonization of the Bahamian Archipelago: a reappraisal. World Archaeol 36(2):422–441

Bicho NF, Haws JA, Davis LG (eds) (2011) Trekking the shore: changing coastlines and the antiquity of coastal settlement, International contributions to archaeology series. Springer, New York, 496 p

Bonsall C, Tolan-Smith C (eds) (1997) The human use of caves, BAR international series 667. Archaeopress, Oxford

Busch L (1994) A glimmer of hope for coastal migration. Science 263:1088–1089

Butzer KW (2008) Challenges for cross-disciplinary geoarchaeology: the intersection between environmental history and geomorphology. Geoarchaeology 101:402–411

Chippendale C, Tacon PSC (1998) The archaeology of rock art. Cambridge University Press, Cambridge/New York, 373 p

Collina-Girard J (1989) Prehistory and Coastal Karst area: Cosquer Cave and the "Callenques" of Marseilles. Karstologia 27:27–40

Culver DC, Pipan T (2009) The biology of caves and other subterranean habitats. National Speleological Society, Huntsville, 254 p

Culver DC, Pipan T, Schneider K (2009) Vicariance, dispersal and scale in the subterranean fauna of karst regions. Freshwater Biol 54:918–929

Daenekas J, Iliffe TM, Yager J, Koenemann S (2009) *Speleonectes kakui*, a new species of Remipedia (Crustacea) from anchialine and sub-floor caves on Andros and Cat Island, Bahamas. Zootaxa 2016:51–66

Dorn RI, Whitley DS, Cerveny NV, Gordon SJ, Allen CS, Gutbrod E (2008) The rock art stability index: a new strategy for maximizing the sustainability of rock art. Heritage Manage 1(1):37–70

Fitzpatrick SM (ed) (2004) Voyages of discovery: the archaeology of islands. Praeger, London

Fitzpatrick SM, Keegan WF (2007) Human impacts and adaptations in the Caribbean Islands: an historical ecology approach. Earth Env Sci T R So 98:29–45

Fitzpatrick SM, Nelson GC (2008) From limestone caves to concrete graves: 3000 years of mortuary practice in Paluan Archipelago, Polynesia. Int J Osteoarchaeol 18:439–457

Fleming TH, Raley PA (2009) Island bats: evolution, ecology and conservation. University of Chicago Press, Chicago, 549 p

Ford B (2011) The archaeology of maritime landscapes. Springer, New York

Gamble DW, Dogwiler T, Mylroie JE (2000) Field assessment of microclimatology of tropical flank margin caves. Climate Res 16:37–50

Goldberg P, Sherwood SC (2006) Deciphering human history through the geoarchaeology of cave sediments. Evol Anthropol 15:20–36

Gusick AE, Faught MK et al (2011) Prehistoric archaeology underwater: a nascent subdiscipline critical to understanding early coastal occupations and migration routes. In: Bicho NF (ed) Trekking the shore: changing coastlines and the antiquity of coastal settlements, Interdisciplinary contributions to archaeology. Springer Science and Business Media, LLC, New York, pp 27–50

Hassan FA (1978) Sediments in archaeology: methods and implications for paleoenvironmental and cultural analysis. J Field Archaeol 5(2):197–213

Hayward MH, Roe PG, Cinquino MA, Alvarado-Zayas PA, Wild KS (2009) Rock art of the Caribbean. University of Alabama Press, Tuscaloosa

Hendey QB, Volman TP (1986) Last interglacial sea levels and coastal caves in the Cape Province, South Africa. Quaternary Res 25:189–198

Heyden D (2005) Rites of passage and other ceremonies in caves. In: Brady J, Prufer K (eds) The Maw of the earth monster: mesoamerican ritual cave use. University of Texas Press, Austin, pp 21–34

Hummelinck PW (1991) De Rockstekeningen van Aruba, the prehistoric rock drawings of Aruba. Uitgeverij Presse-Papier, Utrecht, 228 p

Hutson AM, Mickleburgh SP, Racy PA (2001) Microchiropteran bats. International Union for Conservation of Nature and Natural Resources (IUCN), Cambridge, UK, 272 p

Iliffe TM, Hart CW, Manning RB (1983) Biogeography and the caves of Bermuda. Nature 302:141–142

Johnson LL, Stright M (1991) Paleoshorelines and prehistory: an investigation of method. CRC Press, Boca Raton, 256 p

Kano Y, Kase T (2008) Diversity and distributions of the submarine-cave Neritiliidae in the Indo-Pacific (Gastropoda: Neritimorpha). Org Divers Evol 8:22–43

Keegan WF, Fitzpatrick SM, Sealey KS, Lefebvre MJ, Sinelli PT (2008) The role of small islands in marine subsistence strategies: case studies from the Caribbean. Hum Ecol 36:635–654

Kornicker LS, Iliffe TM (1995) Ostracoda (Halocypridina, Cladocopina) from an anchialine lava tube in Lanzarote, Canary Islands. Smithson Contr Zool 568:1–40
Lace MJ (2012) Anthropogenic use, modification and preservation of coastal cave resources in Puerto Rico. J Isl Coast Archaeol 7(3):378–403
Lee G (1991) Rock art and cultural resource management. Wormwood Press, Calabasas, 72 p
Maire R, Tourte B, Jailet S, Despain J, Lans B et al (2009) Geomorphic and archaeologic features of coastal caves in the Madre de Dios Archipelago, (Patagonia, Chile). In: Proceedings of international congress of speleology, International Union of Speleology, I. Kerrville, TX, pp 516–521
Martin JB, Wicks CM, Sasowsky ID (eds) (2010) Hydrogeology and biology of post-Paleozoic carbonate aquifers. Proceedings of the symposium on karst frontiers, Karst Waters Institute, special publication no. 7. Karst Waters Institute, Leesburg, 212 p
Mason OK (1993) The geoarchaeology of beach ridges and cheniers: studies of coastal evolution using archaeological data. J Coastal Res 9(1):126–146
Mercer HC (1896) The hill caves of Yucatan: a search for evidence of man's antiquity in the caverns of Central America. Zephyr Press, Teaneck, 183 p
Miracle P (2005) Late mousterian subsistence and cave use in Dalmatia: the Zooarchaeology of Mujina Pecina, Croatia. Int J Osteoarchaeol 15:84–105
Mylroie JE, Mylroie JR (2007) Development of the carbonate island karst model. J Cave Karst Stud 69: 59–75
Mylroie JE, Mylroie JR (2009) Caves as geologic indicators, Kangaroo Island, Australia. J Cave Karst Stud 71:32–47
Mylroie JE, Mylroie JM (2011) Void development on carbonate coasts: creation of anchialine environments. Hydrobiologia 667:15–32
Mylroie JE, Carew JL, Moore AI (1995) Blue holes: definition and genesis. Carbonate Evaporite 10(2): 225–233
Mylroie JE, Mylroie JR, Nelson CN (2008) Flank margin cave development in telogenetic limestones of New Zealand. Acta Carsologica 37:15–40
Northup DE, Lavoie KH (2001) Geomicrobiology of caves: a review. Geomicrobiol J 18:199–122
Pastoors A, Weniger G-C (2011) Cave art in context: methods for the analysis of spatial organization of cave sites. J Archaeol Res 19:377–401
Pessis A-M, Strecker M and Bahn P (1999) Dating and the earliest known rock art. Oxbow Books Limited Oxford, 200 p
Petit S, Pors L (1996) Survey of columnar cacti and carrying capacity for nectar – feeding bats on Curacao. Conserv Biol 10(3):766–775
Prufer KM (2005) In: Brady J, Prufer K (eds) Maw of the earth monster: mesoamaerican cave use. University of Texas Press, Austin
Pyatt FB, Wilson B, Barker GW (2005) The chemistry of tree resins and ancient rock paintings in the Niah Caves, Sarawak (Borneo): some evidence of rain forest management by early human populations. J Archaeol Sci 32:897–901
Reid BA (ed) (2008) Archaeology and geoinformatics: case studies from the Caribbean. University of Alabama Press, Tuscaloosa, 234 p
Rissolo D (2005) Beneath the Yalahua: emerging patterns of ancient Maya ritual cave use from northern Quinatana Roo, Mexico. In: Brady J, Prufer KM (eds) In the maw of the earth monster: mesoamaerican cave use. University of Texas Press, Austin
Rodríguez-Durán A (2009) Bat assemblages in the West Indies: the role of caves. In: Fleming TH, Racey P (eds) Island bats: evolution, ecology, and conservation. University of Chicago Press, Chicago, pp 265–280
Roe PG (2009) The mute stones speak: the past present and future of Caribbean rock art research. In: Hayward M, Atkinson L, Cinquino MA (eds) Rock art of the Caribbean. University of Alabama Press, Tuscaloosa, pp 198–239
Romero A (2009) Cave biology: a life in darkness. Cambridge University Press, Cambridge, UK/New York, 291 p
Rouse IC (1986) Migrations in prehistory: inferring population movement from cultural remnants. Yale University Press, New Haven, 202 p
Sanz N (ed) (2008) Rock art in the Caribbean: towards a serial transnational nomination to the UNESCO world heritage list. UNESCO world heritage papers. Paris, no. 24
Sasowsky ID, Mylroie JE (2004) Studies of cave sediments: physical and chemical records of paleoclimate. Springer Publishing, Dordrecht, 329 p
Schaffer WC, Carr JS, Pateman MP (2012) Lucayan-Taino burials from Preacher's Cave, Eleuthera, Bahamas. Int J Osteoarchaeol 22(1):45–69
Shackleton JC, van Andel TH, Runnels CN (1984) Coastal paleogeography of the central and western Mediterranean over the last 125,000 years and its archaeological implications. J Field Archaeol 11(3): 307–314
Stone AJ (1995) Images from the underworld: Naj Tunich and the tradition of Maya cave painting. University of Texas Press, Austin, 304 p
Stone AJ (2005) A cognitive approach to artifact distribution in caves in the maya area. In: Brady J, Prufer KM (eds) In the maw of the earth monster: mesoamaerican cave use. University of Texas Press, Austin
Whitley DS (ed) (2001) Handbook of rock art research. Left Coast Press, Walnut Creek, 863 p
Wilkens H, Iliffe TM, Oromi P, Martinez A, Tysall TN, Koenemann S (2009) The corona lava tube, Lanzarote: geology, habitat diversity and biogeography. Mar Biodivers 39:155–167
Wilson WL (1994) Morphology and hydrology of the deepest known cave in the Bahamas: Dean's Blue Hole, Long Island. In: Boardman MR (ed) Seventh symposium on the geology of the Bahamas. Bahamian Field Station, San Salvador Island, p 21

Winters JH (2009) Rock art of the Bahamian Archipelago. In: Hayward MH, Atkinson LE, Cinquino MA (eds) Rock art of the Caribbean. University of Alabama Press, Tuscaloosa, pp 13–21

Woodward JC, Goldberg P (2001) The sedimentary records in Mediterranean rockshelters and caves: archives of environmental change. Geoarchaeology 16(4):327–354

Yuwono JSE (2009) Late Pleistocene and mid-Holocene coastal and inland interaction in the Gunung Sewu Karst, Yogyakarta. IPPA Bull 29:33–44

Coastal Cave and Karst Resource Management

6

Michael J. Lace and John E. Mylroie

Abstract

Human impacts to coastal karst landscapes encompass a broad range of examples, including small and large scale construction, resource mining, and recreational and agricultural uses. The geographic definitions of coastal zones vary widely from one area to another. Coastal resource management, preservation and restoration efforts and their respective operational plans, are equally and understandably as diverse and in some instances overlapping or conflicting in their intended scope, legal frameworks and practical application. The very structure of coastal karst and associated carbonate aquifers introduces unique problems in terms of geologic stability, extraction of hydrocarbon reserves, environmental preservation, and water resource quantity and quality. These problems incorporate additional complexity to the definition of physical boundaries, resource documentation, and the design and implementation of management plans. Many of the case studies presented in Part II illustrate some of the regional karst resource management approaches applied to specific littoral settings. This chapter examines several examples of coastal karst resource utilization and outlines potential models for predicting geologic stabilities, identifying sustainable land uses and approaching preservation challenges associated with such landforms.

6.1 Coastal Karst Resource Management Approaches

The benefits and challenges of implementing a universal system for systematically classifying coastal areas have been discussed in other venues (Finkl 2004; Brommer and Bochev-van der Burgh 2009) and together with widespread coastal management and restoration approaches (Clark 1996; Green 2009; McConney et al.

M.J. Lace (✉)
Coastal Cave Survey, West Branch, IA 52358, USA
e-mail: michael-lace@uiowa.edu

J.E. Mylroie
Department of Geosciences, Mississippi State University, Mississippi State, MS 39762, USA
e-mail: mylroie@geosci.msstate.edu

M.J. Lace and J.E. Mylroie (eds.), *Coastal Karst Landforms*, Coastal Research Library 5, DOI 10.1007/978-94-007-5016-6_6, © Springer Science+Business Media Dordrecht 2013

Fig. 6.1 What is sustainable development in a coastal setting?

2003) will not be addressed in detail within this chapter. For a more comprehensive treatment of coastal hazard management, see the companion volume in this coastal research library series by Finkl (2013). Similarly, general approaches to cave and karst preservation, management and restoration are not addressed in detail but the reader is directed to seminal referenced works in these fields (Haslett 2009; Jones et al. 2003; Van Beynen 2011; Werker and Werker 2006). Yet, given the limited presence of proactive preservation in the face of increasing anthropogenic modification of many coastal zones, a discussion of the challenges faced in the design and implementation of effective models of coastal cave and karst resource management is warranted.

The following sections offer perspectives on sustainable resource management: (1) coastal karst resource documentation, (2) legal protections and (3) resource access, education and interpretation. The remainder of the chapter examines additional case studies and preventative modeling approaches specific to the potential pitfalls and consequences of land use patterns, small and large scale development and incumbent modification of coastal karst. As this chapter illustrates, the definition of sustainable land uses in coastal karst settings can prove problematic as they are often site-specific and may not readily conform to standardized criteria (Fig. 6.1).

Karst areas introduce a range of complexities to the management of coastal areas with considerations specific to these structures, including unique geologic stabilities and aquifer vulnerabilities (Bear et al. 1999) that directly influence resource preservation strategies (Fleury et al. 2007; Gillieson 1996). In spite of these complexities, coastal development in many areas continues and includes a wide range of land uses from minimal subsistence agriculture (Fig. 6.2a) to complex infrastructure and municipal scale expansion within the context of complex coastal geomorphologies (Fig. 6.2b). For additional examples of coastal modification and management issues, readers are also directed to the companion volume (no. 3) in this coastal research library series by Cooper et al. (2012) for a more expansive discussion of shorelines within this context. As discussed in Chap. 4, mechanisms of void development (e.g. cave formation) specific to coastal settings generate distinctive structures in abundance. Dissolutional voids in this setting comprise not only those that have been revealed by cliff retreat or platform erosion but also voids that are not visible

Fig. 6.2 (**a**) Basic construction and agricultural use on an uplifted karstic terrace on the Northwestern coast of Haiti. (**b**) Commercial construction overlying a karst limestone escarpment of the north-central coast of Puerto Rico

or quantifiable in the absence of remote sensing techniques, rendering the management of such unseen coastal resources problematic.

6.1.1 Resource Documentation

As with any management plan, reconciling the physical and conceptual boundaries of a coastal resource is a critical first step, preceding detailed resource documentation and the design and implementation of successful initiatives. Defining the geographic boundaries of a coastal karst landform can be approached by three distinct, and often non-overlapping, perspectives: (1) geological/geomorphological; (2) geoarchaeological/cultural; and (3) coastal zone resource management. As discussed in Chap. 4, defining the relevant geomorphologic boundaries of coastal resources, for example, island structure, platform area and associated hydrogeology, has direct implications in modeling the sequence of geologic events that shaped its past and present form. Similarly, the previous chapter illustrated that the very definition of what constitutes a coastal landform influences our perceptions of past anthropogenic uses of coastal areas as well as associated geoarchaeological interpretations similarly can influence present coastal resource management and preservation approaches applied to such cultural sites. Thus, cultural landscapes in a coastal setting may not readily conform to either a geologic delineation of the landform structure or the coastal zone management boundaries of a given protected area and vice versa. Clearly, an integrated conceptual view of a coastal landform that incorporates multiple perspectives should prove useful but in practice is infrequently applied to coastal settings. Similarly, combining a range of systematic geophysical mapping techniques can prove effective in coastal karst modeling (ESRI 2007), such as LIDAR, microgravity assays, GPR and standard field mapping methods as shown in the following examples. Such methods can further support long term resource monitoring as another key component of karst resource management.

6.1.2 Legal Protections

In many coastal settings, government agencies are charged with the responsibility of managing both terrestrial and marine resources. A wide range of legal statutes are utilized as a framework to support specific cave resource management approaches. In some instances, practical necessity has prompted many government agencies to outsource management of protected resources to private foundations or land trusts – institutions which in some cases can offer greater continuity, broader management capabilities and at times have proven less susceptible to economic limitations or shifting political currents faced by publicly funded government initiatives. Examples of NGOs engaging in coastal karst management include the Bahamas National Trust, which oversees a wide range of natural resource areas (including karst). In parallel, the Antiquities, Monuments and Museums Corporation (AMMC) which oversees Bahamian cultural resources, often associated with caves. Such partitioned management is not uncommon with separate governmental ministries charged with environmental or cultural resources, respectively, even though both may employ similar protection strategies applied to the same coastal zone. The Republic of Haiti utilizes multiple ministries to manage aspects of environmental resources with several standing legal statutes. As in many countries however, continuity of legal enforcement of such statutes through political transition has proven challenging in the past.

Independent private agencies can also play an important role. The Carmabi Foundation (Caribbean Marine Biology Institute) is an NGO contracted to manage significant karst resources on the island of Curaçao as well as engaging in coral reef health monitoring in the broader region. The Puerto Rico Land Trust (as discussed in Chap. 9) has acquired and effectively manages a series of significant karst preserves in coastal areas and in the island interior. Thus, both developed and developing nations often share complex coastal zone management issues involving multiple agencies (private and governmental) charged with overseeing resources within the same karstic landforms, potentially leading to redundant or conflicting management approaches (Kueny and Day 2002).

6.1.3 Resource Access, Education/Interpretation and Preservation

The concept of caves as "underground wilderness" environments has gained traction in resource management in recent years, offering a more accurate definition of cave environments within a context that supports a range of preservation approaches. Cave management plans span a broad spectrum of preservation approaches, supporting a range of recreational cave uses (Figs. 6.3 and 6.4), scientific

Fig. 6.3 Recreational use of undeveloped or altered coastal cave sites. (**a**) Pristine cave site, Dalmation coast, Croatia. (**b**) Anthropogenic modification of a coastal cave (Cala Figuera, Mallorca). (**c**) Modification (note large excavated trench for building material extraction) and modern tourism use of a pre-ceramic aged archaeological site within a karstic eolian calcarenite dune (Cueva del Indio) on the northern coast of Puerto Rico

research and associated land uses specific to each coastal karst resource as determined by the respective management entities. Recent qualitative approaches to modeling sustainable land uses in karst environments also show promise in shaping effective long-term management of such complex landforms but an established, universal assessment protocol remains elusive (Watson et al. 1997; Van Beynen et al. 2012).

In sharp contrast to a pristine underground wilderness model, cave commercialization has been invoked (although controversial) in some karst preservation approaches (Gurnee 1967; Huppert et al. 1994), ranging from small-scale community-based ecotourism strategies incorporating cave and karst features to full scale recreational development. Commercial cave development is not restricted to any specific coastal landform or cave type and can take many

Fig. 6.4 Examples of commercial cave development. (**a**) Hawaiian Lava tube (Kaumana Caves), (**b**) Cueva Del Arte (Mallorca, Spain), (**c**) Hato Cave (Curaçao), (**d**) Grotte Azzurra (Capri) – ca 1900s postcard, (**e**) Yucatan cenote – Chaak Tun Caverns. (**f**) Harrisons Cave (Barbados) tram trail

forms (Fig. 6.4); examples include Grotte Marie-Jeanne (southern peninsula, Republic of Haiti), to moderate or large-scale commercial cave development, such as Green Grotto (Jamaica), Hato Cave (Curaçao), Rio Camuy Cave (Puerto Rico), Cueva Del Arte (Mallorca), Grotta Azzurra (Capri), Sea Lion Cave (USA – see Chap. 14) and Harrison's Cave (Barbados –

Chap. 10) as natural economic resources (in some cases comprising significant components of the national GDP) within broader sustainable management models (CRSTD 2008).

Cave commercialization has often proven effective in supporting resource education through interpretive programs stressing the complexities and importance of diverse cave ecosystems which in turn support future resource protection efforts. However, preservation caveats certainly apply as many sites with unrestricted or unmanaged visitor access have seen significant negative impacts that have proven devastating to archaeological and historical resources, cave biodiversity, hydrology and even significantly altering cave microclimate, resulting in a cascade of negative effects to cave environments and the species they support. Clearly, a universal resource management protocol may not be tenable as access to sensitive karst landscapes requires a balance between education, research and site preservation – specific to each setting and flexible enough to respond to an ever-changing status.

The following section explores ways of assessing potential pitfalls and remediation costs associated with coastal development on karst. It also illustrates the power and utility of expansive coastal cave inventories, derived from field exploration and refined by multidisciplinary analyses, as key resource management and karst modeling tools.

6.2 Modeling Geologic Instabilities in Coastal Karst

Coastal cliffs and terraces comprise a predominant portion of the world's coastlines (Emery and Kuhn 1982; Migon 2010). Coastline changes are the result of a dynamic interplay of natural and anthropogenic influences that are intimately linked to geological processes and climate change in coastal settings (Fitzpatrick et al. 2006; Harff et al. 2007). Consequently, many carbonate coastal zones can benefit from detailed karst resource and stability assessments in the face of expanding residential and commercial development on dynamic littoral landscapes.

As with karst in continental areas, land use issues in island karst are centered around three principal factors: (1) landform stability; (2) water quantity and quality; and (3) environmental preservation.

6.2.1 Landform Stability

As presented in Chaps. 3 and 4, there are major differences between island karst and continental karst. In terms of land stability, carbonates in tropical and subtropical locations are commonly eogenetic, with less diagenetic maturity, and therefore potentially less strength, than telogenetic carbonates in continental interiors. The development of dissolutional voids in the subsurface of islands, which can constitute a collapse or subsidence risk, follows different rules than for continents. In many cases, such as flank margin caves, the location of caves can be reasonably predicted, especially if the coastal sea-level history is known so that the site of past fresh-water lens margins can be identified. In the Bahamas, the development of banana holes in last interglacial strand plains creates a situation where the region of risk can be identified by geology (specifically, the strand plain facies location), but the site-specific risk within that region can be difficult to establish (where is the actual roofed banana hole that has not yet collapsed?). Large progradational collapse features such as blue holes in the Bahamas, indicate that very large collapse events can occur, and that their location is extremely difficult to predict in advance. Given that it is now believed the majority of blue holes (away from the bank margin) are collapses into conduit flow systems at depth in the Bahama Banks (Chap. 4), in this case the planning and prediction of karst collapse more closely resembles that for continental interior karst areas.

Geophysical techniques have been used in the Bahamas with varying degrees of success. Microgravity was demonstrated to be capable of locating small, shallow voids (Kunze and Mylroie 1991), however the technique is labor intensive in the field, especially as a reconnaissance tool as survey lines must be cleared and measured. Better success has occurred with ground-penetrating radar (GPR), as many caves are shallow, and the youth of the Bahamian limestones means there is minimal soil and clay to attenuate the signal (Wilson et al. 1995). Wilson et al. (1995) also developed a karst hazard report for San Salvador Island, in an attempt to quantify the geologic hazards presented by karst features. The report initially characterized the landforms of the island as plains, beaches, ridges, and inland water bodies. The location of known karst features was then catalogued as to landform specificity, and as to median size. From this work, the authors were able to generate a subsidence and collapse risk factor.

As noted in Chap. 4, sinkhole development, and therefore their collapse potential, is different in eogenetic carbonate islands from that found in continental interiors in telogenetic rocks. In the Bahamas, with thin to no soil, cover collapse sinkholes are almost non-existent. However, Barbados has a thick soil cover over its extensive limestone surface, and cover collapse sinkholes can occur there. In both localities, cave collapse sinkholes are common, especially true for the stand plain banana hole country of the Bahamas. True dissolutional sinkholes are also rare in the Bahamas, as the youth and eogenetic nature of the limestone has lead to pit cave development instead; pit caves could be considered an exaggerated mode of sinkhole development, where the dissolution is mostly downward and very little laterally. On Bermuda (Mylroie et al. 1995), vadose flow on the limestone/basalt contact created meandering cave streams that undermined the limestone, leading to numerous large progradational collapse features. The key point here is that based on the volume of meteoric recharge, a large conduit should not have developed. But because the vadose flow was perched, a relatively small stream, as it meandered, could create a broad void span that was mechanically prone to failure. Once collapse began, the vadose stream was able to dissolve and mechanically transport away the accumulated collapse debris, maintaining accommodation space so that the collapse could continue to prograde upward (collapsed rock takes up 40 % more space than the rock did when in place, so collapse generally ceases unless the surface has been reached or material is removed, see White (1988) for a full discussion of cave collapse). Large closed contour depressions, as discussed in Chap. 4, are mostly constructional in origin.

Hydrologically, the epikarst inflow in eogenetic carbonates is dispersed across the entire carbonate outcrop, and not specifically directed to joints as in continental telogenetic limestones. If a significant soil cover exists, predicting the meter scale bedrock topography can be quite difficult. Loading such a landscape with a building structure, in the absence of proper site investigation, can result in differential compaction and building failure (Fig. 6.5).

As noted by the CIKM, karst development is very island specific. Knowledge of the island's geologic history is critical to risk assessment. The age and strength of the rock, the amount of soil, the development of vadose flow along the limestone/non-limestone contact, the tectonic rate, are all important factors to determine where the voids are, and what risks they represent.

6.2.2 Water Resource Quantity and Quality

Water quantity. As described in Chap. 3, flow within a fresh-water lens is quite different from the aquifer setting in continental interiors. The presence of underlying and adjacent marine water makes the fresh-water lens vulnerable to salt water contamination. This problem is true for all marine coasts and islands in any type of rock or sediment material. In carbonates, the development of dissolutional flow paths can accentuate the difficulties in utilizing the resources of the fresh-water lens (Farrell and Boyce 2007).

Two simple examples can provide perspectives on the pitfalls of water extraction in islands. In most continental aquifers, if one had a water-producing interval of 20 m, one would drill and set the extraction pipe close to the bottom of the aquifer, so as to be able to extract the greatest amount of water. If this approach is taken with

Fig. 6.5 Road cut, northern Puerto Rico, as seen in 1992 when newly made. The overlying grass-covered surface is a uniform slope, but the road cut reveals that the bedrock beneath has a jagged form produced by karstification, with a terra rossa soil overlying it. That soil will compact on the right side, while the bedrock to the left will not, leading to differential compaction if the site was loaded by a building or other construction. Note that test boring could strike pinnacles and not troughs (or vice versa), creating a false impression of depth to bedrock

a fresh-water lens, marine water from below is almost immediately taken up the well, and the aquifer is salt contaminated. Less obvious is what happens if one drills only a few meters into the fresh-water lens, so as to avoid the previous problem. As the well is pumped, a depression in the water table, called a *cone of depression*, is formed. This cone of depression creates the slope that delivers water from the farther reaches of the aquifer to the well. The problem is that the fresh-water lens is floating as a buoyant body on the underlying marine water. Because the difference in density of fresh water compared to marine water is one part in 40, for each meter downward a cone of depression forms, the underlying marine water rises up 40 m, a process called *upconing* (see Chap. 3 for details). Successful extraction of water from the fresh-water lens requires skimming the top of the lens at multiple locations and multiple times to prevent upconing. On Guam, where the northern part of the island acts as a carbonate cover island, wells are sunk into the part of the fresh-water lens that has ramped up onto the volcanic basement, a region called *parabasal water*, which can be strongly pumped as there can be no upconing from the volcanic rocks below (Jocson et al. 2002).

In carbonate rocks, dissolution through time results in the aquifer undergoing self-modification, permeability increases even as net porosity decreases, as the remaining porosity is organized into flow paths called touching vug permeability (Vacher and Mylroie 2002). The longer a carbonate rock has held a fresh-water lens, the more permeable the lens becomes. Initially this increase might seem like a good thing, the better aquifers tend to be the more permeable ones. In carbonate islands, however, this is a bad thing. As explained in Chap. 3, a fresh-water lens forms because for meteoric recharge to reach the sea, a gradient must be established from the island interior to the sea. If the aquifer has low permeability, this slope will be steep so as to drive the water coastward. If the aquifer is highly permeable, the slope will be very gentle. Because of the buoyant nature of the fresh-water lens, the more the water piles up to make a steep gradient, the more the lens extends below sea level in the 40–1 ratio. Therefore, as the carbonates hosting the fresh-water lens become more permeable, the lens thins over time. Storage is lost, and less water is available for extraction. In the Bahamas, the thickest fresh-water lens is commonly in Holocene sand bodies near the coast (Wallis et al. 1991), because although the porosity is high, commonly over 30 %, it is not organized as in the adjacent older Pleistocene rock, and

Fig. 6.6 (**a–d**) Fais Island. (**a**) Outcrop of Late Cenozoic reef facies, flashlight 12 cm high for scale. (**b**) Aerial view of a portion of Fais Island, looking southwest; the impermeable reef flat, and the rock peninsulas that cross the reef flat visible to left and right. (**c**) The west – facing cliff of the peninsula to the *left* in image **b**, showing uplifted flank margin caves. Person in *white oval* for scale (**d**) Fossil spur and groove topography that mimics karst. Note person in *white oval* for scale

the lens is consequently thicker, despite its near-coast position. On a siliciclastic island, such as Nantucket, dissolutional modification does not occur over time, but in the Bahamas and Bermuda (Vacher 1988), the older the carbonate rock, the thinner the fresh-water lens.

Fais Island is part of the Kingdom of Yap, within the Federated States of Micronesia in the western Pacific. The island is an uplifted atoll 220 km east of Yap, the only uplifted island in a large archipelago of atolls. The ~320 inhabitants use rainwater collection as their source of drinking and cooking water. Typhoons blow off the building roofs, and ENSO conditions lead to drought, both of which create a water crisis on the island. The work described here is presented in detail in MacCracken et al. (2007) and Mylroie et al. (2008).

Fais Island is 1.2 km wide and 2.9 km long, with a maximum elevation of 40 m, composed entirely of Late Cenozoic limestones (Fig. 6.6a). Because of the water resources issues for the island population, a study was done to determine if a karst analysis of the island would allow a better water recovery strategy than just a simple assumption of a uniform fresh-water lens. The work was done in late spring, when the maximum negative tides of the year are experienced, so that selective water discharge from the lens, if it occurred, would be identifiable. The island is surrounded by a well-cemented reef flat, except in a few areas where rocky peninsulas cross the reef flat and extend to deeper water (Fig. 6.6b). Field work demonstrated that uplifted flank margin caves were found only in the rocky peninsulas, and not in the back-beach cliff areas inland of the reef flat (Fig. 6.6c). Fresh-water discharge was also only found in these rocky peninsulas, directly beneath the areas where uplifted flank caves were found. The reef flat displayed almost no fresh-water discharge, indicating it was acting as an aquitard or aquiclude. The only exception was an embayment on the northeastern side of the island, filled with Holocene sand. As has been reported earlier, these Holocene sand bodies have a high porosity but an undeveloped permeability, and so hold water in a thick lens. This sand-filled embayment has a dug well lined with a stone wall, and the water present is fresh. It is located some distance from the island village, and the water was not considered as clean as rainwater, so this resource was not utilized. Long linear troughs on the island surface, considered anecdotally to be collapsed caves, were recognized as uplifted fossil spur and groove structures inherited from a lagoonal origin (Fig. 6.6d).

The model that was developed from these observations is that the fresh-water lens does not discharge uniformly to the coast, because of the inhibiting nature of the well-cemented reef flat. Where rocky peninsulas cross the reef flat, the rock provides a flow matrix that allows preferential fresh-water discharge. The uplifted flank margin caves are excellent proxies for modern lens flow, as they themselves had formed at an earlier time when they were at sea level and the lens discharged through them. These karst interpretations allowed for a better strategy to provide fresh water to the island population.

Water quality. Carbonate islands can have a myriad of flow systems, from simple diffuse flow through a porous matrix, as in the Holocene sands described above, or high permeability flow in touching vug conditions, as in older Pleistocene carbonates, or conduit flow within the carbonate mass, as in large banks, or a perched flow on underlying non-carbonates. In addition, if the rock has been fractured, those fractures can also provide efficient flow paths. These many possible flow paths create a major water quality issue in carbonate islands. The porous media flow will be slow, with maximum storage. The touching vug flow system will be moderate in flow velocity, with some storage. The conduit and fracture flow pathways move rapidly, but with low storage. If a contaminant is introduced into the fresh-water lens, where and when that contaminant reappears depends on which flow paths are taken (they are not mutually exclusive), and how the aquifer is being managed (will water wells draw flow to them?). Moran and Jenson (2004) were able to demonstrate on Guam how water flow from injection wells and sinkholes through the carbonate aquifer took different simultaneous pathways with very different arrival times and destinations. Davis and Johnson (1989) on San Salvador Island, Bahamas, found that the fresh-water lens was partitioned by interior hypersaline water bodies, and that favored flow paths had distorted the lens. These examples demonstrate that water quality issues on carbonate islands require a full understanding of the CIKM and the local geology (White et al. 2006). Contaminant issues are heightened by the fact that on islands, there are not many groundwater resource alternatives. One cannot run a pipeline to an adjacent aquifer. Watershed contamination, however is not limited solely to island settings as continental coastal settings face the same difficulties and require the same in-depth understanding of the karst hydrogeology specific to coastal aquifers (Bonacci and Roje-Bonacci 1997; Harmon and Wicks 2006; Metcalfe et al. 2010).

6.2.3 Environmental Preservation and Biodiversity

Islands can be near to continents, part of archipelagoes, or distant in the far ocean. They contain ecologies that have evolved in isolation, and the impact of human colonization has been devastating. First, humans place a stress on the existing fauna and flora; second, they import competing or exotic species that undermine the existing ecology, particularly in geophysically constrained island settings (see Chap. 12 for a good example). Caves can act as refugia for some species (as discussed in Chap. 5), and thereby assume significant ecological importance on islands and in submerged continental karst (Bakran-Petricioli and Petricioli 2008), requiring site-specific resource protections.

Fig. 6.7 (**a**, **b**) Impacts of coastal karst instability on Barbados. (**c**) Development on a karstic coastal terrace Barbados

6.3 Additional Examples of Karst Management and Resource Utilization in Coastal Settings

6.3.1 Barbados: The Arch Cot Collapse

Similar to other carbonate-overlain islands in the Caribbean region, Barbados is composed of series of well-defined uplifted reef terraces. Each of these carbonate escarpments has been found to be intensely karstic with dense dissolutional cave development expressed as a variety of cave structures. As similarly illustrated in the coastal limestone escarpments of Puerto Rico (Chap. 9), construction on such terraces on Barbados is widespread with the concomitant risk of geologic instability (Chap. 10). In August of 2007, a catastrophic collapse of a large cave segment directly beneath a residential area (Fig. 6.7) resulted in complete structural failure of a residence built above the main chamber and the tragic loss of a family of five. The immediate area has since been evacuated in the wake of continuing karst landscape instability.

Fig. 6.8 Natural Bridge of Aruba. (**a**) A residual cave and natural bridge structure. (**b**) Fissure development adjacent to collapse area. (**c**, **d**) Seaward and landward views of collapsed natural bridge

6.3.2 Aruba Natural Bridge: Structural Failure of a Coastal Landmark

Coastal karst landforms frequently function as important economic resources in terms of tourism development which is not surprising as coastlines the world over harbor some of the most stunning vistas. The Natural Bridge on the northern coast of Aruba (Andicuri Bay) was touted as one of the largest such structures in the Caribbean, spanning more than 40 m and standing 7 m above the energetic Atlantic surf; it continues to serve as a popular tourist destination. Formed within Pleistocene-aged coral limestone overlying loosely consolidated, weathered basalt, the structure collapsed on September 2nd, 2005 (Fig. 6.8). A similar collapse was also recently recorded at Waverly Beach Cave (New Zealand) in June of 2012. Fortunately, no injuries were associated with either event. A second Aruban bridge remnant (known as "The Baby Natural Bridge") adjacent to the former Natural Bridge remains intact however contemporary fissures (prominently visible in satellite imagery) have recently formed (Fig. 6.8b) within the segment of cliffline that once separated the two bridges, raising the potential of future instability and collapse events. This recurrent coastal structure (the "sea arch") has also been recorded on Curaçao (e.g. Sheta Boka), Barbados (Chap. 10) and other islands in the Caribbean region and elsewhere (Chap. 1). Evolution of this landform is consistent with emerging theories of coastal gully genesis (called "*bokas*" in the Netherlands Antilles and "*calas*" on Mallorca) which in many cases involve the complex interplay of fluvial incision intersecting preexisting dissolutional voids (or conversely, the creation of the incision allows mixing dissolution in the incision walls) formed by the freshwater lens within coastal limestone terraces (Fig. 6.9).

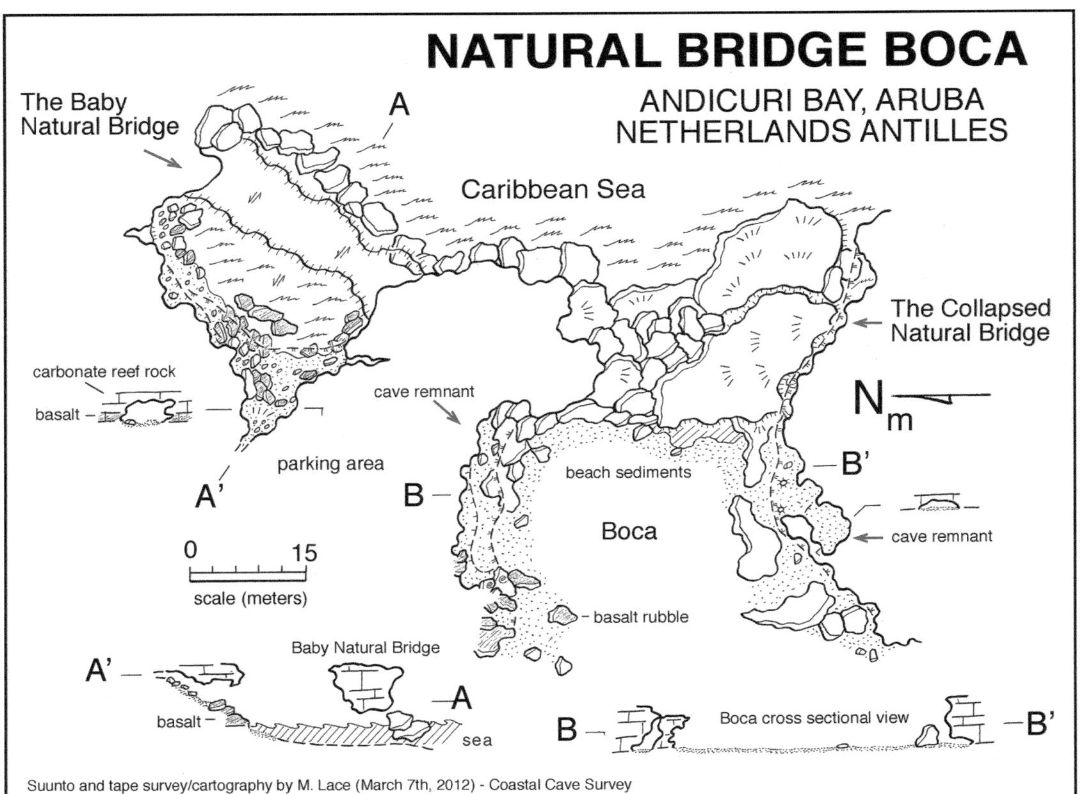

Fig. 6.9 Map of Natural Bridge (Aruba) and associated coastal landform

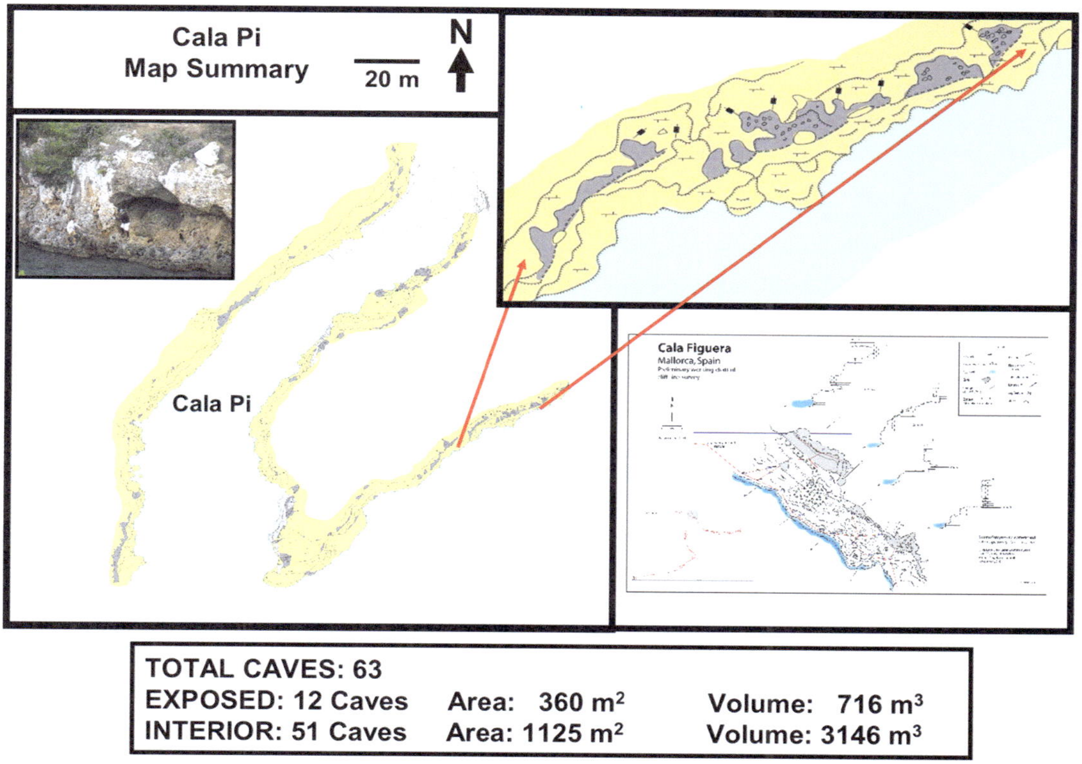

Fig. 6.10 Coastal landform mapping and quantitative density analysis of void development (Mallorca, Spain)

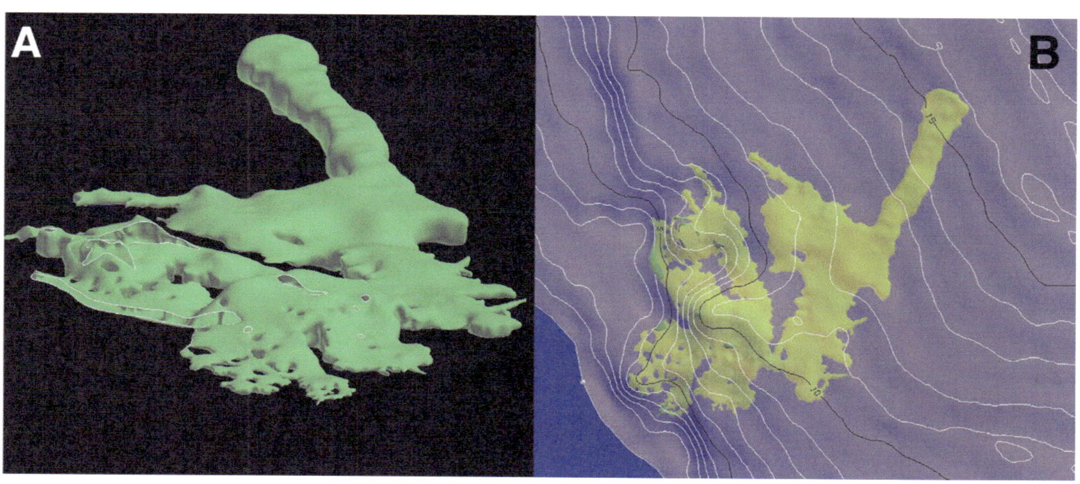

Fig. 6.11 (**a**) Computer generated 3D map of Salt Pond Cave, Long Island, Bahamas, with (**b**) a topographic contour overlay of associated coastal landform (Lascu 2005)

6.3.3 Mallorcan Coastal Karst: Applied Modeling of Void Development in Coastal Karst

The eogenetic coastal karst of the Balearic Islands is associated with complex patterns of speleogenesis, coastal aquifer structure and associated land uses (Chap. 13). In addition to geologic stability assessments, coastal karst landform mapping and quantitative analysis of associated void development can be applied to the study of speleogenetic processes (as discussed in Chap. 4) and to modeling petroleum reservoir structure and porosity preserved within deeply buried carbonates, as illustrated in the following example from the Mallorcan coast.

Analysis of cave development associated with protected coastal inlets (i.e. "calas") composed of Messinian-Tortonian carbonate cliffs was used to define speleogenetic controls within a differential coastal cliff retreat model (Mylroie et al. 2012). Several geophysical criteria were incorporated into a quantitative assessment of void development on varying scales (Fig. 6.10). The approach further supported 3D modeling of macroporosity and microporosity of preserved structures within deeply buried carbonate reefs associated with petroleum reservoirs, similar to mathematical modeling of flank margin cave development reported by Labourdette et al. (2007) and Lascu (2005) (Fig. 6.11). Thus, eogenetic coastal karst can serve as a quantitative model for diagenetically mature paleo-coastal structures with implications for petroleum resource utilization.

In summary, coastal karst landforms present significant and complex challenges to effective management, resource utilization and preservation. Inherent to any sustainable approach is the need for a detailed conceptual and practical understanding of coastal karst processes borne of detailed exploration and documentation of these evolving landscapes.

References

Bakran-Petricioli T, Petricioli D (2008) Habitats in submerged karst of the eastern Adriatic coast – Croatian national heritage. Croat Med J 49:455–458

Bear J, Cheng A, Sorek S, Oazar D, Herrera I (eds) (1999) Seawater intrusion in coastal aquifers: concepts, methods and practices, vol #14, Theory and applications of transport in porous media series. Springer-Verlag, New York

Bonacci O, Roje-Bonacci T (1997) Sea water intrusion in coastal springs: example of the Blaz spring, Croatia. Hydrolog Sci J 42(1):89–100

Brommer MB, Bochev-van der Burgh LM (2009) Sustainable coastal zone management: a concept for long-term and large-scale coastal evolution. J Coastal Res 25(1):181–188

Clark JR (1996) Coastal zone management handbook. CRC Press, Boca Raton, 694 p

Cooper J, Pilkey AG, Orrin H (2012) Pifalls of shoreline stabilization. Coastal research library series, vol 3. Springer Publishing, Dordecht, 131 p

CRSTD (2008) Development of a strategic business management model for the sustainable development of heritage tourism products in the Caribbean. Caribbean regional sustainable tourism development programme report, Barbados, 102 p

Davis RL, Johnson CR (1989) Karst hydrology of San Salvador. In: Mylroie JE (ed) Proceedings of the fourth symposium on the geology of the Bahamas. Bahamian Field Station, Port Charlotte, pp 73–81

Emery K, Kuhn G (1982) Sea cliffs: their processes, profiles and classification. Geol Soc Am Bull 93(7): 644–654

ESRI (2007) GIS for caves and karst. GIS best practices. www.esri.com/library/bestpractices/caves-karst, 51 p

Farrell D, Boyce S (2007) Water resources management in coastal aquifers in the Caribbean. Eos. Trans. AGU proceedings 88(23), Jt. Assembly Supplement (abstract HS1F-03)

Finkl CW (2004) Coastal classification: systematic approaches to considering the development of a comprehensive scheme. J Coastal Res 20(1):166–213

Finkl CW (2013) Coastal hazards, Coastal research library series. Springer, Dordrecht, 800 p, (in press). ISBN 978-94-007-5233-7

Fitzpatrick SM, Kappers M, Kate Q (2006) Coastal erosion and site destruction on Carriacou, West Indies. J Field Archaeol 31(3):251–262

Fleury P, Bakalowicz M, de Marsily G (2007) Submarine springs and coastal aquifers: a review. J Hydrol 339:79–92

Gillieson D (1996) Caves: processes, development and management. Blackwell Publishing, Malden, 324 p

Green DR (2009) Coastal zone management. Telford Publishing, London, 392 p

Gurnee RH (1967) Conservation through commercialization: Rio Camuy development project. B Natl Speleol Soc 29(2):27–71

Harff J, Hay WW, Tetzlaff DM (2007) Coastline changes: interrelation of climate and geological processes. Geol Soc Am Spec Pap 426, 214 p

Harmon RS, Wicks CM (eds) (2006) Perspectives on karst geomorphology, hydrology and geochemistry: a tribute volume to Derek C Ford and William B White. Geological Society of America, Boulder, 306 p

Haslett SK (2009) Coastal systems. Taylor and Francis Publishing, Routledge, 216 p

Huppert GH, Burri E, Forti P, Cigna A (1994) Effects of tourist development on caves and karst. Cave Conservationist 13(3):13–22

Jocson JMU, Jenson JW, Contractor DN (2002) Recharge and aquifer responses: Northern Guam lens aquifer Guam, Mariana Islands. J Hydrol 260:231–254

Jones WK, Hobbs HH, Wicks CW, Currie RR, Hose LD, Kerbo RC, Goodbar JR, Trout J (2003) Recommendations and guidelines for managing caves on protected lands, Special publication no. 8. Prepared for the National Cave and Karst Research Institute, National Park Service. Karst Waters Institute, 81 p

Kueny J, Day MJ (2002) Designation of protected karstlands in Central America: a regional assessment. J Cave Karst Stud 64(3):164–174

Kunze AWG, Mylroie JE (1991) Use of gravity techniques to detect shallow caves on San Salvador Island, Bahamas. In: Bain RJ (ed) Proceedings of the 5th symposium on the geology of the Bahamas. Bahamian Field Station, Port Charlotte, pp 139–149

Labourdette R, Lascu I, Mylroie JE, Roth M (2007) Process-like modeling of flank margin caves: from genesis to burial evolution. J Sediment Res 77: 965–979

Lascu I (2005) Speleogenesis of large flank margin caves of the Bahamas. M.Sc. thesis. Department of Geosciences, Mississippi State University, Mississippi State, 218 p. http://library.msstate.edu/etd/show.asp?etd=etd-05102005-132949

MacCracken RS, Jenson JW, Heitz LF, Rubenstein DH, Mylroie JE (2007) Water resources analysis of Fais Island, Federated States of Micronesia. Technical report no. 111, Water and Environmental Research Institute of the Western Pacific, University of Guam, p 41

McConney P, Pomeroy R, Mahon R (2003) Guidelines for coastal resource co-management in the Caribbean: communicating the concepts and conditions that favour success. Caribbean Conservation Association (CCA), 55p

Metcalfe CD, Beadows PA, Gold Bouchot G, Metcalfe TL, Li H, van Lavieren H (2010) Contaminants in the coastal karst aquifer system along the Caribbean coast of the Yucatan Peninsula, Mexico. Environ Pollut 159:991–997

Migon P (ed) (2010) Geomorphological landscapes of the world. Springer publishing, Dordrecht, 375 p

Moran DC, Jenson JW (2004) Dry trace of groundwater from Guam International Airport and Harmon Sink to Agana Bay and Tumon Bay. Water & Environmental Research Institute of the Western Pacific, University of Guam technical report 9

Mylroie JE, Carew JL, Vacher HL (1995) Karst development in the Bahamas and Bermuda. In: Curran HA, White B (eds) Terrestrial and shallow marine geology of the Bahamas and Bermuda, Geological Society of America, Boulder, special paper no. 300. Geological Society of America, Boulder, pp 251–267

Mylroie JE, Mylroie JR, Jenson JW, MacCracken RS (2008) Fresh-water lens anisotropy and flank margin cave development Fais Island, FSM. In: Freile D, Park L (eds) Proceedings of the 13th symposium on the geology of the Bahamas and other carbonate regions, pp 135–139. Reprinted in speleogenesis and evolution of karst aquifers, 2011, Issue 11. http://www.speleogenesis.info/content/

Mylroie JE, Mylroie JR, Kambesis PN, Lace MJ, Larson EB, Owen AM, Sumrall JB (2012) Quantitative description and coastal mapping of flank margin caves on Mallorca island, Spain. Geologia, special issue. In: Onac BP, Fornos JJ (eds.) Proceedings of the National Science Foundation workshop. Sea level changes in MIS 5E, from observations to predictions. Palma de Mallorca, Spain, pp 42–44

Vacher HL (1988) Dupuit-Ghyben-Herzberg analysis of strip-island lenses. Geol Soc Am Bull 100:223–232

Vacher HL, Mylroie JE (2002) Eogenetic karst from a perspective of an equivalent porous medium. Carbonate Evaporite 17:182–196

Van Beynen PE (ed) (2011) Karst management. Springer Publishing, Dordrecht, 489 p

Van Beynen PE, Brinkman R, Van Beynen K (2012) A sustainability index for karst environments. J Cave Karst Stud 74(2):221–234

Wallis TN, Vacher HL, Stewart MT (1991) Hydrogeology of the freshwater lens beneath a Holocene strandplain Great Exuma, Bahamas:. J Hydrol 125: 93–100

Watson J, Hamilton-Smith E, Gillieson D, Kiernan K (1997) Guidelines for cave and karst protection. WCPA Working Group. IUCN – the World Conservation Union. Gland, Switzerland and Cambridge, UK, 63 p

Werker V, Werker J (2006) Cave conservation and restoration. National Speleological Society, Huntsville, 600 p

White WB (1988) Geomorphology and hydrology of karst terrains. Oxford University Press, New York

White S, White N, Middleton G (2006) Report of findings on the proposed Iralalaro hydro-electric power scheme, Timore-Leste. Report submitted by the Haburas Foundation and the Australian Conservation Foundation, 65 p

Wilson WL, Mylroie JE, Carew JL (1995) Quantitative analysis of caves as a geologic hazard, San Salvador Island, Bahamas. In: Proceedings of the 7th symposium on the geology of the Bahamas and other carbonate regions. San Salvador, Bahamas, pp 103–121

Part II

Selected Case Studies in Coastal Cave and Karst Development

Caves and Karst of the Bahama Islands

7

John E. Mylroie and Joan R. Mylroie

Abstract

The Bahama Islands provide the simplest setting for expression of island karst as described by the Carbonate Island Karst Model (CIKM). The rocks are carbonates of mid to late Quaternary age, relief is low, climate is stable, and tectonics non-existent. As a result, there are significant constraints in time (a few thousand years) and space (a few hectares) for cave and karst development. Despite these constraints, cave development is prolific, and caves of large size are common across the entire archipelago. The Bahamas demonstrate the complex interaction between deposition and dissolution, with syndepositional caves forming as the carbonates they are in are still being deposited in immediately adjacent areas. Despite all the time and spatial constraints, the cave variety and morphology is complex and well represents the special ground-water flow conditions and geochemistry that exist within the fresh-water lens. The Bahamas were the site of origin for both the flank margin cave model, and the subsequent CIKM, as the conditions present allowed establishment of the fundamental theoretical controls of cave and karst development in coastal settings.

7.1 Introduction

The development of the Carbonate Island Karst Model (CIKM) was initiated by work begun in The Bahamas[1] in the mid-1970s. That research eventually lead to the development of the flank margin cave model, explained in Chap. 4. When that model was applied to other, progressively more complex carbonate islands, such as Isla de Mona (Chap. 9) and the Mariana Islands (Chap. 13), then the more integrated CIKM was proposed. Because the Bahamas are young, tectonically stable, and often small islands on small platforms, this archipelago provides significant constraints of time and space when considering possible explanations for the cave and karst features present in these beautiful islands. The simplicity of the Bahamian setting allowed the fundamental controls of cave and karst

[1] The official name of the country is The Bahamas; we have dropped the capital T in the chapter as we have found it commonly confuses readers (they assume a new sentence has begun).

J.E. Mylroie (✉) • J.R. Mylroie
Department of Geosciences, Mississippi State University, Mississippi State, MS 39762, USA
e-mail: mylroie@geosci.msstate.edu

M.J. Lace and J.E. Mylroie (eds.), *Coastal Karst Landforms*, Coastal Research Library 5,
DOI 10.1007/978-94-007-5016-6_7, © Springer Science+Business Media Dordrecht 2013

development in carbonate islands to be worked out, so that when more complicated carbonate islands were inspected, the underlying controls could be understood, and the differences more easily explained. The following presentation is drawn from a chapter published in Palmer and Palmer (2009), and updated to the current conditions. This chapter does not present a review of submerged caves, but only representative examples, mostly flank margin caves, of those caves accessible in subaerial conditions.

7.2 Regional Setting

The following information is drawn from a variety of sources, the leading two of which are publications by Carew and Mylroie (1995a, 1997). The Commonwealth of the Bahamas is an extensive archipelago of carbonate islands and shallow platforms in the western North Atlantic Ocean, from 21° to 27°30′ N latitude and from 69° to 80°30′ W longitude. The Bahama Islands comprise a 1,400 km long portion of a NW-SE trending archipelago that extends from Little Bahama Bank off the coast of Florida to Great Inagua Island, just off the coast of Cuba (Fig. 7.1). The archipelago extends farther southeast as the Turks and Caicos Islands, a separate political entity (and still a British colony), and terminates with the submerged Mouchoir, Silver and Navidad banks. The Bahamian archipelago covers 300,000 km^2, of which 136,000 km^2 is shallow bank, and 11,400 km^2 is land as islands (Meyerhoff and Hatten 1974). The banks are generally less than 10 m deep and are bounded by near-vertical sides into abyssal depths of 1,000 m or more. The Bahamas consists of 29 land masses identified as islands, 661 cays (pronounced "keys," usually minor islands), and 2,387 rocks (Albury 1975). The term "island" is used in this chapter to refer to both formal islands and cays.

The islands are predominantly low lying. The topography is dominated by eolianite (dune) ridges that extend up to 20 m on most islands. The highest elevation (63 m) occurs on Cat Island (Fig. 7.1). The northwestern Bahamas consist of islands on two large banks, Great Bahama Bank and Little Bahama Bank. Great Bahama Bank is cut by two deep troughs: Tongue of the Ocean in the center (1,400–2,000 m deep), and Exuma Sound to the east (1,700–2,000 m deep). Little Bahama Bank is separated from Great Bahama Bank by Northwest and Northeast Providence Channels. To the southeast, the islands are on small banks that are separated by deep water (2,000 m to >4,800 m). In many cases, the islands that occupy these banks occupy most of the bank area, such as Mayaguana Island and Great Inagua Island (Fig. 7.1). The northwestern Bahama islands are isolated landmasses that project above sea level from two large carbonate platforms, Little Bahama Bank and Great Bahama Bank.

The Bahamian platforms have been sites of carbonate deposition since at least Cretaceous time, resulting in a minimum sedimentary cover thickness of 5.4 km (Meyerhoff and Hatten 1974) and perhaps as much as 10 km (Uchupi et al. 1971). The Bahamas are tectonically stable, with slow isostatic subsidence of 1–2 m per 100,000 years occurring (Carew and Mylroie 1995a). The geologic literature on the Bahamas is extensive, but the bulk of that literature deals with the flooded carbonate banks and related deep-water environments. With the exception of San Salvador, comparatively little work had been done on the subaerial geology of other Bahamian islands until the 1990s. A thorough overview of Bahamian island geology can be found in Curran and White (1995) and Vacher and Quinn (1997: Chapter 3): , and a variety of field guides to specific islands have bean published by the Gerace Research Centre on San Salvador Island (www.geraceresearchcentre.com).

The climate of the Bahamas ranges from subtropical temperate in the north (18 °C in January to 28 °C in July with 1,355 mm rainfall annually) to semiarid in the south (23.5 °C in January to 28.5 °C in July with 700 mm rainfall annually) (Sealey 1990). Initial European reports indicate that Bahamian islands were once heavily vegetated with mixed tropical broadleaf coppice including mahogany. At present, the northern islands are largely covered by pine barrens with palmetto, but there are regions of limited

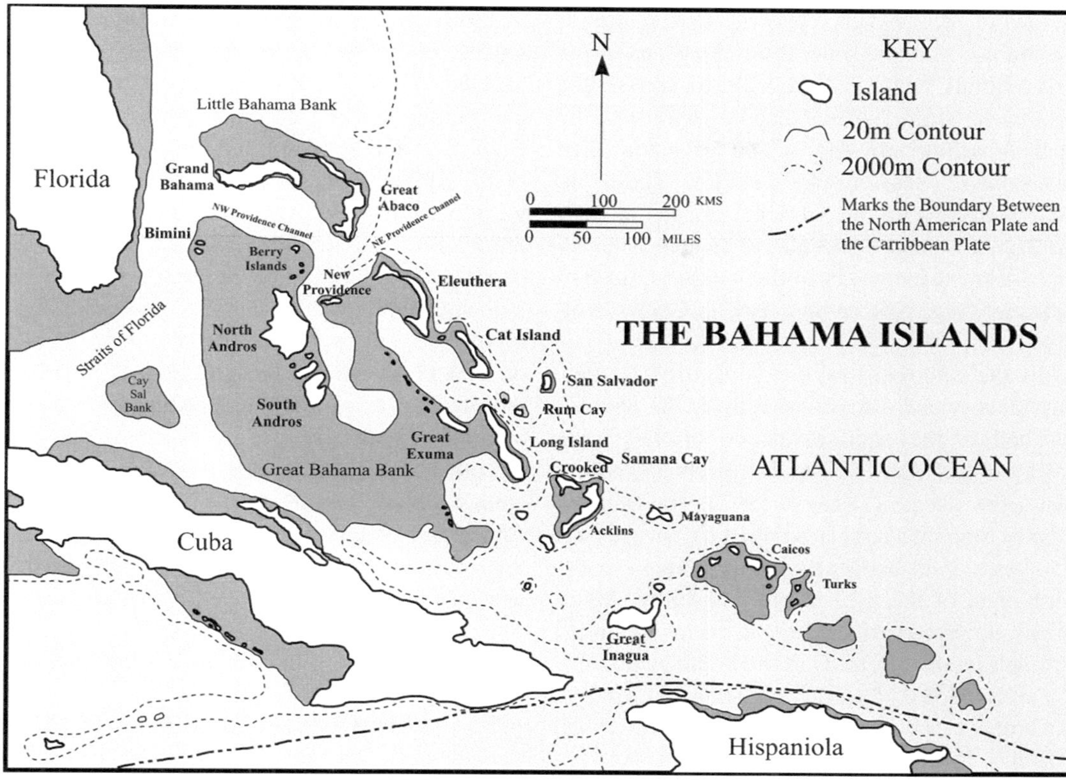

Fig. 7.1 Map of the Bahamian Archipelago

broadleaf coppice. South of New Providence Island the coppice becomes less dense, and tree size declines as the climate becomes drier; on many islands, xeric vegetation and scrub dominate (Sealey 1990). The temperatures of marine waters of the Bahamas average 18 °C in winter and 28 °C in summer. The north equatorial current (Antilles Current) delivers water to the carbonate banks from the southeast. The current diverges and flows northwestward along the eastern margin of the archipelago at 0.6–0.8 kn, and then northwestward through Old Bahama Channel south of Great Bahama Bank at 0.9 kn. The Bahamas are bounded on the west by the Florida Strait, through which the Gulf Stream flows at ~2.5–2.8 kn. (Data from the *Hydrographic Chart of The Commonwealth of The Bahamas*, first edition 1977). The Bahamas are an entirely carbonate environment because the ocean currents isolate the banks from terrigenous sediment from the Greater Antilles to the south and North America to the west (Fig. 7.1). The Bahamas lie within the northeast trade winds, and as a result, the preferential occurrence of islands is on the eastern (windward) side of most banks. These trade winds have influenced the position, shape, and alignment of many of the topographic ridges, but some eolianite ridges are aligned with other wind directions, especially that of the seasonal westerlies associated with fronts from the North American continent.

7.3 Historical Background

The Bahamas are well known as the site of first landfall of Christopher Columbus in the "New World" in 1492. Today, despite a lively debate in the 1980s, it is generally agreed that San Salvador Island, formerly known as Watling's Island and called *Guanahani* by the native Lucayans, was most likely the first landfall (see discussion in

Carew et al. 1995). According to the log of Columbus, when he sailed from the island of his first landfall, many islands could be seen to the southwest. Many "islands" can be seen from hills at the southwestern end of San Salvador when atmospheric conditions are favorable. These "islands" are in fact the refracted images of hills on Rum Cay and Conception Island that lie 35 and 54 km respectively to the southwest of San Salvador (Fig. 7.1), as described in Carew et al. (1995).

In the aftermath of the voyage of Columbus, slavery and disease brought by the Spanish resulted in the rapid extinction of the native Lucayan and Arawak peoples, probably within just a few decades (Keegan 1992). The Bahama Islands remained largely uninhabited for the next 150 years, until British adventurers began sparse settlement of the area in the mid-1600s. Much piracy occurred in the Bahamas, which provoked Spanish retaliatory raids in the archipelago until the early 1700s, when the British began realistic governance out of Nassau on New Providence Island.

When the British colonies in continental America won their independence, a number of British loyalists from the southeastern United States left the colonies and settled in the British-governed Bahamas. Because the size of the land grant from the British Crown depended on family size, including slaves, plantation owners transported their families and slaves to the Bahamas to re-establish their plantations. During this time of British control, additional African slaves were brought in to work the plantations. The soils of the Bahamas could not support long-term production of cotton or other large-scale farming, and the plantations soon began to fail. The Bahamas languished under British inattention, and most of the plantation owners ultimately left. The former slaves, freed by British government decree in 1834, were left behind, and the current population is composed largely of their descendants.

In the late eighteenth and early nineteenth centuries, the economy of these islands was based on agriculture, privateering, and wrecking (salvaging of wrecks, some of which were "induced" by false lighthouses). The Bahamas went through modest boom times and intervening relatively hard times; significant boom times resulted from gun running to the Confederacy during the U.S. Civil War and rum running during Prohibition in the 1920s. Tourism began to flourish when wealthy Americans vacationed in the Bahamas during Prohibition. The Bahamas gained independence in 1973 from Great Britain but remain a British Commonwealth nation. The population is now $\sim$300,000, of which 200,000 live in Nassau on New Providence Island, and another 70,000 in the Freeport area of Grand Bahama Island. The remaining 30,000 people are distributed among the other Bahamian islands, called the "Out Islands" or the "Family Islands" by the natives. Abaco and Eleuthera contain the largest populations outside of Freeport or Nassau; San Salvador hosts about 1,000 people.

7.4 Subaerial Geology of the Bahamas

The exposed rocks of The Bahamas are all mid to late Quaternary carbonates, dominated by eolian calcarenites or eolianites (fossilized carbonate sand dunes) and subtidal facies (including fossil reefs) at low elevations; and solely by eolianites at elevations above 8 m. The highest elevation is 63 m on Cat Island. Paleosols can occur at all elevations. The glacio-eustatic sea-level changes during Quaternary time alternately have flooded and exposed the Bahamian platforms, subjecting them to cycles of carbonate deposition and dissolution, respectively. Significant carbonate deposition has occurred in the past only when the platforms are flooded, as is the case today.

The carbonate sequences of the Bahamas can be viewed as individual packages deposited on each sea-level highstand, separated by erosional unconformities (usually marked by paleosols) produced by each sea-level lowstand (Carew and Mylroie 1995b, 1997). The oldest unit (Fig. 7.2) is the Owl's Hole Formation, consisting entirely of eolianites. This unit represents multiple carbonate deposition events during glacioeustatic platform flooding cycles

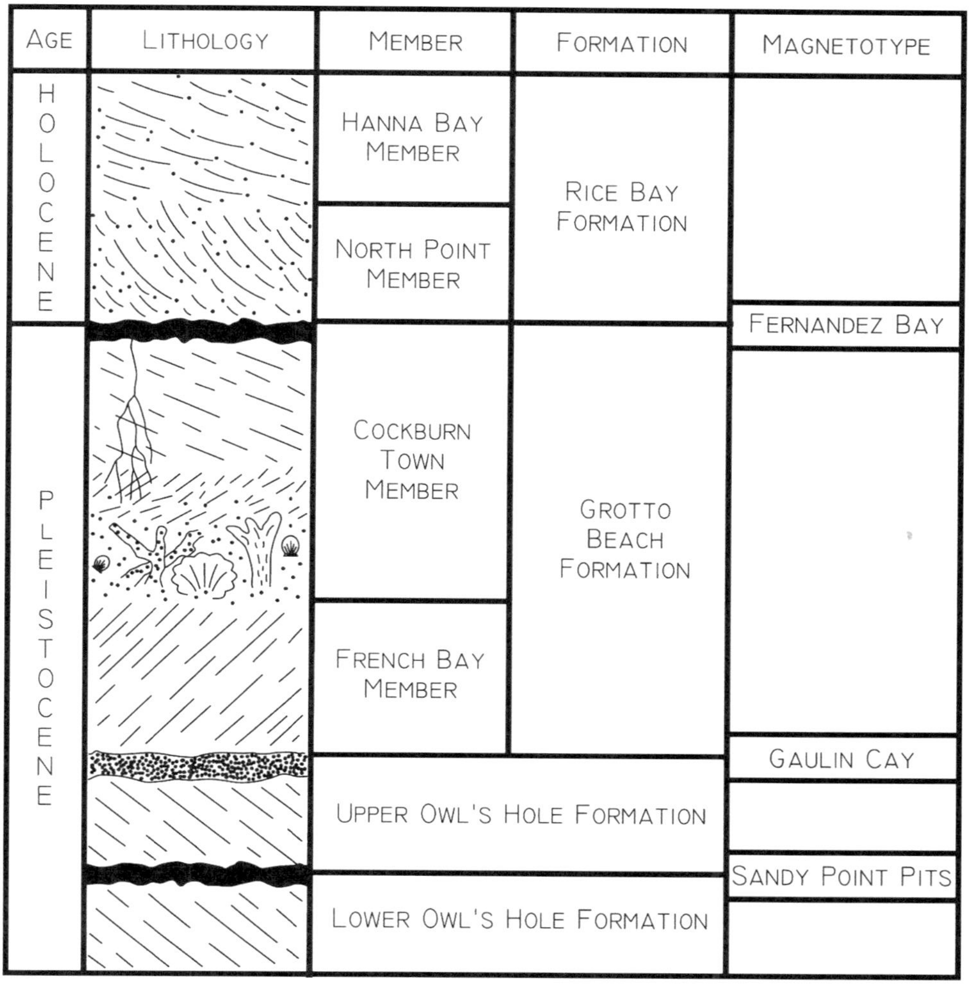

Fig. 7.2 Simplified stratigraphic column for the Bahamas. A detailed column is available in Mylroie et al. (2012)

beginning at least 200,000 years ago, and older. The Owl's Hole contains several paleosols representing emergence of the platforms during glacioeustatic sea-level lowstands. The Owl's Hole Formation eolianites contain most of the dry caves in the Bahamas. The Owl's Hole can be subdivided into members (Fig. 7.2) based on paleomagnetic records of secular variation within the paleosols, to create magnetotypes (Panuska et al. 1999), or by amino acid racemization (AAR) techniques (e.g. Kindler and Hearty 1995). Neither geochronologic technique is useful while in the field, so the Owl's Hole subdivisions are of little use in real time. On Eleuthera Island, however, excellent outcrops allow the Owl's Hole to be subdivided and named in the field (Kindler and Hearty 1995).

Overlying the Owl's Hole Formation and separated from it by a paleosol or other erosion surfaces is the Grotto Beach Formation, which developed during the last interglacial, marine isotope substage (MIS) 5e. MIS 5e had long been reported to have lasted for ~12,000 years in the Bahamas, from 131,000 to 119,000 years ago, based on U/Th dating (Chen et al. 1991), but recent work on diagenetic effects in fossil corals has changed that value to ~9,000 years, 124,000 to 115,000 years ago (Thompson et al. 2011).

This new interpretation reduces the time for all sea-level highstand activities, both depositional and dissolutional, by 3,000 years or 25 %, providing even tighter time constraints than previously believed. The last interglacial was about 6 m higher than present sea level, so subtidal deposits, such as fossil reefs and lagoons, are found above sea level in this unit. Two members occur in the Grotto Beach Formation, the French Bay Member, and the Cockburn Town Member, representing the transgressive (French Bay), and stillstand and regressive episodes (Cockburn Town) of the last interglacial sea-level highstand. There are a number of karst caves developed in the unit, which formed during and immediately after the deposition of the rock, about as fast as caves can form following limestone deposition (the syndepositional caves defined in Chap. 4).

Overlying the Grotto Beach Formation and separated from it by a paleosol or other erosion surface are the rocks of the Rice Bay Formation, deposited during Holocene time. The Rice Bay Formation consists entirely of eolianites and beach rock. Two eolianite suites are recognized, the North Point Member and the Rice Bay Member. Because of its very young age, the Rice Bay Formation appears to lack karst caves; however, some small caves may have developed recently by mixing dissolution. The origin of these caves is being studied.

7.5 Freshwater Lens Hydrology and Karst Processes

This topic was discussed in detail in Chaps. 3 and 4. In any essentially homogeneous body of rock like that of the carbonates forming the Bahamian islands, the freshwater lens floats on underlying, denser seawater that permeates the subsurface (Fig. 7.3). The model for the ideal behavior of such water masses is the Ghyben-Herzberg-Dupuit model. In reality, variations in rock permeability and other factors result in distortion of the ideal lens shape in the Bahamas (e.g. Vacher and Wallis 1992). Nonetheless, the Ghyben-Herzberg-Dupuit model serves as a useful first approximation of the relationship between the freshwater and underlying marine groundwater in an island. In simple carbonate islands like the Bahamas, the largest subaerial caves are the flank margin caves, which developed during a past, higher sea level in the distal margin of the fresh-water lens, under the flank of the enclosing land mass.

The Bahamas extend from positive water budget islands in the north (annual precipitation exceeds evapotranspiration) to negative water budget islands in the south (annual evapotranspiration exceeds precipitation), with the balance line being just north of San Salvador

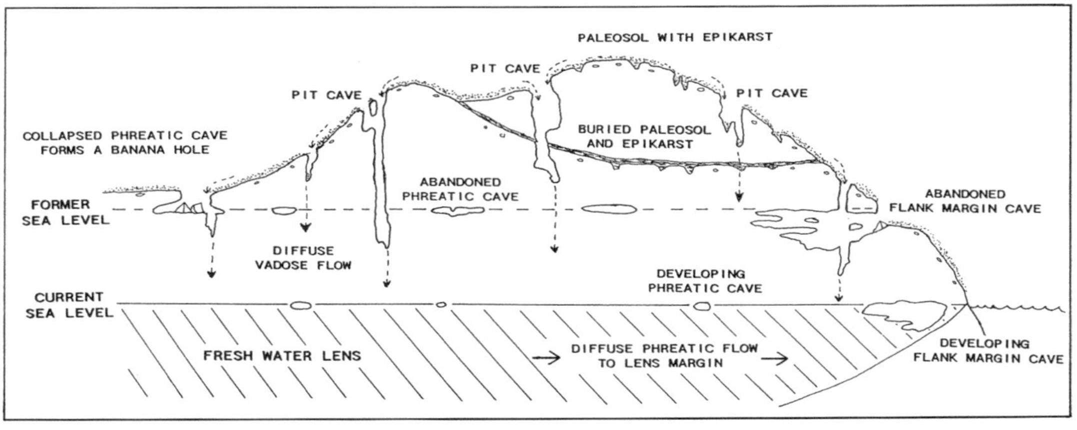

Fig. 7.3 Diagrammatic representation of the fresh-water lens in a Bahamian island, showing the various common cave and karst features, except for blue holes

Island (Fig. 7.1). Despite this gradient, all islands and many cays have a fresh-water lens, as precipitation is episodic and can infiltrate downward to escape evaporation. In the positive water budget islands, interior depressions from fresh-water lakes and swamps, but in the more arid southern Bahamas, the exposed water bodies evaporate, upconing the marine water below. Continued evaporation can drive the water bodies to hypersalinity. Once salinity exceeds 60 ppt, the water bodies cannot support most aquatic species, and cyanobacteria dominate, in some cases forming stromatolites. The fresh-water lens can become partitioned by these saline water bodies, and become isolated within the eolianite ridges (Davis and Johnson 1989). Karst conduits provide marine connections to some interior water bodies, such that their conditions do not reach hypersalinity (Mylroie et al. 1995a). Cave development in the Bahamas does not seem to be influenced by these climatic gradients, which supports the argument, in the case of flank margin caves, that they are hypogenic.

As in all carbonate coasts, and as described in Chap. 2, Bahamian coasts have a variety of coastal karren features (Horowitz and Roberts 2010). The Holocene eolianites in particular provide a natural laboratory to demonstrate how fast these coastal karren are produced, as the rocks cannot contain any karst forms inherited from pre-Holocene times. Studies on San Salvador Island have shown that coastal karren are recycled continuously, as major storms break of large rock slabs, and the karren formation process begins anew (Mylroie and Mylroie 2009).

7.6 Caves of the Bahamas

There is ample evidence that early Lucayan natives entered the caves in the Bahamas prior to the arrival of European adventurers. Artifacts, burials, and most notably, petroglyphs, have been discovered (Winters 2009) (see also Fig. 5.3). Because almost all large or accessible caves in the Bahamas were mined for their bat guano, beginning with Lucayan natives and continuing up into the twentieth century, the only common cave archeological data are the petroglyphs, or rock carvings (Fig. 7.4). Anything left on the cave floor surface, or within excavated sediments, was removed or destroyed by the guano mining. Many caves have been used as hurricane and storm shelters, right up to the present day.

Modern speleology in the Bahamas began not with cavers, but with scientists such as archeologists, or more importantly, bat researchers. The mid-twentieth century bat literature from the Bahamas has proven to be a good initial resource when hunting for caves. Organized cave exploration by cavers began in the 1960s from two different origins. One was exploration of blue holes by cave divers. A historical perspective of that cave diving work is provided in Mylroie et al. (1995b) and Chap. 4, which have a series of references into the blue hole literature. Blue holes, and cave diving, are not a focus of this chapter. The second impetus came from traditional "dry" cavers, mostly American tourists who poked around but did little mapping or publishing of their work. Documentation of caves in the Bahamas began in earnest in the mid-1970s, mostly as a result of the resources and opportunities provided by the Gerace Research Centre (then the CCFL Bahamian Field Station). That work has continued up to present, and a wide collection of cave data are now available for most Bahamian islands. Mylroie and Mylroie (2007) provide a review of the development of caving and cave science in the Bahamas for the last 30+ years. Despite the vast number of banana holes and pit caves that occur in the Bahamas, they will not, with one exception, be described here as they are usually short (under 10 m) in length and depth, and provide a very rudimentary caving experience. The flank margin caves are numerous, and in some cases extensive, and the largest or most significant ones are reviewed here.

Cave length, the usual indicator of cave size, is a vague term in the case of flank margin caves. These cave are hypogenic (Palmer 1991), meaning that they form de-coupled from the surface hydrology, as a result of mixing of fluids within the rock mass to generate dissolution (see Chap. 4). Classic continental examples of hypogenic caves are Carlsbad Caverns and Lechuguilla Cave in New Mexico, or Wind Cave and Jewel Cave in South Dakota. In flank margin

Fig. 7.4 Petroglyphs in Hartford Cave, Rum Cay, Bahamas, directly in front of the person, and on the far right of the image

caves, the mixing is between fresh water and salt water, and the result is caves with low, wide globular chambers that commonly interconnect by small windows between chambers. Survey of these caves consists of running a base line with numerous splay shots to define the chamber dimensions. Survey length, in this setting, is not representative of cave size. Instead, areal footprint has been selected (Mylroie 2007). The use of areal footprint to quantify cave size can be justified by comparing the 1,207 m survey length of Hole In The Wall Cave to the 1,200 m survey length of Hatchet Bay Cave to the 1,030 m survey length of Hamilton's Cave. The areal footprint of those three caves, in the same order, is 3,422, 5,934, and 8,931 m^2, respectively. Cave size for these hypogenic mixing caves is best quantified by measuring their areal footprint. Given that the caves are vertically restricted because of their development in the thin distal margin of the fresh-water lens, areal footprint is a proxy for cave volume. This footprint is then a measure of how much carbonate rock has been removed by dissolutional processes. As shown by Fig. 4.16 in Chap. 4, a rank order plot of Bahamian flank margin caves, used areal footprint as the size measure. The plot shows that the caves fall into three main groups: small (<100 m^2), medium (100–1,000 m^2), and large (>1,000 m^2). These groupings are the result of small chambers forming in the fresh-water lens by mixing dissolution, then their sudden enlargement by intersecting an adjacent chamber, and another enlargement when connected sets of chambers intersect other connected sets. These ideas of how the caves form and grow greatly influence cave discovery and exploration strategies in the Bahamas.

The large flank margin caves of the Bahamas occur preferentially in the eastern Bahamas. Grand Bahama and the Andros Islands lack large caves, mostly because they lack significant eolian ridges to host the caves. New Providence has caves of some size, but commercial development (the island contains >200,000 people, over 70 % of the nation's population) has resulted in some caves being destroyed. The largest cave here was Hunts Cave, located in the interior of the island (Mylroie et al. 1991), and now threatened by quarrying. As New Providence is the hub for communication in the Bahamas, many people

Fig. 7.5 Caves Point East Cave, showing the low, wide chamber configuration of flank margin caves. Person standing in entrance for scale. Note that the cave has been subaerially exposed by a relatively small amount of hill slope retreat; the original slope can be seen on the right hand side of the picture. A parking lot now occupies the foreground of this 1988 picture

pass through New Providence, and the capital city Nassau, on their way elsewhere. Two locations are presented here as they are very representative of Bahamian cave development, and because of their ease of access.

7.6.1 New Providence Island

Caves Point Caves: On the north shore of New Providence, just east of the intersection of West Bay Street and Blake Road (which heads south towards the airport), is a small public park with two very large cave entrances. The location is appropriately called "Caves Point", and the two caves are named Caves Point East Cave and Caves Point West Cave (Mylroie et al. 1991). The East Cave (Fig. 7.5) is one broad chamber with a small pit cave entering near the back wall. It is a classic simple flank margin cave, much wider (22 m) than it is high (3–4 m), and situated just inside the hill that contains it, penetrating 15 m, with an areal footprint of 289 m^2 (Roth 2004). West Cave is similar, 25 m wide, except that in the back wall and floor are openings into a deeper chamber, which is 6 m high as it extends down to the high tide mark. The cave penetrates a total of 25 m into the eolianite hill, with an areal footprint of 359 m^2 (Roth 2004). A local legend states that at very low tide, a passage can be followed under the road to the sea. A small karst cave does exist in the low sea cliff, but no connection was located. Examination of the eolianite ridge containing the two caves demonstrates how the caves were opened by a relatively small amount of hill slope erosion (Fig. 7.5). As a result of developing under the flank of the landmass, these caves are formed without entrances, but such entrances later form fairly easily by surficial erosion processes. The caves developed during the last interglacial, between 124,000 and 115,000 years ago, when sea level was 6 m higher than at present, and waves lapped up against the hill as it does lower down across the road today. The caves, while short, are still voluminous chambers, and they only had 9,000 years to form. Figure 7.5 shows the caves as they looked in 1988. Currently the foreground of that image has a parking lot, and a hotel dominates the top of the hill above the caves, examples of developmental pressures on islands as discussed in Chap. 6.

The Primeval Forest: Following West Bay Street to the west end of New Providence, South Ocean Road leads south, and a side road leads to a new national park in the Bahamas National Trust system, the Primeval Forest. This small park is fenced and protected, but arrangements can be made with the Bahamas National Trust to visit the area. The park only covers 3.3 ha, but contains a mind-boggling array of pit caves, banana holes, flank margin caves, natural bridges, and shallow karst valleys. Over 100 karst features have been identified. None of the caves are very large, and they are developed in lagoonal, beach, and eolian facies of the Cockburn Town Member of the Grotto Beach Formation. They are a good example of syndepositional karst, in that the dissolutional features in the lagoonal facies developed when beach sands prograded over them, allowing the fresh-water lens to extend into these subtidal facies, dissolving flank margin caves and banana holes (Mylroie et al. 2008). The caves are syndepositional because they developed in the same sea-level highstand that created the carbonate deposit hosting them and also elevated the fresh-water lens into that deposit to dissolve the caves (Chap. 4). Earlier it was stated this chapter would not deal with banana holes and pit caves, but the Primeval Forest location on New Providence Island (Nassau), and its protected status as a Bahamas National Trust park, allows access to one of the most densely-packed karst areas in the entire archipelago, hence its inclusion here. Given that the caves are a mix of types, and are all small, no areal footprint information has been developed for them, and the reader is referred to Fig. 7.6 for approximate size relationships. Many of the pit caves, banana holes, and flank margin caves are connected by open collapses, or tubes not negotiable by humans, creating cave complexes such as "A-Survey Cave" (Fig. 7.6).

7.6.2 Abaco Island

Hole in the Wall Cave: At the southeastern tip of Great Abaco Island is a peninsula with a large sea-arch in it, the "Hole In The Wall". A lighthouse occupies the base of the peninsula, and a very rugged road (four wheel drive) heads west and then north to the main Abaco road system. Just west of the lighthouse about 500 m, and north of the road, are several entrances into Hole In The Wall Cave, a large flank margin cave (Fig. 7.7) with spacious rooms, interesting passage connections, and speleothems. The cave contains bats, and shows evidence of guano mining (Fig. 7.8). Its proximity to the road has resulted in a lot of visitation, and graffiti and vandalism are apparent. The cave has at least six negotiable entrances; the most commonly used one, Entrance 1, is the most northeastern entrance, and in 2005 contained a ladder for ease of access. From this entrance, the cave continues to the northeast for 160 m+ as a very linear feature (for flank margin caves) consisting of crawls, chambers and tubes with few side passages. To the southeast of Entrance 1, past a short breakdown crawl, lies the main part of the cave containing several large chambers and a maze of crawls, walking passages and short climbs that trend to the south and southeast from the main chambers. The remaining five entrances are all found in this maze area, as the cave runs into the hillside and the individual passages are breached to the surface. With 1,207 m of linear survey, and a 3,422 m^2 areal footprint, Hole In The Wall Cave is one of the larger caves in the Bahamas. A complete description of the cave and its geology can be found in Walker (2006).

Eight Mile Cave: North of Hole In The Wall Cave, within the boundaries of the Abaco National Park, about 2 km inland from the east coast in the vicinity of Lantern Head, and to the southwest of Eight Mile Rocks, is Eight Mile Cave, the second largest known flank margin cave on Abaco (Fig. 7.9). The cave is very difficult to find without a guide. The cave has 671 m of survey, and an areal footprint of 919 m^2. The cave has three entrances, access is easiest by way of Entrance 1 at the southernmost part of the cave. From Entrance 1, the cave trends north as a large passage with a maze of rooms and passages to the east, and a series of large chambers to the

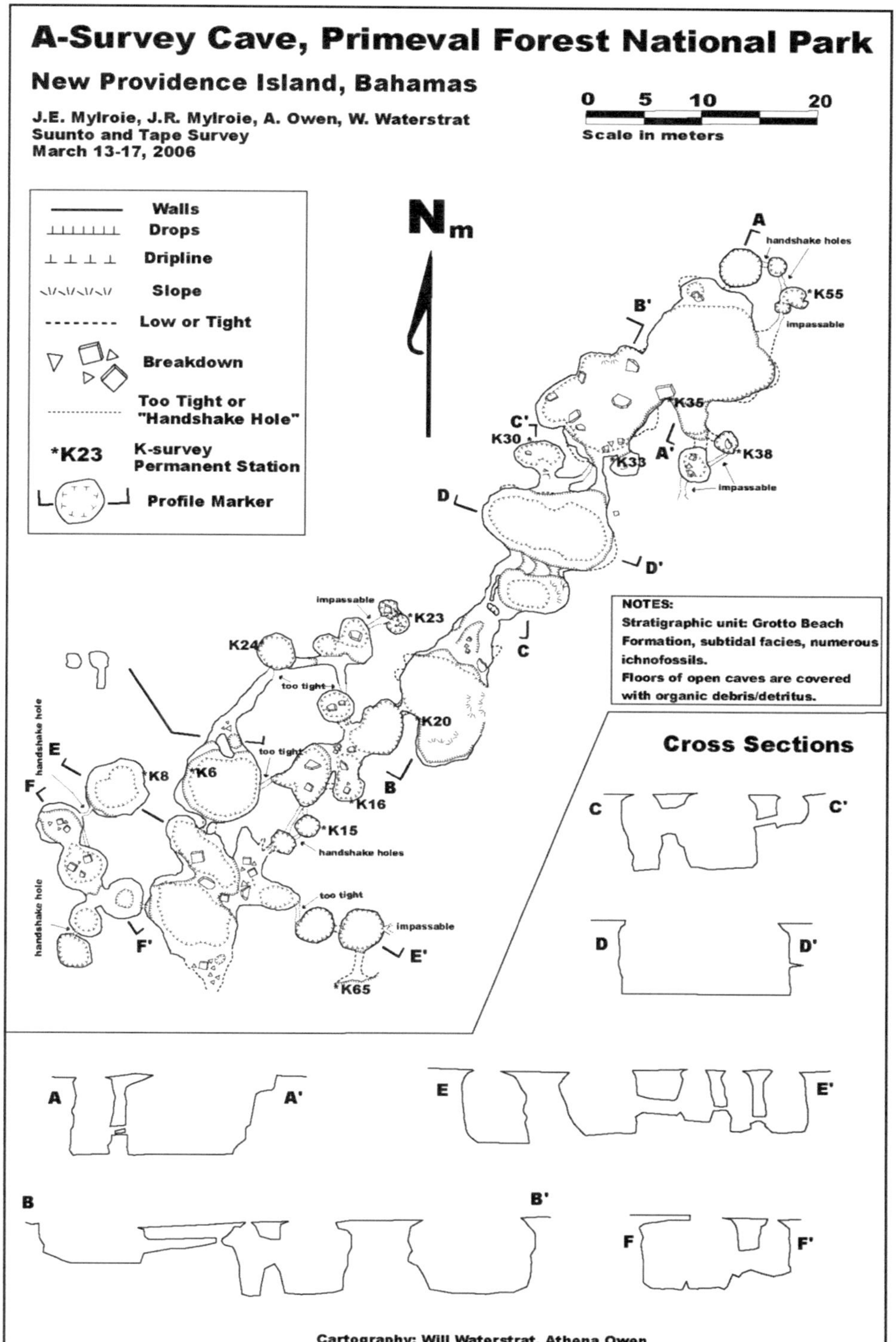

Fig. 7.6 Map of A Survey Cave, Primeval Forest, NEW Providence Island, Bahamas. This cave is a collection of pit caves, bananas holes, and flank margin caves, all interconnected by small tubes and partial collapse zones

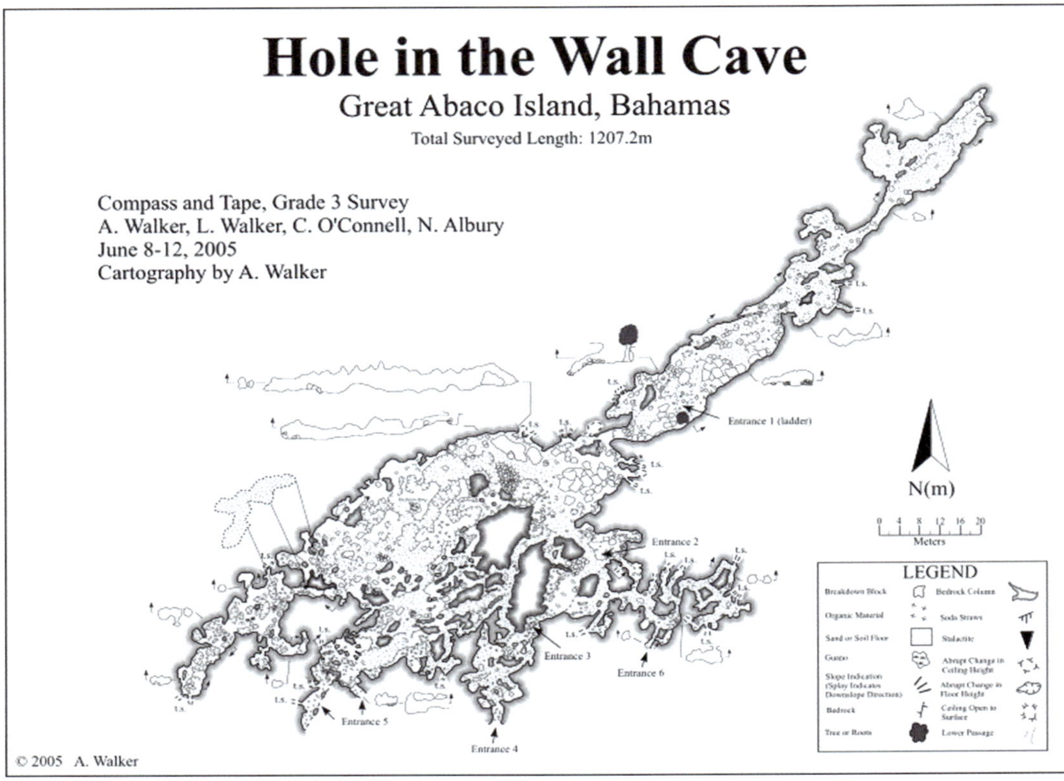

Fig. 7.7 Map of Hole In The Wall Cave, Abaco Island, Bahamas (Walker 2006)

Fig. 7.8 Chamber in Hole In The Wall Cave, Abaco Island, Bahamas. The irregular wall outline and isolated bedrock pillar pattern is typical of flank margin caves. The cave has been mined for guano

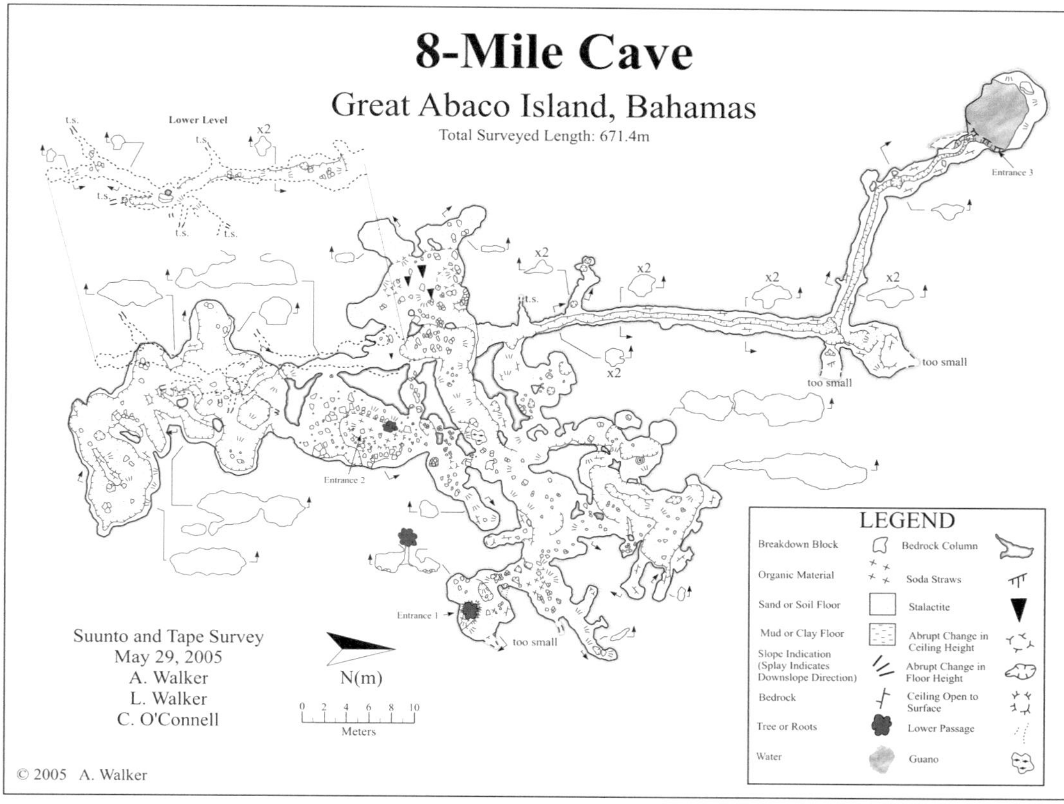

Fig. 7.9 Map of Eight Mile Cave, Abaco Island, Bahamas (Walker 2006)

west, where a roof collapse creates Entrance 2 (Fig. 7.10). Extending northeast from the main cave, a lower-level passage trends east then north to a large collapse feature floored with water, Entrance 3. This passage is remarkable as it has a classic keyhole cross section, with an oval tube at the top, and a slot in the floor (Fig. 7.11). This resembles a phreatic tube with a vadose floor canyon, as is common in continental caves, but here it appears to be dissolutional widening at a favored horizon, with enlargement downward of the cave floor. The "canyon" portion does not display a continuous down-gradient trend or turbulent flow markings typical of vadose flow. The relative paucity of joints and fractures in the young eolianites of the Bahamas results in there rarely being linear passages with floor slots. Eight Mile Cave represents the unusual exception. A complete description of the cave and its geology can be found in Walker (2006).

Other Abaco Caves: A medium-sized flank margin cave can be found in the Long Beach community, in an open field near the road. Long Beach Cave has 428 m^2 of areal footprint, contains crawls and some chambers, and two entrances. It contains flowstone surfaces that have been cut across by a dissolution event. This dissolution of flowstone suggests sea level fluctuations that (1) created the cave by phreatic dissolution in the fresh-water lens margin, (2) dropped to lower the lens, allowing vadose speleothem production, and (3) rose to return the lens to dissolve the speleothems and surrounding rock with renewed phreatic dissolution (Walker 2006).

In the Little Harbour community on eastern Great Abaco are a series of small to medium-sized flank margin caves. One, Hunters Cave, is almost entirely intact, while many other caves have lost most of their seaward walls to reveal

Fig. 7.10 Cave chamber in Eight Mile Cave, Abaco Island, Bahamas, showing ceiling collapse and invasion of the cave by plant roots seeking to use the cave as a shortcut to water and nutrients

Fig. 7.11 Phreatic tube in Eight Mile Cave, Abaco Island, Bahamas, with a slot in the floor. While entirely phreatic in origin, the passage resembles a "keyhole" passage, that is, a phreatic tube with a vadose notch in the floor

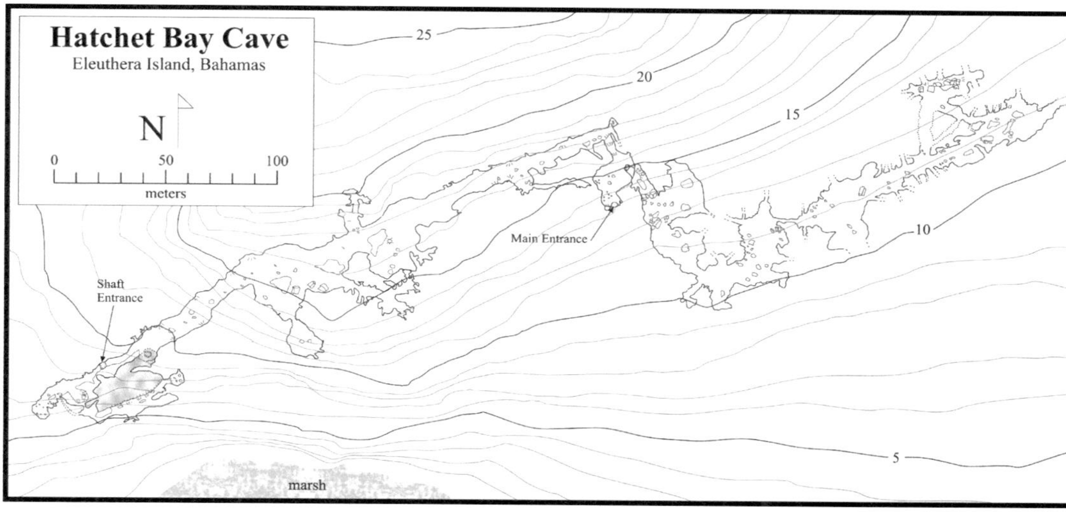

Fig. 7.12 Map of Hatchet Bay Cave, Eleuthera Island, Bahamas, with topographic overlay (Lascu 2005)

large chambers and some subsidiary passages. These caves are described in Walker (2006).

West of Great Abaco Island, across a causeway to Little Abaco Island, are the Cedar Harbour Caves. This series of small caves occupies a significant portion of the small eolianite ridges that host them. The caves are interesting in that they contain a complex of sediment infills, of both terrestrial and marine origin, that indicate a complex history of cave formation, draining, reflooding, and breaching by wave action (Walker 2006).

7.6.3 Eleuthera Island

Hatchet Bay Cave: One of the largest caves in the Bahamas is Hatchet Bay Cave, located north of the Hatchet Bay community on the west side of the main north–south road on Eleuthera (Fig. 7.12). It is marked on the topographic maps of the island, and on many tourist maps as guided tours are sometimes taken there. This notoriety has resulted in the cave having been greatly vandalized, mostly by spray paint, in the larger rooms and passages (Fig. 7.13). Hidden amongst the recent graffiti are names, dates and inscriptions going back to the 1840s. The cave has over 2,000 m of linear survey, and an areal footprint of 5,934 m^2. For such a large cave, it is unusual for having only two entrances, one of which may be artificial, created (or enlarged) to assist guano mining in the 1840s. The Main Entrance (Fig. 7.12) is located approximately in the center of the cave, which trends from east-northeast to west-southwest over a linear trend in excess of 500 m. To the WSW of the Main Entrance the cave is a large tubular passage of very linear dimension, unusual for the Bahamas. To the south side of this passage are a series of large chambers and interconnecting smaller passages, but to the north only one small side passage exists. The cave ends to the WSW in a maze of smaller passages that lead to a lower level that is partially flooded by marine tidal water. In this area, a rectangular shaft leads 5 m to the surface. The walls of this shaft have been clearly squared off by chiseling, but there may have been a small natural entrance here originally, as the hillside is dropping down to lowlands in this area and the cave is just beneath the land surface (Fig. 7.13). To the ENE from the main entrance, the cave character is different. The large, roomy passages and chambers seen to the WSW are now low, broad chambers with complex interconnections. A single fissure passage, with slight airflow, connects to more low chambers and maze passages heading ENE. The ceiling of the cave shows steeply dipping

Fig. 7.13 Chamber in Hatchet Bay Cave, Eleuthera Island, showing wall cusps and pockets, and vadose speleothems. Note also modern spray-paint graffiti

foreset beds that have been laterally crosscut by the horizontal dissolutional surfaces. The source of the wind (a third entrance, possibly, or perhaps barometric in origin) has not been determined. These passages are floored with thick guano, and ceiling heights are commonly 1 m or less. A single large chamber exists.

The difference in the cave passage configuration between the WSW and ENE portions of the cave is caused by a paleosol (Lascu 2005). The paleosol outcrops as a sloping surface along the main tubular passage forming the WSW portion of the cave. This paleosol forms the contact between two eolianites of different ages, the northern eolianite being older. The paleosol inhibited water flow in the fresh-water lens, causing it to ramp up along the contact between the two eolianites. As a result, there are no passages but one small one crossing this contact to the north, and the chambers to the south are large and very high as a result of this ground-water ponding. The extreme linearity of the cave in this region is also a response to the fresh-water lens being constrained to the north by the paleosol. In the ENE portion of the cave, the hillside formed by the younger eolianite has begun to trend more to the east, while the paleosol separating the younger and older eolianites has continued its trend to the ENE. As a result, the margin of the lens, where flank margin cave development is favored, was now separated from the influence of the paleosol. The result is a series of passages and chambers that are very low and maze-like, reflecting the thin, undistorted margin of the lens in this area of the cave. The development of the cave in a large, long ridge explains why the cave has few entrances. The thick soil above the cave may explain the abundant and well-developed speleothems inside the cave (Fig. 7.13). A full discussion of Hatchet Bay Cave and its geology is available in Lascu (2005).

Ten Bay Cave: Located inland and at the north end of Ten Bay, but west of the main road, is Ten Bay Cave (Fig. 7.14). This is also one of the largest caves in the Bahamas, with over 1,000 m of linear survey, and an areal footprint of 5,147 m^2. The cave has numerous entrances,

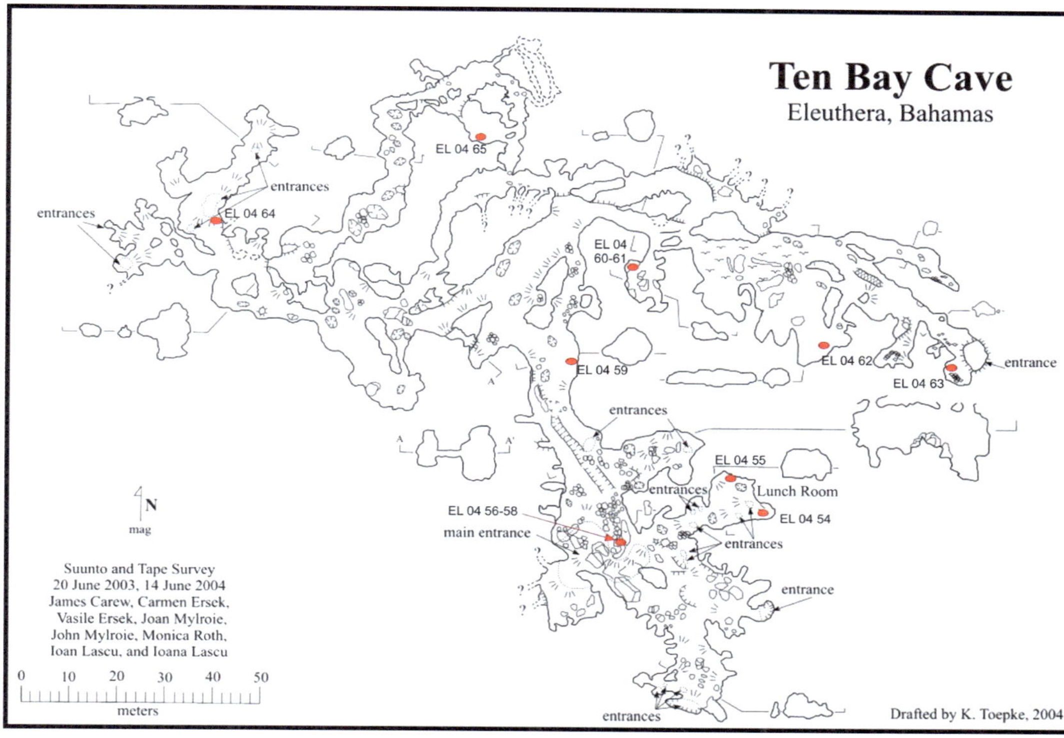

Fig. 7.14 Map of Ten Bay Cave, Eleuthera Island, Bahamas. Code numbers on the map beginning with "EL" mark locations of wall rock samples taken for petrographic examination (Lascu 2005)

ceiling holes, and skylights, and unlike Hatchet Bay Cave to the north, it lies just beneath the surface of the eolianite that encloses it. The thinness of the soil on the rock above may also explain why the cave is almost entirely devoid of speleothems. There are two important entrances used to gain access to the cave. The first is at the extreme east end of the cave, where a natural skylight (as at Hatchet Bay Cave, modified to assist guano mining) allows a careful climb down into large passage. The passage continues to the west, with interconnecting passages and chambers developed along both sides, until a large chamber and passage junction area is found at the approximate center of the cave. To the southeast, a large passage leads to some collapse areas and very large rooms, eventually running into the hillside as a walk-in entrance at the extreme southern end of the cave. This entrance is the second primary way of access into the cave, as a path leads from here to a saline pond. Numerous skylights dot the ceiling. To the west from the central junction, passages and chambers head north and west, all eventually becoming low as they reach the edge of the ridge, with numerous entrances. Low passages loop back to the initial entrance passage to the east. The wall rock of the cave shows spectacular vugular porosity in places (Fig. 7.15).

Ten Bay Cave is an example of a "nose cave" (Lascu 2005; Chap. 4, Fig. 4.13), a flank margin cave that wraps around the end, or "nose", of an eolianite ridge. The two primary entrances are on either side of that ridge, and the cave is constrained to the lens margin. But as the nose of the ridge is approached going west, the margins of the lens began to interact, and cross-connections developed between the caves formed on each side of the ridge, following the lens margin around the ridge nose. In Ten Bay Cave, the unusual passage configuration is the result of self-distortion of the lens as it follows the

Fig. 7.15 Chamber in Ten Bay cave, Eleuthera Island, Bahamas, showing vugular porosity at various scales in the wall rock of the cave

ridge terminus. In Hatchet Bay Cave, the lens was distorted by a paleosol within the ridge. In each case, the cave development was controlled by how the lens shape and position changed under these conditions. A full discussion of Ten Bay Cave and its geology is available in Lascu (2005).

Cow and Bull Caves: These caves are extremely small, but very significant. Located in northern Eleuthera, the Cow and Bull are large boulder-like features that are labeled on the topographic map. They are two of about eight "boulders" in this immediate area. They were once considered to be boulders that had been flung up by tsunami or mega-hurricanes from the east onto the large sea cliffs that face the Atlantic (Hearty 1997). Fieldwork has demonstrated that each boulder has at least one phreatic cave inside it, and that the boulders may be remnant tower karst features, not tsunami or storm rubble (Mylroie 2008). Kindler et al. (2010) provide evidence that the boulders are not remnant tower karst, but are also unlikely to be tsunami deposits either.

7.6.4 Cat Island

Cat Island has a great number of caves, a few large, but many of medium to small size. Two caves are mentioned here because they add new information about the development of caves in the eolianites of the Bahamas. Information on the caves of Cat Island can be found in Palmer (1986), Palmer et al. (1986), Mylroie et al. (2006) and Waterstrat (2007).

Port Royal Cave: This cave is located in a coastal outcrop of eolianite at the north end of Cat Island, just north of the community of Port Royal. The cave has an areal footprint of 597 m^2, which is not extensive. The cave consists of a large chamber, breached to the southwest, trending northwest to southeast. Numerous interconnecting smaller passages trend into the hill to the northeast but all end or loop back to known cave, with a large secondary chamber lying to the southeast, breached to the southwest as is the main chamber. Port Royal Cave sits on a coastal cliff, and is being attacked by wave action. It is being degraded and will eventually be destroyed. To the southeast, beyond the current cave, a notch in the cliff line runs for hundreds of meters, with flowstone and other speleothems attached to the back wall. This notch represents the last remnants of the extension of Port Royal Cave to the southeast. Flank margin caves form in coastal environments, their development dependent on the distal margin of the lens where mixing dissolution works best. These caves are vulnerable to later destruction by wave attack, either on the sea-level highstand that formed them (as with the Cedar Harbour caves on Little Abaco), or on a later sea-level highstand (as at Port Royal Cave). As this wave attack continues, the flank margin caves become modified into hybrid caves by wave action, and may appear to the casual observer to be sea caves. Differentiating sea caves from flank margin caves is important, as flank margin caves contain information about the condition and behavior of past fresh-water lenses. If sea caves are misinterpreted as flank margin caves, or vice versa, geologic, denudation, and hydrologic interpretations may be mistaken. Using caves like Port Royal Cave, Waterstrat et al. (2010) attempted to develop quantitative criteria to separate such cave populations based on cave morphology.

Big Cave (St Francis Grotto): This cave was initially discussed in Chap. 4 because of its unusual position. It is a small phreatic cave that has the general appearance of a flank margin cave. It has an areal footprint of only 92 m^2. It represents a major problem for the flank margin cave model, in that the cave sits at 55 m elevation on Mt Alvernia, the highest elevation in the entire Bahamian Archipelago. In fact, the cave itself is at a higher elevation than any other topographic feature in the Bahamas. Mt Alvernia is famous for the small hermitage that sits at the peak (Mylroie et al. 2006). The cave was used as a place of worship while the hermitage was being built. Big Cave has all the features of a phreatic origin – ceiling pockets, curvilinear walls, isolated bedrock pillars, etc. It also lacks any vadose or turbulent flow bedrock sculpture. If the cave formed by the flank margin model, then sea level would have been at an elevation of at least 55 m. There is no evidence anywhere on earth of a glacioeustatic sea level at such a high elevation (it would require melting almost all the ice on the planet) in the Quaternary. How then, to create such a phreatic cave? The best answer, unproven so far, would be to have a paleosol near the crest of Mt Alvernia that perched a local water table, and the cave is a result of bananas hole process (mixing of fresh vadose and phreatic water), not a flank margin cave. However, the banana hole theory has just been revised (Chap. 4), so uncertainty remains. The cave's size is in the range of the larger known banana holes, and it is small as regards the majority of flank margin caves. No good explanation has been put forth to explain this cave.

7.6.5 San Salvador

Lighthouse Cave: Lighthouse Cave is located southeast of the Dixon Hill Lighthouse on the northeast corner of San Salvador Island (Mylroie and Carew 1994). The cave has an areal footprint of 1,378 m^2 (Fig. 7.16), and is one of the best-known flank margin caves in the Bahamas. This notoriety is the result of the Gerace Research Centre being located on San Salvador Island, such that many researchers visit the island, and the cave has been investigated for biological, geological, and geochemical information for more than three decades. Lighthouse Cave has three entrances, all clustered on the central southwest side of the cave, where the cave begins to emerge from the side of a steep eolianite ridge. The cave trends northeast as a large chamber but slopes down both to the south and to the northwest into passages at or in the intertidal zone. Phreatic features dominate the walls and ceiling (Fig. 7.17). The southern passages trend away from the central chamber, parallel to the hillside, and end. The passages to the northwest loop and interconnect in a complex pattern, rejoining the central chamber a number of times. The abundance of passages at current sea level, partially filled with tidally-responsive marine water, is very unusual for Bahamian flank margin caves. The southern end of Hatchet Bay Cave is another example of

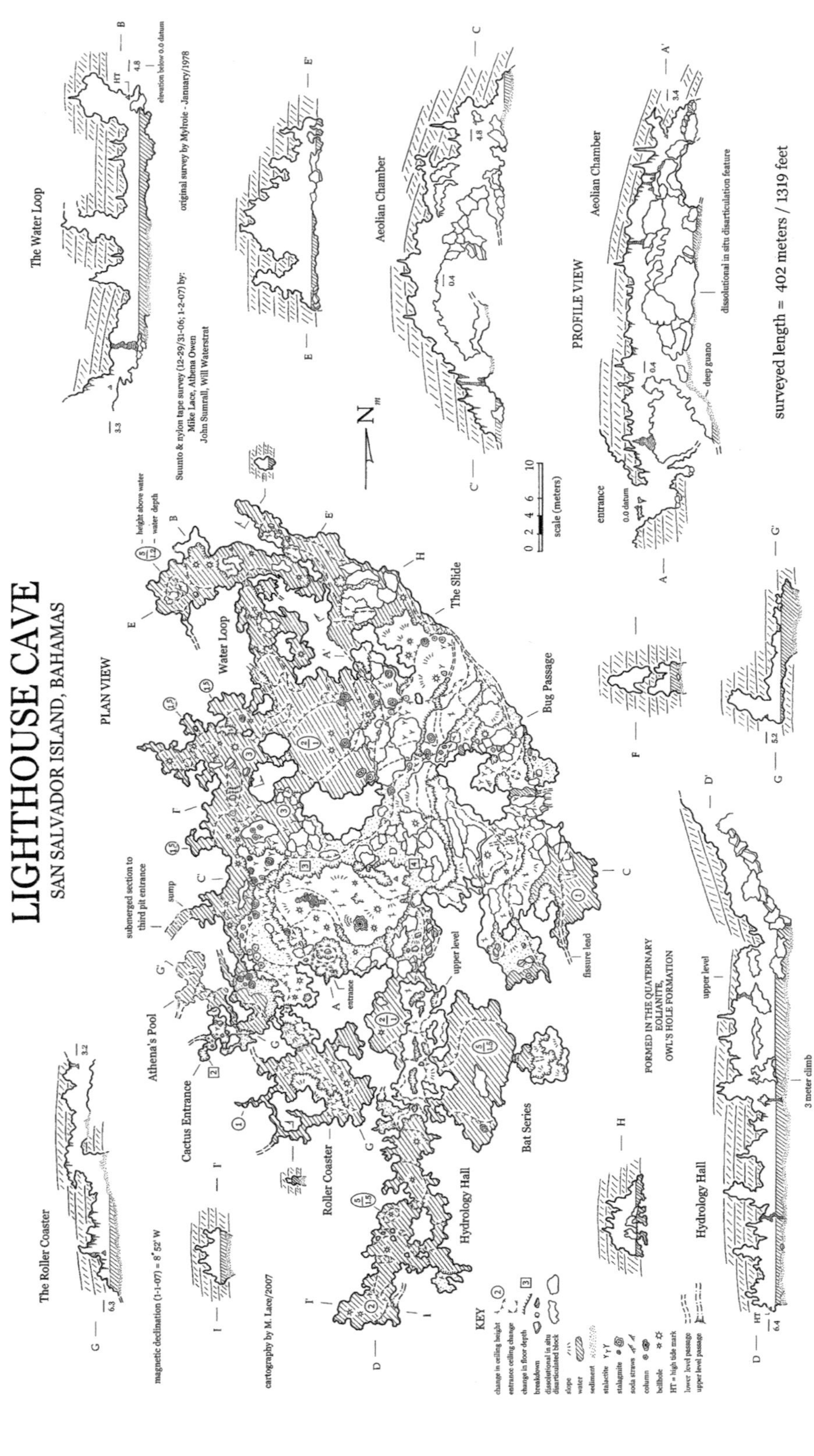

Fig. 7.16 Map of Lighthouse Cave, San Salvador Island, Bahamas. Note that north is to the *right*

Fig. 7.17 Dissolutional pockets and cusps, Lighthouse Cave, San Salvador Island, Bahamas. Note eolian bedding dipping *left* to *right*

such passages. The passages in Lighthouse Cave contain many bellholes, which were the subject of a detailed study (Dogwiler 1998) that concluded they were phreatic, as opposed to condensation corrosion, features (see also Birmingham et al. 2010).

The vast majority of flank margin caves in the Bahamas are dry, a result of their having formed in a fresh-water lens elevated on the last interglacial sea-level highstand. The extensive passages at sea level in Lighthouse Cave, as well as in the southern end of Hatchet Bay Cave (and a few other caves in the Bahamas) do not fit a last interglacial model of origin. It may well be that the sea-level portions of these caves are paleokarst from an earlier (pre-last interglacial) sea-level highstand. The subsidence rate of the Bahamas is 1–2 m per 1,000 years (Carew and Mylroie 1995a). Features, such as fossil reefs and flank margin caves, formed on earlier sea-level highstands are believed to have subsided below the levels of modern terrestrial observation. However, these sections of phreatic cave found at modern sea level today in Lighthouse Cave and elsewhere could be such remnants. No geochronologic data are available to confirm or refute this supposition.

Other Caves: As noted above regarding the extensive amount of research that has occurred on San Salvador Island as a result of the presence of the Gerace Research Centre, more is known about the caves and karst of San Salvador than any other Bahamian Island. However, only Lighthouse Cave falls in the large cave category. At least 22 other flank margin caves, in the 640 m^2 to 14 m^2 areal footprint size range, have been mapped on the island. Notable among these are Altar Cave, Beach Cave, Garden Cave, and Majors Cave. Alter Cave is one of the largest syndepositional caves in the Bahamas.

7.6.6 Long Island

Salt Pond Cave: Located in the middle of Long Island, to the west of the main road just north of the community of Salt Pond, is Salt Pond Cave. The cave has an areal footprint of 3,901 m^2 (Fig. 7.18). The cave has a large hillside walk-in entrance on its west side, and some collapse entrances immediately nearby. The initial chamber is large, with short side passages to the north. The east wall of this chamber has a single 3 m diameter opening that leads eastward into a large

Fig. 7.18 Map of Salt Pond Cave, Long Island, Bahamas. Note how the large inner (*right*) chamber has a single short passage connecting it to the outer (*left*) chamber. The inner chamber formed when the mixing front in the outer chamber stepped inward into the fresh-water lens. Trending northeast from the inner chamber is a large tubular passage (see Fig. 7.19) that ends abruptly in a bedrock wall. This wall represents the location of the mixing front when sea level fell at the end of the last interglacial

Fig. 7.19 Large tube in Salt Pond Cave, Long Island, Bahamas, ending in a bedrock wall (See caption for Fig. 7.18)

second chamber. This inner chamber has small side passages to the south and northwest, but to the northeast a long tubular, passage trends for over 50 m north-northeast before ending in a blank bedrock wall (Fig. 7.19). The sudden end of the passage represents the position of an inward migrating mixing front when the sea level dropped, the cave drained, and phreatic dissolution ceased. The wide horizontal size of the inner and outer chambers, with their relatively restricted vertical range, clearly demonstrates the development of the cave in the thinning distal margin of a past fresh-water lens. The geology of Salt Pond Cave is discussed in Mylroie et al. (1991) and Lascu (2005).

Hamilton's Cave: This cave is located in the Hamilton community, just east of the main road, between Deadmans Cay to the north and Clarence Town to the south. It has a linear survey record of 1,030 m, and an areal footprint of 8,931 m^2, making it one of the largest caves in the Bahamas. The cave trends southeast to northwest, following the side of the eolianite ridge that contains it (Fig. 7.20). To the northwest, the main entrance is provided by a large collapse that truncates the cave from Hamilton's Annex Cave, which consists of two large rooms with a few short side passages at the extreme northwest end. From the collapse entrance, the cave winds southwest, then east as a series of very large passages, some of which contain immense flowstone columns. Numerous ceiling skylights exist in the southwest trending portion. The cave has a small (1×2 m) connection to the southeastern portion of the cave, which shows more passage complexity and interconnections than the northwestern part of the cave. Many dead-end passages exist here, some with T-like configurations that come close to, but do not connect with, other nearby passages (Fig. 7.21). These connection failures are expected in a mixing dissolution cave without conduit or turbulent flow. Hamilton's Cave is described in greater detail in Mylroie et al. (1991) and Lascu (2005).

7.6.7 Crooked Island

1702 Cave: At the north end of Crooked Island, a few kilometers east of Pitts Town Landing, is a large cave called 1702 Cave (Fig. 7.22). The cave has an areal footprint of 3,957 m^2. From its northernmost point, the cave initiates as isolated

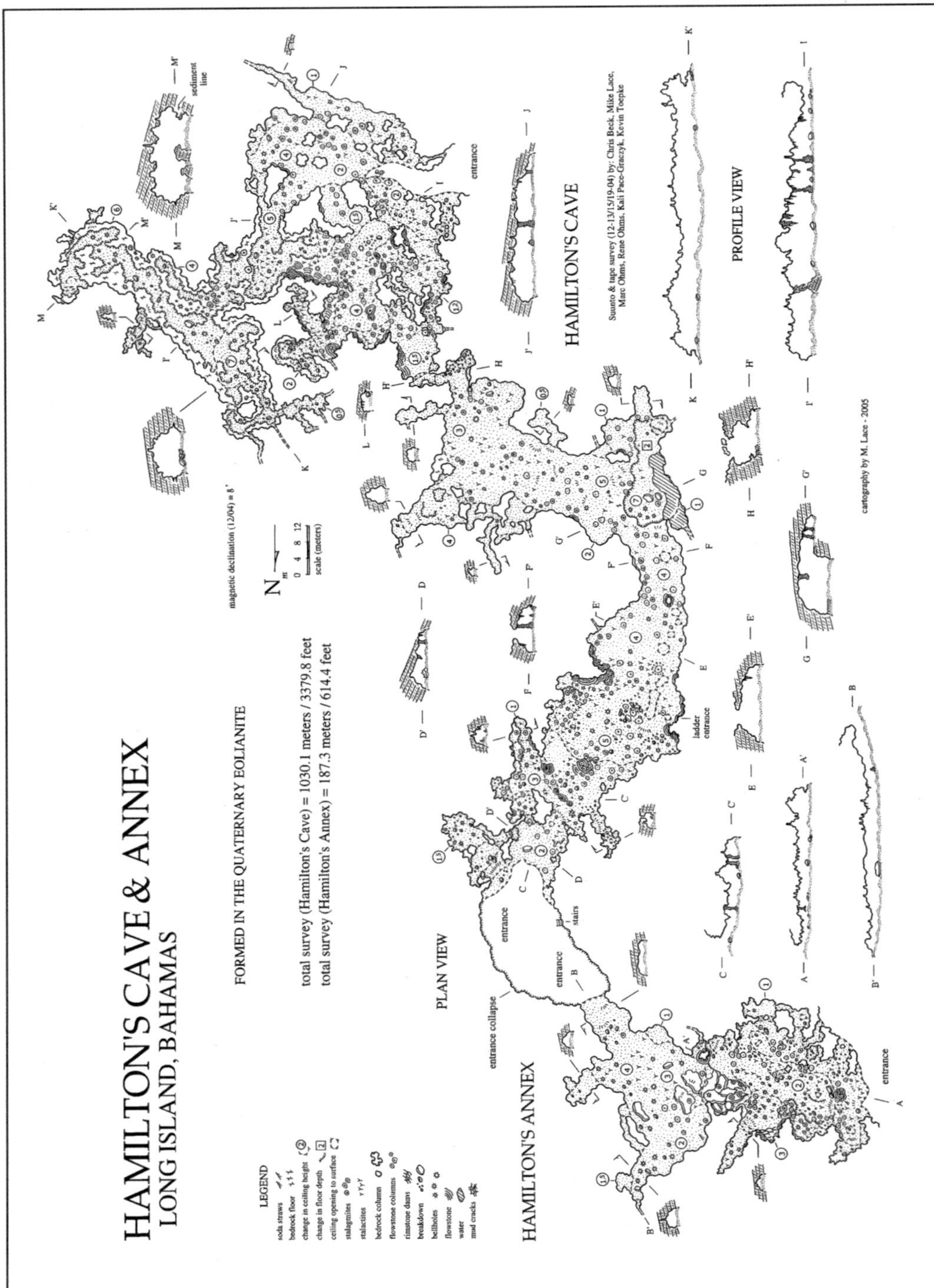

Fig. 7.20 Map of Hamilton's Cave, Long Island, Bahamas. Note that north is to the *left*

Fig. 7.21 Large phreatic pocket in Hamilton's Cave, Long Island, Bahamas. Note the eolian foreset beds dipping *upper left* to *lower right*, while the axis of the phreatic pocket is horizontal, cutting across the foreset beds. The *brown* to *white* transition in the wall color represents the elevation of guano, now mined away

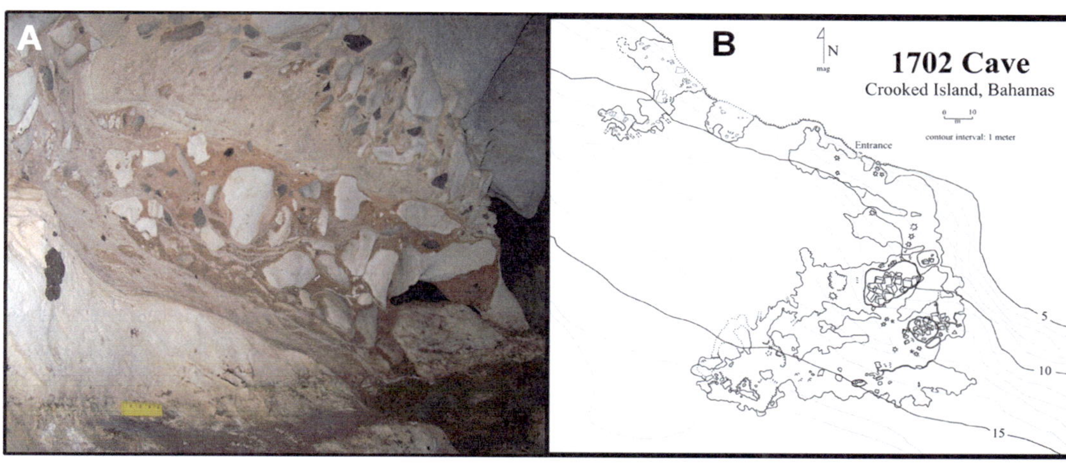

Fig. 7.22 1702 Cave, Crooked Island, Bahamas. (**a**) Paleosol infill material in 1702 Cave. Phreatic dissolution has cut smoothly through the paleosol matrix, included clasts, and wall rock of a remnant bedrock pillar. Such clean-cut dissolutional surfaces reflect the geochemical power of mixed-water dissolution. *Rectangle* in *lower left* is 10 cm long for scale. (**b**) Map of 1702 Cave. Note how the cave follows the hillside morphology, bending as the fresh-water lens margin that formed the cave would have curved following the paleocoastline

hillside pockets, then the main cave trends southeast and then south just inside the ridge flank as large passages and chambers. The change in cave direction to the south mimics a change in ridge orientation on the surface. The southern portion of the cave has chambers that intersect a paleosol that separates an overlying younger eolianite from the older eolianite that contains

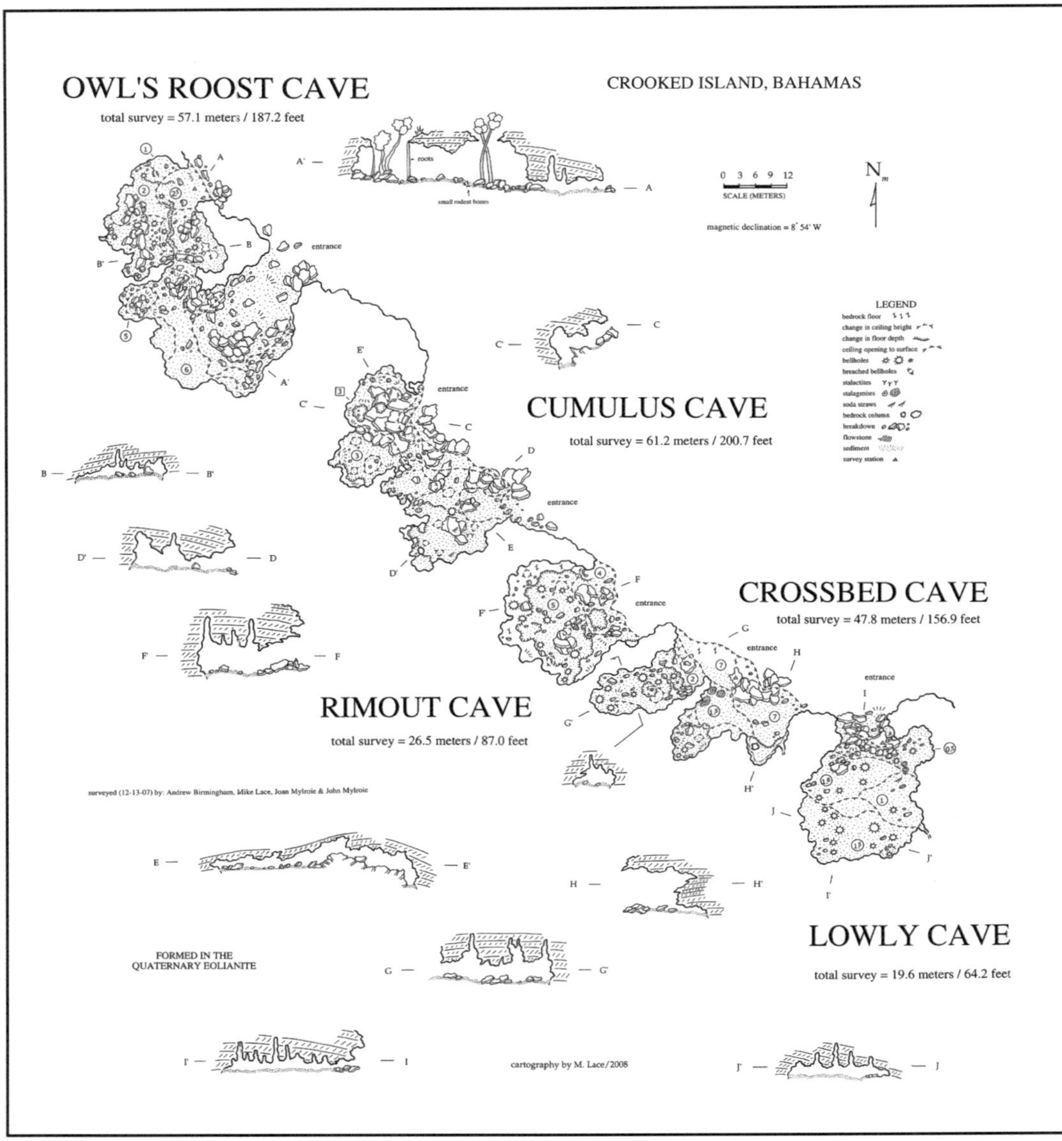

Fig. 7.23 Maps of a series of small caves east of 1702 Cave. The caves display the "beads-on-a-string" pattern (Chap. 4) common to flank margin cave development

most of the cave. The cave has intersected an old, paleosol-infilled dissolution pit (pit cave), and cut a dissolution surface through both bedrock and paleosol as a smooth surface (Fig. 7.22). A description of the cave and its geology can be found in Lascu (2005).

Continuing along the ridgeline to the southeast, six more flank margin caves, of small to medium size, are located. They appear to be of the same approximate size (~200 m^2), and equally spaced (Fig. 7.23). Their size and spacing relationships may reveal information about how the fresh-water lens discharges, and how mixing dissolution creates and enlarges voids in the distal margin of the lens.

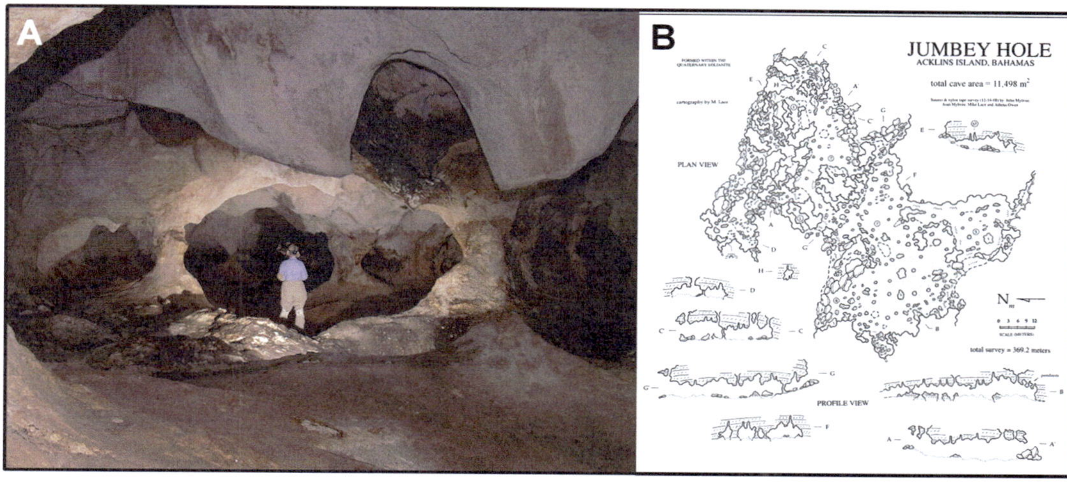

Fig. 7.24 Jumbey Hole, Acklins Island, Bahamas (**a**) Image of the phreatic cave morphology in the cave. (**b**) Map of Jumbey Hole

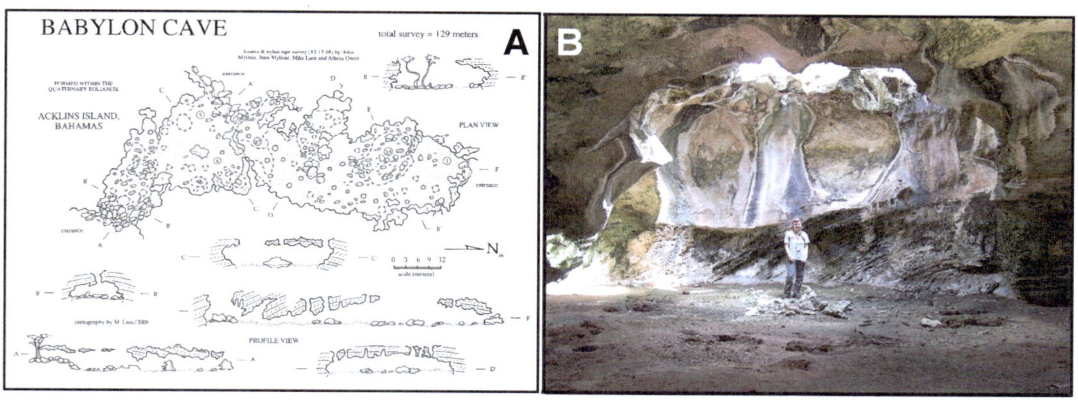

Fig. 7.25 Babylon Cave, Acklins Island, Bahamas. (**a**) Map of Babylon Cave. (**b**) Chamber illuminated by daylight

7.6.8 Acklins Island

Acklins Island is actually an extension of Crooked Island, separated by a shallow, narrow channel. The island follows the edge of the bank east and south, and like Long Island, Cat Island, Eleuthera Island, and Abaco Island, all to the north, is a very long, linear piece of land on the windward side of its bank. Reconnaissance work to find, catalog and map caves on Acklins Island has just begun, but a few significant caves have been located. At the north end of the island, east of the village of Chesters, is Harbour Cave, a large single chamber cave. About one-third of the way south, in the vicinity of Goodwill, are a series of caves in one section of dune ridge, the largest of which is Jumbey Hole (Fig. 7.24), a very large chamber with numerous bedrock pillars, alcoves and mazy areas, especially to the north. Far to the south, near Salina Point, is another collection of flank margin caves, some extend to sea level and have wide, guano-floored pools, such as Nibbles Cave and Wishing Well Cave. Nearby Babylon Cave is highly degraded, with daylight illuminated rooms and passages (Fig. 7.25).

7.6.9 Other Bahamian Islands

Great Inagua Island contains a number of small flank margin caves east of Mathew Town, the single village on the island. North, and east past Palacca Point, is a very large flank margin cave that has not yet been mapped. Little Inagua Island is uninhabited and little is known of caves there. North and South Andros Islands have small flank margin caves (see Carew et al. 1998 for South Andros data), and Mangrove Cay, located between North and South Andros, is rumored to have a large bat cave. Caves are also known on Exuma Island from recent bat research, but have as yet not been mapped. Mayaguana Island also has caves, again not mapped. Rum Cay has several flank margin caves, the most famous of which is Hartford Cave, known for its extensive petroglyphs (Fig. 7.4). Work continues in the archipelago to document the subaerial caves of the Bahamas.

7.7 Summary

The Bahamas are a 100 % carbonate environment. These carbonates are eogenetic – youthful, diagenetically immature, with high primary porosity. They do not behave like limestones from continental interiors. The ground-water environment is dominated by the fresh-water lens. As a result, cave and karst development proceeds differently than that observed in continental interiors. The Carbonate Island Karst Model, or CIKM, takes into account these factors, as well as the control of sea level, and hence the fresh-water lens, by glacioeustasy. The Bahamas are the simplest possible aspect of the CIKM. There are no tectonic influences, and no non-carbonate rock to produce allogenic recharge, or to partition the lens. Despite these simplifications, and the short time intervals of sea-level highstands in which speleogenetic processes could act, the Bahamas have a wide variety of caves of significant complexity. That these caves formed in a 9,000 year time window makes them even more impressive. The Bahamas have been famous for years for their blue holes and large, extensive underwater cave systems. Hopefully this chapter has effectively introduced the reader to the abundant and impressive dry caves of the Bahamas.

7.7.1 Field Research Logistics

The Bahamas are a convenient country to visit and to conduct cave research. Nassau, on New Providence Island, is the capital of the country, and the hub of travel. One can fly from Europe and Latin America directly to Nassau, and avoid the travel and paperwork problems of entering and departing from the United States. It is not possible, except by charter, to fly amongst the islands themselves. So travel time and cost is elevated by the need to go into Nassau and out again to visit a new Bahamian island. Scientific research cannot be done in the Bahamas without a government permit. Otherwise, cave access is generally easy, with most landowners willing to let visitors go in their caves.

References

Albury P (1975) The story of the Bahamas. St Martin's Press, New York, 294 p

Birmingham AN, Lace MJ, Mylroie JR, Mylroie JE (2010) Bell hole origin: constraints on developmental mechanisms, Crooked Island, Bahamas. In: Martin JB, Siewers FD (eds) Proceedings of the 14th symposium on the geology of the Bahamas and other carbonate regions, pp 18–30. Gerace Research Centre, San Salvador. Reprinted in Speleogenesis and Evolution of Karst Aquifers, 2010, Issue 10 (http://www.speleogenesis.info/content/)

Carew JL, Drost DM, Sealey NE, Mylroie JE (1995) Refracted images of Bahamian islands, and possible implications regarding the first landfall of Columbus. Bahamas J Sci 2(3):29–33

Carew JL, Mylroie JE (1995a) Quaternary tectonic stability of the Bahamian Archipelago: evidence from fossil coral reefs and flank margin caves. Quat Sci Rev 14:144–153

Carew JL, Mylroie JE (1995b) A stratigraphic and depositional model for the Bahama Islands. In: Curran HA, White B (eds) Terrestrial and shallow marine geology of the Bahamas and Bermuda. Geological Society of America Special Paper 300, Boulder, pp 5–31

Carew JL, Mylroie JE (1997) Geology of the Bahamas. In: Vacher HL, Quinn TM (eds) Geology and hydrogeology of carbonate islands. Elsevier Science Publishers, Oxford, pp 91–139

Carew JL, Mylroie JE, Schwabe SJ (1998) The geology of South Andros Island Bahamas: a reconnaissance report. Cave Karst Sci 25(2):57–66

Chen JH, Curran HA, White B, Wasserburg GJ (1991) Precise chronology of the last interglacial period: ^{234}U-^{230Th} data from fossil coral reefs in the Bahamas. Geol Soc Am Bull 103:82–97

Curran HA, White B (eds) (1995) Terrestrial and shallow marine geology of the Bahamas and Bermuda, Geological Society of America Special Paper 300, Boulder, 344 p

Davis RL, Johnson CR Jr (1989) Karst hydrology of San Salvador. In: Mylroie JE (ed) Proceedings of the fourth symposium on the geology of the Bahamas, Port Charlotte Florida, Bahamian Field Station, Gerace Research Centre, san Salvador, Bahamas, pp 73–81

Dogwiler TJ (1998) Analysis of bell hole morphology and distribution: a tool for evaluating formational processes. M.Sc. thesis, Mississippi State University, 106 p

Hearty PJ (1997) Boulder deposits from large waves during the last interglaciation on north Eleuthera Island, Bahamas. Quat Res 48:326–338

Horowitz MH, Roberts TM (2010) Geomorphic zoning and eogenetic karst on limestones within the supratidal environment: San Salvador, Bahamas. Geologia 55(1):17–27

Keegan WF (1992) The people who discovered Columbus: the prehistory of the Bahamas. The University of Florida Press, Gainesville

Kindler P, Hearty PJ (1995) Pre-Sangamonian eolianites in the Bahamas? New evidence from Eleuthera Island. Mar Geol 127:73–86

Kindler P, Mylroie JE, Curran HA, Carew JL, Gamble DW, Rothfus TA, Savarese M, Sealey NE (2010) Geology of Central Eleuthera, Bahamas: a field trip guide. Gerace Research Centre, San Salvador, 74 p

Lascu I (2005) Speleogenesis of large flank margin caves of the Bahamas, M.Sc. thesis, Department of Geosciences, Mississippi State University, Mississippi State, MS, 218 p

Meyerhoff AA, Hatten CW (1974) Bahamas salient of North America: tectonic framework, stratigraphy, and petroleum potential. Am Assoc Petrol Geol Bull 58:1201–1239

Mylroie JE (2008) Late Quaternary sea level position: Bahamian carbonate deposition and dissolution cycles. Quat Int 183:61–75

Mylroie JE, Carew JL (1994) A field trip guidebook of lighthouse cave, San Salvador Island, Bahamas, Bahamian Field Station, San Salvador, Bahamas, 10 p

Mylroie JE, Mylroie JR (2007) Development of the Carbonate Island Karst model. J Cave Karst Stud 69:59–75

Mylroie JE, Mylroie JR (2009) Coastal eogenetic karren of San Salvador Island. In: Gines A, Knez M, Slabe T, Dreybrodt W (eds) Karst rock features Karren sculpturing. Karst Research Institute/ZRC Publishing, Ljubljana, pp 475–485

Mylroie JE, Carew JL, Sealey NE, Mylroie JR (1991) Cave development on new providence Island and long Island, Bahamas. Cave Sci 18(3):139–151

Mylroie JE, Carew JL, Vacher HL (1995a) Karst development in the Bahamas and Bermuda. In: Curran HA, White B (eds) Terrestrial and shallow marine geology of the Bahamas and Bermuda, Geological Society of America Special Paper 300, Boulder, pp 251–267

Mylroie JE, Carew JL, Moore AI (1995b) Blue holes: definition and genesis. Carbonates Evaporites 10(2):225–233

Mylroie JE, Carew JL, Curran HA, Freile D, Sealey NE, Voegeli VJ (2006) Geology of Cat Island, Bahamas: a field trip guide. Gerace Research Centre, San Salvador Island, 43 p

Mylroie JE, Mylroie JR, Owen AM, Waterstrat WA (2008) Cave and karst inventory of the Primeval Forest, New Providence Island, Bahamas: unexpected discoveries. In: Freile D, Park L (eds) Proceedings of the 13th symposium on the geology of the Bahamas and other carbonate regions, Gerace Research Centre, San Salvador, Bahamas, pp 107–118

Mylroie JE, Carew JL, Curran HA, Godefroid FM, Kindler P, Sealey NE (2012) Geology of New Providence Island, Bahamas: a field trip guide. Gerace Research Centre, San Salvador, 57 p

Palmer RJ (1986) Preliminary studies of speleogenesis on Cat Island, Bahamas. Cave Sci 13(2):79–82

Palmer AN (1991) Origin and morphology of limestone caves. Geol Soc Am Bull 103:1–21

Palmer RJ, McHale M, Hartlebury R (1986) The caves and blue holes of Cat Island. Cave Sci 13(2):71–78

Palmer AN, Palmer MV (eds) (2009) Caves and karst of the USA. National Speleological Society, Huntsville, pp 348–353

Panuska BC, Mylroie JE, Carew JL (1999) Paleomagnetic evidence for three Pleistocene paleosols on San Salvador Island. In: Curran HA, Mylroie JE (eds) Proceedings of the ninth symposium on the geology of the Bahamas and other carbonate regions. Bahamian Field Station, San Salvador Island, Bahamas, pp 93–100

Roth MJ (2004) Inventory and geometric analysis of flank margin caves of the Bahamas. MSc thesis, Department of Geosciences, Mississippi State University, 117 p

Sealey NE (1990) The Bahamas today. MacMillan Education Ltd., London, 120 p

Thompson WG, Curran HA, Wilson MA, White B (2011) Sea-level oscillations during the last interglacial highstand recorded by Bahamian corals. Nat Geosci 4: 684–687

Uchupi E, Milliman JD, Luyendyk BP, Brown CO, Emery KO (1971) Structure and origin of the southeastern Bahamas. Am Assoc Petrol Geol Bull 55:687–704

Vacher HL, Quinn TM (eds) (1997) Geology and hydrogeology of carbonate islands. Elsevier Science Publishers, Amsterdam, 948 p

Vacher HL, Wallis TN (1992) Comparative hydrogeology of fresh-water lenses of Bermuda and Great Exuma Island, Bahamas. Ground Water 50:15–20

Walker AD (2006) Bahamian cave and karst geodatabase, and GIS analysis of San Salvador Island, Bahamas. M.Sc. thesis, Mississippi State University, 81 p

Waterstrat WJ (2007) Morphometric differentiation of flank main caves and littoral, or sea caves. M.Sc. thesis, Mississippi State University, 201 p

Waterstrat WJ, Mylroie JE, Owen AM, Mylroie JR (2010) Coastal caves in Bahamian eolian calcarenites. Differentiating between sea caves and flank margin caves using quantitative morphology. J Cave Karst Stud 72:61–74

Wilson, WL (1994) Morphology and hydrology of the deepest known cave in the Bahamas: Dean's Blue Hole, Long Island. In: Boardman MR. (ed) Seventh symposium on the geology of the Bahamas. Bahamian Field Station, San Salvador Island, Bahamas, p 21

Winters JH (2009) Rock art of the Bahamian Archipelago. In: Hayward MH, Atkinson LE, Cinquino MA (eds) Rock art of the Caribbean. University of Alabama Press, Tuscaloosa, pp 13–21

Tafoni Development in the Bahamas

8

Athena M. Owen

Abstract

Tafoni are cuspate pseudokarst weathering features with sloping sediment covered floors, and overhangs and projections, occurring in many rock types. Tafoni appear to be polygenetic. Tafoni have been confusingly defined in many ways: variations in size, rock type, and formation mechanisms. This study addresses tafoni in Quaternary eolian carbonates of the Bahamas to help better define the term. Both large (meters) and small (decimeters) tafoni were studied and comparisons were made between tafoni occurring in Pleistocene and Holocene aged rock units. The differentiation between the main cave types on these islands: tafoni, flank margin caves, and sea caves, is important as all three cave types form in the same area, but flank margin and sea caves can be used as paleo-sea level indicators, while tafoni cannot. Small tafoni were measured and studied from cliff faces and various cultural locations and show a growth rate of 0.022 m^3/year; and may amalgamate to form larger tafoni, which grew at faster rate of 0.65 m^3/year as a result of that amalgamation. Petrographic analysis was done to help identify tafoni-forming mechanisms; results revealed no evaporites present, removing crystal wedging as a formation mechanism in this environment, while other data indicated wind erosion as the primary mechanism.

8.1 Introduction

What are tafoni? The term *tafoni* (*tafone* is the singular) is used widely in the geologic literature; however, it is poorly defined. The literature review for this project used several papers from peer reviewed journals, other articles, and textbooks, and each provided a different definition of the term. The Glossary of Geology, which may represent the ultimate authority in this case, defines tafone as: "tafone (ta-fo'-ne) (a) A Corsican dialect term for one of the natural cavities in a honeycomb structure, formed by *cavernous weathering* on the face of a cliff in a dry region or along the sea shore. The hole or recess may reach a depth of 10 cm, and is

A.M. Owen (✉)
Department of Geosciences, Mississippi State University, Mississippi State, MS 39762, USA
e-mail: amo58@msstate.edu

M.J. Lace and J.E. Mylroie (eds.), *Coastal Karst Landforms*, Coastal Research Library 5,
DOI 10.1007/978-94-007-5016-6_8, © Springer Science+Business Media Dordrecht 2013

explained as the result of solution of free salts in crystalline rock following heating by insolation. (b) A granitic or gneissic block or boulder hollowed out by cavernous weathering. Pl: tafoni" (Neuendorf et al. 2005: 655). The original reference to tafoni comes from the island of Corsica around 1882 (Tucket 1884, in Brandmeier et al. 2010), the term comes from the Corsican word "tafonare" meaning to be perforated; the term is morphological in origin and not based on genesis (Wilhelmy 1981, in Rögner 1988). The scientific literature has provided much contrast to this definition, varying in terms of shape and size of the void, mechanism of formation, orientation, climate, and rock type. Given that tafoni are described as being formed by cavernous weathering, they appear to meet the definition of pseudokarst, that is, "A karst-like terrain having closed depressions, sinking streams, and caves, but produced by processes other than dissolving of rock" (Neuendorf et al. 2005, p. 523).

Features believed to be tafoni have been described in the Bahamas from Abaco, San Salvador, Cat, Long, New Providence (Walker 2006; Walker et al. 2008; Owen 2007) and Crooked Islands. Tafoni features in the Bahamas can resemble typical island karst caves produced by dissolutional processes, their analysis in this setting is important for a number of reasons. Certain cave types, such as flank margin caves, are used as paleo sea-level indicators in carbonate islands. If tafoni caves are present, and have been misidentified as flank margin caves, then past sea-level interpretations will be in jeopardy. Additionally, flank margin caves are used as proxies for past fresh-water lens configuration and behavior, so misidentification of tafoni as flank margin caves would make such lens interpretations incorrect. Tafoni can also be confused with sea caves, and create an impression of a sea-level position where none existed.

Study of these features is also important for the interpretation of the term itself. For example if these cave features are identical in physical description to tafoni caves found in non-karstic rocks, but their mechanism of formation is by carbonate dissolution, do they represent a new category of karst cave? Given that tafoni are rarely reported from carbonate rocks, are there unique aspects of the Quaternary carbonate eolianites of the Bahamas that facilitate tafoni development (Owen 2007)?

8.2 General Geologic Overview of the Bahamas

The Bahamian Archipelago is a northwest-southwest trending island chain, 1,400 km long, extending from the east coast of Florida to just off the coast of Cuba to the southwest, and then continuing southeast towards the Turks and Caicos Islands (Carew and Mylroie 1997 and references therein). The Bahamas show no evidence of active tectonics Carew and Mylroie (1995a). San Salvador Island, where the majority of the tafoni research was done, is one of the outermost islands in the Bahamian Archipelago, located on the east side of the Bahamian island chain (Fig. 8.1). The island is composed of young limestones, Pleistocene and Holocene in age. Included in these limestones are eolianites, beach rock, fossilized coral reefs, beach facies, and paleosols (Carew and Mylroie 1995b). San Salvador also has a wide variety of karst features, including flank margin caves, banana holes, blue holes, lake drains, and pit caves (Mylroie 2005). Additional information about the geology and caves of the Bahamas is presented in greater detail in Chap. 7.

8.3 Tafoni Background

Tafoni have been described and defined in many ways. The lack of an exact definition, and the overlap of other terms such as honeycomb and alveoli, along with the various weathering techniques believed to go with each of these terms, makes it very difficult for researchers to recognize the differences between these features (Turkington 1998). Honeycombs are described as having a cell-like structure present in a rock or soil (Fig. 8.2a, b), while alveoli is defined as "a space or cavity" (Neuendorf et al. 2005). McBride and Picard (2004) differentiate

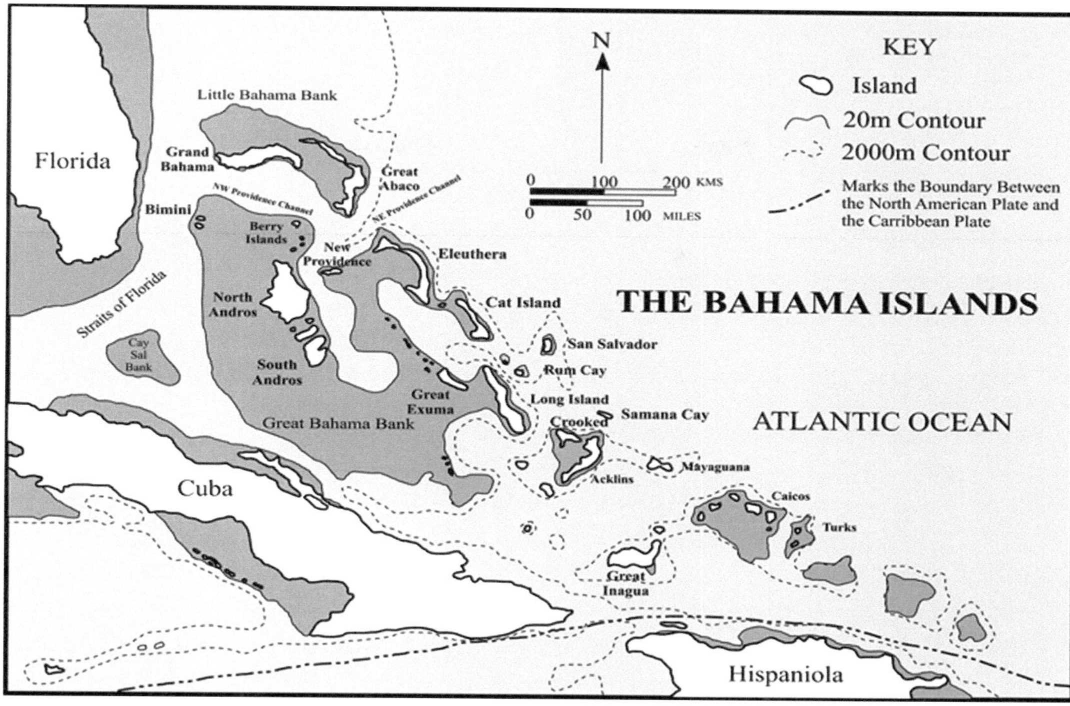

Fig. 8.1 A map of the Bahamian Archipelago (Carew and Mylroie 1995a)

honeycomb from tafoni by the fact that tafoni can have solitary occurrences, where honeycombs must be in clusters. Tafoni have also been defined as a weathering feature resulting from honeycomb weathering (Maroukian et al. 2010). The simplest and most inclusive definition of tafoni comes from Campbell (1999), stating that tafoni are small holes or caves in rock that result from some type of weathering (Fig. 8.2c, d). The definitions have been linked by many to size, shape, rock type, and by some to climate, although the word seems to be used in the current literature to represent any form of weathered hole in a rock face or boulder. Sidewall tafoni are distinguished by, and occur on, near-vertical or vertical surfaces; tafoni occurring at ground level are known as basal tafoni (Wilhelmy 1964 in Turkington 1998).

The Glossary of Geology, as previously noted, defines tafoni as having a depth of 10 cm, however, tafoni size as discussed in the literature is quite variable. Tafoni are described as meter-scale openings with overhangs and projections (McBride and Picard 2000). Mellor et al. (1997), also describes tafoni at this large scale, stating tafoni to be several cubic meters in volume (as seen in Fig. 8.2c, d).

Tafoni are believed by many (Turkington 1998; Turkington and Phillips 2004; Hacker 2003; Maroukian et al. 2010) to be the result of cavernous weathering or honeycomb weathering. Cavernous weathering is the result of chemical and mechanical processes on a cliff face, resulting in the removal of the grains resulting in large hollows. Honeycomb weathering is the result of chemical process, resulting in numerous pits in the rock surface. Honeycomb weathering is common in tuffs and sandstones, often occurring in arid regions (Neuendorf et al. 2005:305). Brandmeier et al. (2010) uses the term tafone weathering synonymously with cavernous weathering. Tafoni have been located in several climates and locations ranging from Antarctica to Mars (Cooke et al. 1993; Brandmeier et al. 2010). The most common climates and settings in which tafoni are found are arid climates and along

Fig. 8.2 (**a**) Cavernous weathering showing honeycomb or aveoli, clastic rocks, west coast of North Island, New Zealand. (**b**) Cavernous honeycomb weathering on Kanmantoo schists, Kangaroo Island, Australia. (**c**) Tafone in the Remarkable Rocks Granite, Kangaroo Island, Australia; note overhanging walls and spherical interior void. (**d**) Tafone in gneiss, Margaret River, Western Australia (Photo by J.E. Mylroie)

coastal cliff faces. According to Rögner (1988), the term tafoni was coined in the Mediterranean subtropic zone, and must be developed in that or similar climate conditions or the tafoni are said to be no longer active, and represent paleoclimate indicators, showing when the climate was similar to that of the Mediterranean subtropic zone. Tafoni occurring in dry arid environments are inactive, and represent wetter paleoclimates (Rögner 1988).

While it is agreed upon that tafoni are caused by some form of weathering, the actual mechanism is disputed. Some of the proposed methods of tafoni development are salt transport, salt weathering, case hardening, core softening, flaking and scaling, and increased air circulation (Brandmeier et al. 2010; Huinink et al. 2004; Rögner 1988). The salt weathering technique is the most commonly supported mechanism in the literature, with case hardening and core softening also well supported. Perhaps the mechanism is dependent on the actual rock type present, and the climate in which it is located, and tafoni caves are therefore polygenetic in origin.

The majority of the literature on tafoni is on the proposed methods of formation of tafoni, the rate at which they develop, and the reasons for the location of their occurrence. The development of evaporite minerals, especially halite, along rock faces is the most supported mechanism for tafoni development. This process is believed to occur when evaporation takes place and salt water is drawn up through capillary action through the rock, the water is then evaporated from the rock surface, causing the salt to crystallize. The water reaches the deepest area of the tafone first, resulting in the most salt crystallization occurring in this interior area. This results in the back of the cave weathering faster, inducing the concave shape. The weathering effect is also increased as the moisture content goes up, causing minerals in the rocks to swell (Hacker 2003 and references there in). Hacker's (2003) research supported this method of tafoni development, indicating higher concentrations of salt in the walls and interior openings within the tafoni than on the floors and in the surrounding rocks. In arid regions, the tafoni development is also caused by halite; however, the salt moves in the wind and crystallizes in the tafoni caves after transport. This is supported by McBride and Picard's (2000) work on desert tafoni, in which tafoni did not develop more than 10 km from the salt source in a playa. Brandmeier et al. (2010) cites salt wedging as the primary formation mechanism, however they link the location of formation to localized faults systems, and rock cleavage. Their work in Corsica also showed the major salts present to be calcium and sodium sulfates as opposed to halite, even in coastal areas.

Flaking and scaling is another mechanism proposed to support the formation of tafoni. Flaking and scaling is induced by salt crystallization. Flaking is defined as the weathering of rocks parallel to the surface resulting in flakes 1–10 mm thick, whereas scaling results in pieces 1–10 cm thick. The evaporation of moisture from the rock containing a large load of salts creates salt horizons that result in pressure, and promotes flaking and scaling. Well-developed tafoni have a higher rate of flaking and scaling than lesser developed tafoni. The rate of flaking and scaling also seem to be dependent on moisture and/or precipitation present (Rögner 1988).

The idea of case hardening is that the outside of the rock is often harder than the inside; this could be from differences in composition, or by some other phenomenon (Mottershead and Pye 1994). The hardening results in an increased resistivity to weathering. If the crust of the rock is damaged in some way, the inside of the rock can come in contact with the air and moisture, which possibly results in the formation of tafoni. There have been many cases where tafoni have developed in areas where case hardening has occurred, however, there are also several cases where tafoni occurred, but the case hardening did not (Huinink et al. 2004). The idea of core softening is also based on rock strength differences, and happens in much the same way as case hardening. The core is believed to become softened by leaching of minerals out of the core, due to moisture content (Huinink et al. 2004 and references there in). The core is softened and again the outside is damaged and the tafoni development occurs at that point. This action seems to happen at some tafoni locations but not all of them (Huinink et al. 2004).

Hacker (2003) showed the tafoni development could also be related to microclimate changes. After looking at the distribution of the tafoni, there was a noticeable lack of tafoni occurring in areas that were never exposed to direct sunlight. Dew points were recorded inside and outside the tafoni caves, showing the dew point to be higher by an average of 1.2 °C inside the tafoni cave compared to outside the caves. A similar comparison was done for temperature, and the variance was found to be much lower, with a mean of 0.1 °C (Hacker 2003). Rögner's (1988) study shows that climate can have a limited effect on the rate of flaking and scaling.

It has also been said that wind circulation could promote the development of the tafoni caves. Increased wind circulation inside the tafoni would promote higher drying rates and therefore increase the rate of salt crystallization along the inside of the tafoni caves (Huinink et al. 2004 and references there in). Brandmeier et al. (2010) found tafoni distributions in some areas to align with prevailing wind directions, but this was not believed to be the primary forming mechanism. It is still in dispute if wind really promotes the tafoni growth or not. Some of this dispute is fueled by the lack of a specific definition of tafoni; it is hypothesized that wind could play a factor in small openings, but probably not in large openings (Huinink et al. 2004).

Another question is to what extent karst processes are considered to be chemical weathering, or cavernous weathering. Tafoni are not listed in karst texts and papers as being part of the

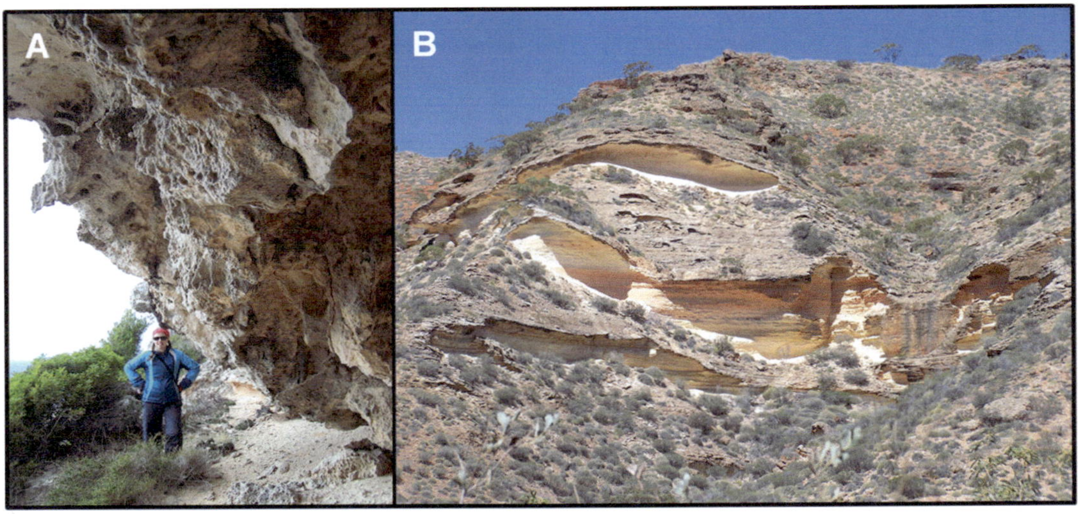

Fig. 8.3 Tafoni are rarely reported from carbonate rocks. (**a**) Tafone in quaternary eolianites, southwest coast of Mallorca. (**b**) Tafoni in Miocene limestones, Exmouth, Western Australia (Photo by J.E. Mylroie)

karst landform regime, so by exclusion tafoni are not karst features. This exclusion is most likely the result of most tafoni being the product of weathering in non-carbonate rocks. If cavernous weathering produces tafoni in carbonate rocks, are they karst features or not? Or, a broader question, can carbonate rocks host this type of pseudokarst?

Following the second definition given in the Glossary of Geology, tafoni could be considered to only occur in granitic or gneissic rocks granite (Fig. 8.2c, d) and tafoni have commonly been described in granite, (Hacker 2003; Guglielmin et al. 2005; Viles 2005; McBride and Picard 2000, 2004; Matsukura and Tanaka 2000; Maroukian et al. 2010; Brandmeier et al. 2010); however, the term tafone(i) has been used to describe openings resulting from weathering in sandstones (Turkington 1998; Souza-Egipsy et al. 2004; Sunamura 1996), greywacke, arkoses, tuffaceous conglomerates, (Sunamura 1996), metamorphosed conglomerate (Hacker 2003) limestones (both oolitic and with calcareous sands) (Sunamura 1996; Norwick and Dexter 2002; Rögner 1988), breccia (Campbell 1999), rhyolite tuff (Cooke et al. 1993; Stoffer 2004), dolomite (Norwick and Dexter 2002; Rögner 1988) siltstone (Norwick and Dexter 2002) and flysch deposits (Mellor et al. 1997). Although tafoni have been cited in the literature as occurring in limestone and other carbonates, very little else is described about their occurrence in carbonate rocks. It may be that they occur preferentially in eogenetic limestones (Fig. 8.3).

The shape of tafoni is another point of disagreement in the field. The shape of tafoni have been described as an opening with arch-shaped overhung entrances, inner walls that are concave in shape, and a smooth slightly sloping floor that is debris covered (Matsukura and Tanaka 2000; Brandmeier et al. 2010). Tafoni have also been described as simple hemispherical cavities (Sunamura 1996). These descriptions of the shape are specific while others are vague, for example, the definition describing the shape as a large cave which can reach several meters in size (Huinink et al. 2004). Tafoni have also been described as an opening which has projections and overhangs (McBride and Picard 2000). This version is perhaps the most general description of the tafoni shape without being overly vague. The emphasis on overhangs and projections, especially in the entrance area of tafoni, is an indication that these cave-like features form after a protective outer layer or coating has been breached, allowing weathering processes to work on a weaker interior (Fig. 8.2c). It is also possible that interior core weakening is occurring as opposed to exterior case hardening.

Fig. 8.4 (**a**) Tafone present in a 15 year-old road cut in 3,000 year old rock, at Alligator Point, Cat Island Bahamas. Note overhanging projection and the sediment accumulation near the bottom. Flashlight is 12 cm tall. (**b**) Tafone present in 5,000 year old rock, on North Point, San Salvador Island Bahamas (Photo by J.E. Mylroie)

There is also much debate about the location of tafoni cave occurrences. Hacker (2003) found that tafoni occur more commonly in bedrock areas of structural and/or composition irregularities. Also indicated by this study was that tafoni growth was controlled by joints in the rocks. Joints formed locations where tafoni development could be initiated, or the joints limit the growth of tafoni caves started elsewhere on the rock face (Hacker 2003; Mellor et al. 1997). Fault systems and rise of salts by capillary action have also been stated as controls on tafoni development (Brandmeier et al. 2010). Bedding patterns, topography, specifically slope and aspect, as well as spring or seepage presences, have also been credited as locations of tafoni origin points (Mellor et al. 1997).

Large amounts of research have also been done on the rates of tafoni development. Tafoni develops at different rates in different rock types (Norwick and Dexter 2002). Sunamura (1996) developed a one-dimensional model used to explain coastal tafoni growth rates, and showed that tafoni development decreases exponentially due to the rock weakening, as a result of the weathering. In the Bahamas, coastal tafoni present in the Holocene rocks must have formed since the last sea level highstand showing the age of the tafoni to be less than 5,000 years, indicating rapid tafoni formation (Owen 2007). It was also shown that there is a lag time in tafoni development, during which the rock surface hardens and must be breached for the tafoni formation to begin (Sunamura 1996). This lag time is different for different areas and environments. The lag time was shown mathematically for the area along the Japanese coast, in the sandstone present there, to be 35 years (Sunamura 1996). Preliminary data in the Bahamas support this kind of rapid tafoni development, as shown in Fig. 8.4. In Northeastern Arizona, the area is composed of sandy dolomite and sandstones, and the lag time is approximately 1,000 years, based on the lack of tafoni in pioneer buildings present in the study area (Norwick and Dexter 2002). The study done on the tafoni in Israel, the tafoni formed by flaking, showed a rate of tafoni growth to be 0.2–0.3 mm/year (Rögner 1988). In certain settings siderophile bacteria may aid in tafoni development by providing additional erosion of biotite and sulfide or minerals that could be present within the tafoni (Brandmeier et al. 2010).

It was also been shown that many previous studies commonly underestimated the growth rate of the tafoni (Sunamura 1996). Brandmeier et al. (2010) states that tafoni weathering is the most effective weathering process that operates on plutonic rocks, working at an order of magnitude faster than other weathering processes in the same area on bare surface rocks. The presence of tafoni in road cuts only 15 years old (Fig. 8.4a) as well as larger tafoni present in 5,000 year old Holocene rock in the Bahamas (Fig. 8.4b) argues for rapid development in the climate and carbonate rocks present there (Owen 2007). It appears that rates will vary by rock type, climate, forming mechanism and other variables.

For the purpose of this Bahamian research, tafoni will be defined as openings in Quaternary eolian carbonates in the size range of decimeters to meters that have a concave and somewhat irregular shape with overhangs and projections, which are the result of some weathering mechanism. The hypotheses for this study were selected to address the issues resulting in confusion in the tafoni literature (Owen 2007).

On San Salvador Island, Bahamas, potential tafoni were found to exist in two main types, a large-scale type and a small-scale type. The large-scale tafoni are present along naturally occurring sea cliffs, along various coastlines on the island, including both sides of North Point. The small-scale potential tafoni occur mainly in cultural features such as road cuts, quarries, and buildings. From this point forward within this chapter, these features are referred to as tafoni (Owen 2007).

8.4 San Salvador Tafoni

The tafoni on San Salvador were surveyed using the compass and tape method, and the approximate location of the tafoni were recorded by GPS. Locations selected for study include: North Point, The Bluff, The Gulf, Watling's Quarry, and a road cut known as Hole 12 Road Cut (Fig. 8.5). All locations were selected because they were logistically easy to access, and had a cliff feature that contained tafoni-like features. North Point was selected due to the fact that it is Holocene in age. The Gulf and the Bluff were selected because they were similar to North Point in morphology, but are Pleistocene in age. Both cultural features were selected because the date of the artificial cliff formation is known. Hole 12 Road Cut was selected because it was along a breezy hillcrest, and was developed in rocks that are Pleistocene in age. Watling's Quarry was selected because of its sheltered bowl shape, and being developed in rocks that are Pleistocene in age.

The tape and compass survey methods were used to measure the distance, inclination, and azimuth of tafoni position and configuration, while additional measurements of height, width, and depth were taken. The survey and mapping provided detailed information about the morphology of the tafoni caves as well as provide some idea of the volume of material which has been weathered away. Large tafoni measurements were divided between the currently-roofed portion of the void, and to characterize what appeared to be collapsed, or the unroofed portions of the void.

Hypotheses tested during the examination of tafoni on San Salvador include first of all that tafoni are forming on San Salvador Island. Secondly the study wanted to quantify that large tafoni in different aged rocks are not morphometrically different. Thirdly the formation of the Bahamian tafoni was examined to see if larger-scale tafoni are correlated to the formation of small tafoni and to test the idea that the formation of tafoni is related to mineral crystallization, and/or wind and air erosional processes once a resistant surface layer has been breached (Fig. 8.4).

For morphometric analysis to be completed on the tafoni, cave maps were created for each of the large scale tafoni, using the typical tape and Suunto compass method as noted above. The data were later reduced in Compass (Fish 2006), the line plot was then converted in to a bitmap, and opened in Xara x Software, (Xara Group Limited 2001) where drafting of the cave maps took place, final maps were converted to jpegs for measurement (Fig. 8.6).

Measurements used to compare the morphology of the different types of tafoni are shown in Fig. 8.6a. Measurements were done digitally using a software program known as Image J (Rasband 2007), this is a program, developed by the Research Services Branch of the National Institutes of Health (NIH) for measurements of biomedical research related images. The program is used to calculate area, distances, angles, perimeter, and other properties, based on an input of known distance.

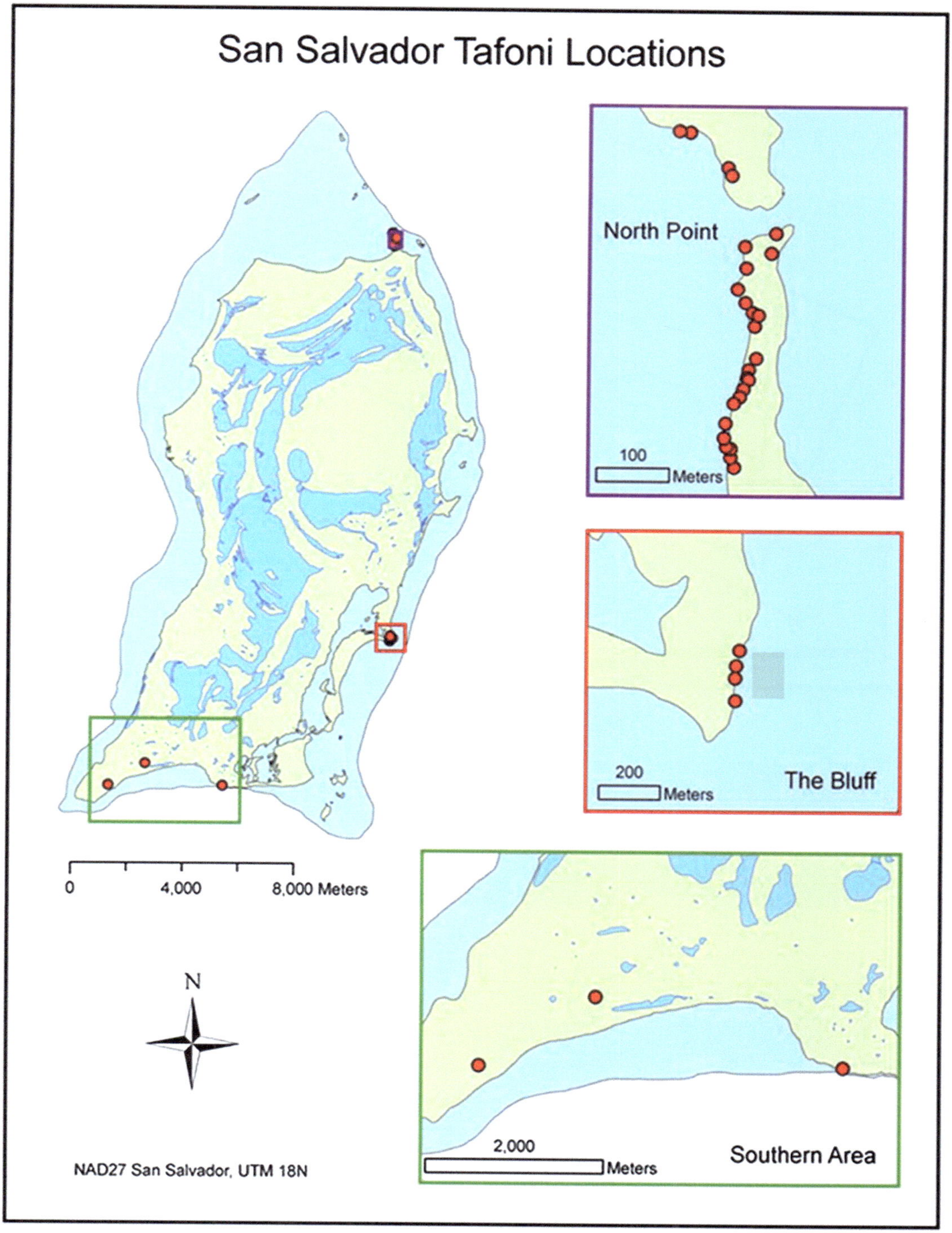

Fig. 8.5 A map of San Salvador Island, showing the locations of tafoni mapped and measured

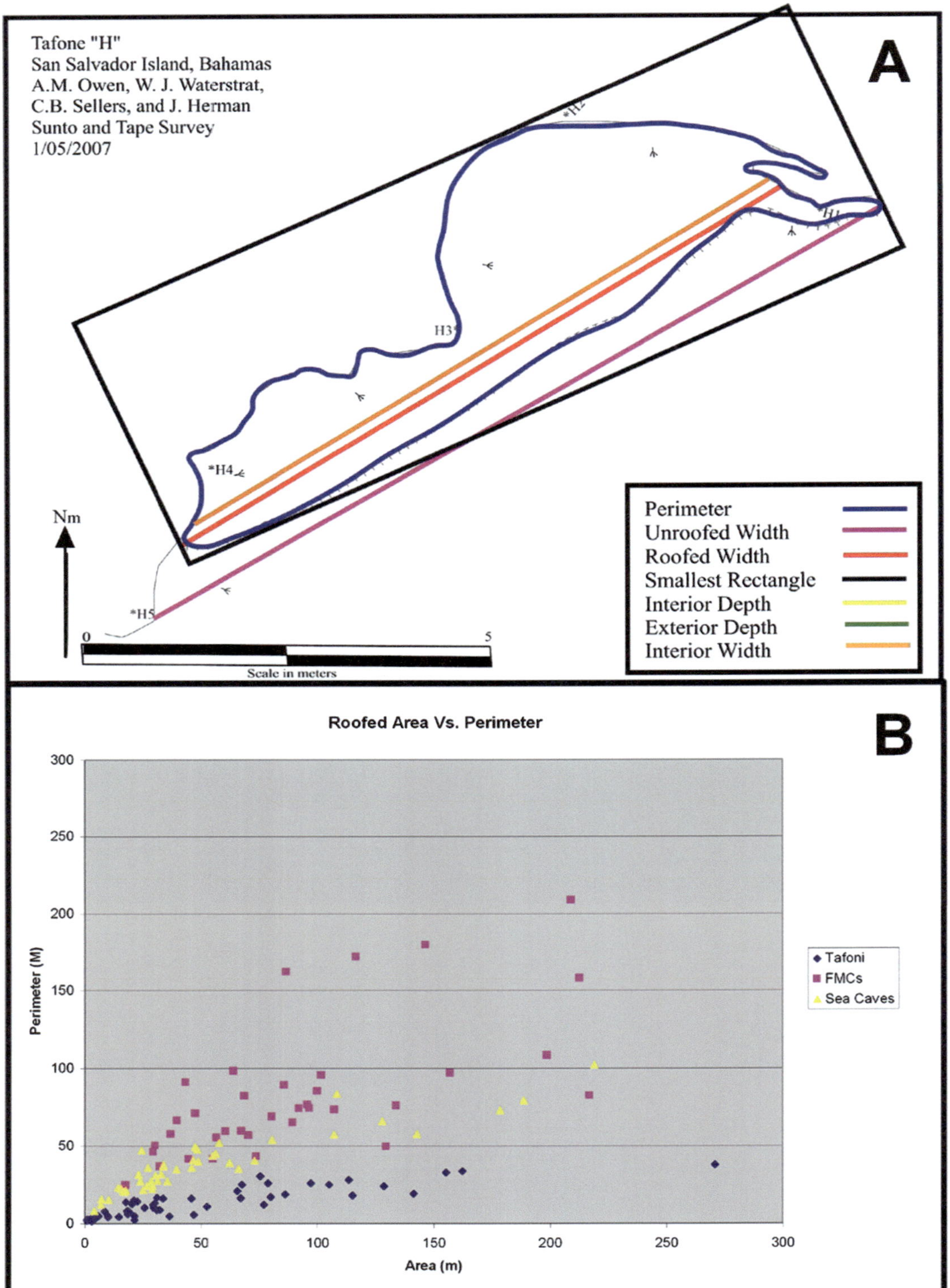

Fig. 8.6 (**a**) A tafone map showing the measurements taken for morphometric analysis of the tafoni maps. Area was the area contained within the perimeter of the tafone. (**b**) Plot of area to perimeter ratios for flank margin caves, sea caves, and tafoni, showing plot separation based on void wall complexity

Measurements were taken for each interior inside the dripline (edge of the roof where meteoric water drips), and for the unroofed area around each dripline. For every cave mapped, the measurements taken were the actual area, perimeter, depth, width (Fig. 8.6a). The area, perimeter, long, and short sides of a rectangle, were recorded for the smallest rectangle that could contain each interior cave area and unroofed area. Many of the cave maps had multiple driplines, whose measurements were taken along each dripline and along the cave wall that was encompassed by that dripline. The unroofed area of each dripline was determined by the projections around the dripline area, which divided that area by line of sight from the next dripline. Measurements were taken from the point of each projection in a straight line across to the other projection and then followed the back wall to determine area, perimeter, and rectangle measurements. In a few cases this straight line approach did not include the entire cave. In these cases the measurements were extended the smallest amount possible off one of the projections along the line of sight until the entire cave was included in the measurement. The smallest rectangle was determined by rotating the image until the smallest rectangle was found, this usually occurred when the long axis to the cave was along one of the long axis of the rectangle (Fig. 8.6a) (Owen 2007).

From the measurements taken from both the tafoni themselves and those measured from the cave maps, ratios were calculated and statistical analyses were performed comparing the small tafoni from the road cut and the quarry, and the tafoni themselves forming in rock types of different ages. The measurements used were selected because they could be applied to both large and small tafoni, and they could be done quickly and simply and still allow for characterization of the voids. Morphometric analysis, using area to perimeter, as well as other ratios have been proven to be a successful method to differentiate these cave types by Waterstrat (2007), Owen (2007), Roth (2004), Lace (2008) and Waterstrat et al. (2010). Measurements were taken for both the roofed and unroofed portions of the tafoni to look at how the cliff lines retreat as the tafoni develop.

Ratios were calculated for each cave, and were used to normalize the cave data, making the caves comparable regardless of size. Ratios used will be referred to by the abbreviation as listed: area vs. perimeter (A/P), short axis over the long axis of the smallest rectangle to enclose the box (S/L) ratio, entrance width vs. the greatest enclosed interior width (EW/IW) for both roofed (R) and unroofed (UR), percent of area the cave that the cave filled was also calculate by dividing the area of the cave by the area of the box (Box Fill %) and height data were also measured, at the highest point in each large tafoni.

Previous workers (Waterstrat et al. 2010) had attempted to differentiate between flank margin caves, sea caves, and tafoni by using area to perimeter ratios and other cave-measurement metrics. That work, building on the differentiation of sea caves and flank margin caves by Waterstrat (2007) and Lace (2008), initially seemed to indicate the three cave types could be differentiated from typical cave maps. Further analysis demonstrated that sea caves and flank margin caves may not be differentiated by cave metrics, but that tafoni could be differentiated from the other two cave types (Fig. 8.6b).

8.4.1 Small Tafoni

Small scale tafoni were measured in two main locations on San Salvador Island, at Watling's Quarry, and at nearby road cut, Hole 12 Road Cut (Fig. 8.7). The first location, Watling's Quarry, is a limestone quarry located near the south central portion of the island, the quarry was approximately 35 years old at the time of measurement (center dot of the lowest inset in Fig. 8.5). This location has three cliff faces that were left exposed when the quarry was abandoned. The second location is known as Hole 12 Road Cut, due to its location near the 12th hole of a once-proposed golf course. This site is located farther south than Watling's Quarry. The road cut was also approximately 35 years old at the time of measurement (left dot in the lowest inset of Fig. 8.5).

Fig. 8.7 Modern tafoni. (**a**) Watlings Quarry, San Salvador Island, about 5 years after initial excavation, showing few tafoni. (**b**) Hole 12 Road Cut, San Salvador Island, about 35 years after excavation, showing small tafoni being measured (Photo by J.E. Mylroie)

Fig. 8.8 A series of four photos showing smaller tafoni merging in the breakdown blocks present within the large Tafone Q, North Point, San Salvador (Photo by J.E. Mylroie)

At this location there are two exposed cliffs. The measurements in these locations were done by numerous teams of two, one measuring and one recording the data. The measurements included the maximum exterior and interior heights, the depth, and the maximum exterior and interior widths, and were converted into the ratios listed above. It was considered possible that the smaller tafoni form independently and continue to grow into each other resulting in the larger tafoni. This idea is also supported by the presence of tafoni forming in the breakdown present in a larger tafone, for example tafone Q (Fig. 8.8).

Other observations in the individual larger tafoni show multiple deeper points within the tafone; is possible that these multiple deeper points are perhaps analogous to the small tafoni forming at one time along the cliff face, at one time individuals that are now still consolidating (Fig. 8.9). The small tafoni have occurred in walls that have not retreated in their 35 year existence, while the large tafoni found in the sea cliffs occur where the original cliff face has eroded back. The cliff face tafoni were compared by the following ratios, Interior width (IW) over entrance width (EW), entrance height (EH) over interior height (IH), entrance height (EH) over entrance width (EW), interior height (IH) over interior width (IW), the EH/IH ratio over depth (D), The IH/IW ratio over depth (D), the EH/EW over depth (D), and the EW/IW ratio over depth (D) (Fig. 8.10).

8.4.2 Tafoni Morphometric Comparisons

Results from this part of the study determined, for tafoni to form in the rocks, those rocks must be of sufficient strength to support the void. This interpretation is supported by observations from

Fig. 8.9 Showing the locations of possible smaller tafoni present in a larger tafone, North Point, San Salvador (Photo by J.E. Mylroie)

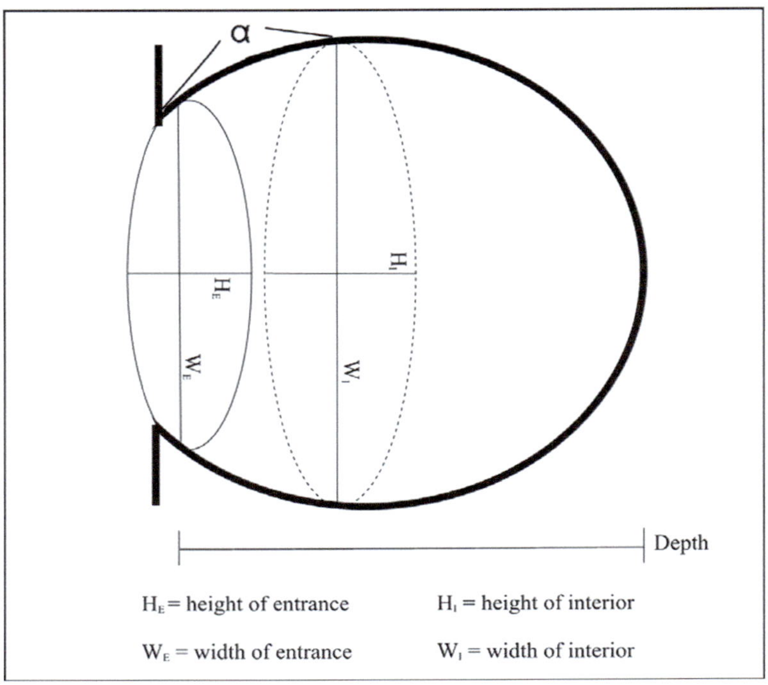

Fig. 8.10 A three dimensional view, idealized diagram showing the measurements taken within the small tafoni

Hanna Bay, where collapse of the voids is evident in poorly lithified 3,000 year-old eolianites, preventing the development of large tafoni along these cliffs. The rocks also appear to need to be a certain distance away from the sea spray and wave action for tafoni development to occur or continue; otherwise the sea spray cements the rock enough that tafoni weathering processes

Fig. 8.11 (**a**) The Graham's Harbor side of North Point, with tafoni present along the side and along Cut Cay to the north. Note the low wave energy and calm conditions, typical for this location. (**b**) The Rice Bay (Atlantic) side of North Point, note the increase in wave energy on this side of the Point, the grey cemented dune surfaces, which limit tafoni development (Photo by J.E. Mylroie)

Fig. 8.12 The Bluff along the southeastern coast of San Salvador island, with the beach and wave cut bench present isolating the cliff face from casual sea spray, with tafoni features present on the sloping cliff face

cannot operate. A majority of the tafoni in this study occur along North Point and Cut Cay, along the protected Graham's Harbor side (Fig. 8.11a) of the peninsula. On the Rice Bay side, open to the Atlantic, where there is greater wave action (Fig. 8.11b), there are no tafoni along Cut Cay and only two tafoni present along the very end of North Point are high on the cliff face compared to the Graham's Harbor side, above the reach of routine sea spray (Owen 2007).

Along The Bluff (small boxed point midway along the east side of San Salvador in Fig. 8.5), an additional Pleistocene location, a truncated wave cut bench has formed with a beach, and the sea cliff is set back from current normal wave activity (Fig. 8.12). This wave cut bench and beach provides the necessary protection from normal sea spray and wave action, similar to the wave cut bench present in Holocene eolianites along Graham's Harbor at North Point (Fig. 8.11a). As a result, normal wave action cannot routinely send sea spray to the sea cliff, which allows tafoni to form. The wave cut bench in both locations is the result of episodic storms, which lack the duration to create a sea spray cementation of the sea cliffs. In The Gulf, a Pleistocene location at the south end of San Salvador (right hand dot in the lower box of Fig. 8.5), tafoni were only found in a protected cliff which is perpendicular to the direction of the major sea cliff direction. Along the sea coast itself sea caves are prominent and there are no tafoni. The protected area has a cliff face formed by wave action, which only reaches this site during storm events. The cliffs containing the tafoni are located here and do not receive sea spray under normal conditions. The key point is that wave erosion occurs often enough in these locations to cut and cliff the side of the eolianites, but not often enough to create a significant amount of sea spray cementation.

Ratios from the small tafoni were used to compare each of the three cliff faces at Watling's Quarry to each other, and the two cliff faces at the Hole 12 Road Cut to each other. Then the Quarry faces and the Road Cut faces were compared to each other. The small tafoni were then compared to the large tafoni using the following ratios: (1) exterior width over interior width for both the roofed and unroofed portion, then (2) depth vs. exterior width over interior width again for both the roofed and unroofed portions.

F tests were run on each data set to show the amount of variance, this information was used to determine the best *t* test to use with the data set. Two tailed t tests were used to make the comparisons; the tests were run for highly varied data sets, due to large variations present. All the variables were tested at the 95 % confidence interval. Graphical comparisons were also made comparing the different measurements and ratios for both the small and large tafoni, testing for morphological differences as the tafoni developed.

Table 8.1 Small tafoni cliff face comparisons results

	Cliff faces			
	Watling's quarry faces			Hole 12 road cut
Ratio compared	A vs. B	A vs. C	B vs. C	D vs. E
IW/EW			*	
EH/IH	*		*	*
EH/EW	*	*	*	
IH/IW	*	*	*	
EH/IH vs. D				
IH/IW vs. D				
EH/EW vs. D				*
EW/IW vs. D				

*indicates significance ($p < 0.05$ or less) to the 95 % confidence interval

The null hypothesis for each cliff face comparison was that the ratios for the cliff faces would not be significantly different. In Watling's Quarry, for the ratios of EH/IH over D, IH/IW over D, EH/EW over D, and EW/IW over D, failed to reject the null hypothesis because these ratios were not significant (p value < 0.05) between the cliff faces compared. For the comparison of cliff face A to B, the null hypothesis was also accepted for the IW/EW ratio. The remaining three ratios EH/IH, EH/EW, and IH/IW, the null hypothesis was rejected with each of this ratios being found to be significantly different (p value < 0.05). The comparison of cliff faces A to C found only the EH/EW and IH/IW ratios to be significantly different (p value < 0.05), while the null was accepted for the IW/EW and EH/IH, ratios. All four remaining ratios were found to be significantly different (p value < 0.05) when comparing cliff faces B to C. In the Quarry all three comparisons showed the ratios of EH/EW and IH/IW to be significantly different (p value < 0.05) (Table 8.1).

The cliff faces present in Hole 12 Road Cut had few ratios that rejected the null hypothesis, the two ratios for which the null was rejected were the EH/IH and EH/IW over D; for the remaining ratios the null was accepted (Table 8.1). None of the ratios were found to be significantly different (p value < 0.05) in all four cliff face comparisons, however, EH/IH, EH/EW,

Table 8.2 Small tafoni location comparison results

Ratio compared	Watling's quarry vs. hole 12 road cut
IW/EW	*
EH/IH	*
EH/EW	
IH/IW	*
EH/IH vs. D	*
IH/IW vs. D	*
EH/EW vs. D	*
EW/IW vs. D	*

*indicates significance ($p < 0.05$ or less) to the 95 % confidence interval

and IH/IW were significant in three out of the four comparisons. The comparisons between cliff faces A and C, and between D and E had the fewest ratios that rejected the null hypothesis at two each.

The data for each of the locations, Watling's Quarry and Hole 12 Road Cut, were combined and then compared to each other using the same eight ratios as the individual cliff faces. The null hypothesis was again that the two locations would not be significantly (p value < 0.05) different in any of the ratios. The result of the *t* test showed these cultural locations to be significantly (p value < 0.05) different in every ratio except EH/EW, for this ratio the null was failed to be rejected, for all others it was rejected, see Table 8.2.

8.4.2.1 Small vs. Large Tafoni

The data from the Watling's Quarry and Hole 12 Road cut were then compiled to comprise the dataset for the small tafoni. These data were then compared to the entire dataset of the large tafoni in all the ratios that were present in both datasets. This resulted in two different ratios, EW/IW, and D over EW/IW, those were then compared to both the equivalent Roofed (R) and Unroofed (UR) tafoni ratio. The descriptive statistics for all of these ratios are shown in Table 8.3. For these comparisons the null hypothesis was that these ratios would not be significantly different (p value < 0.05). The null was accepted only for the comparison of unroofed tafoni EW/IW

Table 8.3 Descriptive statistics for large vs. small tafoni

Ratio	Average	Standard deviation	Variance
UR EW/IW	0.987	0.047	0.002
R EW/IW	0.998	0.0154	0.002
Sm. tafoni EW/IW	0.983	0.047	0.006
UR D over EW/IW	3.455	2.458	6.042
R D over EW/IW	1.649	1.234	1.524
Sm. tafoni D over EW/IW	0.125	0.169	0.028

Table 8.4 Large vs. small tafoni results

Ratio compared	Significant
UR tafoni EW/IW ratio vs. Sm. tafoni EW/IW ratio	
R tafoni EW/IW ratio vs. Sm. tafoni EW/IW ratio	*
UR D over UR EW/IW vs. Sm. tafoni D over EW/IW	*
R D over R EW/IW vs. Sm. tafoni D over EW/IW	*

*indicates significance ($p < 0.05$ or less) to the 95 % confidence interval

ratio vs. the small tafoni EW/IW ratio. For the other three comparisons done the ratios were significantly different (p value < 0.05) and the null was rejected (Table 8.4).

8.4.2.2 Rock Samples

Rock samples were collected in two consecutive tafoni, one sample set in Pleistocene eolianites and one sample set Holocene eolianites, then additional samples were collected from the unweathered rock between each set of tafoni, also collected was the sediment accumulating along the bottom of each tafone. Samples were collected high and low in each different tafone and in one location in the unweathered rock. After the sample location was selected the five samples were collected in each site, with the orientation noted for both upward and outward directions. The samples were collected in a four cubic centimeter progression, starting with the outer surface and working back approximately 20 cm, so to show the progression of any variation present in the rock. The samples were then wrapped and bagged, with two silicon gel packs included to

Fig. 8.13 Photo showing the accumulation of sediment in a Pleistocene tafone near the Gulf. Note the additional accumulation near the pencil and along the wall to the left of the picture

remove moisture from the air and to preserve any halite or other evaporite crystals which may have been present.

After returning to the sample collection locations approximately 6 months after the original sample collection took place, strong evidence was present to suggest wind erosion as a new mechanism for tafoni formation. Sampling resulted in centimeter-scale holes in the outcrops with a known date of origin. When the sampling locations were revisited the sampling holes within the tafoni themselves were already partially filled with as much as 6 cm of additional sediment in both the Holocene and Pleistocene locations. The crust sample holes located between the sampled tafoni, however, contained very little sediment. The sediment commonly appeared to be predominantly accumulating on the same sides of the hole, along the north to northeast sides. This was especially evident in both the sample locations in a Holocene aged tafone on Cut Cay. This tafone also is the most exposed of the sampled tafoni. This result could show the effect of winds, and wind direction on the tafoni development (Figs. 8.13 and 8.14).

8.4.2.3 Thin Sections

Thin sections were created from the samples of four different tafoni, two Pleistocene and two Holocene in age. Thin sections were created for areas both high and low in each tafone. Thin sections were also prepared from the non-weathered area between each set of tafoni. When the thin sections were made the samples were impregnated due to their brittle nature, then cut using water sensitive methods (i.e. cut in oil), orientated in both the out and up directions, and covered with cover slips to prevent reaction of any evaporities to the moisture in the air. Thin sections were analyzed to determine the percentage of pore space, cement, and allochems present in the rocks, while checking for evaporites. The percentages of the allochems, porosity, and cement was determined by a 100 point count, and ground truth with one thin section being counted to 400 points.

The point count of all the thin sections showed no evaporatite grains in the Holocene or the Pleistocene aged rocks, either inside the tafoni themselves or in the protective crust present between the tafoni. The Holocene rock result displayed the

Fig. 8.14 The sample site of the crust between the two Pleistocene sampled tafoni, note the lack of sediment compared to the previous figure

lowest porosity and the highest percent cement, with average percentages of 17.2 and 25.4, respectively. The allochem average percentage for the Holocene-aged thin sections was 57.4, lower than the both the Pleistocene and crust samples. The Pleistocene samples had the exact opposite result, with an average porosity of 25.9 and 10 % average cement. The allochem percentage was higher than the Holocene at 64.1 %. The crust is exactly in the middle in all three categories with porosity percent of 23.5, allochems at 62.9 %, and an average cement percentage of 13.6.

The results of the thin section analysis found allochems that were consistent with previous work done in the Bahamas from the Holocene (Schwabe 1992; Carney and Stoyka 1993; White 1995; Hutto and Carew 1984; Stowers 1988) with presence of the foraminifera, coral fragments, mollusks, and algae. The fact that no evaporites were found in any of the samples prepared for thin section removes salt wedging as a forming mechanism for the tafoni present in the Bahamas.

The idea that wind plays a roll in the development of tafoni is expressed by the sample holes. The regular alignment of sediment present in the sample holes could show the effect of winds, and wind direction on the tafoni development, especially because the less wind-protected holes from tafone Y on Cut Cay, showed more sediment alignment that the other holes. The accumulation of this sediment also shows the rapidness of the process, accumulating 6 cm in 6 months. The wind is enough to align the sediments present in the tafoni, however, it is not enough to remove the sediments. Based on this idea and the bowl-like shape of Watling's Quarry, which inhibits the wind, it does not take a lot of wind to promote the tafoni development once the crust is breached.

The porosity values were problematic, as they were lower in the Holocene eolianites than in the Pleistocene eolianites. Earlier workers, cited above, had found porosity in the Holocene eolianites to be greater than in the Pleistocene eolianites, as vadose cementation had progressed to a greater extent in the older eolianites. The explanation turned out to be the method of thin section preparation. Water cutting of the eolianites, the technique used by previous workers, had washed out biofilms and attached $CaCO_3$ material; with oil cutting, the biofilms and their associated carbonate material were preserved. This preservation created a lower porosity in the Holocene eolianites when the thin sections were point counted. In the older Pleistocene eolianites, the biofilms had decayed away, releasing the attached carbonate material, which was lost upon rock cutting for thin sections. A full discussion of this unexpected outcome can be found in Owen et al. (2010); this unexpected result was a by-product of the attempt to preserve any evaporate minerals present in the rock.

8.5 Tafoni from Other Islands

In addition to San Salvador Island, in the Bahamas, tafoni have also been noted from Long Island (Wilson 1992), Abaco Island (Walker 2006; Walker et al. 2008), New Providence Island (Owen 2007), Cat Island (Owen 2007), and Crooked Island (this study).

8.5.1 Long Island

Long Island Bahamas has two tafoni sites of particular interest because they are not associated directly with cliff formation resulting from coastal wave action. The first site is in an area called The Crevice (Figs. 8.15 and 8.16). At this location, a deep-seated void has progradationally collapsed to the surface under the flank of an eolian ridge. As a result, a cliff developed on the side of the ridge, which removed the usual protective calcrete crust. A very large void is developed in this ridge cliff. It has the typical appearance of a tafoni cave, and if so, is the largest known tafoni cave in the Bahamas. The second feature is Dean's Blue Hole (Wilson 1992), the deepest blue hole in the world at 202 m depth (Fig. 8.17). This blue hole is another progradational collapse feature that also has collapsed upward through the side of an eolian ridge. These exposed cliffs, which lack a calcrete crust, contain numerous shallow caves that appear to be tafoni (Fig. 8.17). Because some of these voids contain vertical rock structures hanging from the ceiling, similar in appearance to weathered stalactites, these caves were initially identified incorrectly as flank margin caves (Wilson 1992). However, close examination show that these vertical rock structures are not stalagmites, but vadose pit infills, exposed by weathering of the less well indurated rock material around them. In addition, the caves are

Fig. 8.15 The Crevice, Long Island, Bahamas, a large tafone, present along the cliff that resulted from the collapse of a dune ridge. The dune collapse resulted in the breeching of the calcrete crust, promoting tafoni development (Photo by J.E. Mylroie)

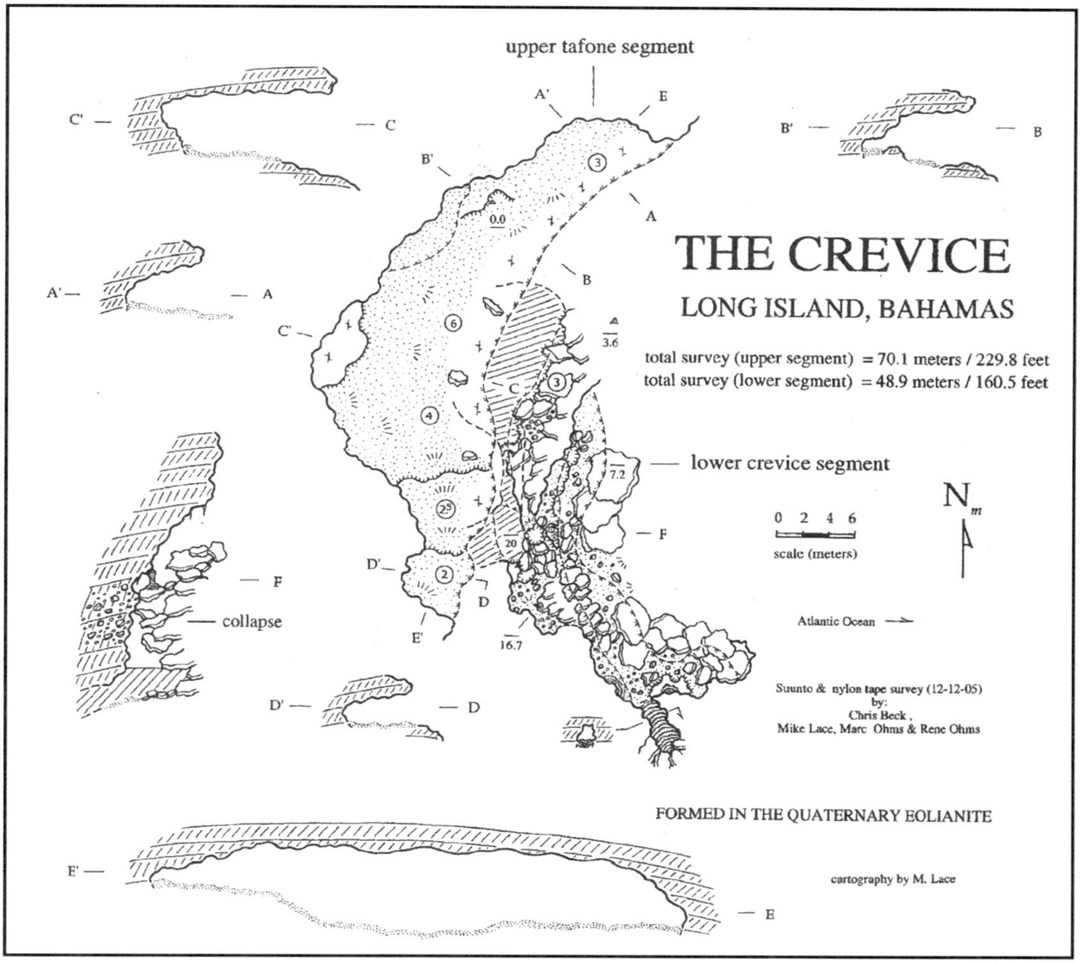

Fig. 8.16 Map showing The Crevice (Courtesy of M. Lace)

at numerous horizons, not typical of flank margin cave development.

From the Long Island examples it appeared that the tafoni are initiated by the cliffing caused by collapse of large voids at depth, which removed the calcrete crust, and left vertical cliffs with the soft eolianite material exposed to outside weathering. As a result, tafoni development occurred. The Crevice especially is illustrative, as it is in an interior location and coastal processes have no influence at this site. Deans Blue Hole, while in a lagoonal setting, lacks significant wave energy. Cliffing of an eolian ridge by collapse into the blue hole created another vertical face, exposing the soft interior of an eolianite ridge. These Long Island examples demonstrate that a key factor in tafoni development in Quaternary eolianites: some mechanism to cliff the eolianite without subsequent cementation so that the soft and poorly indurated interior can be exposed to wind and sun. The verticality of cliffs limits meteoric cementation, protected lagoons limit sea spray cementation.

8.5.2 Abaco Island

Tafoni caves, on Abaco Island were originally described by Walker (2006) (Fig. 8.18). Tafoni caves were originally located and believed to

Fig. 8.17 Deans Blue Hole, Long Island, Bahamas, the deepest blue hole in the world, with tafoni present above in the cliff created by progradation of the blue hole. The tafoni show pit infill features originally misinterpreted as stalactites (Photo by J.E. Mylroie)

Fig. 8.18 Great Abaco Island tafone, at 10.8 m this tafone is at the lowest elevation among the PITA caves, named "Cave J" (Walker 2006)

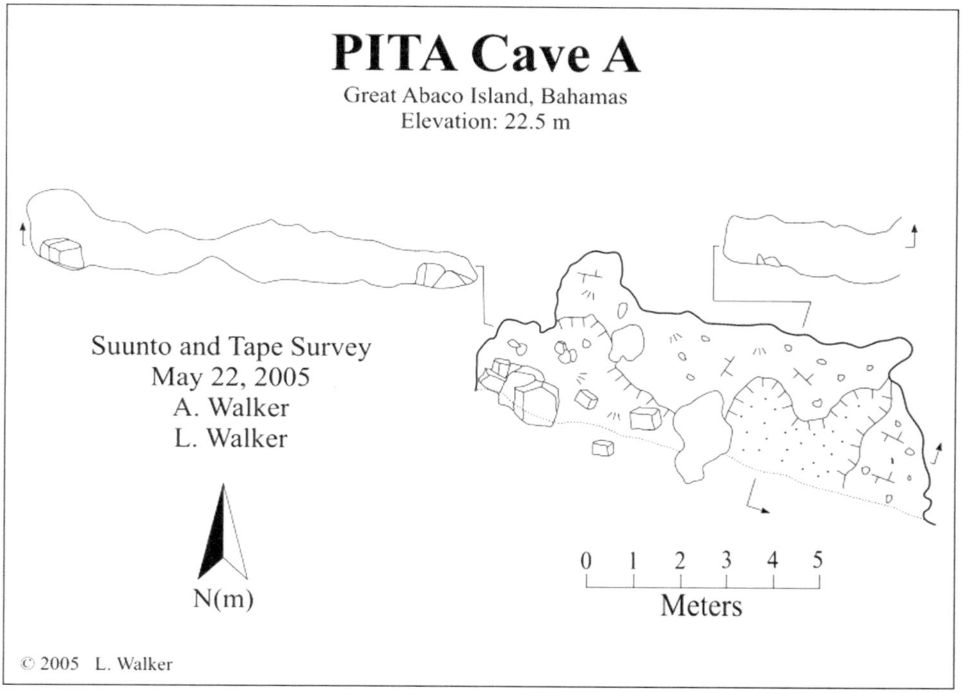

Fig. 8.19 Map of PITA Cave A, Great Abaco Island, Bahamas (Walker 2006)

be flank margin caves formed at 20 m above current sea level. This was problematic since the Bahamas are tectonically stable. According to the flank margin cave model, it would be necessary to have a previous sea level highstand approximately 20 m higher than current sea level to form these caves. However, modern dry flank margin caves were formed during the only sea level highstand known to be higher than modern sea level, the MIS 5e highstand, resulting in flank margin caves forming at approximately 6 m above current sea level (see Chap. 7 for a full discussion). After further investigation the caves were found not to be in a continuous elevation horizon, and were found to be lacking in all the features common to flank margin caves, such as phreatic dissolutional surfaces, bell holes and stalagmites. Therefore, they appeared not to be representative of flank margin caves. The caves did look erosional in formation, and are now believed to be tafoni (Fig. 8.19) (Walker 2006; Walker et al. 2008).

There are 14 known tafoni present on Abaco Island and are thought to be inactive, screened by vegetation and in a very degraded state. They are believed to have formed during the MIS 5e highstand when the +6 m higher sea level cliffed an already existing dune, starting the tafoni erosional process. The dune was large enough that sea spray could not reach the upper reaches of the cliff, and sea spray cementation did not occur at those heights. This situation allowed tafoni to develop. The 115,000 years since that highstand have allowed enough time for rainfall to cement the cliff and the tafoni are no longer undergoing erosional processes, and have become relict features. Once sea level fell at the end of the last interglacial, active cliffing stopped, and the cliff gradually cemented with meteoric water and stabilized.

8.5.3 New Providence Island

Tafoni from New Providence to date are only described in buildings and other cultural features. These features again attest to the rapidity of the tafoni-forming process. Norwick and Dexter

Fig. 8.20 Tafoni present in an old road cut, now found in the parking lot of a Bahamian government building, New Providence Island (Photo by J.E. Mylroie)

(2002) showed a lag time of 1,000 years for tafoni development in Arizona, based on the lack of tafoni present in the blocks of building. The tafoni present in the buildings on New Providence, only several hundred years or less old, show that tafoni form more rapidly in the Bahamas, in building blocks made from eolianites (Owen 2007).

Tafoni present on New Providence were located in numerous cultural features, including road cuts (Fig. 8.20) and buildings (Fig. 8.21). There are two areas of buildings, both of which are historic, with sawn eolianite blocks that show clear evidence of weathering. This weathering occurs low, closer to the ground, along these building which implicates wind grain transport in these tafoni-like features (Owen 2007).

8.5.4 Cat Island

The tafoni on Cat Island are similar to that present on the other Bahamian islands, and are probably reacting to similar situations. Two high relict tafoni were studied and appear to be similar to the tafoni present on Abaco, high on a cliff line outside the sea spray range on the south coast of the island. The event that resulted in the cliff face where the tafoni are present was probably a product of the last sea level highstand, MIS 5e, ~125 ka. The coastally active tafoni on Cat Island also seem to forming under the same restrictions that limit tafoni development on San Salvador, that is, in protected environments away from sea spray, with some wind interaction.

On Cat Island, like San Salvador, tafoni were present in various locations, and included tafoni high along cliff faces (Fig. 8.22), and in building and other cultural features (Fig. 8.23). The high tafoni present along a large cliff face (Fig. 8.24), are significantly removed from present sea level and covered partially by vegetation, similar to the Abaco situation. They are relict from MIS 5e cliffing of the eolianite hill. These tafoni were mapped and measured, in the same methods used on the San Salvador tafoni.

8.5.5 Crooked Island

Recent work on Crooked Island has located high, inactive and abandoned tafoni as seen on

Fig. 8.21 Tafoni present low on the side of a building on a busy street, in Nassau, on New Providence Island. Note the development preferentially close to the ground (Photo by J.E. Mylroie)

Fig. 8.22 White Mountain, on Cat Island, represents active coastal tafoni. The entire mountain is friable and unsafe for traverse up the cliff face (Photo by J.E. Mylroie)

Cat and Abaco Islands. One example at 20 m elevation, as seen in Fig. 8.25, is similar to Figs. 8.18 and 8.24, being fronted by vegetation, and showing modification by collapse, indicators of its likely MIS 5e age, when waves eroded the base of the eolianite hill containing the tafone, creating the open face of unweathered eolianite. Holocene sediment now protects this cliff from active wave attack, allowing senescence to occur.

Fig. 8.23 Tafoni present in Mt. Alvernia monastery masonry work, on Cat Island. Note that the eolianite blocks weather back faster than the original mortar

8.6 Summary

Tafoni form in very specific environments in the Bahamas. The large tafoni form along eolianite cliff faces, recently produced, that are inland, or coastal but which do not typically receive sea spray. Sea spray promotes cementation and slows erosion rates. In the absence of sea spray the cliffed surface is made of carbonate grains with small amounts of vadose meniscus cements, and these grains are therefore vulnerable to wind erosion. Small tafoni can form in artificial cliff-like features such as quarries and road cuts, or even in buildings, where they are again protected from rapid meteoric cementation by their verticality, and their development can continue. The large tafoni may form as small tafoni and erode in to each other producing large surface areas, increasing the rate of erosion. The large tafoni also seem to have a depth limit, developing deeper as the cliff supporting them breaks down – the "unroofed" portions described earlier, such that their depth varies around a value of a few meters.

Tafoni were originally hypothesized for this study to be formed from salt wedging or other evaporate crystal wedging; however, the lack of evaporates present in the petrographic analysis rules this out as a mechanism for formation of tafoni on San Salvador. The data collected implicate wind erosion as a possible mechanism, although it also shows that the wind amounts do not need to be strong enough to actually remove the sediment, just to promote the weathering of the surface, which can be seen in Watling's Quarry, and in the four large sampled tafoni. Both of these mechanisms have been suggested by previous authors. Tafoni are probably polygenetic, forming by several methods given different environments, with different methods needed for promoting the mechanical weathering process, producing morphometrically similar results.

Tafoni were found to be forming on San Salvador Island, as well as other islands in the Bahamas, including Cat, Abaco, New Providence, Crooked and Long Island. 34 large-scale tafoni were mapped, 32 from San

Fig. 8.24 Coastal relict tafone, high on a cliff on Cat Island, similar to that found on Abaco Island at the PITA caves location (Photo by J.E. Mylroie)

Salvador Island, and two from Cat Island, both in the Bahamas; Abaco (Walker 2006), and Long Island (Wilson 1992) both had previously documented occurrences of tafoni. Tafoni were photographed in buildings and cultural features from New Providence and Cat Island. These tafoni were not measured or included in the analysis.

Four hundred and ninety small tafoni were measured from two cultural features, Watling's Quarry and Hole 12 Road Cut, on the south end of San Salvador Island. Comparisons in these measurements show that there is no pattern in how small tafoni form when comparing the entrance and interior measurements, however, patterns do exist when the depth was added to the ratios, comparing the tafoni in three dimensions. The formation of large-scale tafoni is correlated to the formation of small tafoni.

There was no evidence found during this study that traditional cave dissolutional processes are supporting the formation of tafoni in this location. Therefore, the tafoni features on the island represent traditional pseudo-karst features, and not a new form of traditional karst or "pseudo-pseudo karst".

Fig. 8.25 Tafone at 20 m elevation on the northeast side of Crooked Island, Bahamas. The tafone is walled on its open side by vegetation, and contains scattered breakdown blocks, indicative of its abandoned state and original genesis during the MIS 5e sea-level highstand ~125 ka (Photo by J.E. Mylroie)

References

Brandmeier M, Kuhleman J, Krumrei I, Kappler A, Kubik PW (2010) New challenges for tafoni research. A new approach to understanding processes and weathering rates. Earth Surf Process 36:839–852

Campbell SW (1999) Chemical weathering and associated with tafoni at Papago Park, Central Arizona. Earth Surf Process Landf 24:271–278

Carew JL, Mylroie JE (1995a) Quaternary tectonic stability of the Bahamian Archipelago: evidence from Fossil Coral Reefs and Flank Margin Caves. Quat Sci Rev 14:145–153

Carew JL, Mylroie JE (1995b) Depositional model and stratigraphy for the quaternary geology of the Bahama Islands. Geologic Society of America Special paper 300, Terrestrial and shallow marine geology of the Bahamas and Bermuda. Geologic Society of America, Boulder, Colorado, pp 5–32

Carew JL, Mylroie JE (1997) Geology of the Bahamas. In: Vacher HL, Quinn TM (eds) Geology and hydrogeology of carbonate Islands developments in sedimentology, vol 54. Elsevier, Burlington, pp 91–139

Carney C, Stoyka GS (1993) Depositional history and diagenesis of a Holocene Strand Plain, Sandy hook, San Salvador, Bahamas. In: White B (ed) The proceedings of the sixth symposium on the geology of the Bahamas. Bahamian Field Station, San Salvador, pp 35–45

Cooke RU, Warren A, Goudie AS (1993) Desert geomorphology. UCL Press, London

Fish L (ed) (2006) Compass Project Manager (version 5.6.12.19.32), Fountain Computer Products. http://fountainware.com/compass

Guglielmin M et al (2005) Biotic and abiotic processes on granite weathering landforms in a cryotic environment, Northern Victoria Land, Antarctica. Permafr Periglac Process 16:69–85

Hacker A (2003) Controls of tafoni development in Castle Rocks, Idaho. In: Proceedings of the 16th annual Keck Research symposium in geology, Beloit

Huinink HP, Pel L, Kopinga K (2004) Simulating the growth of tafoni. Earth Surf Process Landf 29: 1225–1233

Hutto T, Carew JL (1984) Petrology of Eolian Calcarenites, San Salvador Island, Bahamas. In: Teeter JW (ed) The proceedings of the second symposium on the geology of the Bahamas. Bahamian Field Station, San Salvador, pp 197–207

Lace MJ (2008) Coastal cave development in Puerto Rico. J Coast Res 24:508–518

Maroukian H, Leonidopoulou D, Skarpelis N, Stournaras G (2010) Effects of lithology, mineralogy and weathering on particle size variability of sediments in the coastal environments of Livada Bay in SE Tinos Island, Greece. J Coast Res 26:167–175

Matsukura Y, Tanaka Y (2000) Effect of rock hardness and moisture content on tafoni weathering in the granite of Mount Doeg-Sung, Korea. Physical geography. Geografiska Annaler series A. Blackwell Publishing Limited, Malden, pp 59–67

McBride EF, Picard MD (2000) Origin and development of tafoni in Tunnel Spring Tuff, Crystal Peak, Utah, USA. Earth Surf Process Landf 25:869–879

McBride EF, Picard MD (2004) Origin of honeycombs and related weathering forms in Oligocene Macigno Sandstone, Tuscan coast near Livorno, Italy. Earth Surf Process Landf 29:713–735

Mellor A, Short J, Kirkby SJ (1997) Tafoni in the El Chorro area, Andalucia, southern Spain. Earth Surf Process Landf 22:817–833

Mottershead DN, Pye K (1994) Tafoni on coastal slopes, South Devon, UK. Earth Surf Process Landf 19: 543–563

Mylroie JE (2005) Coastal caves. In: Culver DC, White WB (eds) Encyclopedia of caves. Elsevier/Academic Press, Amsterdam, pp 122–127

Mylroie JE, Carew JL (1995) Karst development on carbonate Islands. In: Budd DA, Saller AH, Harris PM (eds) Unconformities and porosity in Carbonate Strata, vol 63, AAPG Memoir. American Association of Petroleum Geologists, Tulsa, Oklahoma, pp 55–76

Neuendorf KKE, Mehl JP, Jackson JA (2005) Glossary of geology, 5th edn. American Geological Institute, Alexandria, p 23, 523, 655

Norwick SA, Dexter LR (2002) Rates of development of tafoni in the Moenkopi and Kaibab formations in Meteor Crater and on the Colorado Plateau, northeastern Arizona. Earth Surf Process Landf 27:11–26

Owen AM (2007) Tafoni caves in quaternary carbonate eolianites: examples from the Bahamas. Masters thesis, Mississippi State University, p 187. http://library.msstate.edu/etd/show.asp?etd=etd-05142007-143443

Owen AM, Kirkland BL, Mylroie JE, Fratesi SE (2010) Serendipitous observations of organic matter in meteoric calcite cement in Holocene eolian calcarenites, San Salvador Island, Bahamas. In: Martin JB, Siewers FD (eds) Proceedings of the 14th symposium on the geology of the Bahamas and other carbonate regions. Gerace Research Centre, San Salvador Island, Bahamas, pp 200–208

Rasband W (ed) (2007) Image J (Java 1.5.01-09). National Institutes of Health USA. http://rsb.info.nih.gov

Rögner K (1988) Measurements of cavernous weathering at Machtesh Hagadol (Negev, Israel) a semiquantitative study. In: Imeson AC, Sala M (eds) Geomorphic processes in environments with strong seasonal contrasts, vol I, Hillslope processes. Catena, Cremlingen/Destedt

Roth M (2004) Inventory and geomorphic analysis of Flank Margin Caves of the Bahamas. Master's thesis, Mississippi State University, 134p

Schwabe SJ (1992) The petrology of Bahamian Pleistocene Eolianites and Phreatic Dissolution Caves: implications for Late Quaternary Island Development. Master's thesis, Mississippi State University, 177p

Souza-Egipsy V, Wierzchos J, Sanchos C, Belmonte A, Ascaso C (2004) Role of biological soil crust cover in bioweathering and protection of sandstones in a semi-arid landscape (Torrollones De Gabarda, huesca, Spain). Earth Surf Process Landf 29: 1651–1661

Stoffer P (2004) Desert landforms and surface processes in the Mojave National Preserve and Vicinity. USGS Open File Rep 2004–1007

Stowers RE (1988) Stratigraphy and geochronology of Pleistocene Carbonates: Sandy Point Area Southern San Salvador Island, Bahamas. Master's thesis, Mississippi State University, 103p

Sunamura T (1996) A physical model for the rate of coastal tafoni development. J Geol 104:741–748

Turkington AV (1998) Cavernous weathering in sandstone: lessons to be learned from natural exposure. Q J Eng Geol 31:375–383

Turkington AV, Phillips JD (2004) Cavernous weathering, dynamical instability and self organization. Earth Surf Process Landf 29:665–675

Viles H (2005) Self-organized or disorganized? Toward a general explanation of cavernous weathering. Earth Surf ProcessLandf 30:1471–1473

Walker LN (2006) The caves, karst, and geology of Abaco Island Bahamas. Master's thesis, Mississippi State University, 241p

Walker LN, Mylroie JE, Walker AD, Mylroie JR (2008) The caves of Abaco Island, Bahamas: keys to Geologic Timelines. J Cave Karst Stud 70(2):108–119

Waterstrat WJ (2007) Morphometric differentiation of Flank Margin Caves and Littoral or Sea Caves. Master's thesis, Mississippi State University, 200p

Waterstrat WJ, Mylroie JE, Owen AM, Mylroie JR (2010) Coastal caves in Bahamian eolian calcarenites: differentiating between sea caves and flank margin caves using quantitative morphology. J Cave Karst Stud 72:61–74

White KS (1995) An imprint of Holocene transgression in quaternary carbonate eolianites on San Salvador Island, Bahamas. In: White B (ed) Geological Society of America Special Paper 300, Terrestrial and shallow marine geology of the Bahamas and Bermuda. Geological Society of America, Boulder, Colorado, pp 125–138

Wilson WL (1992) Beyond the blue: Dean's Blue Hole Long Island, Bahamas, Exploration and Scientific Study of the Deepest Blue Hole in the World. Subsurface Evaluations, Inc., Winter Springs, Florida

Xara x Software (version 1.0) (2001) Xara Group Limited Hemel Hempstead, UK. http://www.xara.com/us/

Coastal Caves and Karst of the Puerto Rican Islands

9

Michael J. Lace

Abstract

Puerto Rico, in addition to numerous associated smaller islands and cays, forms a complex archipelago exhibiting a wide range of coastal geomorphologies influenced by the interaction of regional tectonics, lithology and sea level fluctuations. Coastal landforms in this setting also display the influences of multiple karst and pseudokarst processes. Expressions of coastal karst in this setting include caves formed by littoral erosion (sea caves), cliff retreat (talus caves and fissure caves) dissolution from processes associated with a freshwater lens (flank margin caves), coastal karren and dissolution pipes. The effects of a combination of Quaternary glacioeustatics and tectonic uplift are illustrated by multiple distinct cave horizons, or discrete elevations associated with past sea-level stillstands, ranging from 0 to more than 40 m above mean sea level (msl). While systematic research into the speleogenesis, biospeleology and archaeology of coastal caves and karst of the Puerto Rican islands has historically been sporadic in its geographic coverage and limited in its scope, recent efforts focusing on these fascinating landforms have revealed an expansive inventory of over 400 coastal cave sites to date with significant modern shoreline and paleoshoreline karst areas remaining to be studied.

9.1 Geophysical Overview of the Puerto Rican Islands

The islands of Puerto Rico in conjunction with the western flank of the Virgin Islands to the east were once defined as the Puerto Rico Bank, including over 240 islands and cays associated with the Puerto Rico platform (Heatwole et al. 1981). This chapter deals with coastal karst within the Commonwealth of the Puerto Rican islands, geographically extending from Isla Monito in the west to Isla Vieques at the eastern edge of the Greater Antilles and encompassed by the windward Atlantic shoreline to the north and the leeward Caribbean coastline to the south (Fig. 9.1). The main island of Puerto Rico and its adjacent smaller islands include 720 km of coastlines which display a variety

M.J. Lace (✉)
Coastal Cave Survey, West Main Street 313 1/2,
West Branch, IA 52358, USA
e-mail: michael-lace@uiowa.edu

M.J. Lace and J.E. Mylroie (eds.), *Coastal Karst Landforms*, Coastal Research Library 5,
DOI 10.1007/978-94-007-5016-6_9,

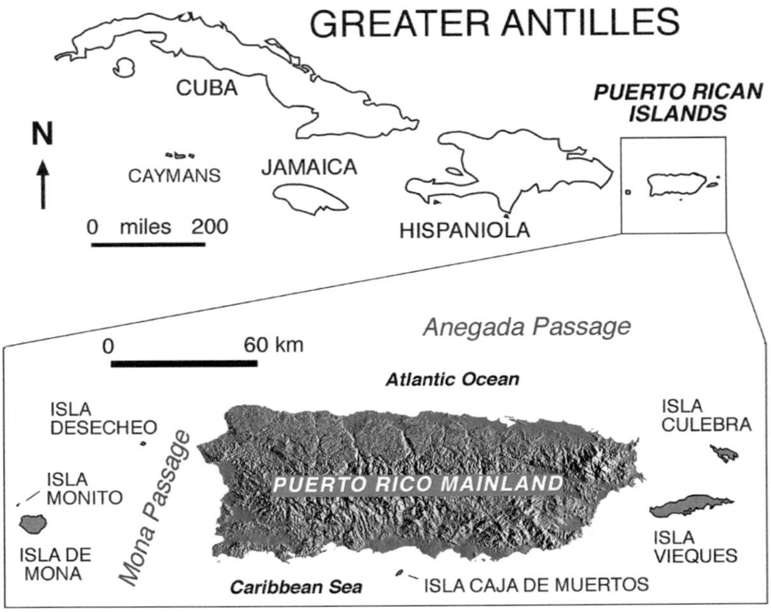

Fig. 9.1 Map of Greater Antilles with inset area detail map of the Puerto Rican Islands. Shaded relief basemap of Puerto Rico mainland produced by sidelong airborne radar (SLAR) 1:200,000 scale – courtesy of the USGS (Bawiec 1999)

of landforms from expansive alluvial coastal plains to rugged Miocene and Quaternary-aged carbonate exposures harboring a diverse inventory of karst features within a contemporary microtidal environment (Kaye 1959a; Lundberg and Taggart 2007; Monroe 1968, 1974). Located at the eastern extreme of the Greater Antilles, the island structure in the Puerto Rico group conform to a range of simple to complex carbonate island karst models (CIKM) according to Mylroie and Mylroie (2007). The structural diversity of these island platforms has contributed to the range of coastal karst expressions in Puerto Rico.

Tectonic evolution of the Puerto Rican islands. The Puerto Rican mainland is associated with the contact between the Caribbean and North American plates and bounded by a series of deep marine trenches and faults which have played a pivotal role in area island platform genesis. The turbulent waters of the Mona Passage, reaching depths of 1,000 m, lie between Isla de Mona and the western coast of the Puerto Rican mainland. The Caja de Muertos fault is located off the southern coast separating Isla Caja de Muertos from the mainland (Garrison 1969; Kaye 1957). The Puerto Rico Trench reaches depths of 8,500 m in the Anegada Passage off the northern shore of the island. Tectonic activities associated with these deep marine fault structures within the context of regional plate and microplate boundaries form a complex model of tectonic evolution for the Puerto Rican islands with concomitant effects on coastal cave development (Gestel et al. 1999; Mann et al. 2005; Reid et al. 1991; ten Brink et al. 2009).

Seminal fieldwork examining coastal karst in Puerto Rico by Briggs, Guisti, Kaye, Monroe and Seiders spanned the 1950s and early 1980s (see References). These works persist to date as invaluable and durable resources in coastal research in this setting though recent works offer more comprehensive views of the geology and karst hydrology of Puerto Rico and the adjacent Virgin Islands (Renken et al. 2002) and area coastal morphologies (Guisti 1978; Hernandez Santana et al. 2002; Hubbard et al. 2008; Morelock et al. 2010). Karst resources of the following coastal regions on the Puerto Rican islands are examined in detail in this chapter, incorporating current cave inventories that encompass the mainland coastlines and those of the smaller, but equally significant, islands offshore.

9.2 Coastal Morphology and Karst Resources of the Puerto Rico Mainland

The Puerto Rican mainland measures 176 km east to west and 62 km north to south with over 500 km of shoreline – significant segments of which are composed of karst landscapes (Lugo et al. 2001; Monroe 1960). Coastal carbonate stratigraphy on the mainland includes Quaternary eolian calcarenites as well as Miocene and Cretaceous-aged subtidal limestone (Kaye 1959a). Elevations rise to 1,338 m asl in the exposed volcanic core of the island interior. To date, over 190 coastal caves have been documented in the western, northern and southern coastal sections of the mainland (Lace 2008) in addition to other coastal karst features with additional reports of isolated submerged cave structures offshore (Miller and Kambesis 2009). Coastal cave types identified to date include sea caves, flank margin caves, talus caves and fissures (Table 9.1) which range from intact to partially or significantly denuded and/or overprinted (hybrid) structures.

9.2.1 Northern Coastal Karst Zone

The windward northern coastline of Puerto Rico naturally harbors the majority of the Quaternary eolian calcarenite exposures on the mainland (Monroe 1974). Reaching elevations over 10 m above msl, these lithified carbonate (eolianite) dunes discontinuously stretch from the northwest point 128 km eastward to a point west of the Atlantic confluence of the Rio Grande de Loiza (Fig. 9.2a). Numerous embayments have formed within the eolianitic ridges—what Kaye (1959a) described as "lunate resonating basins" (Fig. 9.2b) along with solutional pans and tidal terraces (Fig. 9.10). Heavy weathering accompanied by energetic littoral and bioerosional processes of the shoreline dune surfaces (i.e. micro and macro-pitting) has also generated dense areas of coastal karren (see Chap. 2). Shear limestone cliffs also dominate the northwest coast and also harbor a significant portion of the recorded caves in this coastal zone.

Coastal Cave Development within these structures is prolific with over 50 caves documented in coastal Quaternary eolianites to date. Another 70 north coast caves lie within adjacent Miocene-aged limestones (e.g. the Aymamon and Aguada limestones), which form littoral escarpments reaching elevations of up to 37 m (Fig. 9.2c). Sea cave development is

Table 9.1 Summary of coastal caves recorded in the Puerto Rican islands

PR island	Island platform type[a]	Number of total caves recorded	Coastal cave types expressed[b]	Range of total cave areas[c] (m^2)	Total island area (km^2)
PR mainland	Composite-complex	>190	tf, lf, fm,sc, tc	13–4,357	8,897
Isla de Mona	Simple	>200	fm, sc, tf, tc, cc	10–235,000	55
Isla Monito	Simple	nd	nd	nd	0.17
Isla Vieques	Composite	16	fm, sc	4–144	135
Isla Caja de Muertos	Simple	2	fm, sc	103–888	1.54
Isla Desecheo	Composite	nd	nd	nd	1.2
Isla Culebra	Non-carbonate	0	No documented karst features	na	28

[a]As determined by CIKM (Mylroie and Mylroie 2007)
[b]Cave types: *tf* tectonic fissure, *lf* littoral fissure, *fm* flank margin cave, *sc* sea cave, *tc* talus cave, *cc* coralloid cave, nd not determined, *na* not applicable
[c]Total floor area determined by quantitative analysis of cave maps (ImageJ software, National Institutes of Health – NIH)

Fig. 9.2 (**a**) Array of Quaternary eolianite dunes with sea arch development-northern coast, Puerto Rico mainland. (**b**) Northern coastal dune ridge breached by Atlantic Ocean wave front (Photo by WH Monroe, 1965 USGS photo archives). (**c**) North coastal limestone escarpment with cave development at discrete elevations (+7 and +35 m msl). Note wave cut bench at an elevation of <1 m

prominent in this energetic littoral environment but numerous examples of breached flank margin caves are also prominent.

9.2.2 Southern Coastal Karst Zone

The southern interior of the Puerto Rican mainland also harbors a significant karst region, known as the southern karst belt (Beck 1974; Miller and Kambesis 2009). However, in contrast to other coastlines of the Puerto Rican mainland, the leeward Caribbean coast in the south harbors no eolianite dune ridges or extensive fringing carbonate marine terraces characteristic of the northern shoreline. Thickly-bedded to chalky, poorly indurated exposures of Miocene-aged limestone form extensive shore cliffines (Fig. 9.3a). A series of complex submerged reef structures and canyons (with sporadic associated cave development) extend to the edge of the faulted Caja de Muertos shelf (Beach and Trumbull 1981; Trumball and Garrison 1973) to where depths plunge into the Venezuelan basin. A range of cave types and other shoreline features (such a sea arches and sea stacks) also occur in this setting (Fig. 9.3b–d).

Coastal Cave Development in the South Shoreline cave development is confined to discontinuous outcrops of Miocene-aged Ponce limestone located at current shoreline and higher elevations consistent with paleoshoreline settings. Cueva Murcielagos (Fig. 9.4) is an example of a flank margin cave in such a paleoshoreline exposure. As observed in other coastal areas in Puerto Rico, the cave serves as a significant bat roost. Littoral erosion of seaward exposures has also resulted in a number of sea caves as well as exposure and progressive denudation of flank margin voids (Fig. 9.3c–d).

9.2.3 Western Coastal Karst Zone

In contrast to the northern and southern coastlines, the western coastal zone is characterized by more extensive alluvial coastal plains with widely distributed and varied coastal karst segments (Mattson 1960; Renken et al. 2002). Heavily modified cliff exposures of Aymamon limestone contain isolated eolianite outcrops and exposed terra rosa paleosol deposits more than 1 m thick. Isolated outcroppings of karstic Aguada limestone have also been observed (Lace 2008).

Coastal cave and karst development on the western coast is comparatively more limited in its extent and associated with a variety of carbonate lithologies, including discontinuous fringing marine terraces, eolianites, Miocene-aged limestones, such as the Aymamon and Cibao formations, as well as isolated coastal exposures of karstified Cretaceous limestone (Monroe 1974). Both sea caves and flank margin caves compromise the majority of the recorded caves in this coastal zone. Cuevas Cofresi, however, are unusual examples of fissure cave development within one of very few coastal exposures of Cretaceous-aged limestone in Puerto Rico (Lace 2012). The caves consist of a network of more than 300 m of maze-like passages exhibiting phreatic morphology and the influence of structural control, consistent with examples of cave development within inland Cretaceous-aged limestone (Miller 2004). The fissure caves harbor small but persistent bat populations and show evidence of pre-contact use and historic guano mining in the nineteenth century (Giles and Carrero 1918).

9.3 Coastal Morphology and Karst Resources of the Adjacent Islands

The smaller adjacent islands display diverse coastal morphologies as well, ranging from periodically submerged alluvial banks to volcanic and uplifted carbonate platforms. The following examples focus on those island platforms with significant recorded coastal karst resources.

9.3.1 Isla de Mona and Isla Monito

Geologic Setting. Isla de Mona (or Mona island) is a Neogenic carbonate platform encompassing

Fig. 9.3 (**a**) Cave development within limestone exposures of the southern coast. (**b**) Sea stacks on the southwestern coast. (**c**) Littoral cave structure – southern coast. (**d**) Flank margin cave development on the southern coast

an area of 55 km^2 within the turbulent Mona Passage with a rectangular shape of 11 km in length (east to west) and 5 km in width - north to south (Fig. 9.5a) (Rodriguez et al. 1977). The distinctive island geomorphology and geography have produced a unique and intensely karstic landscape that has shaped an equally complex ecosystem (Aron 1973). Regional tectonic activity has uplifted the platform, resulting in elevations up to 90 m above msl. These same influences along with minor fracture development have resulted in significant cliff retreat, generating sheer cliffs

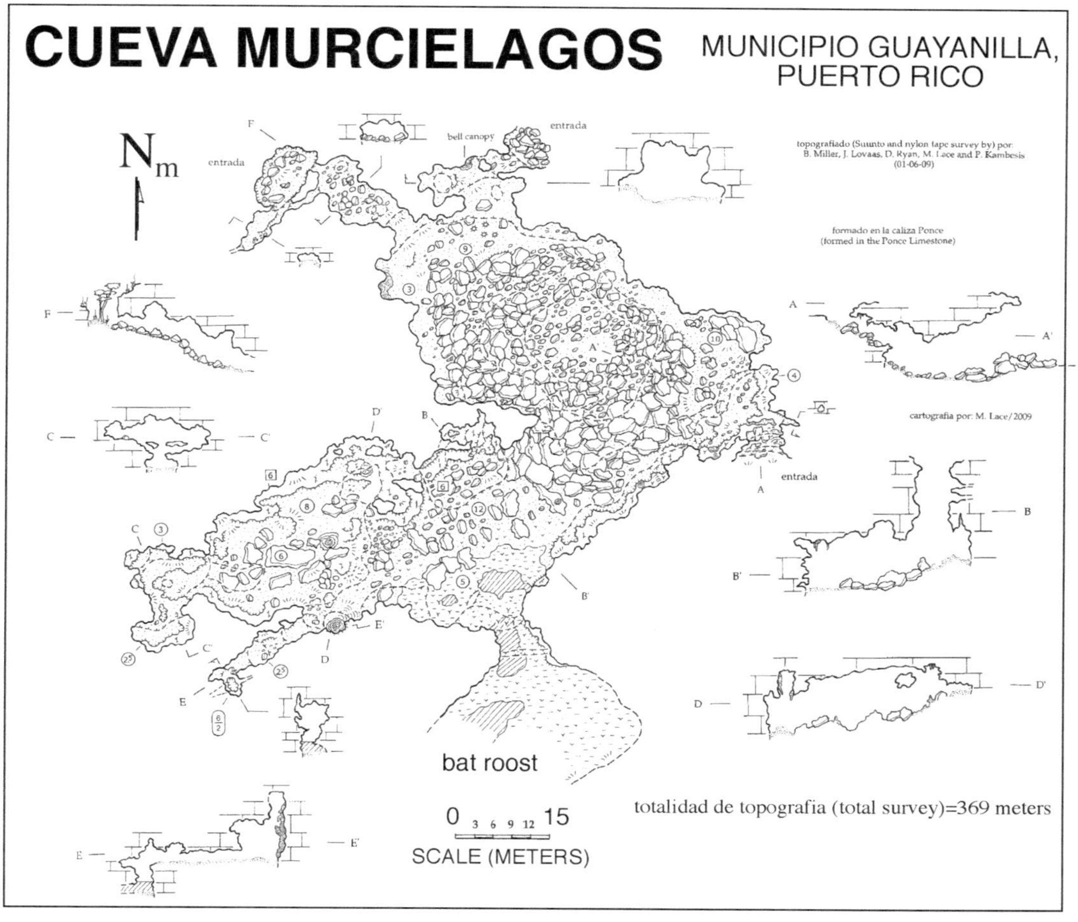

Fig. 9.4 Map of Cueva Murcielagos, a flank margin cave formed within Miocene-aged limestone on the southern coast

that range from 20 to 80 m in height (Fig. 9.5c). Coastal carbonate stratigraphy includes exposed Pleistocene and Mio-Pliocene reef structures as well as a predominant Miocene-aged dolomite (Mona Dolomite) platform base overlain by a limestone veneer (Lirio Limestone) of varying thickness, ranging from 1 to 30 m, across a north to south transect (Briggs and Seiders 1972; Gonzalez et al. 1997). On the south and southwest side of the island, a 3–6 m high Pleistocene fossil reef abuts the cliff line, forming a narrow coastal plain. Offshore, a modern reef partially encircles the shorelines - separated from the west, east and southern coastlines by intermittent shallow lagoons (Fig 9.5a). A series of parallel grooves within the submerged platform slope beyond the modern reef are thought to be littoral erosion (rather than constructional) features as documented on other islands in the region (Kaye 1959b). While a number of phreatic karst features and collapsed dissolutional caves have been documented in the rugged island interior, the majority of expressed cave structures are dissolutional voids concentrated along the intensely karstic island perimeter (Frank et al. 1998a, b) (Fig. 9.5b–d).

Coastal Cave Development. More than 200 caves have been systematically documented on Mona so far but thorough exploration of more remote coastal and inland areas remains incomplete (Kambesis and Lace 2009). Flank

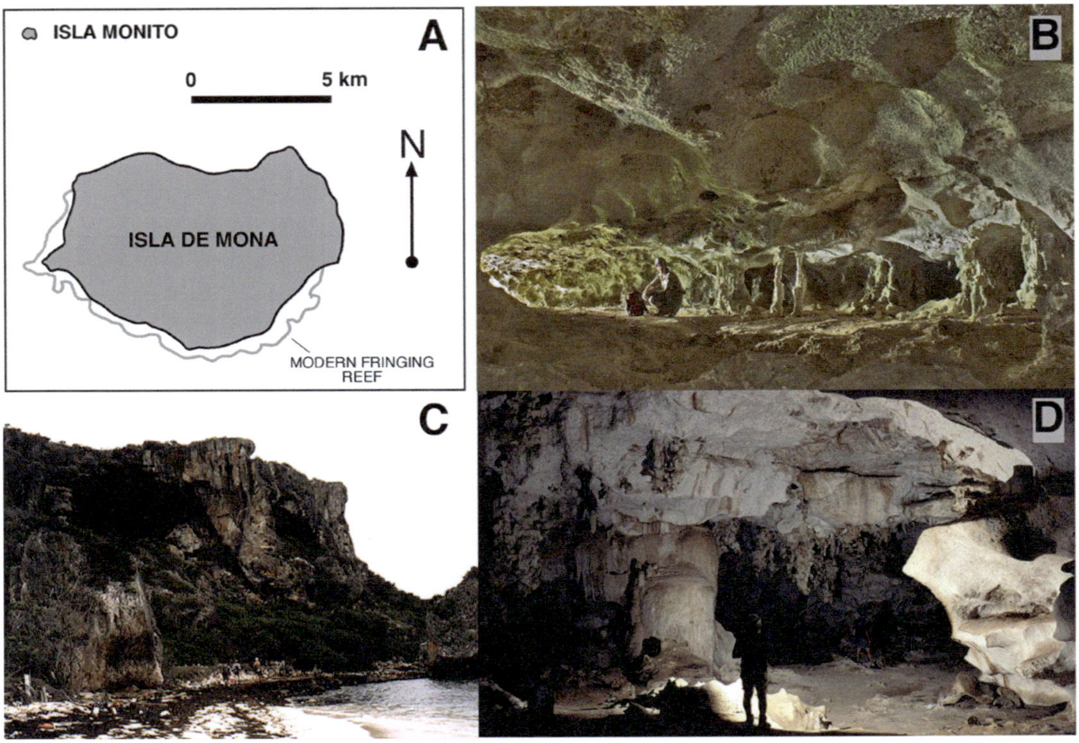

Fig. 9.5 (**a**) Area map of Isla Mona and Isla Monito. (**b**) Morphology of flank margin cave development, Cueva Esqueleto, west coast (Photo by D. Bunnell). (**c**) Coastal Cave entrance on eastern shoreline of Isla de Mona. (**d**) Cave passage in Cueva Chito, eastern coast

margin caves comprise the overwhelming majority of the recorded cave inventory on Mona, including Sistema Faro which is the most extensive documented flank margin cave in the world to date (see Chap. 4, Fig. 4.5). They are not conduits that move groundwater but rather are relict mixing chambers that mark the location of a paleo halocline, i.e. the interface between the saltwater and the freshwater lens (Wicks and Troester 1998). Similar to complex flank margin cave development in the Marianas (see Chap. 13), the caves have formed at a variety of discrete elevations (or "cave horizons") within the Lirio limestone and the underlying Mona Dolomite, indicating a combination of tectonic uplift and changes in sea level spanning its geologic past and showing dissolutional overprinting indicative of repeated cycles of vadose and phreatic conditions. The majority of the larger dissolution structures are associated with the limestone/dolomite contact and can extend laterally (parallel to the cliff facies) for several hundred meters with comparatively limited inland extent (ranging from 5 to 257 m) (Fig. 9.6). In addition to flank margin caves, phreatic pits, talus caves (formed within displaced segments of retreating cliff facies), constructional coralloid caves and sea caves have also been identified. Littoral erosion of the sheer perimeter of nearby Isla Monito has also been noted (Kaye 1959b) but to date no caves have been mapped on this small remote island (with an area of merely 0.17 km^2) which shares a similar morphology to Isla de Mona as a simple model of an uplifted carbonate platform.

Cultural cave resources of Isla de Mona. The spectacular geomorphology of Isla de Mona is not the sole source of its uniqueness and importance in coastal karst studies of the

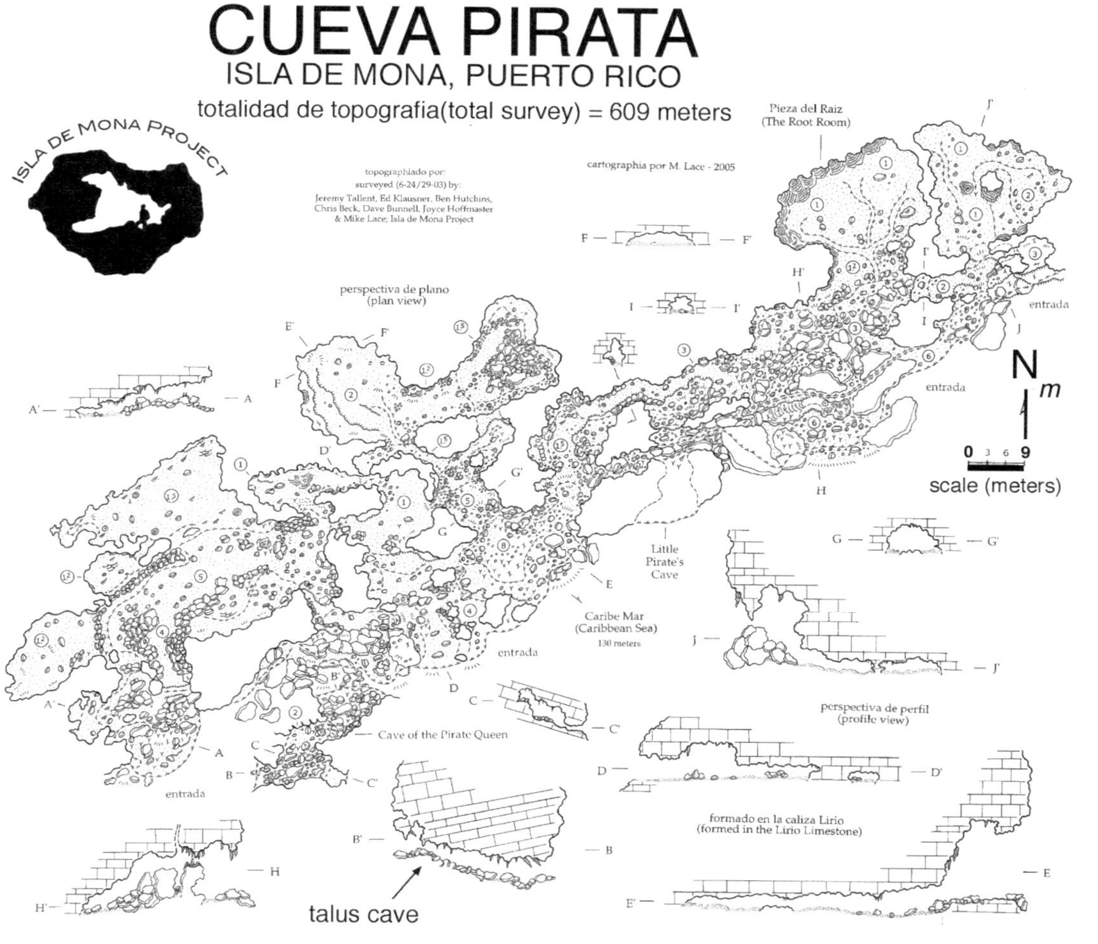

Fig. 9.6 Map of Cueva Pirata and adjacent cave segments – a flank margin cave modified by progressive cliff retreat, Isla de Mona

region. In addition to harboring past and present endemic faunal species, many of the caves of Mona have also provided a physical setting for anthropogenic use of these structures including the archaic to pre-contact (or pre-colonial) use by indigenous peoples, spanning several thousand years, through post-contact nineteenth century commercial guano mining by European interests (Wadsworth 1973; Frank 1998). Earlier works defined indigenous migrations and presence in the Caribbean region and in the Mona Passage within the context of an established general model yet did not specifically address Taino activity in detail on Isla de Mona (Rouse 1992). Alegria (1983) defined ceremonial plazas on the arid rocky plateau of the island as well as the Puerto Rican mainland and Davila (2003) offered a more complete overview of archaeologic record on Isla de Mona, including field research performed within some of its caves. Similarly, overviews of indigenous rock art distribution in the region, while citing fieldwork on Mona, implied that the study of cave rock art in this setting was "complete", noting a total of 4 cave sites harboring rock art (Hayward 2009). However, research by the Isla de Mona Project, spanning the past 13 years, has demonstrated that the work is far from complete as over 20 cave sites harboring a diverse inventory of rock art forms have recently been documented, some distinct from those found elsewhere in Puerto Rico (see also Chap. 5).

Isla de Mona and Isla Monito were named a natural national landmark in 1975 by the U.S. Department of the Interior due to its complex and delicate ecology. The islands and the surrounding reefs were designated a protected marine environment in 1986. The Mona Island Nature Reserve is the largest such protected marine area in Puerto Rico. The site is managed by the Puerto Rican Department of Natural Resources (Departamento Recurso Naturales y Ambientales – DRNA).

9.3.2 Isla Caja de Muertos

Geologic Setting. Located 8 km SE of the south-central coast and measuring 2.75 by 1.75 km in area, Isla Caja de Muertos (Fig. 9.7a) was originally proposed to be composed of a series of faulted blocks – some volcaniclastic in origin and some composed of Miocene-aged limestones (Kaye 1957). Tectonic uplift of the platform is associated with the Muertos Trench which extends S-SW from the southern coast of the Puerto Rico mainland and west of Caja de Muertos (Beach and Trumbull 1981). Similar to Isla de Mona and Isla Monito, Cayo Morrillito is a flat-topped, uplifted carbonate platform rising several meters above msl (Fig. 9.7c).

Coastal Karst Resource and Cave Development. Only a few coastal caves have been thoroughly documented on the island to date (Monroe 1974). Cave resources on Caja de Muertos include intact flank margin caves (Fig. 9.8), reports of denuded flank cave structures that ring the rocky southern tip of the island and sea caves formed within a shallow, intermittent Pleistocene-aged reef terrace (Fig. 9.7b, d). Cueva Almeida is the largest cave documented on the island with a vertical extent of 26 m. It is a flank margin structure, formed within the principal limestone escarpment on the island. Structural control of the uplifted limestone terrace and its role in shaping the hydrodynamic structure of the freshwater lens, and influencing subsequent void development, is reflected in the prominent joint-control patterns illustrated on the cave map (Fig. 9.8a). Cueva Almeida is one of two caves commercially mined for guano-rich cave sediments in the nineteenth century (Gile and Carrero 1918) with a small modern bat population. Local history associated with the coastal caves of Caja de Muertos also includes potential archaeologic (i.e. Taino) cave use (Dubelaar 1995) as well as historic activities, ranging from seventeenth century pirate activity. Adjacent Cayo Morrillito, though smaller in land area, also harbors expressions of apparent littoral cave development (Fig. 9.7c). Additional cave segments are likely to be found on these platforms in the course of additional fieldwork.

The nature reserve, which includes both Caja de Muertos and Cayo Morrillito, is managed by the Puerto Rican Department of Natural Resources (Departamento Recurso Naturales y Ambientales - DRNA). While some of the caves are accessible to island visitors, access to some karst segments of the island is prohibited due to sensitive nesting areas for endemic fauna.

9.3.3 Isla Vieques

Geologic Setting. Isla Vieques (Fig. 9.9a) also represents a complex island platform example composed of inland Cretaceous-aged volcanics with intermittent coastal overlays of alluvial sediments, igneous exposures, recemented accumulations of beach rubble (Fig. 9.9d) and Cenozoic Miocene aged carbonates (Fig. 9.9b), generally referred to as the Puerto Ferro Limestone (USGS 1982; USN 2001). Isolated segments of modern fringing reef structures have also been described along the southern coast. While igneous shoreline exposures offer poor substrates for extensive or durable void development, the fringing limestone was found to harbor numerous examples of coastal caves (Fig. 9.9b, c).

Coastal Cave Development. While cave and karst features on Vieques have been poorly documented, recent exploration has documented

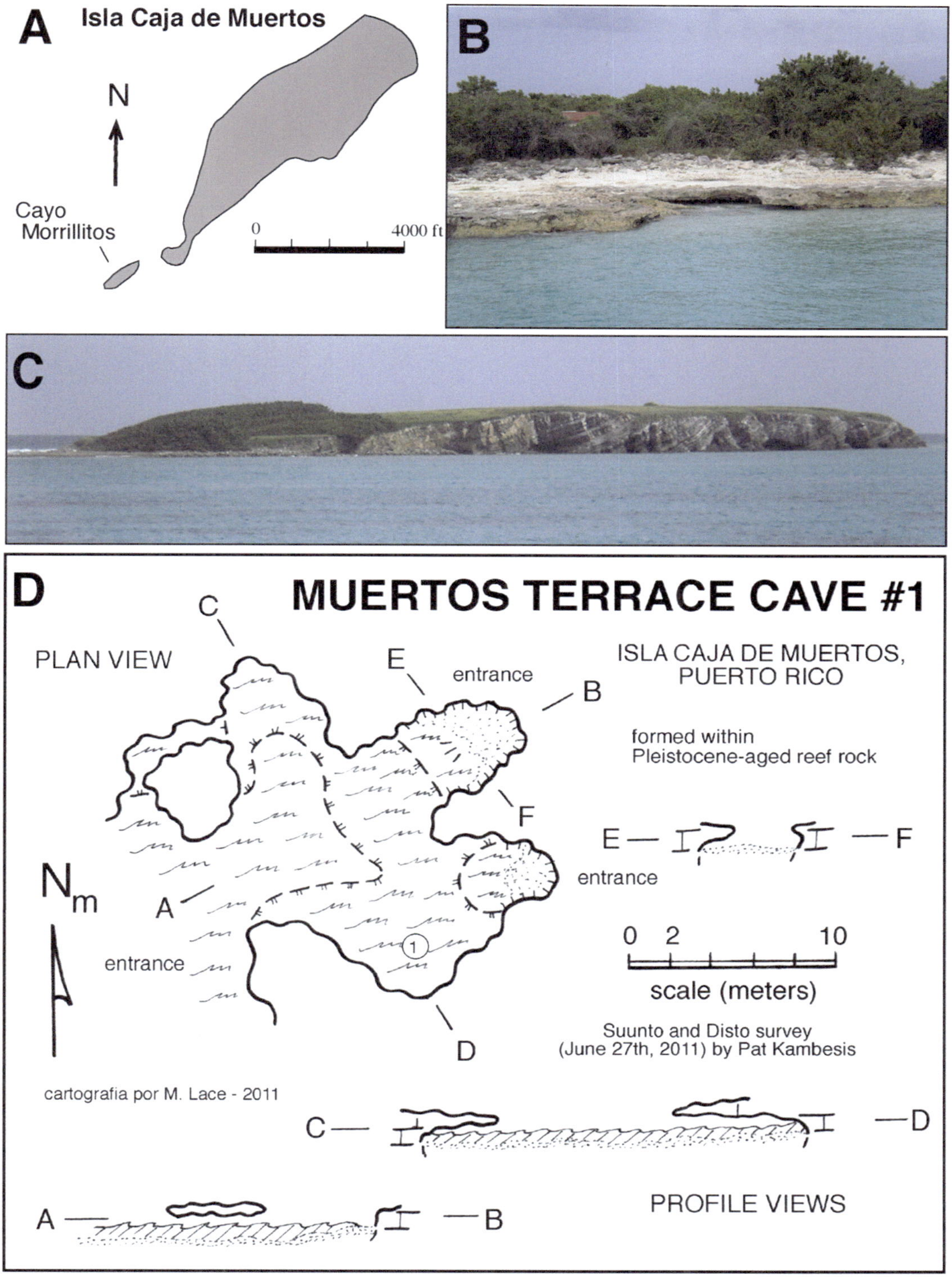

Fig. 9.7 (**a**) Map of Isla Caja de Muertos and Cayo Morrillito. (**b**) Littoral cave development within a shallow Pleistocene reef terrace, Caja de Muertos. (**c**) Cayo Morrillito shoreline. Noted bedding angle. (**d**) Map of sea cave formed within reef terrace

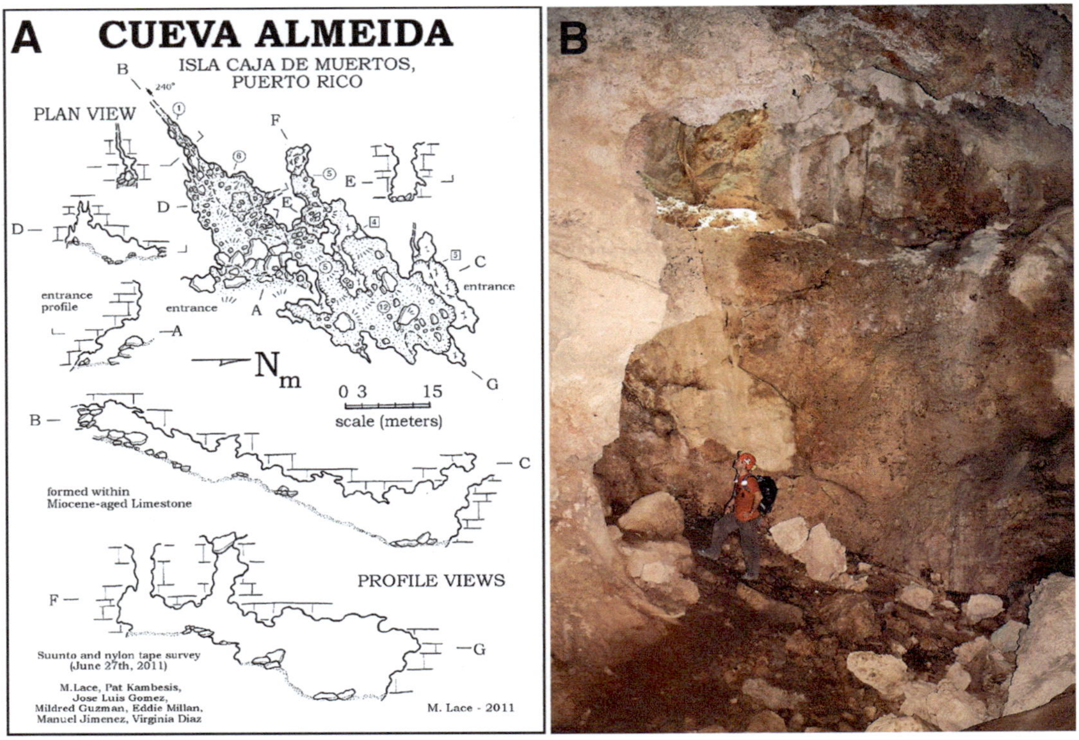

Fig. 9.8 (**a**) Map and (**b**) photo of Cueva Almeida – a flank margin cave formed within an uplifted Miocene – aged limestone terrace

16 coastal caves associated with carbonates located along the southern shoreline. The majority of these structures can be classified as sea caves. Littoral erosion and associated cave development at or near the current shoreline has been influenced by minor faults or unconformities in littoral cliff exposures, resulting in a range of cave structures – some expressed as elongate fissures extending up to 10 m inland from the cliff faces. Apparent mechanisms of coastal speleogenesis in this setting also include an example of mixing zone dissolution. Cueva Navio is a flank margin cave consisting of a single partially submerged tidal chamber (Fig. 9.9e).

Isla Vieques is one of the least developed inhabited islands in the Puerto Rico chain. Consequently, traditional human impacts are comparatively limited in many coastal areas which are now designated as protected wildlife areas (USFWS 2007). Anthropogenic overprinting of karst terrain on the eastern portion of Vieques is perhaps more unique than other island platforms in the region as long-term naval bombardment training targeted inland and coastal areas; this training was discontinued in 2000, however, such long-term land uses may have altered significant karst features on Vieques (USN 2001). On a positive note, numerous concrete bunkers constructed on Vieques during WWII now serendipitously serve as roosts for several bat species (Rodriguez-Duran 2009). While overall coastal cave development is limited on Vieques due to limited carbonate exposures in comparison to the shorelines of the Puerto Rico mainland, further exploration will no doubt reveal additional karst features. In-depth study and impact assessment of some areas, however, must wait for completion of proposed residual unexploded ordinance remediation.

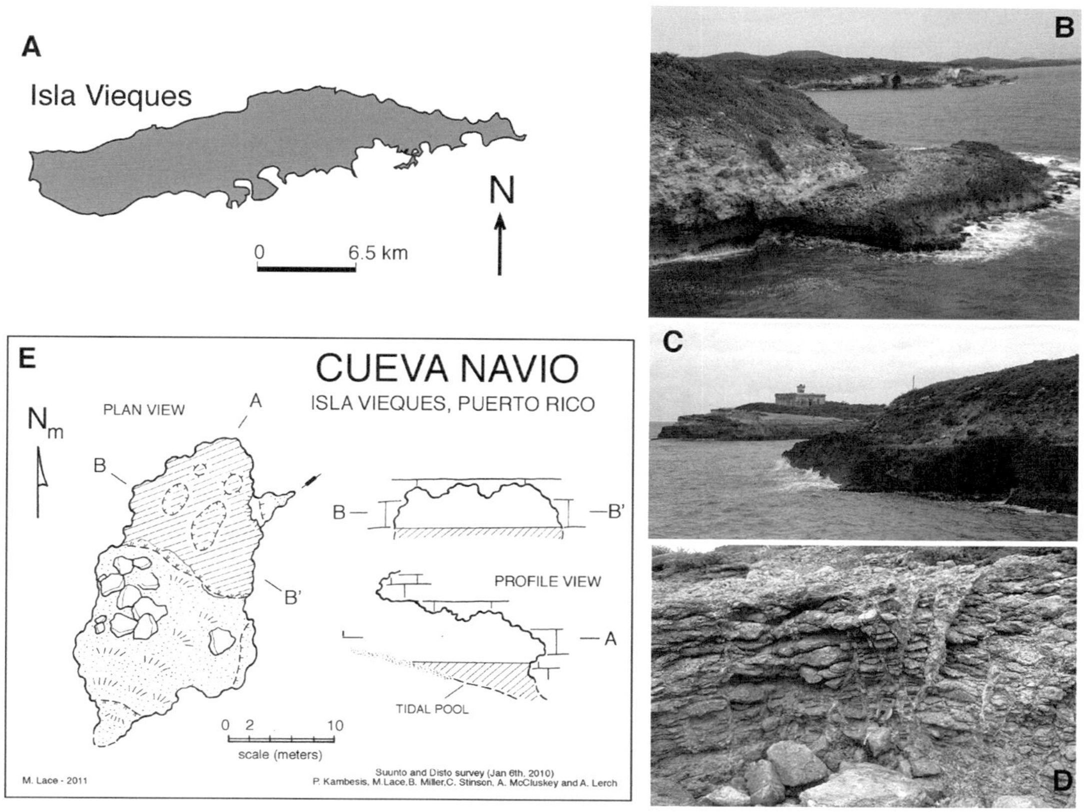

Fig. 9.9 (**a**) Map of Isla Vieques. (**b**) Photo Isla Vieques southern coast. (**c**) Photo of Puerto Ferro limestone exposures, southern coast (Note: Eighteenth century lighthouse in background). (**d**) Igneous beach rubble recemented with calcite (photo by B. Miller). (**e**) Map of Cueva Navio – a flank margin cave formed within coastal limestone

9.3.4 Isla Desecheo

Geologic Setting. Located 21 km off the NW shore of Puerto Rico within the Mona Passage. Isla Desecheo is a non-carbonate platform fringed with intermittent shallow carbonate marine terraces up to 1 m thick. The platform exhibits signs of tectonic modification consistent with its location on the great southern Puerto Rico fault zone. The island is thought to be formed within the Cretaceous to lower Tertiary Rio Culebrinas formation, consisting primarily of volcanic sandstone, breccia and mudstone (Seiders et al. 1972). There are reports of microkarst features formed within these terrace structures as well as some sea cave development associated with moderate-energy wave activity acting on the calcarenous terraces and the adjacent volcanoclastic exposures (Wetmore 1918). Evidence for Quaternary stillstands have also been described on the Desecheo platform in the form of discrete uplifted terraces located at 2 and 13 m above msl (Seiders et al. 1972), similar to observed on the northern coastline of the PR mainland and the western shoreline of Isla de Mona (Lace and Kambesis 2009).

Coastal cave development and management. While the extent of coastal deposits suitable for cave development such as fringing carbonate marine terrace structures is limited on Isla Desecheo, there are anecdotal reports of small sea caves associated with the contact between coastal carbonates and the sedimentary volcaniclasts (Seiders et al. 1972). None of these karst features, however, have been surveyed to date. The island itself and offshore adjacent coral reefs are designated a natural reserve (also managed by the Puerto Rico DRNA) with no permanent human habitation or man-made structures in place.

9.4 Mechanisms of Coastal Cave Development in Puerto Rico

As illustrated in Table 9.1, the density of coastal cave development does not directly correlate with relative size of coastal caves (i.e. as measured by total cave floor area) or with the total area of an island platform. This is consistent with current models of coastal speleogenesis on carbonate islands (as discussed in Chap. 4) since coastal processes responsible for cave formation can be independent of factors associated with inland karst (Mylroie and Mylroie 2007). Flank margin caves are a predominant expression of coastal cave development that is not limited to the world class example of Isla de Mona but such caves have also been documented on the karstic coastlines of the mainland and smaller islands examined to date (i.e. Isla Vieques and Isla Caja de Muertos).

A combination of Quaternary glacioeustatics and tectonic uplift is illustrated by multiple distinct cave horizons, or discrete elevations associated with mixing zone speleogenesis, ranging from 0 to 35 m above msl across the Puerto Rican islands. Thus, expressions of cave development associated with glacioeustacy have been altered by tectonic uplift of island platforms to yield cave horizons well above any recorded interglacial sea-level stillstands in the region. Further, secondary coastal processes have partially or in some areas completely overprinted coastal cave structures, complicating the interpretation of their speleogenic origins, but providing tangible examples of complex interactions between multiple coastal processes (Lace 2008). Modeling the interplay of speleogenic processes within dynamic, evolving coastal landforms presents unique challenges and while qualitative criteria identifying specific coastal structures successfully lays the groundwork for such analysis, more robust quantitative methodologies are still needed.

Tectonic activity, littoral modification and minor fracturing of coastal carbonate exposures have contributed to extensive spalling and progressive cliff retreat, particularly on the northern coastline of the Puerto Rico mainland and all coastal areas of Isla de Mona. In such a tectonically active setting, talus overprinting of dissolution structures or littoral cavities are alternatively accompanied by varying degrees of overprinting of talus structures by coastal dissolution or erosion processes. The emerging inventory of coastal caves offers a range of structural expressions, documented in detailed cave maps, that reflect the complex evolution of coastal karst landforms in the Puerto Rican islands.

Cueva Golandrinas (Fig. 9.10) presents an interesting case study of cave development within lithified coastal calcarenite dunes of the mainland – subsequently modified by secondary coastal processes. Similar to other caves formed within the network of eolianite dunes of the northern coast, Cueva Golandrinas is a residual flank margin cave that has been exposed and significantly modified by secondary littoral processes. Adjacent cave segments (Fig. 9.10, "remnant A" and "remnant B") were similarly modified yet partially protected from such processes due to their position within the breached dune ridge. These expressions of flank margin cave development still retain residual characteristic cuspate morphologies consistent with dissolution as opposed to speleogenesis solely determined by littoral erosion. Detailed mapping of this coastal landform included mapping all associated cave structures within the physical context of a breached eolianite dune ridge, revealing a plausible sequence of coastal karst and pseudokarst processes. As the Atlantic wave front advanced, breaching the dune ridge and the dissolutional voids within, these flank margin caves were exposed to subsequent littoral erosion followed by progressive destabilization and collapse of the residual dune slopes. Consistent with the approaches of coastal karst analysis outlined in Chap. 4 and illustrated in applied modeling examples in Chap. 6, the coastal site documented in Fig. 9.10 illustrates the utility of integrating a functional

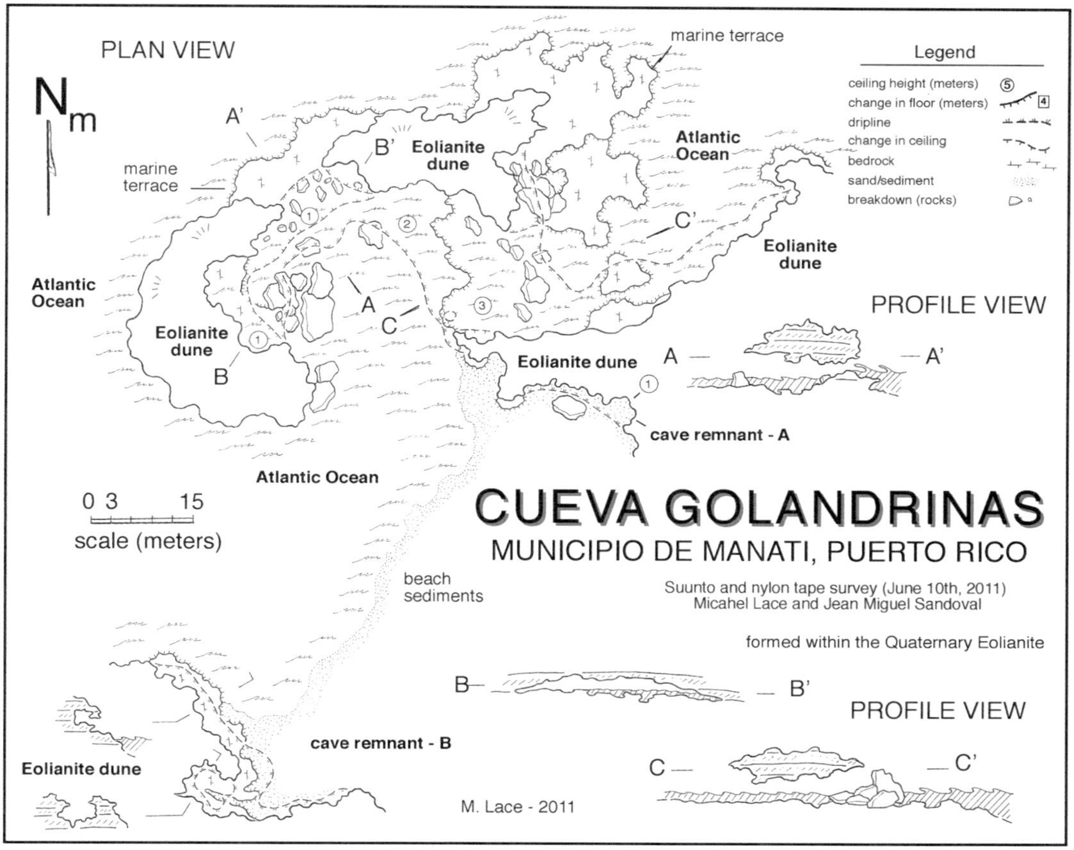

Fig. 9.10 Map of Cueva Golandrinas and associated cave structures formed within a breached Quaternary eolianite dune complex, northern coast of the Puerto Rico mainland

understanding of coastal cave development and modification within the site-specific and regional context of associated coastal landform evolution patterns.

9.5 Coastal Cave Archaeology and Biospeleology in Puerto Rico

The pre-colonial, colonial and modern cultural influences in Puerto Rican islands are imprinted on its coastal karst. Coastal caves in the Puerto Rico shorelines have functioned as fossil repositories as well as contemporary coastal habitats for endemic species (Rodríguez-Durán 2009) and cultural materials (Davila 1988, 1991). While the overall archaeology related to the cultural range of archaic, ceramic through historic periods in Puerto Rico is beyond the scope of this chapter, examples of pre-contact cultural uses of coastal caves are illustrated here (Fig. 9.11a–d). As in other coastal settings (Sect. 5.2), cave structures have served as archaeologic repositories, preserving cultural materials in an otherwise dynamic littoral environment (Lace 2012). Successive migrations of indigenous peoples through the Caribbean basin resulted in the establishment and complex evolution of early societies in Puerto Rico and its neighboring islands (Siegel 1989).

As in other island settings, coastal landforms figured prominently in the introduction and expansion of these cultures across the Puerto Rican islands. For example, Isla de Mona was theorized

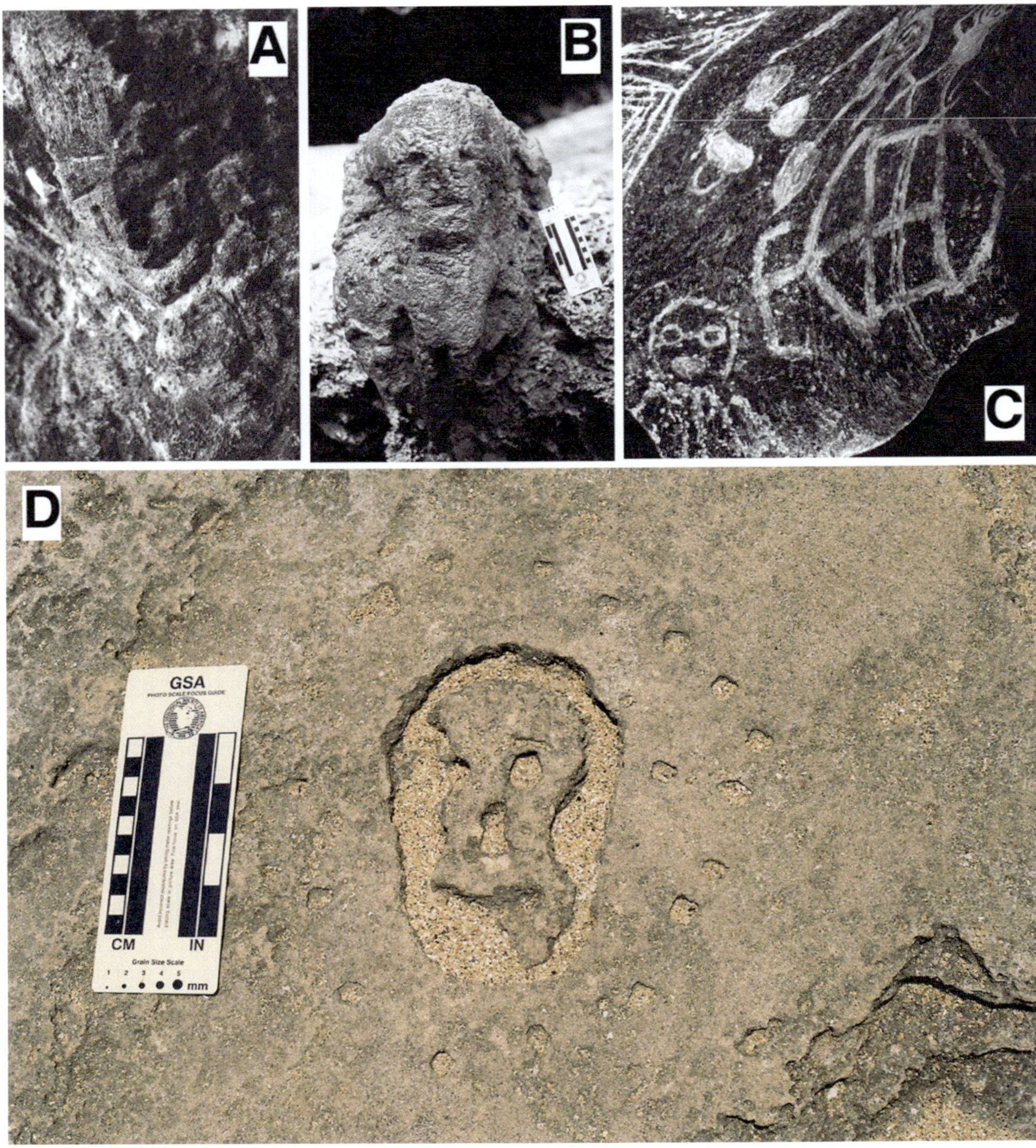

Fig. 9.11 (**a–c**) Pre-contact rock art examples in coastal caves. (**d**) Petroglyph on eolianite dune surface on the northern Purto Rico coast

to have occupied a critical geocultural position as it represented a frontier between distinctive Taino cultural groups while Isla Vieques occupied a similarly critical position at a divergence point in proposed pre-contact population migrations northwest across the Antilles Arc (Rouse 1992). However, more recent interpretations of pre-contact cultural development, utilizing both rock art (Roe and Hayward 2008) and lithic use in cave sites, paint a far more complex cultural landscape spanning 5,000 years bp through post-contact periods (Rodriguez Ramos 2010).

9.6 Coastal Cave Resource Management and Preservation Approaches in Puerto Rico

As in many other localities within the Caribbean region, Puerto Rico continues to experience extensive urban expansion driven by periods of modern population growth, shifting economic climates and changing land use patterns that continue to put inland karst areas under significant pressure (Porter 2010). Littoral areas of the Puerto Rican mainland are not exempt from the same negative effects of encroaching development in coastal municipalities. Several coastal cave sites continue to be profoundly impacted as a direct result of site excavation for future residential and commercial shoreline construction (Fig. 9.12) – a trend seen in many other coastal settings in the region (as discussed in Chap. 6).

As illustrated in this chapter, management approaches applied to coastal and inland cave sites in Puerto Rico vary and include active resource management, in the forms of private commercialization and public natural or historical preserve structures open to the public, as well as passive resource management plans by private land trusts or public agencies in the form of degrees of restricted public access or specific permit use (Diaz 2010; Kambesis 2007; Petrae 1995; Weaver 2009). However, these are exceptions within the inventory of coastal karst features in Puerto Rico.

Fig. 9.12 Modern development and impact to coastal and inland karst landforms, northern shore of Puerto Rico. Image courtesy of the USGS seamless server (http://nationalmap.gov)

Some of these coastal sites are comparatively remote which has contributed to preserving the cultural and natural features within them but it is perhaps unrealistic to assume that physical remoteness of some coastal localities with corresponding limited contemporary human traffic alone will serve as a sufficient long term deterrent to progressive degradation of natural and cultural aspects of coastal resources. Interpretive and educational initiatives are in development and remain critical tools complementing emerging sustainable resource management plans in both coastal and interior karst landscapes.

References

Alegria RE (1983) Ball courts and ceremonial plazas in the West Indies. Yale University Publications in Anthropology 79, pp 1–185

Aron JM (1973) Geology and mineral resources of isla de mona and monito islands: an assessment of their natural and historical resources. Environmental Quality Board, San Juan, PR, 2: B1–B7

Bawiec WJ (1999) Geochemistry, geophysics, mineral occurrences and mineral resource assessment of the Commonwealth of Puerto Rico. U.S. Geological Survey open file online report 98–038. http://pubs.usgs.gov/of/1998/of98-038

Beach DK, Trumbull JVA (1981) Marine geologic map of the Puerto Rico insular shelf, Caja de Muertos area. U.S. Geological Survey Misc. Investigations Series; I-1265, map

Beck BF (1974) Geology and hydrology of the El Convento Cave-Spring System, southwestern Puerto Rico. Int J Speleol 6:93–107

Briggs RP, Seiders VM (1972) Geological Map of the Isla de Mona Quadrangle, Puerto Rico. U.S. Geological Survey Miscellaneous Geological Investigations, Map I-718

Dávila O (1988) Importancia arqueológica de la Cueva del Indio, Arecibo, Puerto Rico. Departmento de Recursos Naturales, Area de Investigaciones Científicas, Divisíon de Asesoramiento Técnico, San Juan

Dávila O (1991) Cueva de las Caracoles, un sitio preceramico de la Isla de Mona, XVII Simposia de los recursos naturales, 13 y 14 de Noviembre de 1991, pp 87–104

Diaz E, Lilyestrom CG (2010) Puerto Rico coastal and estuarine land conservation plan. Puerto Rico Department of Natural and Environmental Resources (DNER), 31 p. http://www.DNRA.gobierno.pr

Dubelaar CN (1995) Recent rock art research in the Caribbean islands, a survey. In: Proceedings of the international association of Caribbean archaeologists, vol 16, pp 399–443

Frank E (1998) History of the guano mining industry, Isla de Mona, Puerto Rico. Nat Speleol Soc J Cave Karst Stud 60(2):121–125

Frank EF, Wicks C, Mylroie J, Troester J, Alexander EC, Carew JL (1998a) Geology of Isla de Mona, Puerto Rico. J Cave Karst Stud 60(2):69–72

Frank EF, Mylroie J, Troester J, Alexander EC, Carew JL (1998b) Karst development and speleologenesis, Isla de Mona, Puerto Rico. J Cave Karst Stud 60(2): 73–83

Garrison LE (1969) Structural geology of the Muertos insular shelf, Puerto Rico, USGS open file report. U.S. Geological Survey, Washington, DC, 9 p

Gestel J, Mann P, Grindlay NR, Dolan JF (1999) Three-phase tectonic evolution of the northern margin of Puerto Rico as inferred from integrated seismic reflection, well and outcrop data. Mar Geol 1616: 257–286

Gile PL, Carrero JO (1918) Bat Guanos of Porto Rico and their fertilizer value. Agr Expt Stat Bull 25, 66p

Gonzalez LA, Ruiz HA, Taggart BE, Budd AF, Monell V (1997) Geology of Isla de Mona, Puerto Rico. In: Vacher HL, Quinn TM (eds) Geology and hydrogeology of carbonate islands, developments in sedimentology. Elsevier Science B.V., Amsterdam, pp 327–358

Guisti EV (1978) Hydrogeology of Karst in Puerto Rico. Geological Society Professional Paper #1012, United States Geological Survey, 65 p

Hayward MH, Atkinson L, Cinquino MA (eds) (2009) Rock art of the Caribbean. University of Alabama Press, Tuscaloosa, 285 p

Heatwole H, Levins R, Byer MD (1981) Biogeography of the Puerto Rico bank. Atoll Research Bulletin no.251, 55 p

Hernandez Santana JR, Barbosa JS, Gonzalez RR (2002) Islas de Puerto Rico y de Culebra: Regularidades geomofoligicas de su relieve de costera y submarino. Investigaciones Geograficas, Boletin del Instituto Geografia (47):7–19

Hubbard DK, Burke RB, Gill IP, Ramirez WR, Sherman C (2008) Coral-reef geology: Puerto Rico and the Virgin Islands. In: Coral reefs of the world, vol I. Springer, Dordrecht, pp 263–302

Kambesis PN (2007) Caves of the Cueva del Indios natural reserve, Puerto Rico. Nat Speleol Soc News 65(10):13–18

Kambesis PN, Lace MJ (2009) Caves of Isla de Mona. In: Palmer AN, Palmer M (eds) Caves and Karst of the USA. National Speleological Society, Huntsville, pp 343–345

Kaye CE (1957) Notes on the structural geology of Puerto Rico. Geol Soc Am Bull 68:103–118

Kaye CE (1959a) Shoreline features and Quaternary shoreline changes, Puerto Rico. Geological Society Professional Paper #317-B, United States Geological Survey, 140 p

Kaye CE (1959b) Geology of Isla de Mona, Puerto Rico and Notes on the Mona Passage, Geological Society Professional Paper #317-C, United States Geological Survey, 178 p

Lace MJ (2008) Coastal cave development in Puerto Rico. J Coast Res 24(2):508–518
Lace MJ (2012) Anthropogenic use, modification and preservation of coastal cave resources in Puerto Rico. J Island Coast Archaeol 7(3):378–403
Lace MJ, Kambesis PN (2009) Coastal speleogenesis in Puerto Rico. In: White WB (ed) Proceedings of the 15th international congress of speleology, vol I. National Speleological Society, pp 509–515
Lugo AE, Castro LM, Vale A, del Mar Lopez T, Prieto EH, Martino AG, Puente-Rolon AR, Tossas AG, McFarlane DA, Miller T, Rodriquez A, Lundberg J, Thomlinson J, Colon J, Schellekens JH, Ramos O, Helmer E (2001) Puerto Rican karst – a vital resource. USDA and Forest Service general technical report WO-65, 100 p
Lundberg J, Taggart BE (2007) Dissolution pipes in northern Puerto Rico: an exhumed paleokarst. Carbonates Evaporites 10(2):171–283
Mann P (2005) Active tectonics and seismic hazards of Puerto Rico, the Virgin Islands and offshore areas, GSA bulletin no. 385. Geological Society of America, Boulder, 299 p
Mattson PH (1960) Geology of the Mayaguez Area, Puerto Rico. Geol Soc Am Bull 71:319–362
Miller TE (2004) Structural control on cave development in Cretaceous Limestone, Southern Puerto Rico. Caribb J Sci 40(2):276–280
Miller TE, Kambesis PN (2009) Puerto Rico. In: Palmer AN, Palmer MV (eds) Caves and Karst of the USA. National Speleological Society, Huntsville, pp 332–338
Monroe WH (1960) Sinkholes and towers in the karst area of north-central Puerto Rico. In: Geology of Puerto Rico, Hawaii, Pacific Islands and Antarctica, Geological Survey Research – short papers in the geological sciences, USGS: B356–360
Monroe WH (1968) High level Quaternary beach deposits in northwestern Puerto Rico. USGS Geol Sur Prof Pap 600-D:140–142
Monroe WH (1974) Karst Landforms of Puerto Rico. Geological Survey Professional Paper #899, United States Geological Survey, 69 p
Morelock J, Ramirez-Martinez WR, Barreto M (2010) Puerto Rico. In: Bird EC (ed) Encyclopedia of the World's coastal landforms. Springer, Dordrecht/Heidelberg, pp 289–294
Mylroie JE, Mylroie JR (2007) Development of the carbonate island karst model. J Cave Karst Stud 69: 59–75
Petrae G (ed) (1995) Barge Morris J. Berman Spill: NOAA's scientific response. HAZMAT report 95–10. Hazardous Materials Response and Assessment Division, National Oceanographic and Atmospheric Administration , Seattle, 63 p
Porter BL (2010) An application and refinement of the karst disturbance index through evaluating variability in island karst disturbance in Puerto Rico. M.Sc. thesis, Department of Geography and Geology, Western Kentucky University, 168 p
Reid JA, Plumley PW, Schellekens JA (1991) Paleomagnetic evidence for late Miocene counterclockwise rotation of north coast carbonate sequences, Puerto Rico. Geophys Lett 18:565–568
Renken RA, Ward WC, Gill IP, Gomez-Gomez F, Rodriguez-Martinez J et al. (2002) Geology and hydrogeology of the Caribbean Islands aquifer system, Commonwealth of Puerto Rico and the U.S. Virgin Islands. U.S. Geological Survey Professional Paper #1419, 148 p
Rodríguez-Durán A (2009) Bat assemblages in the West Indies: the role of caves. In: Fleming TH, Racey P (eds) Island bats: evolution, ecology, and conservation. University of Chicago Press, Chicago, pp 265–280
Rodriguez Ramos R (2010) Rethinking Puerto Rican precolonial history. University of Alabama Press, Tuscaloosa, 267 p
Rodriguez RW, Trumbull JV, Dillon WP (1977) Marine Geologic map of the Isla de Mona area, Puerto Rico. U.S. Geological Survey, Misc. Investigations Series: I-1063 map
Roe PG, Hayward MH (2008) Rocks of ages: petroglyphs, pictographs and identity in Puerto Rico. In: Flore DF, Sanz ID, May SK (eds) Archaeologies of art: time, place, identity. Left Coast Press, Inc, Walnut Creek, pp 51–76
Rouse I (1992) The Tainos: rise and decline of the people who greeted Columbus. Yale University Press, New Haven
Seiders VM, Briggs RP, Glover L (1972) Geology of Isla Desecheo, Puerto Rico, with notes on the great southern Puerto Rico fault zone and Quaternary stillstands of the sea. U.S. Geological Survey Professional Paper #739, 22 p
Siegel P (ed) (1989) Early ceramic population lifeways and adaptive strategies in the Caribbean. BAR International series: 506, University of Alabama Press, Tuscaloosa
ten Brink US, Marshak S, Granja-Bruna JL (2009) Bivergent thrust wedges surrounding oceanic island arcs: insight from observations and sandbox models of the northeastern Caribbean plate, GSA bulletin no. 121. Geological Society of America, Boulder, pp 1522–1536
Trumball JVA, Garrison LE (1973) Geology of a system of submarine canyons south of Puerto Rico. US Geol Sur J Res 1(3):293–299
USFWS (2007) Vieques National wildlife refuge, comprehensive conservation plan and environmental impact statement. United States Department of the Interior – Fish and Wildlife Service, 353 p
USGS (1982) Isla de Vieques, PR, 7.5 Minute Quadrangle, map scale 1:30,000 U.S. Department of the Interior
USN (2001) Description of current conditions report: naval Atlantic fleet weapons training facility Vieques,

Puerto Rico. Report to Department of the Navy, Atlantic Division, Facilities Engineering Command, CTO-031
Wadsworth H (1973) The historical resources of Mona Island. In: Junta de Calidad Ambiental de Puerto Rico, Isla Mona, vol 2, appendix N: 1–37
Weaver PL, Schwagerl JJ (2009) U.S. fish and wildlife refuges and other nearby reserves in Southwestern Puerto Rico. USDA, Institute of International Tropical Forestry general technical report, IITF-40. 122 p
Wetmore A (1918) The birds of Desecheo Island, Porto Rico. The Auk 35(3):333–340
Wicks C, Troester J (1998) Groundwater Geochemistry of Isla de Mona, Puerto Rico. Nat Speleol Soc J Cave Karst Stud 60(2):107–114

Caves and Karst of Barbados

10

Patricia N. Kambesis and Hans G. Machel

Abstract

The island of Barbados is composed of a series of stepped Pleistocene carbonate terraces that overlie Oligocene-Miocene chalk that is draped over deformed deep-sea clastics of Eocene age, the result of Quaternary sea-level fluctuations combined with tectonic uplift. The island is pervasively karstified, as shown by well over 100 caves, hundreds of gullies, and more than 2,800 sinkholes. Barbados has five types of cave: (1) epigene caves formed by fresh-water dissolution; (2) flank margin caves formed by mixing-zone dissolution; (3) littoral caves formed by marine corrasion; (4) 'mechanical caves' formed by mass movement of bedrock; and (5) hybrid (polygenetic) caves that formed when an initial cave type and morphology was overprinted by another cave-forming process. Epigene caves are found in the upland interior of the island, yet stream sinks are rare and most springs occur along the coast. Flank margin caves are found on the modern and paleo-coastlines, and along many of the gullies. Calcite speleothems are developed almost exclusively in these two types of caves. Littoral and mechanical caves occur on the modern coastlines. However, the majority of the island's caves are of the hybrid variety. Flank margin caves are the most common cave type converted to the hybrid or polygenetic state, as they are readily breached and modified by cliff retreat and littoral processes.

The caves of Barbados have been important to humans from the earliest times of inhabitation of the island. When the sugar cane industry was

P.N. Kambesis (✉)
Department of Geosciences, Mississippi State University, Mississippi State, 39762, USA
e-mail: pk249@msstate.edu

H.G. Machel
Department of Earth and Atmospheric Sciences, University of Alberta, Edmonton, AB, Canada T6G2E3
e-mail: Hans.machel@ualberta.ca

M.J. Lace and J.E. Mylroie (eds.), *Coastal Karst Landforms*, Coastal Research Library 5,
DOI 10.1007/978-94-007-5016-6_10,

at its height, caves served as storage and as hideouts for escaped slaves. In modern times most caves on Barbados remain undeveloped, except for Animal Flower Cave at the north coast and Harrison's Cave in the highlands, which are developed for tourism. In 2007, the collapse of a large cave in the densely populated capital Bridgetown cost a family of five their lives.

10.1 Introduction

Barbados is located in the Atlantic Ocean approximately 125 km east of the Lesser Antilles volcanic island arc (Fig. 10.1). The island has a north–south extent of ~34 km and an east–west extent of ~23 km at its widest point, with 97 km of coastline that envelopes a total land area of ~431 km^2 (Inniss et al. 2001). The surface of the island is characterized by gently sloping terraces of Pleistocene carbonates (locally called 'coral rock' or 'coral stone') separated by cliffs that parallel the coasts. The Upper Coral Rock terrace ranges in elevation from about 180 to 240 m above sea level, the Middle Coral Rock terrace ranges in elevation from about 60 to 90 m, whereas the Lower Coral Rock terrace is only a few meters to tens of meters above sea level. The highest point on the island, Mt. Hillaby at 340 m above sea level, is in an area of Paleogene outcrops called the Scotland District (Fig. 10.1). The physiography of the island is the result of glacioeustasy combined with more or less continuous tectonic uplift of about 5 cm/100 years on average over the last one million years, punctuated by spurts of increased rates of uplift that created the major cliffs that now border the main terraces, which are called First High Cliff and Second High Cliff (Fig. 10.1, Speed 1983; Taylor and Mann 1991; Schellmann and Radtke 2004).

Barbados has a tropical monsoon climate, as defined by the Köppen climate classification, with a wet season that runs from June to November/December and a dry season from December/January through May. Over the last 20–30 years global climate change has tended to shift the end of the rainy season by 3–5 weeks almost into February, and the end of the dry season has moved by about the same amount. Annual precipitation ranges between 1,000 and 2,300 mm (Inniss et al. 2001). The island lies within the path of the northeast trade winds with average monthly temperatures ranging between 21 and 31 °C, depending on the season.

Archaic Amerindians were the first inhabitants of Barbados, as recently documented by archeological artifacts that date back to approximately 2710–2300 B.C., consistent with the earliest known periods of human migration through the Caribbean (Fitzpatrick 2011). Around 1200 A.D. the Arawak inhabitants were thought to have been displaced by invading Carib Indians from South America. Their population eventually declined and were recorded as absent at the time of the arrival of Europeans. The Portuguese explorer Pedro a Campos, who briefly visited the island in 1536 but did not claim the island for his country, supposedly coined the name Los Barbados which translates as the 'Bearded Ones' after the indigenous fig trees whose long, hanging aerial roots have a beard-like appearance. In 1625 Captain John Powell claimed Barbados for England and 2 years later British colonists arrived and established the island's first European settlement in what is now known as Holetown, the second largest city in the island after the capital Bridgetown. The British colonists cleared much of the island's forest and initially planted tobacco and cotton. In the mid-1600s sugar cane began to replace other crops. Sugar cane agriculture was labor intensive and as a consequence, landowners imported large numbers of African slaves to do the work. The sugar cane industry became very successful and continued to prosper after the abolition of slavery in 1834. As a result of social strife and pressures, Barbados' black majority gradually gained more access to the political process. England granted Barbados self-government in 1961, and in 1966

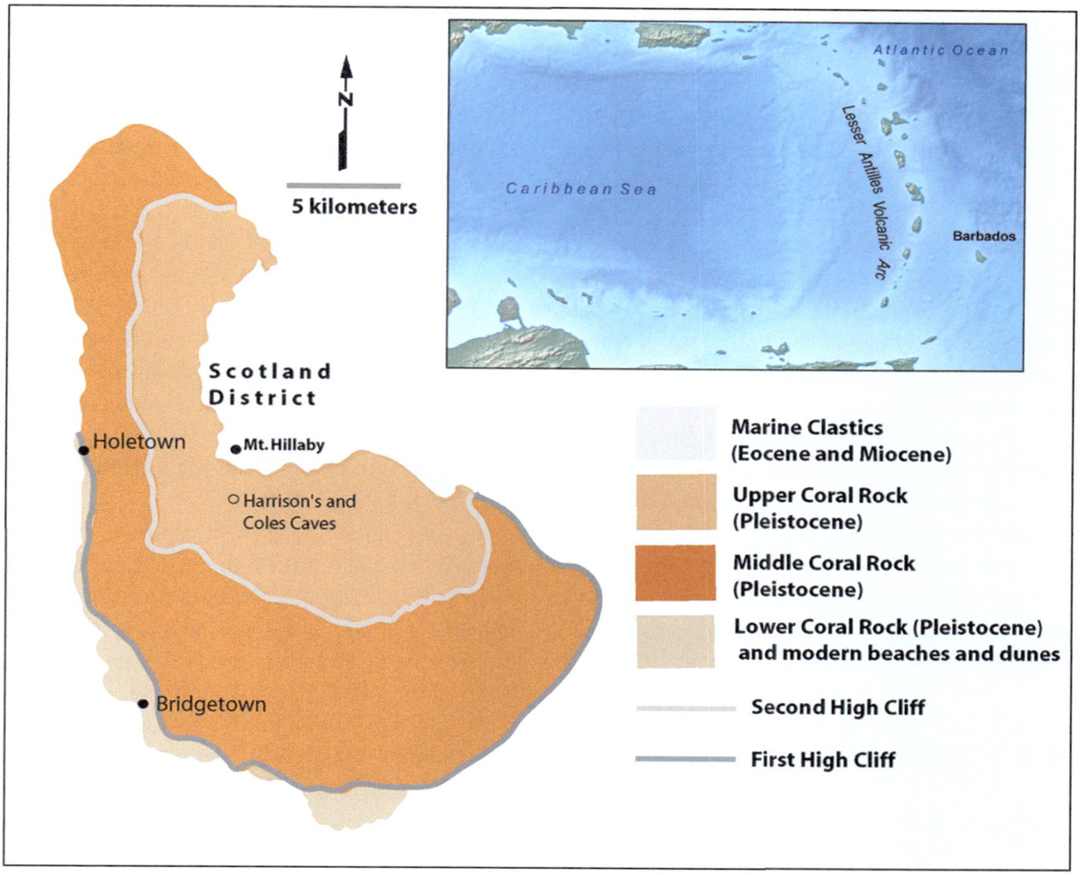

Fig. 10.1 Map of Barbados, geologic units and other reference features (Modified from Machel 1999). *Inset map*: ESRI Caribbean map template

the island became an independent nation and a member of the Commonwealth. After the Second World War the sugar cane industry went into decline. Tourism is now the predominant industry on the island.

10.2 Geologic Setting

Barbados is an uplifted carbonate island located on a forearc bulge known as the Barbados Ridge, which extends for several hundred kilometers approximately 125 km east and parallel to the Lesser Antilles volcanic island arc. The island is the highest point of an accretionary prism complex and is one of the few places in the world where an active accretionary prism is subaerially exposed (Cook and Abbott 2011; Machel 1999, 2011; Speed 1983, 2002, 2012). The forearc bulge is situated above the subduction zone between the descending South American Plate to the east and the overriding Caribbean Plate to the west.

The 'basement' of the island consists of structurally complex marine clastics of Eocene age, which crop out in an erosional window on the east-central side of the island, called the Scotland District (Fig. 10.1). This section is overlain by Miocene chalks, marls and radiolarites of the Oceanics Group. The chalks form a regional aquitard that is missing in a few, relatively small locations. The clastics host oil and gas that extrude in a number of natural seeps and sustain a small local oil industry (Machel 1999, 2011).

The Paleogene and Neogene basement is overlain by up to ~100 m thick Pleistocene-aged platform and reef limestones, which cover about 80 % of the surface of the island today. The limestones form three major terraces that decrease in age from the Scotland District toward the shorelines (Fig. 10.1), each consisting of multiple smaller-scale sedimentary sequences or subterraces that formed in response to eustatic sea level variations, and that have been used to calibrate the global oxygen isotope stages in considerable detail (e.g. Schellmann and Radtke 2004). Sedimentologically each terrace consists of back reef, reef, and fore reef zones that are typical of modern Caribbean reefs (James 1972; James et al. 1977; Humphrey 1997). The oldest limestones are mid-Pleistocene and up to ~800 ka in age, the First and Second High Cliffs (Fig. 10.1) are about 500 and 120 ka in age, respectively. A dolomitic outcrop of Pleistocene limestone is found locally at Golden Grove (Machel and Burton 1994).

10.3 Hydrogeology

The hydrogeology of Barbados is controlled by the contact between the Pleistocene reef limestones and the underlying Oligocene-Miocene chalk that forms a regionally extensive aquitard (Senn 1946; Banner et al. 1990). Recharge enters the limestone aquifer at discrete points from streams flowing off of nearby non-carbonate rocks (Upchurch 2002). Autogenic recharge infiltrates through the vadose zone to the aquifer. Where the aquitard lies above sea level, groundwater flows along the base of the limestone in underground streams. Along the coasts where the aquiclude lies below sea level phreatic freshwater accumulates as a wedge-shaped lens overlying saltwater that has intruded into the aquifer.

10.4 Sinkholes and Gullies

Barbados is classified as a 'composite island' according to the Carbonate Island Karst Model (CIKM) (Mylroie and Mylroie 2007), as clastics are surficially exposed in the northeast part of the island (Figs. 10.1 and 10.2). Both epigene and hypogene karst can form in this island type. Allogenic and autogenic modes of recharge typify epigene karst and as a consequence, the karst is directly coupled to surface hydrology and is associated with sinking streams, karst windows, and springs (Palmer 1991). Hypogene karst forms from conditions that are decoupled from surficial hydrologic processes (Palmer 2011). This type of karst is present in and influenced by coastal areas where fresh/salt water mixing occurs, accentuated by tidal pumping and eustatic sea-level changes (Vacher and Mylroie 2002). The creation of flank margin caves is the main expression of this type of karstification.

Epigene karst features in Barbados consist primarily of sinkholes (dolines) and gullies (dry valleys) (Fig. 10.3). A sinkhole inventory documented 2,830 sinkholes on the island based on remote sensing and ground survey data (Wandelt 2000). Day (1983) found sinkhole abundance to be 9.47/km^2, with the highest density at the 100–150 m elevation, but highly variable regionally both along and transverse to the coral rock terraces. The sinkholes of Barbados come in two forms: large interfluvial sinkholes located between gullies, and small shafts occurring within the gullies (Day 1983). The interfluvial sinkholes tend to be filled with low permeability soil that impedes infiltration of water into the aquifer, whereas the shafts act as conduits that transmit large volumes of water downward at times of heavy rain (Jones and Banner 2003).

Barbados has hundreds of gullies (dry valleys) of variable lengths and depths. They are distributed more or less evenly on the moderate to steep slopes of the carbonate terraces (Fig. 10.3) and many contain cave remnants along their perimeters (Machel et al. 2012). The gullies drain toward the northern, western, and southern coastlines and resemble a network of narrow creek beds in the interior highlands, merging progressively to fewer and wider channels down slope and seaward (Machel et al. 2012). The gradient of most gullies is 2° on average, and some are up to ~30 m deep (Fermor 1972). Most gullies are dry much of the year, although

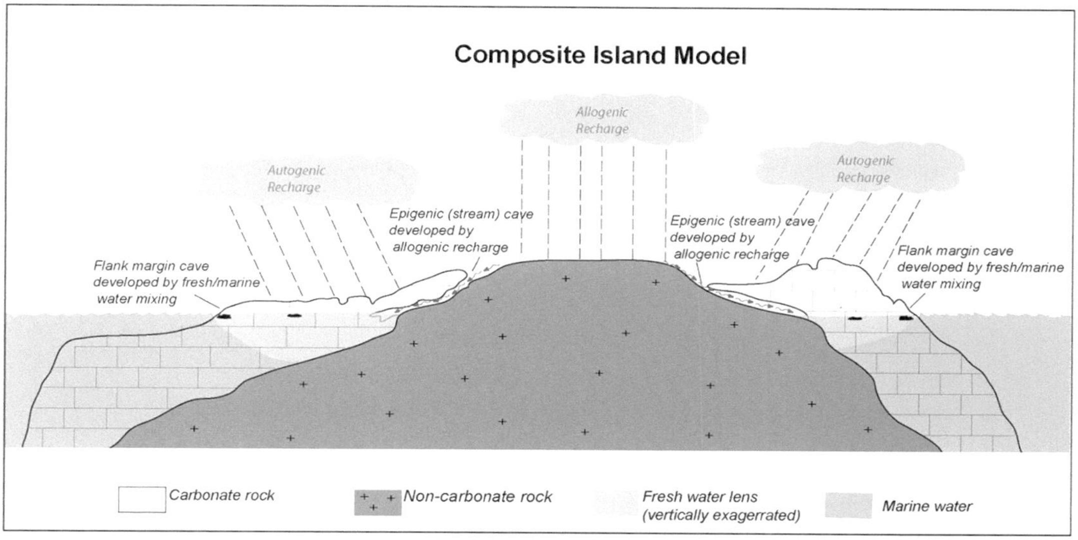

Fig. 10.2 Composite Island Model – from Carbonate Island Karst Model (CIKM) for Composite Island where carbonate and non-carbonate rocks are exposed at the surface resulting in allogenic and autogenic recharge of the fresh water lens. Mixing of fresh and marine water forms flank margin caves. Allogenic recharge at the carbonate/non-carbonate contact forms epigenetic (conduit flow) caves (Modified from Mylroie and Mylroie 2007)

some contain ephemeral streams. During heavy rain events most gullies carry floodwaters, which can quickly become torrents that can lead to considerable property damage and even loss of life. When a tropical depression drenched the island in August 1986, walls were toppled, countless houses were flooded, many cars and even several houses were swept out to sea on the west coast.

The origin of the gullies in Barbados has been the subject of considerable controversy (Schomburgk 1848; Harrison and Jukes-Browne 1890; Tricart 1968; Fermor 1972; Machel 1999; Mylroie et al. 2010; Machel et al. 2012). The abundant and in parts massive speleothems found along many gully walls (Fig. 10.4), along with huge boulders with sizes of up to ~200 m^3 that are scattered on the floors of some gullies, suggest an origin from stream cave dissolution and subsequent roof collapse (Machel 1999). However, the widest passages of the three large stream caves of Barbados (Harrison's, Cole's, Springhead, Arch Cave) are not nearly as wide as the widest gullies. Also, the speleothems that now line the gully walls indicate that the gullies have not been enlarged significantly by surface erosion from their hypothesized pre-collapse state, otherwise the speleothems would have been eroded away. These considerations render the hypothesis of an origin of the gullies solely by stream cave collapse unlikely. Rather, the gullies most probably formed from the succession of the following processes: (1) downward erosion via surface drainage; (2) flooding during a sea level rise, giving rise to flank margin cave development in the gully walls; (3) speleothem formation in these caves upon another sea level drop that placed the caves in the vadose zone, and further widening of the gullies from surface drainage, which may have opened some of the caves through gully wall retreat; (4) possible further exposure of caves from partial marine erosion of the gully walls during another sea level rise; and (5) final exposure of the gully through tectonic uplift, resulting in wide gullies with opened flank margin caves that are lined with speleothems (Mylroie et al. 2010; Machel et al. 2012). This rather complex scenario appears to be the only one compatible with all geologic and geomorphologic observations, tectonic uplift,

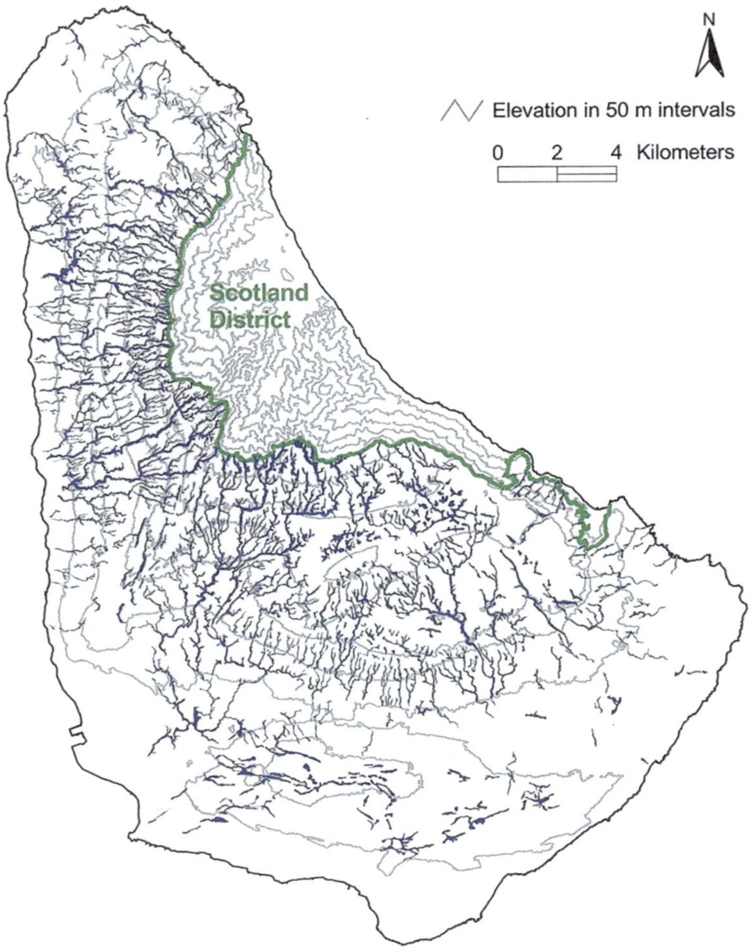

Fig. 10.3 Distribution of sinkholes and gullies. Most of the gullies are located in the Upper Coral Rock terraces which have moderate to steep slopes. Most of the sinkholes occur in flat areas and near the second high cliff (Modified from Jones and Banner 2003)

and the multiple eustatic sea-level fluctuations throughout the Pleistocene (Speed 1983; Taylor and Mann 1991; Schellmann and Radtke 2004).

While we advocate this scenario for most gullies in the Lower and Middle Coral Rock terraces of Barbados, such as Sailor's Gully (Fig. 10.4), at least some gullies appear to have a polygenetic origin. In these cases the upstream portion started out as a stream cave, which was 'converted' into a gully after its roof collapsed, and where flowing water excavated the extension of the gully toward the coastline by surface drainage. Jack-in-the-Box gully, which is located close to Harrison's and Coles' Caves in the Upper Coral Rock Terrace, appears to have formed in this manner. It contains evidence of a collapsed stream cave in its highest portion (narrow width, relatively shallow, decorated walls, boulders from the collapsed cave roof), yet the gully appears significantly widened and deepened by mixing zone dissolution and gully wall retreat farther downstream past the Second High Cliff in the Middle Coral Rock terrace.

10.5 Caves

Barbados has flank margin caves, stream caves, littoral caves (sea caves), mechanical caves (fissure and talus), and hybrid (polygenetic) caves. Speleogenetic factors that controlled formation of these cave types are a function of chemical

Fig. 10.4 Sailors Gully, Barbados. The gully has been developed for road traffic. (**a**) Map shows a series of small breached flank margin caves located along the length of the segment of gully. (**b**) Typical morphology of caves which form on the perimeters of gullies. (**c**) Weathered calcite speleothems from a cave in Welchman's Gully. (**d**) Welchman's Gully near Harrisons Cave

dissolution, mechanical erosion, mass movement, or a combination thereof. Most caves in Barbados appear to be flank margin caves overprinted by other cave forming processes, thus they are hybrid caves.

10.5.1 Epigene Caves

Stream caves form when there is recharge of meteoric water into soluble bedrock. If recharge occurs directly on a limestone surface it is autogenic, and concentration of the water into discrete streams commonly occurs in the subsurface. A catchment that contains non-carbonate and carbonate bedrock will result in allogenic recharge. In this case the streams typically originate in the non-carbonate section and sink once they encounter the carbonate bedrock. Dissolutional conduits formed as a result of epigenic recharge typically display dendritic drainage patterns (Palmer 1991). The conduits carry water to discrete points of discharge in the form of springs or seeps.

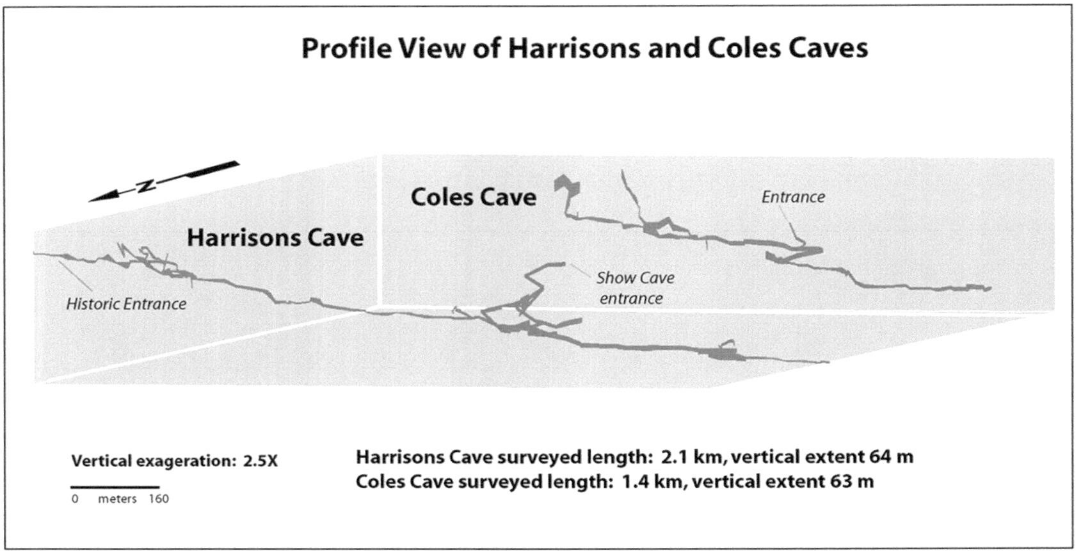

Fig. 10.5 Profile views of Harrisons and Coles Caves. Note the distinct gradient on both caves. The upstream sections of both caves are located under gullies that are a kilometer from the border of the Scotland District, and near the highest elevation point of the island

Harrison's Cave and Cole's Cave (Fig. 10.5) are the longest and best-known stream caves in Barbados, each with close to 2 km of dendritic/branching passages. They begin approximately 1 km south of the edge of the Scotland District in the Upper Coral Rock terrace and host a variety of speleothems. Harrison's and Cole's Cave are recharged from sinkholes on the surface and from direct input of water at the upstream end of Harrison's Gully near Sturgis (Groves 1994). The underground streams recharge a freshwater lens that is located close to the coast.

10.5.2 Flank Margin Caves

Flank margin caves form near the seaward margin of a freshwater lens where the water is brackish and commonly undersaturated with respect to calcite (e.g. Machel and Burton 1994; Mylroie and Mylroie 2007). These caves have no 'entrances' or connections to fresh water streams, unless the latter tunnel into the caves upon uplift at a later time. Flank margin caves of Barbados have been documented within reef, back-reef and fore-reef limestones (Machel et al. 2012).

Golden Grove Cave and Slave Cave (Fig. 10.6) display morphologies that are typical of flank margin caves. The chambers are irregular in shape with cuspate walls. Most flank margin caves have a relatively low height-to-footprint ratio. In this regard the low ceiling and wide profile of Golden Grove Cave (Fig. 10.6a) typifies development in the distal margin of the freshwater lens. In contrast, Slave Cave (Fig. 10.6b) has high height-to-footprint ratios, with a large dome-shaped room. This probably reflects tectonic uplift during formation while the freshwater lens remained in about the same position. Furthermore, the cave wall morphologies are partially controlled by rock facies type. For example, flank margin caves that formed in back-reef rubble, such as in Berwick Cave (Fig. 10.7a), display subdued dissolutional wall morphologies because of the large grain size of the limestone. This is because the back-reef limestones consist mainly of *Acropora cervicornis* and *Porities porites* fragments that mask small-scale dissolution

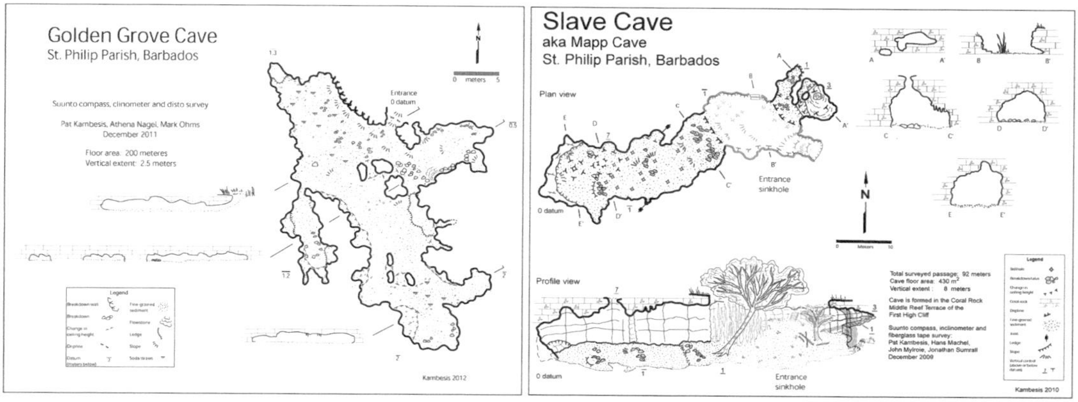

Fig. 10.6 Flank margin caves of Barbados. Map of Golden Grove Cave (*left*) displays the limited vertical extent of the cave and its globular morphology. Slave Cave (*right*) has a vertical extent of 8 m. The cave is segmented at the entrance where it has been breached by ceiling collapse

features (Fig. 10.7b). However, larger-scale features such as bedrock pillars and bellholes can be formed (Fig. 10.7c), as is commonly the case in flank margin caves. At Cove Bay a series of unusual caves is developed in the Miocene chalk just above the high water mark (Fig. 10.8). The chalk is highly fractured, allowing for chemically active fluid flow and diagenesis in this unit. Partial dolomitization was facilitated by cold, at times methane-bearing seawater (as indicated by stable isotope data) along a fracture network (Machel et al. 2011). At other times the hypogene pore water ascending through the fracture network dissolved limestone or dolostone but did not form dolomite. Tectonic uplift led to dissolution in the fracture network under freshwater-seawater mixing zone conditions. Eventually the fractures were exploited by meteoric water, producing a localized freshwater lens within the fracture set. Finally cliff retreat exposed the cave(s) to wave energy. Thus some, if not most caves in the Miocene chalks of Barbados have a polyphase, hybrid origin (Sumrall 2013). The relatively greater hardness of the dolomitized chalk probably provided for enhanced mechanical resistance that allowed the karst voids to survive over about one million years, i.e., since the emergence of the oldest parts of Barbados.

10.5.3 Littoral Caves

Littoral caves (sea caves) are pseudokarstic features, meaning they mimic dissolutional caves but are produced by a different process. Littoral caves can be formed by corrasion, which is erosion by waves aided by sand that acts as an abrasive (Bunnell 2004; Mylroie 2007), by biological action on lithologic or structural heterogeneities of the bedrock (Bunnell 2004), from focusing of wave energy on specific coastal locations (Waterstrat et al. 2010), or from a combination thereof. The erosional forces commonly result in irregular, broken cave wall surfaces where the rock is fractured or jointed, while the cave walls are often rounded and smoothed by the swirling motion of broken rock and cobbles (Bunnell 2004).

The north coast of Barbados displays very high wave energy and sea caves are common (Fig. 10.9). It is unclear, however, whether the genesis of these caves was initiated by erosive wave action or by dissolutional void formation inside the cliff. Visual inspection of some of these caves revealed morphological features of flank margin caves, and there is no connection to stream caves. Thus, it appears possible if not likely that at least some of the caves along the north coast are hybrid, in that they originated

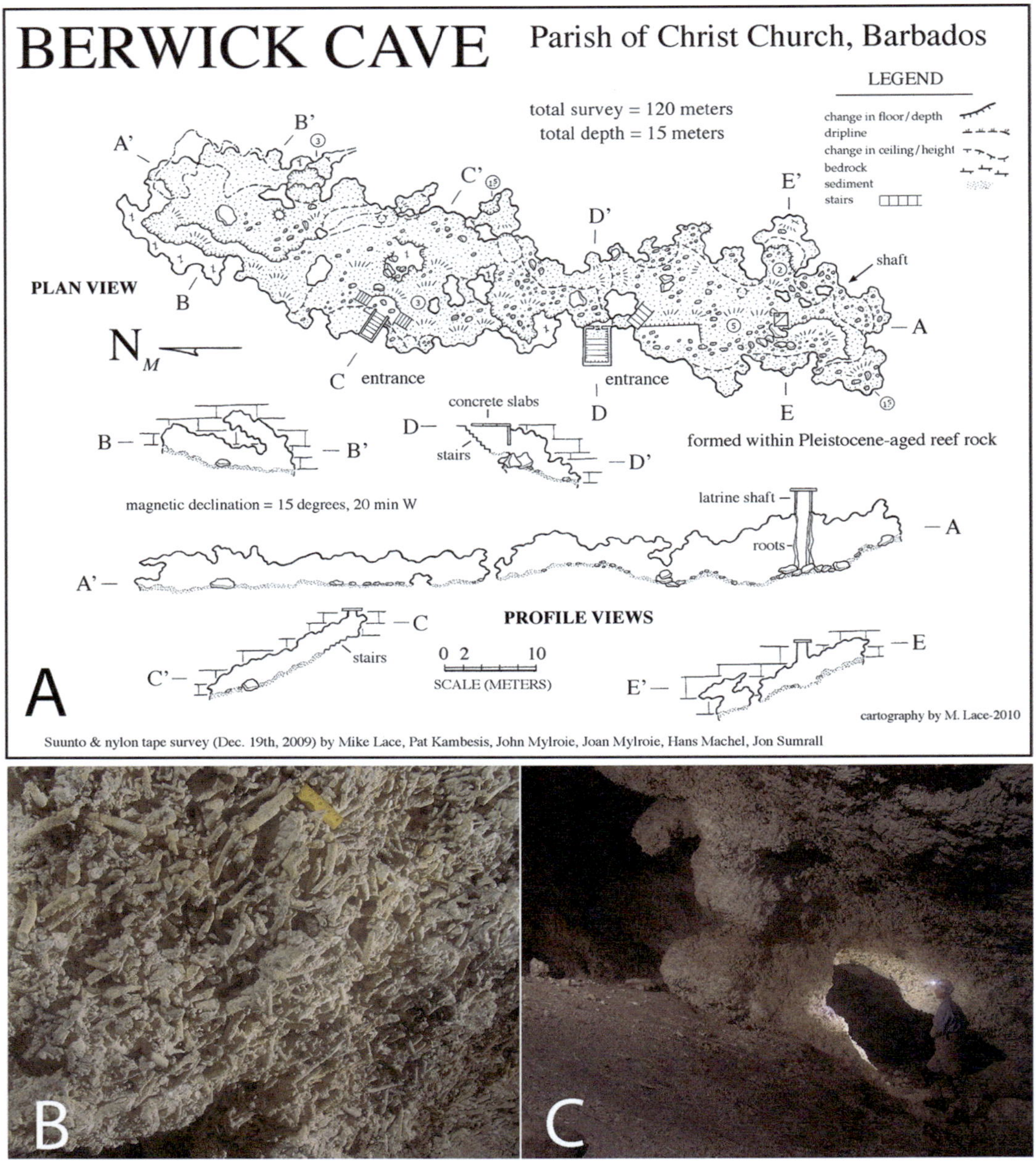

Fig. 10.7 (**a**) Map of Berwick Cave. (**b**) The coarse-grained nature of the *Porities* facies masks smaller dissolutional features. (**c**) Large scale dissolutional features (such as the pillar) are not affected by the *Porities* facies (Berwick Cave) (Photos: John Mylroie)

as flank margin caves that were exposed by cliff retreat and are now overprinted by littoral corrosion. Quantitative morphologic analysis of cave passage layouts has shown promise for differentiating cave types in carbonate coastal environments (Mylroie 2007; Lace 2008; Waterstrat et al. 2010). However, the quantitative criteria for this type of analysis are still under development and have not been applied to this cave type in Barbados.

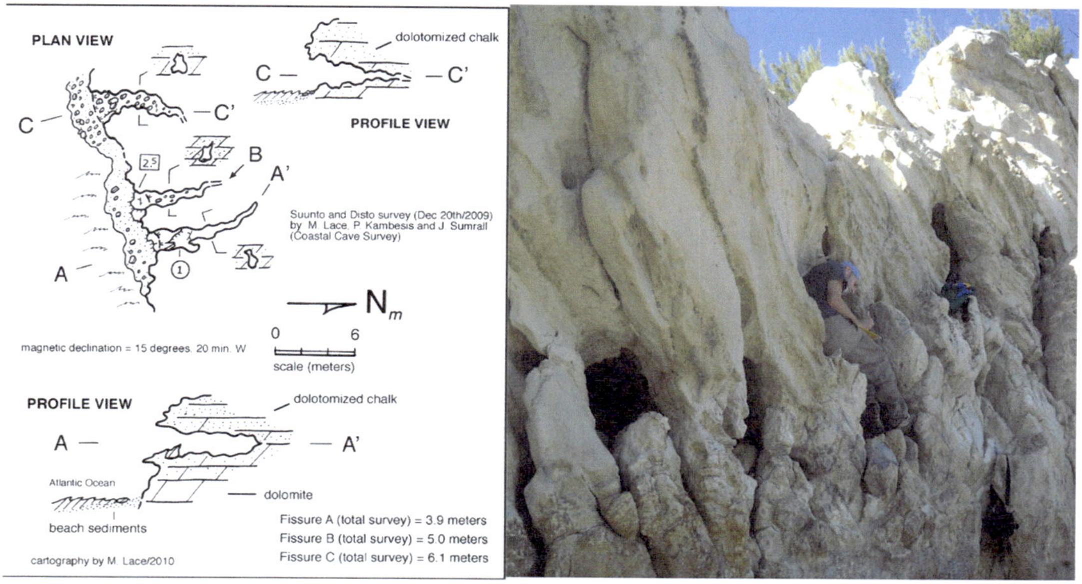

Fig. 10.8 Flank margin caves formed in Miocene-age dolomitic chalks. Photograph shows series of small, narrow cave entrances. Note well-developed jointing

10.5.4 Mechanical (Talus and Fissure) Caves

Mechanical caves result from the mass movement of bedrock and are not dependent on rock type. Caves formed by rock slides and collapses are called talus caves (Fig. 10.10a) and consist of openings between and under rocks that have randomly fallen into a pile, commonly at the bases of a cliff. Fissure caves are formed by mechanical widening of a single or series of joints or fissures along escarpments and cliff lines. Linear passage trends and canyon-like passages with triangular-shaped cross sections are typical of fissure caves (Fig. 10.10b). The cliff line below Sam Lord's Castle contains a variety of talus and fissure caves (and some small flank margin caves). Mass movement can overprint all cave types.

10.5.5 Hybrid (Polygenetic) Caves

Hybrid caves are the result of cave-forming processes changing over time, such that one cave type is overprinted by another. This is a common phenomenon in Barbados where the hydrological conditions have changed repeatedly by the interplay of glacioeustatic sea level variations and tectonic uplift. As a result, flank margin caves are commonly exposed by wave action and cliff retreat, which subjects these cave to littoral processes, as appears to be the case for Animal Flower Cave (Fig. 10.9). A genetically similar cave is 'Indian Castle' aka Mt. Brevitor Cave (Fig. 10.11), which is located within in the Second High Cliff in the northern part of the island. This cave has a large domal room that was cut open on two sides, i.e., the cliff face to the west and the gully face to the south where a gully met the paleocoastline, attesting to cliff and gully wall retreat by corrasion. The cave itself shows most attributes typical for flank margin caves, i.e., there are no connections to a stream cave, the walls have cuspate morphologies, there are pillars and large bell holes. As in the case of Golden Grove Cave and Slave Cave, the high domal shape probably is due to a 'spurt' of tectonic uplift while the freshwater lens and mixing zones were relatively stable (Machel et al. 2012).

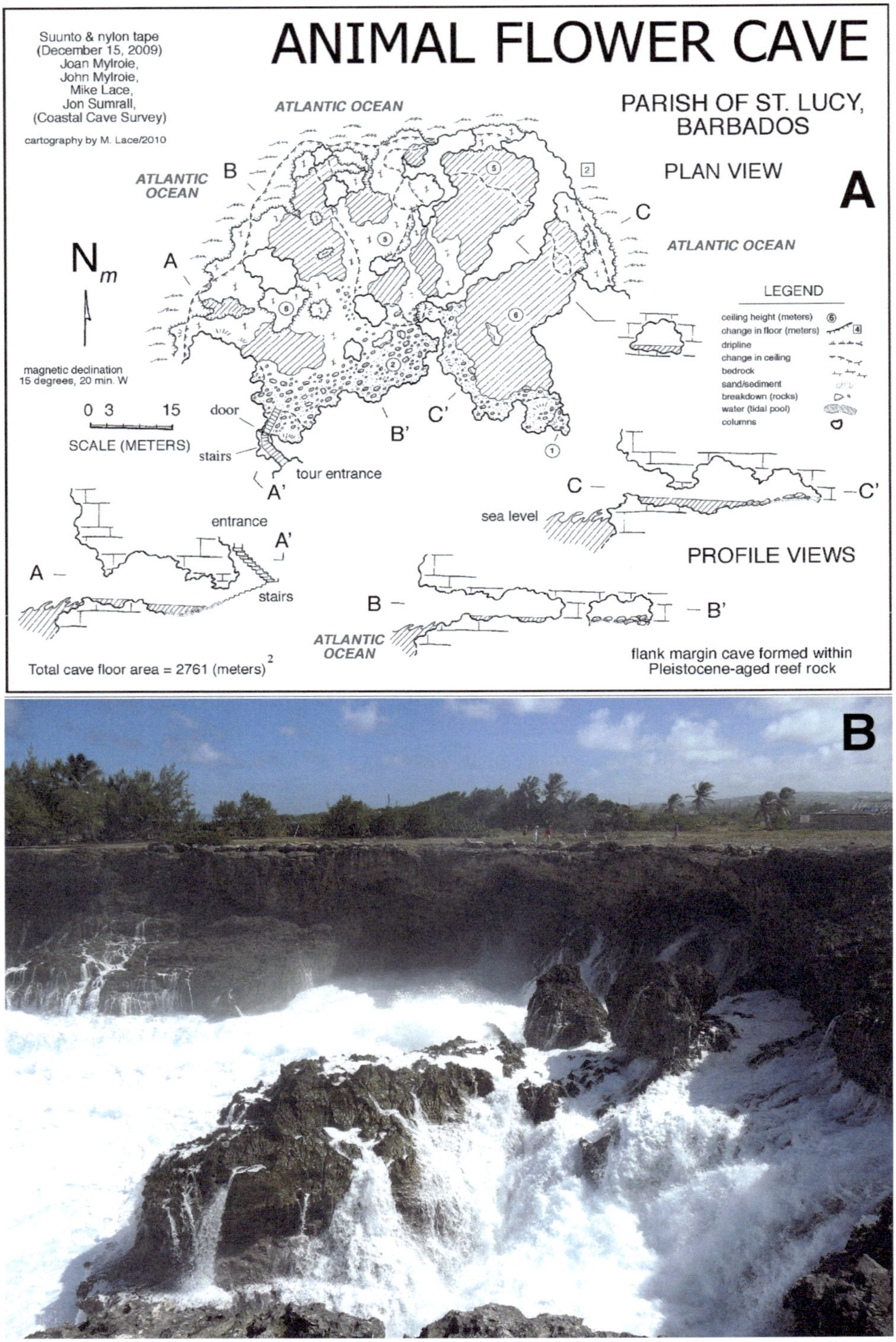

Fig. 10.9 Animal Flower Cave, North coast of Barbados. (**a**) Cave map displays flank margin cave morphology. (**b**) Photograph shows the wave energy at Animal Flower Cave that is typical for the North Coast of Barbados. Wave energy breaches the flank margin cave and overprints it with the effects of littoral processes, modifying it to a hybrid cave

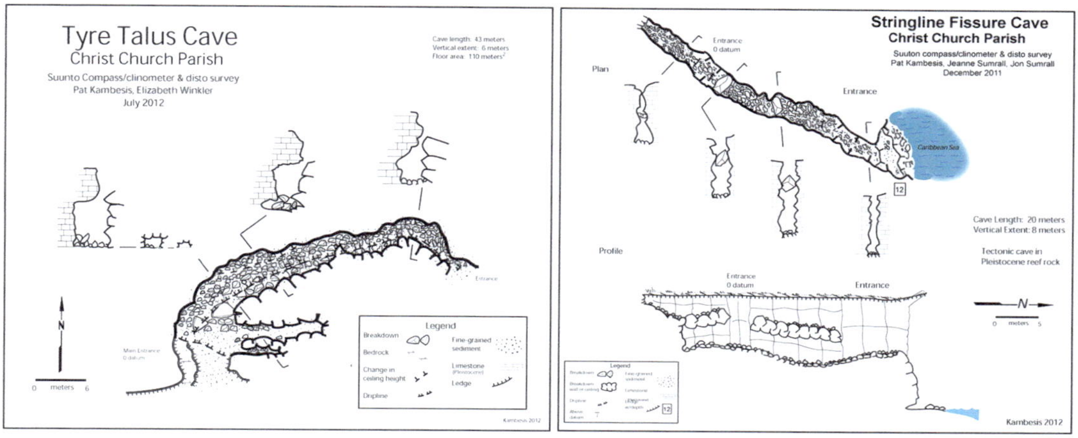

Fig. 10.10 Maps of mechanical caves. (**a**) Tyre Cave, located on the cliffs below Sam Lords Castle, is a talus cave. (**b**) Stringline Fissure, located on the south coast in the parish of Christ Church is a fissure cave

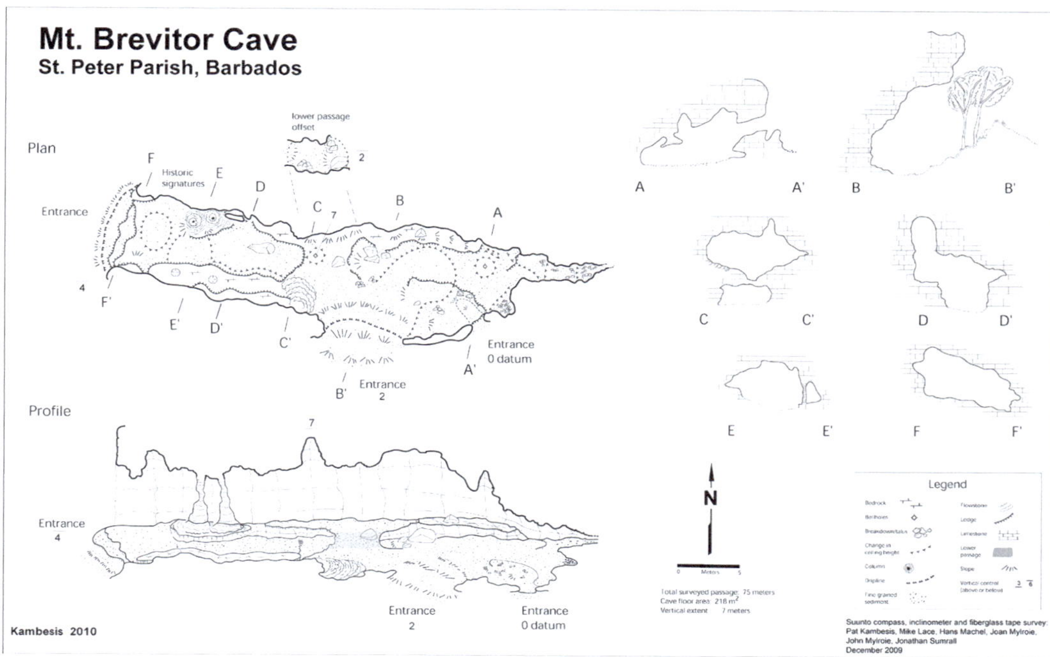

Fig. 10.11 Map of Mt. Brevator Cave is a hybrid cave located within an interior cliff. The cave had a flank margin genesis and was eventually breached by cliff erosion. It was likely overprinted by wave energy during a rise in sea level

10.6 Caves and People

The caves in Barbados have been used for various purposes throughout prehistoric and historic times all the way to the recent. Reports about their use by people are scant and scattered, however, and very few of the many caves on the island are identified in previous publications (Schomburgk 1848; Gurnee and Gurnee 1991).

In prehistoric times, some caves were used by the Amerindian and/or Arawak cultures.

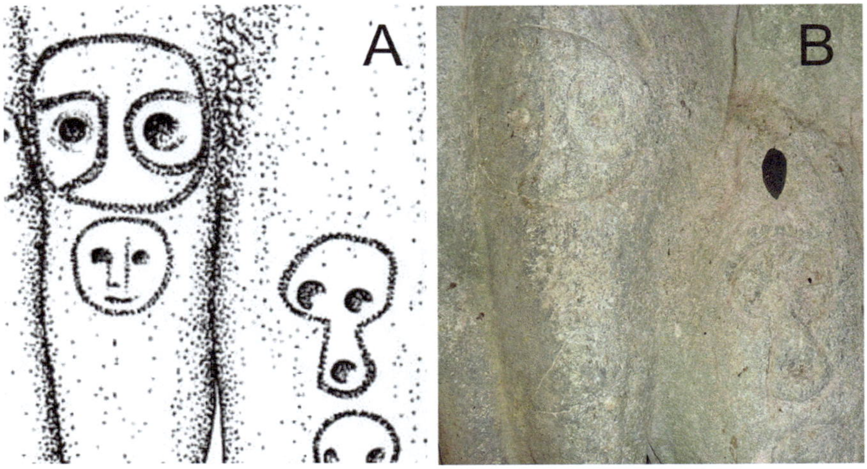

Fig. 10.12 (**a**) Drawings (Drewitt 1997) and (**b**) Photograph of some petroglyphs in Springhead Cave. Leaf for scale (5 cm from top to bottom)

Springhead Cave, one of only four known stream caves of Barbados (located on private land on the Springhead plantation, thus accessible only with special permission), contains the only petroglyphs known from Barbados near the entrance of the largest chamber, which is about 24 × 12 m in dimension. These artifacts depict faces that were carved into flowstone and stalactites, and were documented in a series of drawings by Dewitt (1997). One of these drawings is here reproduced, along with a recent photograph of the exact same petroglyphs (Fig. 10.12). Their date(s) of origin are currently unknown as no other pre-contact cultural materials have been recovered from the site and no direct dating of the rock art has been performed, although some speculate that they date at least to the Troumassoid (650–1600 A.D.) or Suazoid (1000–1450 A.D.) cultures (Drewitt 1997). There are no indications that these cultures actually lived in Springhead Cave, but they surely lived in other caves, as shown by other artifacts (Schomburgk 1848). Apparently some caves were used at least for temporary shelter, storage, or ritual activities. A well-known cave of this type is 'Indian Castle' in the northern part of the island, also called Mt. Brevitor Cave (Fig. 10.11).

In historic times, after settlement by the English and throughout the plantation period of sugar cane production, some caves were used for storage of materials. More notoriously, fugitive slaves sought shelter in some of the hidden caves that are located in the higher terraces and/or in gullies under dense vegetation, as shown by multiple findings of burnings, pottery, fishbones and the like, dating back to the eighteenth century. The Barbados National Museum contains exhibits of this period. One cave in the south-eastern part of the island, aptly named "Slave Cave" (also "Mapps Cave", Fig. 10.6b), served as a prison for escaped slaves. This is the cave in which Bussa was once held. Bussa was an African-born Barbadian slave who led a slave uprising, known as "Bussa's Rebellion". At one time Bussa commanded some 400 freedom fighters. He was killed in battle in 1816, and the rebellion failed. Bussa is revered in Barbadian history as a national hero. In 1985 the Bussa Emancipation Statue was unveiled on a prominent highway intersection.

In recent times, no inhabitants of the island are known to live in caves, with isolated exceptions. The infamous Winston Hall, who escaped from a maximum security prison three times, managed to hide out in a cave for several years without being discovered. Eventually he was cornered and shot by a policeman in 2004 (Hall 2007). Another person lived in a cave until a few years ago in the northwesternmost part of the island. Nearby villagers provided him with food and other necessities on a regular basis. Yet another man, a former employee of an expensive hotel complex, is currently living in a cave, to which he retreated after falling on hard times.

Examples of caves currently exhibiting human impact include Animal Flower Cave (Fig. 10.10) at the northernmost tip of the island, and Harrison's Cave and Cole's Cave (Fig. 10.5) in the central part of the island. The former is a flank margin cave that was cut open by cliff retreat, currently slightly above fair weather sea level and accessible to tourists via a stair case at low tide and calm seas to view cave morphologies and sea anemones ('animal flowers'). When the sea is high and rough, the cave is closed for visitation.

Harrison's Cave is the longest stream cave in the island with over 2.5 km of passages. It was developed for tourism in the 1970s and opened in 1981 for visitation by the public (Groves 1994). Visitors enter the cave via a trolley that takes them through several hundred meters of a paved road. The largest room is about 20 m in height, and the cave is relatively well decorated with various calcite speleothems.

Nearby Cole's Cave is nearly as long as Harrison's but muddier and less decorated. Cole's is not developed for tourism but quite popular with recreational cavers. Branching off Cole's Cave via a very narrow and dangerous sump is the almost unknown Shedith's Cave, discovered by a group led by the famed Danish speleologist Ole Sorensen in 1969. A dive in this cave almost cost one member of the team his life, which is documented in a gripping narrative by Reil (2000).

Some caves have been used in recent times for 'industrial' purposes, such as excavation of marl, temporary storage of goods, or garbage. One small cave that opens up to the sea into a small pool at the coastline by Dawlish at the south coast, accessible only via a steep staircase and hidden from view. Known as "Bachelor's Cave", it has served as a popular dating spot. Almost every year someone somewhere in Barbados reports a small cave collapse on or near his or her property, which contribute to the now well over 2,800 known sinkholes that dot the island (Wandelt 2000). In most cases damage to infrastructure, such as houses and roads, is minor. In one case, however, this was not so.

Before dawn in the early morning hours of August 26th, 2007 the unthinkable happened: a large cave collapsed at Arch Cot Terrace in Brittons Hill, a densely populated area of Bridgetown, the capital of Barbados, taking down about one third of an apartment building (Fig. 10.13a, b) and burying five members of a young family. Frantic rescue attempts were in vain and the entire family perished.

The cave under the ill-fated house was quite large with a maximum height of about 16 m (Fig. 10.13c), and it reached to within about 60 cm of the surface at its shallowest. The cave had been known for several decades, during which many children and adults had played in or visited the cave. At certain times this cave was used as a garbage dump and/or for storage of explosives. In addition, the coral rock that makes up the cliff in this area is separated into large blocks by deep-reaching joints and at least one fault, which provided access to copious amounts of rainwater that had weakened the rock over time. Also, most of the rock matrix around the cave is composed of stick-shaped coral fragments (*Acropora cervicornis*, *Porites porites*) that are embedded in only loosely cemented carbonate sand. This kind of rock disintegrates within bare hands and crumbles in a matter of months when exposed to the surface. The potential pitfalls of construction associated with such a karst matrix is discussed in greater detail in Chap. 6.

In early 2007, prior to the collapse event, construction work began for another building on the adjacent lot, where heavy machinery was used to excavate the site. This caused serious vibrations in the entire neighborhood. Minor ground movements in a radius of about 100–120 m around the construction site were reported over the following months. These movements caused uneven settling of pavements, cracks in walls, buckling of floor tiles, and two holes opened up in the cave roof, one of them large enough to swallow construction equipment. Work was temporarily halted, yet resumed after a cave survey was done,

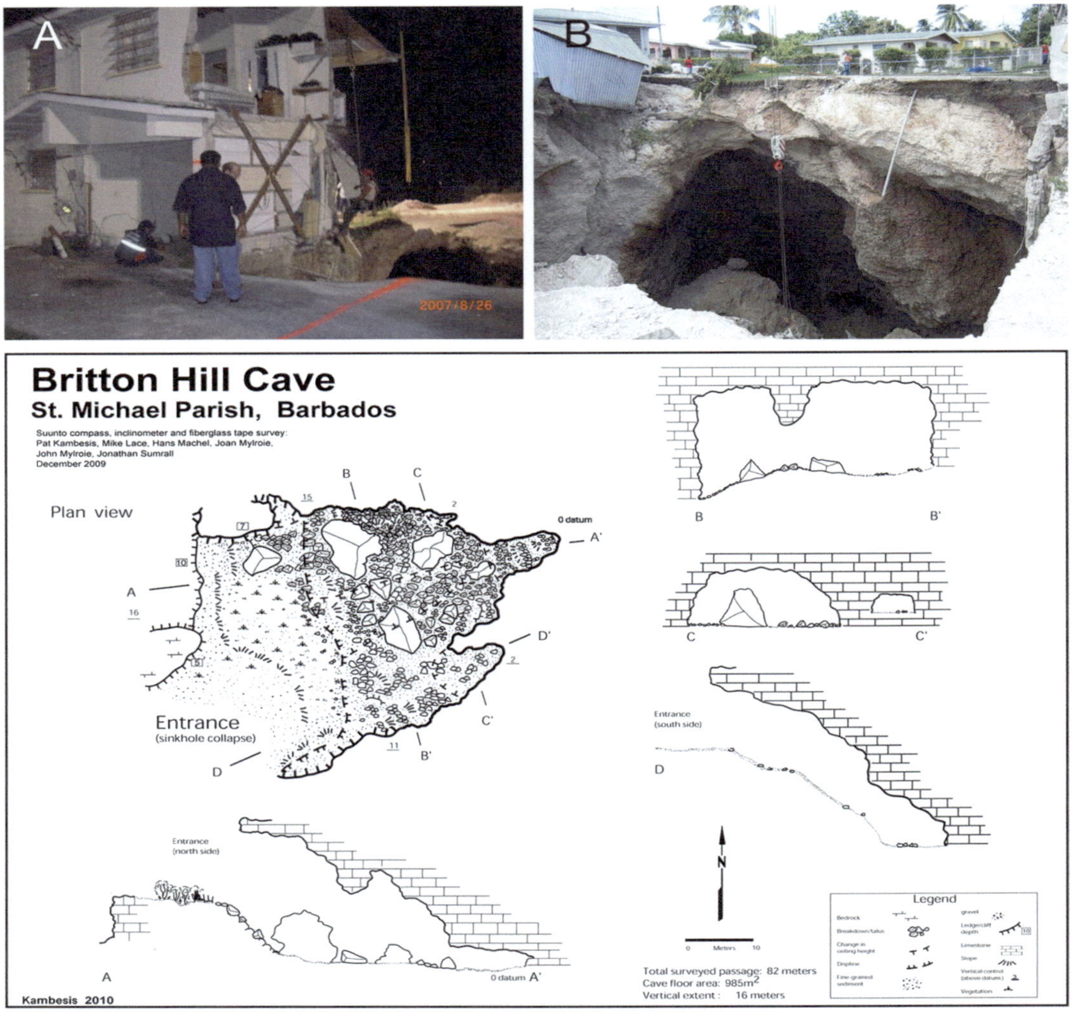

Fig. 10.13 Arch Cot/Brittons Hill Cave collapse. (**a**) View southwestward, showing the southern part of the ill-fated house that remained standing, which was later bulldozed. Note that the cave reached within about ~0.6 m of the surface where the roof was narrowest (*above date stamp*). (**b**) View eastward the day after, the collapsed house just off the field of view to the right. Planar surfaces in the cave walls are joints that dissected the rock into large blocks. (**c**) Map of Arch Cot/Brittons Hill Cave

the main findings of which included that a fault line went right through the cave roof. When the cave collapsed a few days later, the roof broke along this very fault line, shearing off the part of the overlying house.

A thorough Government-commissioned investigation was undertaken by technical experts in the months following the cave collapse, followed in turn by a Coroners Inquest that lasted several months and concluded in December 2011. In her verdict, the Coroner deemed the cave collapse an "avoidable accident" and cited several reasons. Civil and perhaps criminal litigation are now pending or under consideration. On the positive side, legislators and civil servants are now working to improve regulations, building codes, and emergency response measures to avert further tragedies of this kind.

References

Banner JL, Wasserburg GJ, Chen JH, Humphrey JD (1990) Carbonate deposition, diagenesis, and hydrology on Barbados, West Indies: uranium-series evidence. Geol Soc Am Prog 22:A88

Bunnell D (2004) Littoral caves. In: Gunn J (ed) Encyclopedia of caves and karst. Fitzroy Dearborn, New York. ISBN 1-57958-399-7

Cook T, Abbott L (2011) Travels in geology: Barbados: ascending the sea-level staircase. Earth 56(12):54–59. http://www.earthmagazinelorg

Day M (1983) Doline morphology and development in Barbados. Ann Assoc Am Geogr 73:206–219

Drewitt P (1997) The Spring Head petroglyph cave: a sample excavation. J Barbados Museum Soc XLII:50–58

Fermor J (1972) The dry valleys of Barbados. Trans Inst Br Geogr 57:153–165

Fitzpatrick SM (2011) Verification of an archaic age occupation on Barbados, southern lesser Antilles. Radiocarbon 53(4):595–604

Groves C (1994) Geology of Barbados and Harrisons Cave area: a study of the environmental factors in Harrisons Cave, Barbados, West Indies. National Speleological Foundation, p 9–17

Gurnee R, Gurnee J (1991) Caves of Barbados, Western Indies. Closter, National Speleological Foundation, Closter, N.J., pp 8–16

Hall WA (2007) Hall Gunned Down in Barbados, 'Bout Damned Time' Witnesses Claim. "http://worldofwinston.blogspot.ca/2007/01/hall-gunned-down-in-barbados-bout.html"

Harrison JB, Jukes-Browne AJ (1890) The geology of Barbados. Bennett Bros., Printers, Salisbury

Humphrey JD (1997) Geology and hydrogeology of Barbados. In: Vacher HL, Quinn TM (eds) Geology and hydrogeology of carbonate islands, vol 54, Developments in sedimentology. Elsevier, Burlington, pp 381–406

Inniss V, Oderson D, Singh A (2001) The Government of Barbados State of the Environment Report 2000, GEO Barbados, United Nations Environment Programme for the Ministry of Physical Development and Environment, Barbados, 112p

James NP (1972) Late Pleistocene reef limestones, northern Barbados, West Indies. Unpub. PhD dissertation, McGill University, Montreal, 282p

James NP, Stearn CW, Harrison RS (1977) Field guidebook to modern and Pleistocene reef carbonates. Barbados, W.I. In: Third international symposium on coral reefs series, Miami, 30 p

Jones IC, Banner JL (2003) Estimating recharge thresholds in tropical karst island aquifers: Barbados, Puerto Rico and Guam. J Hydrol 278:131–143

Lace MJ (2008) Coastal cave development in Puerto Rico. J Coast Res 24(2):508–518

Machel HG (1999) Geology of Barbados: a brief account of the Island's origin and its major geological features. Barbados Museum and Historical Society, St. Michael, 52p

Machel HG (2011) The geology of Barbados – a little paradise in its own right. In: Carrington CMS (ed), Preserving paradise. Barbados Museum and Historical Society, Bristol, UK, pp 13–51

Machel HG, Burton EA (1994) Golden Grove dolomite, Barbados: origin from modified seawater. J Sediment Res A64:741–751

Machel HG, Sumrall JB, Kambesis PN, Mylroie JR, Mylroie JE, Lace MJ (2011) Episodic dolomitization by methane-bearing seawater in Barbados, West Indies. In: 14th Bathurst meeting of carbonate sedimentologists, University of Bristol. Abstract Book, p 39, 12–14 July 2011

Machel HG, Kambesis PN, Lace MJ, Mylroie JR, Mylroie JE, Sumrall JB (2012) Overview of cave development on Barbados. In: Kindler P, Gamble DW (eds) Proceedings of the 15th symposium on the geology of the Bahamas and other carbonate regions, Gerace Research Centre, San Salvador, Bahamas, pp 96–106

Mylroie JE (2007) Caves surveys, cave size and flank margin caves. Compass Tape 17(4), Issue 60

Mylroie JE, Mylroie JR (2007) Development of the carbonate island karst model. J Cave Karst Stud 69: 59–75

Mylroie JR, Mylroie JE, Sumrall JB, Machel HG, Kambesis PN, Lace MJ (2010) Origin of Barbados gully system: overprinting of dissolution and surface erosion. In: The 15th symposium on the geology of the Bahamas and other carbonate regions, Gerace Research Centre, San Salvador, Bahamas, Abstracts with Program, pp 27–28, 17–21 June 2010

Palmer AN (1991) Origin and morphology of limestone caves. Geol Soc Am Bull 103:1–21

Palmer AN (2011) Distinction between epigene and hypogenic maze caves. Geomorphology 134:9–11

Reil R (2000) Narrative # 6 – cave diving "Unlucky Dive # 7". http://ronreil.abana.org/dive.shtml

Schellmann G, Radtke U (2004) A revised morpho- and chronostratigraphy of the late and middle Pleistocene coral reef terraces on southern Barbados (West Indies). Earth Sci Rev 64:157–187

Schomburgk RH (1848) History of Barbados: comprising a geographical and statistical description of the island; a sketch of the historical events since the settlement; and an account of its geology and natural productions. Longman, Brown, Green and Longmans, London

Senn A (1946) Geological investigations of the groundwater resources of Barbados, B.W.I.: Report of the British Union Oil Company Ltd, 123p

Speed RC (1983) Structure of the accretionary complex of Barbados, I: Chalky Mount. Geol Soc Am Bull 94: 92–116

Speed RC (2002) Field guide to the sub-Quaternary of Barbados. Field guides, 16th Caribbean geological conference, Barbados, 16–21 June 2002, pp 1–83

Speed RC (2012) Geology and geomorphology of Barbados: a companion text to maps with accompanying cross sections, scale 1:10,000. Geol Soc Am (special publication no. 485): Special Paper 491, 63p

Sumrall JB (2013) Relating karst development to island dolomite formation using petrography, geochemistry, and geomorphology. PhD dissertation, Mississippi State University, 162 p

Taylor FW, Mann P (1991) Late quaternary folding of coral reef terraces, Barbados. Geology 19:103–106

Tricart J (1968) Notes géomorphologiques sur la karstification en Barbade (Antilles). Centre de Recherches et Documentation Cartographique et Géographique, Memoirs et Documents 4:329–324

Upchurch SB (2002) Hydrogeochemistry of a karst escarpment. In: Martin JB, Wicks CM, Sasowsky ID (eds) Hydrogeology and biology of post-Paleozoic carbonate aquifers, vol 7, Special publication. Karst Waters Institute, Charles Town, pp 73–75

Vacher HL, Mylroie JE (2002) Eogenetic karst from the perspective of an equivalent porous medium. Carbonate Evaporite 17:182–196

Wandelt B (2000) Geomorphologische Detailkartierung und chronostratigraphische Gliederung der quartaeren Korallanriffe auf Barbados (West Indies) unter besonderer Beruecksichtigung des Karstformenschatzes., Unpublished PhD, thesis, University of Cologne, Germany

Waterstrat WJ, Mylroie JE, Owen AM, Mylroie JR (2010) Coastal caves in Bahamian eolian calcarenites: differentiating between sea caves and flank margin caves using quantitative morphology. J Cave Karst Stud 72:61–74

11 Cave Development and Patterns of Caves and Cave Systems in the Eogenetic Coastal Karst of Southern Mallorca (Balearic Islands, Spain)

Angel Ginés, Joaquín Ginés, and Francesc Gràcia

Abstract

Coastal cave patterns can be studied in exceptional conditions along the southern and eastern coast of Mallorca Island owing to the widespread outcrop of the Upper Miocene calcarenites, in which the development of eogenetic karst features started approximately 6 Ma ago, at the end of Messinian period. Some remarkable coastal-karst caves are today accurately surveyed, including among others the labyrinthine Cova des Pas de Vallgornera (more than 73 km in length), the celebrated Coves del Drac (explored by E.A. Martel in 1896 and visited each year by about one million of tourists) and the impressive submerged passages of Cova de sa Gleda cave-system (exceeding in its current state of exploration a length of 13.5 km, as shown by detailed scuba-diving surveys). A three-fold approach to studying the cave patterns characterizing this complex island CIKM-type karst is presented here on the basis of morphometric analysis of cave-segments, individual caves and cave-systems. At every one of these three levels of observation, the most outstanding feature that can be generalized for the vast majority of caves turns out to be the presence of large breakdown chambers, wider than 20 m along their minor axes, that are partially or totally submerged by sea level fluctuation and the rise of the water table in Holocene times.

11.1 Introduction

Interest in coastal karst and coastal caves has increased during the last few decades because of the particular hydrogeological and speleogenetic characteristics, which differ from karst and caves that develop in continental settings. Karst caves developed in coastal areas have been explained as the result of dissolution from the mixing of waters with different chemical compositions, and it is

A. Ginés (✉)
Departament de Biologia, Universitat de les Illes Balears, Palma de Mallorca, Spain
e-mail: agines.gracia@yahoo.es

J. Ginés • F. Gràcia
Departament de Ciències de la Terra, Universitat de les Illes Balears, Palma de Mallorca, Spain
e-mail: xescgracia@yahoo.es

M.J. Lace and J.E. Mylroie (eds.), *Coastal Karst Landforms*, Coastal Research Library 5,
DOI 10.1007/978-94-007-5016-6_11, © Springer Science+Business Media Dordrecht 2013

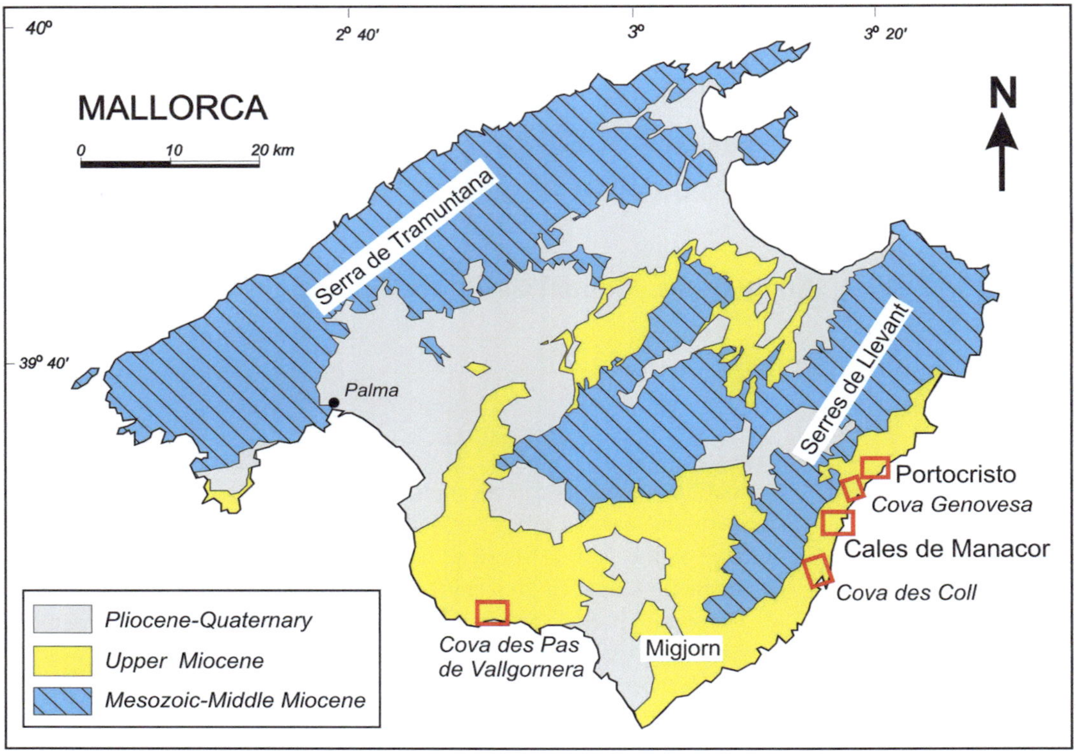

Fig. 11.1 Simplified geological map of Mallorca Island showing the location of the caves and the cave-areas discussed in the text. The Migjorn area is in the *lower right* of the figure

assumed that distinctive cave patterns could be expected in such special geochemical and hydrodynamical environment. These caves commonly exhibit spongework and ramiform development believed to reflect geochemical processes including the mixing of fresh and saline water and hydrodynamics of the freshwater lens.

Knowledge of caves depends on the availability of accurate cave maps. Surveying of caves located in the immediate vicinity of the sea is frequently hampered by the closeness of the water table. Many coastal caves appear drowned in most of their real sections by the postglacial sea-level rise and the consequent rise of the water table. For this reason, cave mapping in these partially-submerged karst systems requires accurate and detailed survey tasks to account not only for the air-filled rooms and passages, but also the water-filled extensions of the caves.

One of the main goals in cave surveying is the disclosure and subsequent explanation of cave patterns. This is especially true for coastal karst because comparisons between cave patterns may reflect different speleogenetic processes, such as the relationship between cave development and glacio-eustasy, the role of breakdown in the evolution of the caves or the bioclimatic, geochemical and structural constraints involved. Since significant information can be obtained using detailed cave maps, a discussion on surveying criteria and standardization of coastal cave maps seems worthwhile (Mylroie 2007; Ginés et al. 2009a).

Cave surveying in the Migjorn region (Fig. 11.1), a coastal karst area located in the southern part of Mallorca Island (western Mediterranean), has a long history of exploration, mapping, and scientific debate on the mechanisms driving cave development, starting in the second half of the nineteenth century. In short, three main periods can be distinguished: (1) the pioneer descriptions of a few celebrated

Migjorn caves (including the Martel map of Coves del Drac, performed in 1896); (2) the conventional-caver mapping of many caves of the region (produced during the 70s to 90s of the twentieth century) that suggested a typical and repetitive pattern made up of apparently isolated collapse chambers that coalesce randomly; and (3) recent detailed mapping by cave divers (from 1995 till present) demonstrating underwater connections between caves and in some cases even with the sea.

11.2 Cave Surveying: A Matter of Accessibility

Producing accurate and detailed cave maps requires overcoming technical problems associated with physical exploration involved with each cave and most importantly traversing the minimum penetrable-size of all cave passages. The availability of cave maps remains a function of exploration, which increases with time. Direct observation, and consequently cave surveying, is limited in some cases by the presence of fluids (e.g. high concentrations of carbon dioxide or, more commonly, water) which require specific exploration equipment. Nonetheless, the most obvious limit for cave surveying is the minimum size of the cavity to explore, whereby many sections of a cave are out of the access of cavers and can only be interpreted through indirect observations.

Inaccessibility accounts for most of the unsatisfactory knowledge of coastal caves and coastal cave-systems in all littoral karst areas as the Holocene rise in sea level flooded these low-lying caves. Furthermore, eogenetic karst is not prone to promote the development of well-structured cave conduits, in spite of its inherent great hydraulic conductivity, impeding the direct observation and mapping of the myriads of tiny interconnected voids that transmit the water toward the sea (the touching vug permeability of Vacher and Mylroie 2002). On the other hand, it is still hard to lay down solid generalizations, because only a small number of locations around the world form the base of the current theoretical approach to coastal karst systems (e.g. Bahamas, Bermuda, Marianas, Mallorca, Puerto Rico, Florida or Yucatan – see accompanying case studies within this volume) and comparison between these coastal locations is still a task to develop in the near future.

11.3 Cave Surveying: A Matter of Methodology and Sampling Strategy

Cave surveying and cave mapping are more than a descriptive task. Although some cave features are striking, many significant features are frequently ignored in the most common cave maps. Selective criteria are applied implicitly when the cavers choose what shape better describes the perimeter of a cave, or when they plot the inside of such a wall-perimeter the most relevant features to enclose in the drawing. Cave maps are the only way to study the pattern of the caves, but these maps can be far from objective representations.

An alternative approach is to assume these facts as the uncertainty bias of a particular "sampling method", thereby avoiding unexpected "graphic artifacts" that would introduce misleading information through the mapping practices. In this way, cave surveying becomes a specific sampling method for collecting accurate field data. For instance, in the case of the Migjorn karst of Mallorca (Fig. 11.1), our current analysis of coastal cave patterns relies on a three-level approach based on the following scale of increasing size and complexity: cave-segments, individual caves and cave-systems.

11.4 Maps of Cave-Segments in the Karst of Southern Mallorca

Detailed analysis of particular cave-segments may make it possible to identify diagnostic features that relate to specific geochemical environments. For example, some irregular chambers may relate to dissolution voids that develop similar to flank margin caves, or definite cave passages may reflect active flow and dissolution processes similar to conduits that develop in telogenetic settings. Additionally, it is important to report significant assemblages of solutional forms, as well as to measure their size and frequency of appearance. Cave maps provide little information about small-scale solution features (e.g. presence or absence of solution notches, scallops, etc.). On larger scales, however, some significant data can be obtained by analyzing chambers and passages in plan view. Cross sections are especially informative because their shapes commonly provide insight on passage genesis. In addition, the cave floor of rooms and passages requires detailed attention when trying to recognize the presence of sediments, speleothem pavements, collapse boulders or *in situ* rock.

In the eogenetic karst of Migjorn, cavers have traditionally focused on the richness in speleothems, the widespread occurrence of breakdown features (boulder heaps and collapsing vaults), and the scarcity of former solutional evidence (that become progressively dismantled by breakdown). This condition is easy to recognize within the whole background of cave surveys produced by Majorcan cavers till present. The vast majority of available maps depict, in detail, the recurrent collapse chambers floored with fallen blocks and speleothems, even below the water table (Fig. 11.2). More recently, however, mapping by cave divers has incorporated specific symbols in order to indicate the presence of conspicuous solutional features (Fig. 11.3), such as certain assemblages of corrosion forms including spongework morphology and single solutional-conduit passages (Gràcia et al. 2005, 2006, 2007, 2009). This information has an additional value because solutional voids appear structured in horizons that are currently located between 1 and 30 m below the water table; being therefore inaccessible for conventional cave exploration.

11.5 Maps of Individual Caves in the Karst of Southern Mallorca

Maps of individual caves in southern Mallorca commonly fail to reflect entire cave systems as new discoveries by cave divers have substantially expanded the old maps. Since most cave maps reflect an "artifact of the current stage of exploration" of the real cave, the new discoveries by cave divers allow the old surveys to be substantially modified. This "expansion" of previous cave maps is especially evident in light of the coastal caves drowned as a result of the Holocene sea-level rise.

In the eogenetic karst of southern Mallorca, excellent examples of caves illustrating the "artificial changes in cave pattern" result from new exploration techniques. For example, the caves surrounding Portocristo harbour (Fig. 11.4) and Cova Genovesa (Fig. 11.5) were previously explored by conventional cavers who produced detailed cave maps that stopped at sea-level pools.

The karstic area around Portocristo encloses two celebrated show caves, Coves dels Hams and Coves del Drac (Fig. 11.4), that reflect the typical trends of caves from the Migjorn karst region (Ginés and Ginés 1989; Ginés 2000a, b). These caves exhibit wandering sets of large breakdown-vault units, 30–50 m in width, connected randomly by local collapses and surrounded by peripheral brackish pools. This pattern is easy to recognize throughout the maps published on these caves over the last century (Martel 1896; Maheu 1912; Faura y Sans 1926; Ginés and Ginés 1992; Ginés et al. 2007). The updated cave surveys (Fig. 11.4) highlight the remarkably non-directional path of Coves del Drac, in spite of its location between the incised creek of Torrent de ses Talaioles and the small bight of Cala Murta. Until now, only some sporadic cave-diver explorations added significant underwater extensions in Coves del Drac (Clarke 1991), although additional passages are expected if underwater surveying continues.

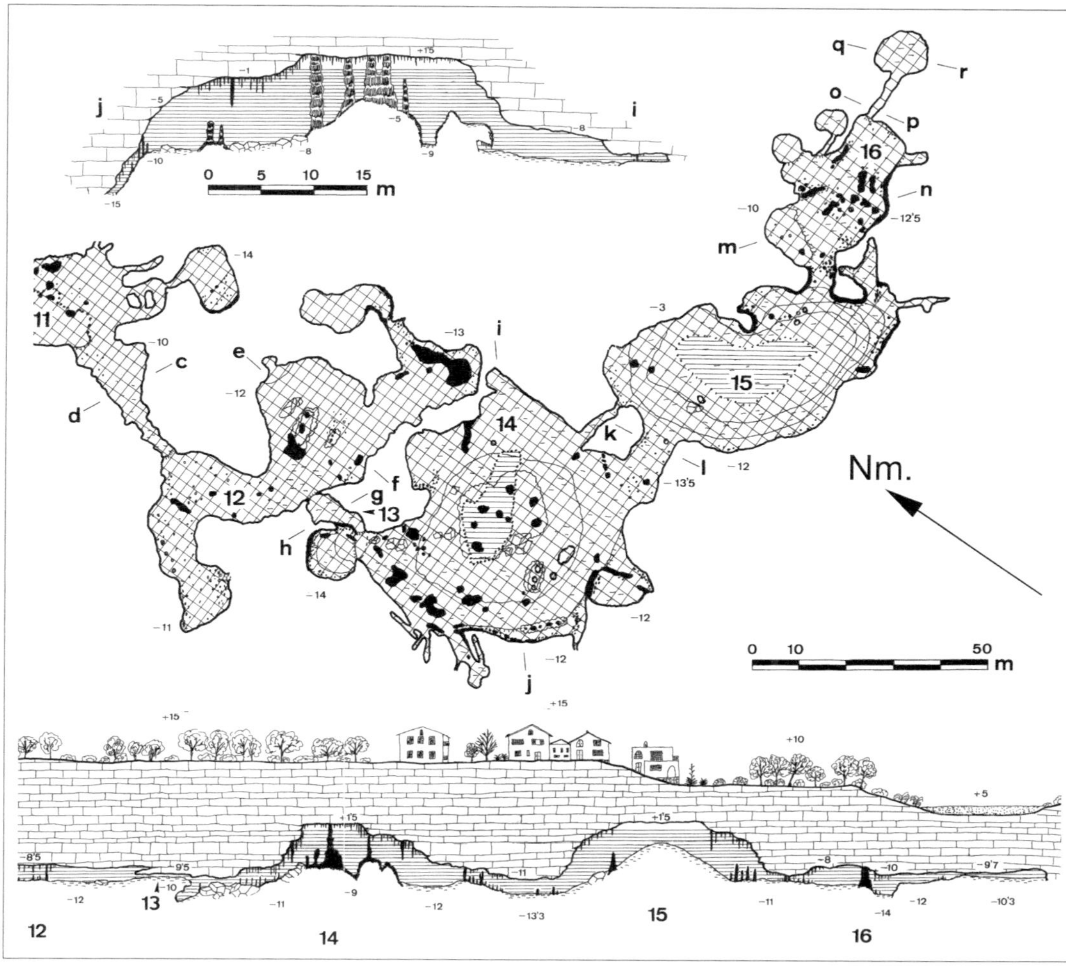

Fig. 11.2 Detailed scuba-diver survey of a sector from Cova Genovesa. The presence of breakdown features and speleothems is easy to recognize with the aid of this kind of underwater mapping (Adapted from Gràcia et al. 2003a)

In addition to the foreseeable discoveries in Portocristo, Cova Genovesa also provides a good example of the striking changes in its plan-view map resulting from underwater exploration and careful mapping (Fig. 11.5). Twenty years ago, the known cave included only two typical collapse chambers near the entrance. Nevertheless, recent mapping has demonstrated a wandering path running toward the Cala Anguila bight, hidden from the "conventional-cavers" below the modern water table (Gràcia et al. 2003a). Such discoveries suggest the previously known caves, namely those accessible without scuba equipment, can be interpreted as just the topmost part of extensive systems of coalescing chambers, whose "solutional roots" are mainly located more than 10 m below the present sea level.

11.6 Maps of Cave Systems in the Karst of Southern Mallorca

The cave systems can be defined as "a collection of caves interconnected by enterable passages or linked hydrologically" or as "a cave with an extensive complex of chambers and passages" (Gillieson 1996). As could be expected, only few of the surveyed caves fit within this

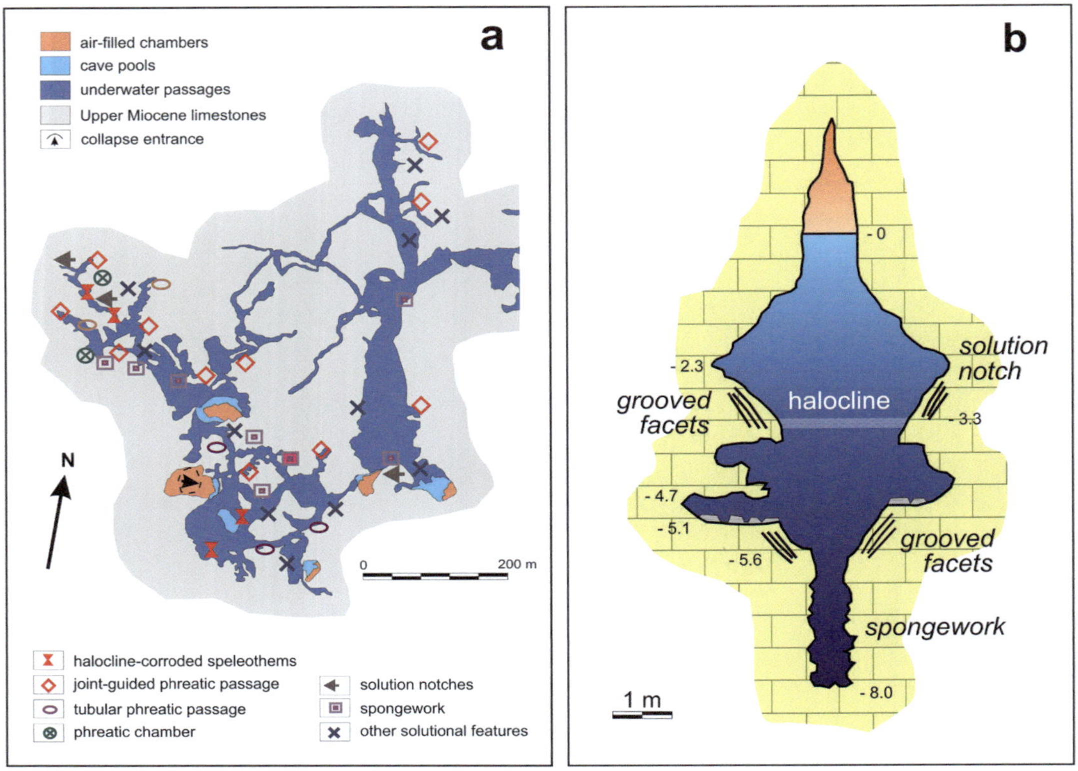

Fig. 11.3 Two examples showing the useful incorporation of specific symbols that indicate the presence of solutional features in underwater cave maps: (**a**) on a plan view survey of a sector from Cova de sa Gleda (Adapted from Gràcia et al. 2007); (**b**) on a cross-section from Galeria Miquel Àngel Barceló, Cova des Pas de Vallgornera (Adapted from Gràcia et al. 2009)

third scale-level of standardized interpretation. Namely, Cova des Coll, Cova dets Ases, Coves de Cala Varques, Cova de sa Gleda and Pirata-Pont-Piqueta system have several connected entrances, and the extensive cave complex of Cova des Pas de Vallgornera that has only one known artificial entrance.

As a result of systematic cave-diver explorations carried out during the last two decades in the sea-level pools of many formerly "well-known" caves from the southern karst of Mallorca, several groups of caves were successfully connected underwater. The exploration of two of them (Cova des Coll and Cova dels Ases) led to a direct opening to the sea, demonstrating a hydrological connection between the caves and the brackish water outlets existing along the coastal line via conduits negotiable by cave-divers. Cova des Coll has today more than 7 km of surveyed passages, of which 5.5 km are submerged (Gràcia et al. 2005), and is distinguished by its rather directional trend (Fig. 11.6). The pattern of the cave includes not only several collapse chambers, but also remarkable joint-guided solutional passages that locally generate mazes of conduits. Cova dets Ases is an array of typical collapse chambers connected with the sea through a shallow underwater passage (Gràcia et al. 1997; Ginés 2000a).

Around the Cales de Manacor karst area (Fig. 11.7) successful explorations provided insight on the pattern of coastal cave-systems of this part of Mallorca. Three cave systems that developed not far from the coast showed no direct drainage toward the sea. Cova de sa Gleda has more than 13 km of submerged chambers and large passages (Gràcia et al.

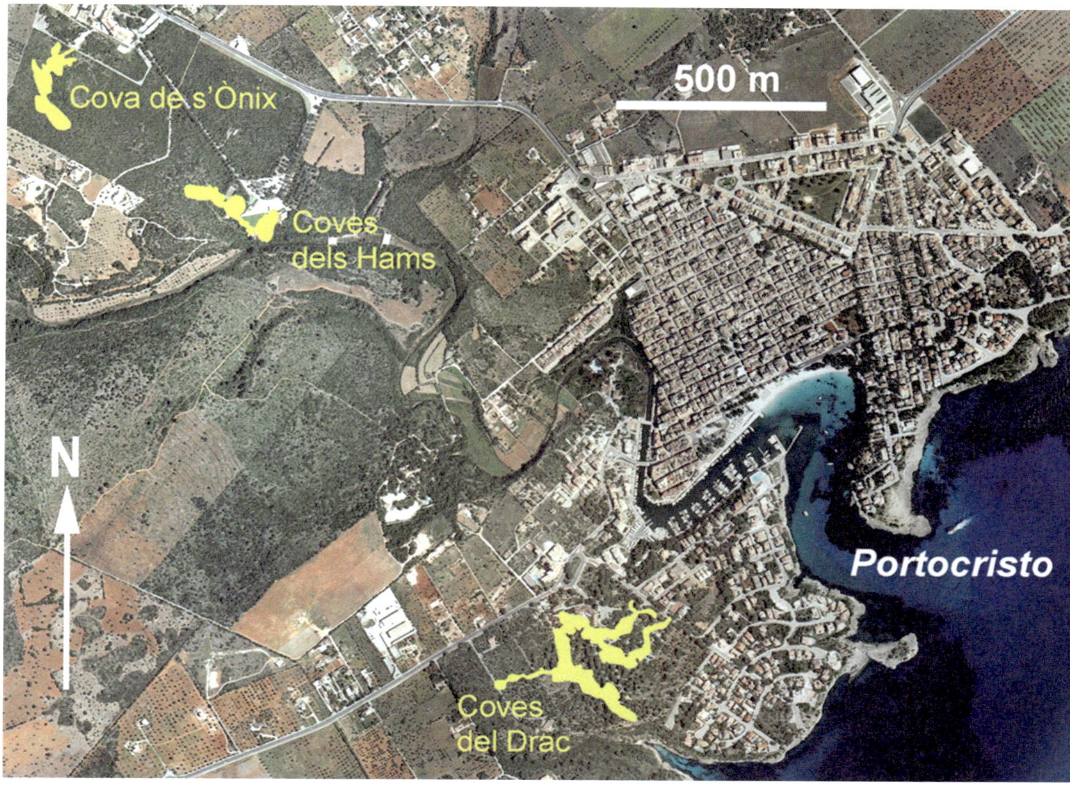

Fig. 11.4 Plotting of the main caves located around Portocristo harbour. Note the wandering trend and the remarkably non-directional path of the Coves del Drac cavern, placed just between the incised creek of Torrent de ses Talaioles and the small bight of Cala Murta visible at the *southern side* of the figure

2007, 2010). Coves de Cala Varques consists of three connected collapse caves that surround the homonymous bight (Gràcia et al. 2000). Pirata-Pont-Piqueta system shows wandering and chaotic paths (longer than 3 km) characterized by the coalescence of nine major collapse chambers (Fig. 11.8), without efficient hydrological connectivity between them, and opened to the surface through three cave entrances (Gràcia et al. 2006). Ramiform patterns are dominant in these three neighboring cave systems (Fig. 11.7).

11.7 Cave-Pattern Statistics: A Matter of Map Availability

Morphometry of caves is a crucial topic trying to explore differences between cave-patterns. Qualitative observations are basic, but they depend excessively on subjective appraisals and rarely could substantiate a statistical approach with enough solid data. Specific measurements of significant features are difficult and tedious to make inside the caves and require too much time. On the other hand, cave maps are the most suitable tool for analyzing the structure of the caves: certainly they are not quite objective representations, but are produced upon measures obtained *in situ* and generally are carried out by authors, not necessarily biased toward any speleogenetic theory, that simply want to describe the path and geometrical development of the caves. For these reasons, we suggest that the most useful criteria in order to implement cave-pattern statistics would be to lay down some general and simple procedures, which being open to comparison between different caves would be based in practice on measures easily to obtain from ordinary cave maps,

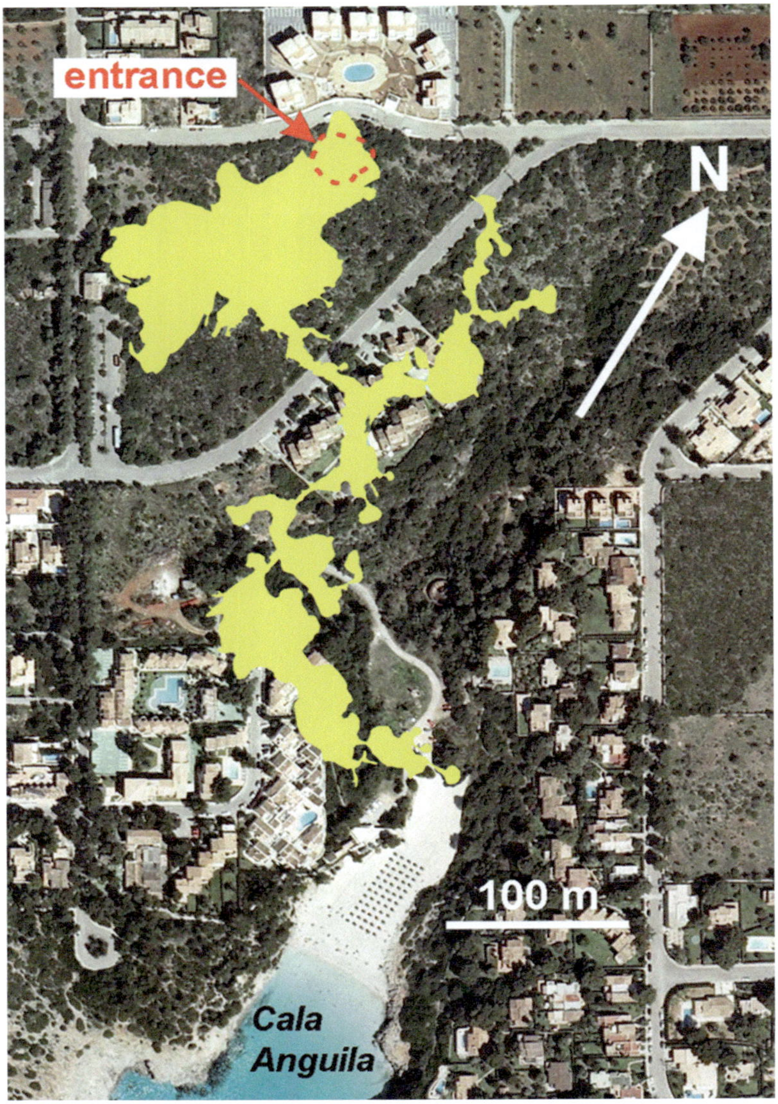

Fig. 11.5 Map of Cova Genovesa plotted over an aerial photograph of Cala Anguila bight (After Gràcia et al. 2003b)

preferably if they are performed with enough detail and accuracy (Fig. 11.2). Let us show some examples of these morphometrical procedures taking as case-studies the cave surveys of Cova Genovesa and Cova des Pas de Vallgornera.

11.7.1 Passage Cross-Section Statistics

Surely one of the most conspicuous features that can be appreciated in the plan view of whatever cave map is the width of the passages and chambers that constitute the entire cave. The breadth of the different sectors of a cave is very easy to measure with the aid of the scale that are plotted on the survey, and at the same time it is quite simple to collect series of width measurements along the imaginary center line that follows the complete pathway of the cave. A good option would be to measure how wide are, in statistical terms, a significant set of transverse cave sections taken at right angles with respect to the main center line route. Furthermore,

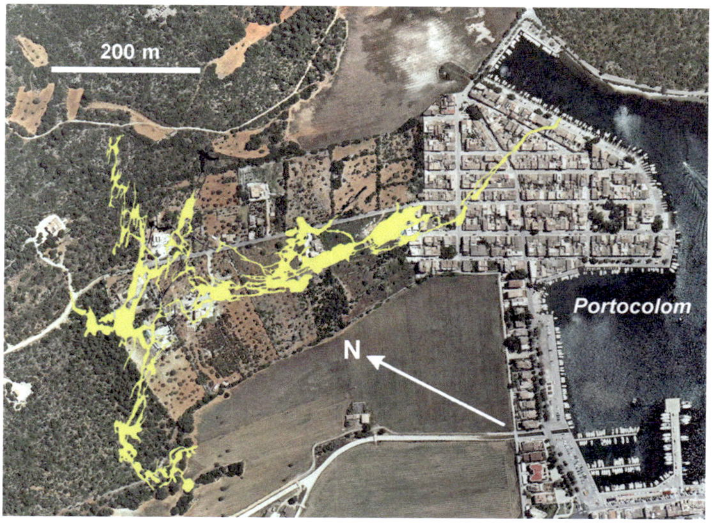

Fig. 11.6 Map of Cova des Coll plotted over an aerial photograph of Portocolom harbour (After Gràcia et al. 2005). The cave connects to the sea in the *upper right*

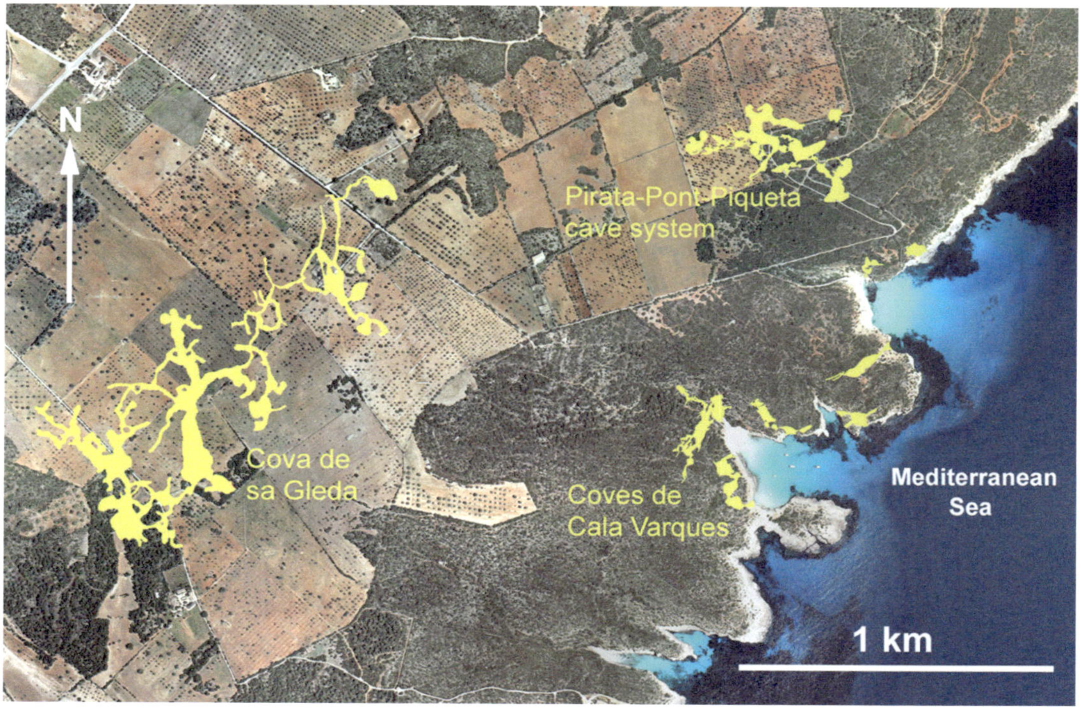

Fig. 11.7 Plotting of the main caves located in Cales de Manacor area. Ramiform patterns are dominant, lacking apparent direct drainage toward the sea (updated after the current explorations performed by the cave-divers of Grup Nord de Mallorca)

in order to improve the validity (even the exhaustiveness) of the set of measurements, each section could follow a definite spatial sequence through the whole center line, for example at a fixed interval. The frequency distribution of the width values would provide a characteristic signature for different caves, probably related to different cave-patterns.

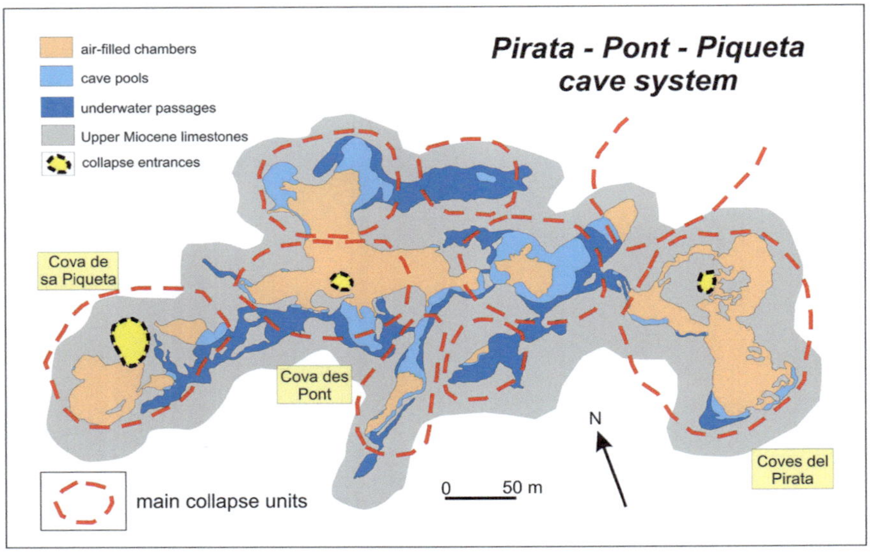

Fig. 11.8 The whole pattern of Pirata-Pont-Piqueta cave system, about 3 km in development, is easily explained as the result of coalescence of nine major breakdown units (Slightly modified from Gràcia et al. 2006)

In the case of Cova Genovesa (see a simplified survey in Fig. 11.5) such a methodology has been applied thoroughly all along the total length of the cave, in the form of 237 cross-sectional measurements spaced between them approximately each 5 m. The results obtained demonstrate that 68 % of the cave is wider than 5 m, and more than 27 % exceed 20 m in breadth. The frequency distribution (Fig. 11.9) points out two significant peaks for the width-classes 20–25 m and 35–40 m, which are clearly related to the extensive collapse breakdown affecting the main part of the cave. This assumption is corroborated by the detailed description of the cave published by Gràcia et al. (2003a, b), that points out the presence of breakdown features over 90 % of the total cave development.

11.7.2 Inferences on Major Cave-Level Development

A rough estimation of the major speleogenetical horizons involved in the evolution of a particular cave can be easily obtained by evaluating the total amount of cave-void that, after the information supplied by its own cave-map, appears distributed at different elevations. For this purpose, a useful method consists simply in drawing a set of horizontal, parallel and equidistant lines across the whole extended profile of the cave (as well as over additional cross-sections if necessary), and measure how much passage length has been intercepted by each leveled line. The most noticeable peaks, that stands out when plotted the frequency distribution corresponding to the amount of void space intercepted within each level, would indicate presumably the location of preferential horizons where karstic erosion have generated more cave.

In the case of Cova Genovesa (Figs. 11.2 and 11.5) two complementary frequency distribution analyses were performed. The first one (Fig. 11.10a) was the simpler result of measuring on the profiles published by Gràcia et al. (2003a) the total amount of cave intercepted at elevation intervals of 2 m. The obtained results show an outstanding maximum of cave-voids placed at 10 m below sea level, as well as several subdued peaks at 18, 8 and −18 m a.s.l.; at the same time, it demonstrates that approximately 40 % of the total surveyed passages and chambers are comprised between 6 and 10 m below the current sea level. On the other hand, the second statistics

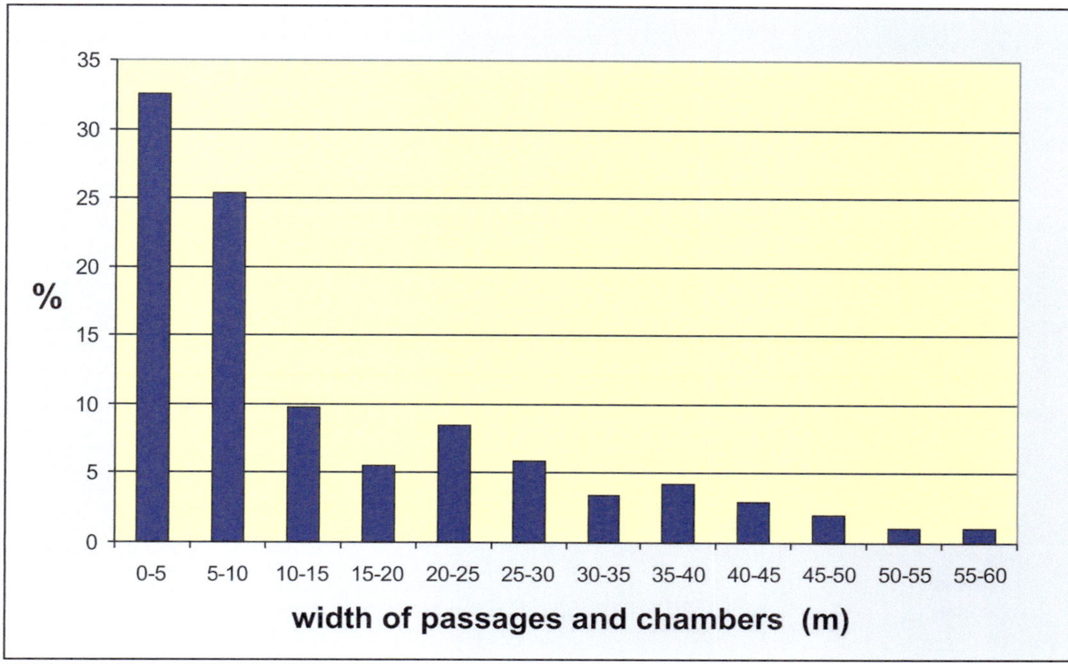

Fig. 11.9 Frequency distribution of cave passage widths from Cova Genovesa. It is worth noting that more than 27 % of the 237 measured cross-sections are wider than 20 m

(Fig. 11.10b) is the result of considering also as measurable voids the space occupied presumably by the great boulder chokes that constitute the floor of the collapse chambers, which are so significant regarding the whole pattern of the cave. The processing of such total cave-void space (real cave plus presumed void occupied by fallen blocks) shows a more smoothed frequency distribution, but the maximum remains clearly the same, at about 10 m below sea level, and indicating only two minor peaks at −2 and −18 m a.s.l. In this way, the statistical data obtained by means of both procedures suggest −10 m a.s.l. as the major horizon of speleogenesis in Cova Genovesa, as well as indicate the existence of a second horizon at −18 m a.s.l., in good agreement with the qualitative observations reported by Gràcia et al. (2003b). The subdued peaks in Fig. 11.10a and the −2 peak from Fig. 11.10b could be caused by the presence of more resistant bedding planes, which probably correspond to the top of the main domes that form some of the greater collapse chambers.

11.7.3 Selection of a Suitable Scale and Grid Size for Comparison Between Cave-Patterns

Whatever comparison between caves is strongly conditioned by the scale of the available surveys, which in turn is tightly dependent on the development and dimensions of the caves or cave-systems. This is particularly true for Cova Genovesa and Cova des Pas de Vallgornera, both located inside the Upper Miocene reef, where they form part of the eogenetic karst of southern Mallorca. In fact, as could be expected for the vast majority of caves developed in the Migjorn coastal karst, they share the same kind of great collapse chambers partially drowned by the current water table, a comparable richness in speleothem abundance and diversity, as well as the presence of several similar features produced by phreatic corrosion. But the difference in their total development is quite significant (2.4 km against 73 km, respectively), and this fact is reflected on the general pattern of both caves

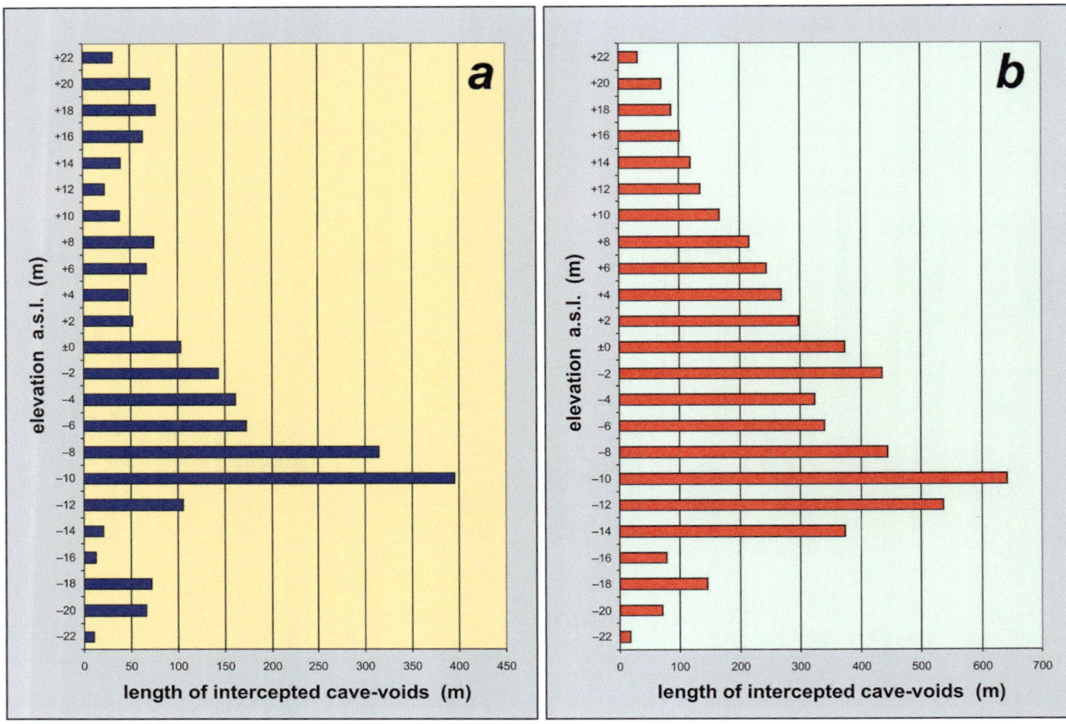

Fig. 11.10 Inference of the main speleogenetic levels in Cova Genovesa as deduced from passage elevation statistics: (**a**) total amount of real cave passages intercepted at elevation intervals of 2 m; (**b**) estimated amount of cave passages at elevation intervals of 2 m including the space occupied presumably by boulder chokes at the floor of the great collapse chambers. Note the outstanding maximum horizon of cave development placed at −10 m a.s.l

as shown by their corresponding plan view. For instance, the map of Cova Genovesa fits better with the ramiform and wandering pattern of Coves del Drac (2.3 km in length), rather than with the complex labyrinthine pattern shown in the complete plan view of Cova des Pas de Vallgornera (Fig. 11.11).

In the case of Cova des Pas de Vallgornera the scale factor illustrates the major reason for the complexity of its pattern, since the cave is huge enough to become conditioned by the different facies characterizing the architecture of the reef inside which the cavern is developed. Namely, whereas in the reef front, being plentiful of coral constructions, a ramiform pattern including collapse chambers and spongework mazes are dominant, in the back reef facies a parallel set of straight joint guided passages is the ruling trend and generates a rather elongated network (Ginés et al. 2009b, c). Conversely, the smaller Cova Genovesa is merely limited to the reef front facies, showing a much simpler ramiform pattern. When the caves to contrast are so different in size, a good option is to observe the maps through a grid and compare, and maybe even measure with the aid of specific descriptors, the enclosed features between significant squares, better than between the entire plan views. A suitable grid size, at least for the caves of the coastal karst of Mallorca, could be formed by squares measuring 200 × 200 m, which means a surface of 4 ha (40,000 m^2, or approximately 10 acres). An example of the fairly diverse perception of several parts of Cova des Pas de Vallgornera as being observed through such a grid is presented in Fig. 11.11, where at least three different cave-patterns related to lithological conditionings are easy to recognize.

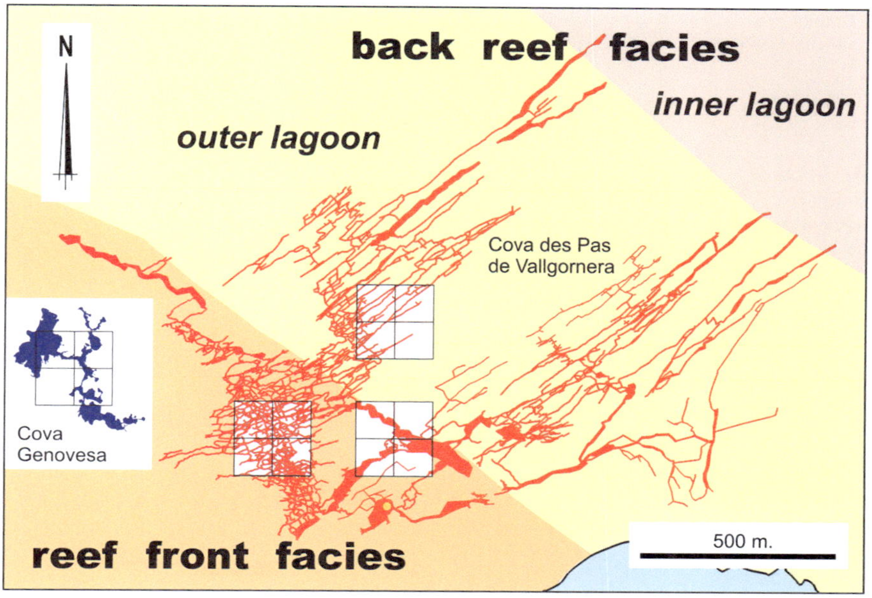

Fig. 11.11 Because cave patterns are strongly conditioned by the scale of observation, they can only be compared through a suitable sized grid. The different facies characterizing the architecture of the Upper Miocene reef of southern Mallorca are reflected in the diverse plan patterns observable in Cova des Pas de Vallgornera (Adapted from Ginés et al. 2009c), whereas the smaller Cova Genovesa (*left-side inset*) only shows the ramiform pattern related to the reef front facies

11.8 Conclusions

Accurate cave maps are essential for the analysis of cave patterns and suggest relevant information such as detailed surveys and complementary data should be included in these maps. This is particularly true in many coastal karst areas, like Yucatan (Mexico) (see Chap. 16), the Gambier Karst (southeastern South Australia) and southern Mallorca (Spain), where specific constraints (e.g. drowned extensions of many cave systems) restrict conventional cavers from access to the lower passages of many caves. The presence of sea-level pools, disrupting the connectivity between caves and hampering the perception of its real pattern, is a general rule in these caves.

Morphometrical approaches to cave patterns are obviously conditioned by speleogenetic hypothesis and at the same time the presence or absence of diagnostic features, as well as the eventual statistical data based on cave surveys, can introduce significant improvements in our perception of the real processes involved in coastal speleogenesis, as well as in computer modeling attempts (Labourdette et al. 2007, see also Chap. 4). For instance, simple statistical measurements on passage cross-sections show that large voids, produced mainly by breakdown collapse, must account for not less than 60 % of the wandering cave development surveyed in Cova Genovesa (Fig. 11.9). The assumption that breakdown is the responsible for the great amount of wide voids in the Majorcan coastal caves is supported by *in situ* observation of boulders occupying the floor of many passages and chambers, as depicted in the subaquatic maps from Gràcia et al. (2003a, 2006, 2010) and sustained furthermore by the conventional-caver knowledge available for the majority of caves in the region, namely about the celebrated Coves del Drac (Ginés 2000a; Ginés and Ginés 2007). The role of vault-collapse in the development of many coastal caves was also documented in Bermuda (Palmer et al. 1977), Yucatan (Beddows 2004) and southeastern South Australia (Grimes 2007).

Fig. 11.12 Tubular phreatic passages are well developed in the less permeable outer lagoon facies of the back reef along the Sector Subaquàtic de Gregal, at Cova des Pas de Vallgornera (Photo Antoni Cirer, Grup Nord de Mallorca)

Recently Mylroie and Mylroie (2007) considered the coastal caves of Bermuda as a particular case in the frame of the CIKM-model: the so-called carbonate-cover island. After our experience in the karst of southern Mallorca, we suggest that, both the composite island and the complex island categories of the CIKM-model could present usually significant breakdown processes along their caves and cave-systems.

The coastal karst caves of Balearic Islands would yield significant contributions to the validation of current successful models concerning development of karst and caves in island environments, namely the eogenetic karst model (Vacher and Mylroie 2002) and the CIKM-model (Mylroie and Mylroie 2007) respectively. Recent studies have demonstrated to what extent typical karst conduits (Fig. 11.12) can grow in eogenetic karsts, as a response to changes in permeability through the complex architecture of the Upper Miocene reef in southern Mallorca; this is the case of the straight passages of Cova des Pas de Vallgornera developed on the inside of the outer lagoon facies. On the other hand, not only southern Mallorca constitutes the higher degree of complexity in the frame of the CIKM-model, as a good example of the complex island category. The Upper Miocene reef in the Balearics is indeed characterized by a good example of simple carbonate island (the smaller island of Formentera) and another typical example of composite island, the karst of southern Menorca.

Our experience in the karst of southern Mallorca (the so called Migjorn region) demonstrates the importance of accurate and detailed cave mapping. Such standards can provide a major significance in the studies of coastal caves. For example, the cave surveys of Coves del Drac (Faura y Sans 1926), Cova Genovesa (Gràcia et al. 2003b) or Cova des Pas de Vallgornera (Gràcia et al. 2009; Merino et al. 2011) are much more than simple description-tools, but instead support inference of speleogenesis (Ginés and Ginés 2007; Ginés et al. 2008, 2009b). In this way, we are suggesting that detailed underwater surveys must to be encouraged, in spite of the incumbent difficulties, with the aim of improving our knowledge on coastal cave morphogenesis (Ginés et al. 2009a). Furthermore, we introduce an open discussion about what data standards could be more advisable regarding future comparisons between coastal cave maps from different coastal settings of the world.

Acknowledgements Thanks are due to the editors aid in the improvement of the manuscript and useful suggestions. The cave-divers of Grup Nord de Mallorca (Pollença) are thanked for their involvement in producing the high quality cave maps and photographs that appeared regularly in different issues of Endins, the journal of the Federació Balear d'Espeleologia. Antoni Cirer has kindly supplied excellent underwater pictures of Mallorcan caves. Joan J. Fornós and Antoni Merino have also contributed in different aspects to this research.

References

Beddows PA (2004) Groundwater hydrology of a coastal conduit carbonate aquifer: Caribbean coast of the Yucatán Peninsula, México. Unpublished PhD thesis. School of Geographical Sciences, University of Bristol, 303p

Clarke O (1991) Diving in Drach. Descent 101:32–33

Faura y Sans M (1926) Cuevas de Mallorca. XIV Cong Geol Intern, Inst Geol España, Gráficas Reunidas, S.A., Madrid, 78p

Gillieson D (1996) Caves: processes development and management. Blackwell, Oxford, 324p

Ginés A (2000a) Patterns of collapse chambers in the endokarst of Mallorca (Balearic Islands, Spain). Acta Carsol 29(2):139–148

Ginés J (2000b) El Karst litoral en el levante de Mallorca: Una aproximación al conocimiento de su morfogénesis y cronología. Unpublished PhD thesis. Universitat de les Illes Balears, Palma de Mallorca, 595p

Ginés J, Ginés A (1989) El Karst en las Islas Baleares. In: Durán JJ, López J (eds) El Karst en España, Sociedad Española de Geomorfología, Monografía 4, Madrid, pp 163–174

Ginés A, Ginés J (1992) Las Coves del Drac (Manacor, Mallorca). Apuntes históricos y espeleogenéticos. Endins 17–18:5–20

Ginés A, Ginés J (2007) Eogenetic karst, glacioeustatic cave pools and anchialine environments on Mallorca Island: a discussion of coastal speleogenesis. Int J Speleol 36(2):57–67

Ginés J, Fornós JJ, Trias M, Ginés A, Santandreu G (2007) Els fenòmens endocàrstics de la zona de Ca n'Olesa: la Cova de s'Ònix i altres cavitats veïnes (Manacor, Mallorca). Endins 31:5–30

Ginés J, Ginés A, Fornós JJ, Gràcia F, Merino A (2008) Noves observacions sobre l'espeleogènesi en el Migjorn de Mallorca: els condicionants litològics en alguns grans sistemes subterranis litorals. Endins 32:49–79

Ginés A, Ginés J, Gràcia F, Fornós JJ (2009a) Cave surveying and cave patterns in the southeastern coastal karst of Mallorca Island (Spain). In: White WB (ed) Proceedings of the 15th international congress speleology, International Union of Speleology, vol 1, Kerrville, pp 488–495

Ginés J, Ginés A, Fornós JJ, Merino A, Gràcia F (2009b) On the role of hypogene speleogenesis in shaping the coastal endokarst of southern Mallorca (Western Mediterranean). In: Klimchouk AB, Ford DC (eds) Hypogene speleogenesis and karst hydrogeology of artesian basins, vol 1, Ukrainian Institute of Speleology and Karstology, Special Paper. Ukrainian Institute of Speleology and Karstology, Simferopol, pp 91–99

Ginés J, Ginés A, Fornós JJ, Merino A, Gràcia F (2009c) About the genesis of an exceptional coastal cave from Mallorca Island (Western Mediterranean). The lithological control over the pattern and morphology of Cova des Pas de Vallgornera. In: White WB (ed) Proceedings of the 15th international congress speleology, International Union of Speleology, vol 1, Kerrville, pp 481–487

Gràcia F, Watkinson P, Monserrat T, Clarke O, Landreth R (1997) Les coves de la zona de ses Partions-Portocolom (Felanitx, Mallorca). Endins 21:5–36

Gràcia F, Clamor B, Lavergne JJ (2000) Les coves de Cala Varques (Manacor, Mallorca). Endins 23:41–57

Gràcia F, Clamor B, Gual MA, Watkinson P, Dot MA (2003a) Les coves de Cala Anguila (Manacor, Mallorca). I: Descripció de les cavitats i història de les exploracions. Endins 25:23–42

Gràcia F, Jaume D, Ramis D, Fornós JJ, Bover P, Clamor B, Gual MA, Vadell M (2003b) Les coves de Cala Anguila (Manacor, Mallorca). II: La Cova Genovesa o Cova d'en Bessó. Espeleogènesi, geomorfologia, hidrologia, sedimentologia, fauna, paleontologia, arqueologia i conservació. Endins 25:43–86

Gràcia F, Clamor B, Jaume D, Fornós JJ, Uriz MJ, Martín D, Gil J, Gracia P, Febrer M, Pons G (2005) La Cova des Coll (Felanitx, Mallorca): espeleogènesi, geomorfología, hidrologia, sedimentologia, fauna i conservació. Endins 27:141–186

Gràcia F, Clamor B, Fornós JJ, Jaume D, Febrer M (2006) El sistema Pirata – Pont – Piqueta (Manacor, Mallorca): geomorfologia, espeleogènesi, hidrologia, sedimentologia i fauna. Endins 29:25–64

Gràcia F, Fornós JJ, Clamor B, Febrer M, Gamundí P (2007) La Cova de sa Gleda I. Sector Clàssic, Sector de Ponent i Sector Cinc-Cents (Manacor, Mallorca): geomorfologia, espeleogènesi, sedimentologia i hidrologia. Endins 31:43–96

Gràcia F, Fornós JJ, Gamundí P, Clamor B, Pocoví J, Perelló MA (2009) Les descobertes subaquàtiques a la Cova des Pas de Vallgornera (Llucmajor, Mallorca): història i descripció dels descobriments, hidrologia, espeleotemes, sediments, paleontologia i fauna. Endins 33:35–72

Gràcia F, Clamor B, Gamundí P, Fornós JJ (2010) El sistema de cavitats Gleda – Camp des Pou (Manacor, Mallorca). Endins 34:35–68

Grimes KG (ed) (2007) Field guide to the caves of the Gambier Karst and nearby areas. CEGSA occasional paper, 10, pp 1–50

Labourdette R, Lascu I, Mylroie J, Roth M (2007) Process-like modeling of flank-margin caves: from genesis to burial evolution. J Sediment Res 77: 965–979

Maheu J (1912) Exploration et flore souterraine des cavernes de Catalogne et des Iles Baléares. Spelunca 8(67–68):1–108

Martel EA (1896) Sous Terre. Cueva del Drach, a Majorque. Ann Club Alpin Franc 23:1–32

Merino A, Mulet A, Mulet G, Croix A, Kristofersson, Gràcia F, Ginés J, Fornós JJ (2011) La Cova des Pas de Vallgornera (Llucmajor, Mallorca). La cavitat de major desenvolupament de les Illes Balears. Endins, 35/Mon Soc Hist Nat Bal 17:147–164

Mylroie JR (2007) Cave surveys, cave size, and flank margin caves. Compass Tape 17(4, 60):1–9

Mylroie JR, Mylroie JE (2007) Development of the Carbonate Island Karst Model. J Cave Karst Stud 69(1):59–75

Palmer AN, Palmer MV, Queen JM (1977) Geology and origin of the caves of Bermuda. In: Ford TD (ed) Proceedings of the 7th international congress speleology, British Cave Research Association, Sheffield, pp 336–339

Vacher HL, Mylroie JE (2002) Eogenetic karst from the perspective of an equivalent porous medium. Carbonates Evaporites 17(2):182–196

Rodrigues – An Indian Ocean Island Calcarenite: Its History, Study and Management

12

Gregory J. Middleton and David A. Burney

Abstract

The remote Indian Ocean island of Rodrigues is primarily of volcanic origin, dating back at least 2.5 (probably 10) million years, with significant deposits of much more recent wind-deposited calcarenite. A distinctive karst has developed in this material, including a stream cave over a kilometer in length, over 30 other surveyed caves, and a range of karstic surface features. The caves have provided excellent repositories for the remains of many endemic vertebrates prior to their extinction, which can in most cases be clearly attributed to human intervention. The same intervention seriously degraded the area ecologically. A restoration, study, tourism and management program in a privately-funded reserve on the karst is now bringing multiple benefits to the island.

12.1 The Island

Rodrigues is a small volcanic island situated in the Indian Ocean approximately 600 km east of Mauritius at about 19°42′S and 63°24′E. Mauritius, Réunion and Rodrigues form the Mascarene Islands (Fig. 12.1). Politically, Rodrigues is part of the Republic of Mauritius, but since 2002 it has enjoyed a degree of administrative autonomy. It is 108 km^2 in area–a little smaller than the uninhabited Hawaiian island of Kaho'olawe–but supports a population of around 40,000 people. It is just over 18 km long, about 6.5 km wide and rises to an elevation of 396 m, at Mt Limon. The economy of the island is based on subsistence agriculture, livestock raising, local fisheries and tourism. The direct depredations of humans and the progressive removal of much of the island's natural vegetation since a tortoise-collecting station was set up there in the mid-1700s (North-Coombes 1971) has led to the extinction of much of the endemic fauna, especially the birds and tortoises, and the severe decline of the endemic flora. Today no area of original natural habitat exists on the island; all of the original plant communities are considered extinct (Strahm 1989).

The bedrock platform on which the island rests has an area of at least 1,650 km^2. Within

G.J. Middleton (✉)
PO Box 269, Sandy Bay, TAS 7006, Australia
e-mail: gregmidd@me.com; ozspeleo@iinet.net.au

D.A. Burney
National Tropical Botanical Garden, 3530 Papalina Road, Kalaheo, HI 96741, USA
e-mail: dburney@ntbg.org

M.J. Lace and J.E. Mylroie (eds.), *Coastal Karst Landforms*, Coastal Research Library 5,
DOI 10.1007/978-94-007-5016-6_12,

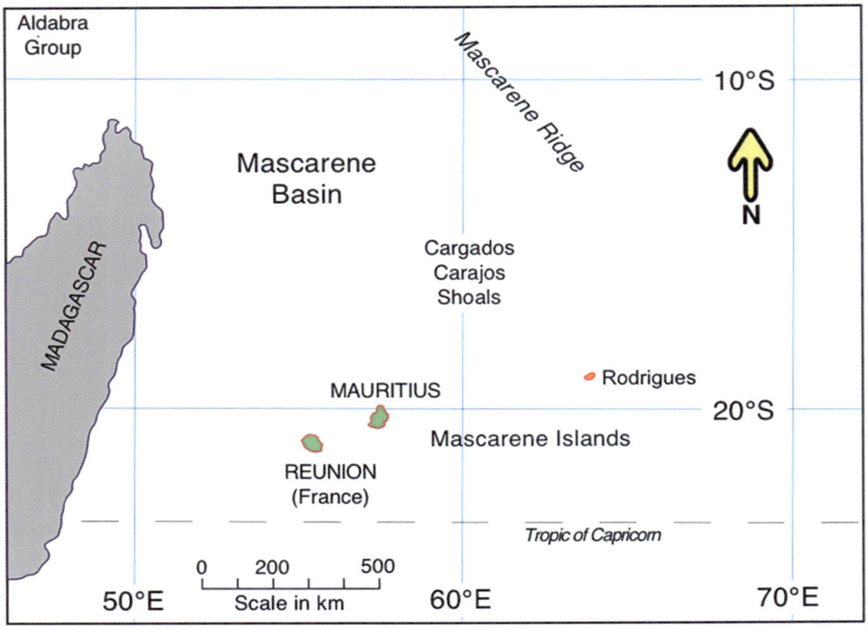

Fig. 12.1 Location of Rodrigues Island within the Mascarene Islands, South-West Indian Ocean

this, a fringing coral reef encloses a large, shallow lagoon (typically less than 1 m deep at low spring tide; Rees et al. 2005), with an area of approximately 240 km^2 (Fig. 12.2). The large platform and lagoon have played a major role in the formation of the island's calcarenites (Braithwaite 1994).

12.2 Igneous Geology

The Mascarenes are entirely of volcanic origin. There seems to be some confusion, however, as to the details of their formation. In accordance with the hot spot theory of volcanic island formation, and given that Réunion is the only one of the islands with a currently active volcano, Rodrigues should be the oldest of the islands. It seems, however that the story is not so simple.

Saddul (2002: 23) summarises his understanding of the situation as: "*... each Mascarene island corresponds to a separate center of eruption associated with an independent magma source. The relative ages of the surface volcanic rocks of the three islands from west to east are given below:*

Island	Oldest rocks	Most recent rocks
Reunion	3 M.Y.	Still active
Mauritius	> 8 M.Y.	20,000 years B. P.
Rodrigues	2.5 M.Y.	1.3 M.Y.

"

This would place Mauritius as the oldest of the islands and Rodrigues the most recent, despite the fact that volcanic activity there ceased first.

McDougall et al. (1965) suggest that Rodrigues "is the summit of a broad but elongated and steep-sided volcano which was built up from the ocean floor by successive eruptions of lava" or that "the submarine platform may be an earlier volcano, beveled by erosion, on which the small volcano of Rodrigues afterwards was built." They provided potassium-argon age determinations on two samples of lava from coastal locations which ranged from 1.30 to 1.58 million years (hence the final figure in the above table). However, the methodology used to date the samples is now considered unreliable, and dated cores obtained by Giorgi and Borchiellini (1998) conclude that the basal basalts have an age of around 10 M.Y., which makes Rodrigues at least as old as Mauritius – and

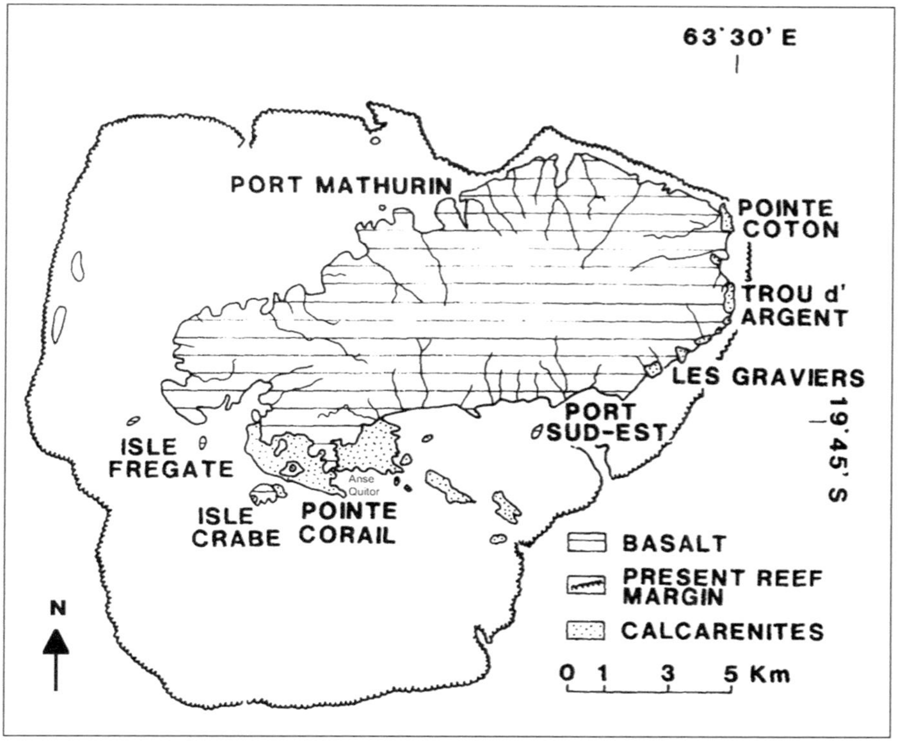

Fig. 12.2 Rodrigues, showing the extent of the present reef and the location of principal calcarenite deposits (After Braithwaite 1994)

this accords better with biological observations (see Cheke and Hume 2008: 20).

Upton et al. (1967) reported in detail on the petrology of the Rodrigues basalts; they observed:

> The shape of the Rodriguez volcano, with the present island standing up from a much more extensive, rather flat-topped rise, suggests that the exposed lavas are the products of a rejuvenation phase long after the formation of the main volcanic edifice.

Rees et al. (2005) cite the absence of elevated Pleistocene reef deposits on Rodrigues as indicating that the island has not been uplifted (presumably since the Pleistocene). Indeed, Upton et al. (1967) see the moderately embayed coastline as suggesting an overall subsidence in the recent history of the island.

12.3 Carbonate Geology

McDougall et al. (1965) observed that limestone forms a discontinuous marginal strip around the eastern and southern coast of the island (Fig. 12.2), lying unconformably on weathered basalt. While they noted that the limestone had previously been described as "coralline" (e.g. by Balfour 1879 who nevertheless provided the first detailed account of the geology of Rodrigues), it is in fact a strongly cross-bedded calcarenite, lacking coral components (Fig. 12.3). They recorded this material to a height above sea level of 62 m, noting eroded sections up to 20 m thick. McDougall et al. described the calcarenite as a fossiliferous packed intrasparite consisting of well-rounded fragments of micrite and various

Fig. 12.3 Eroded section through calcarenite, Canyon Tiyel, showing obvious cross-bedding. The rock, formed by the consolidation of wind-blown calcareous particles, probably dates back to a previous interglacial period

foraminifera, algae and occasional gastropod and echinoderm fragments. This micro-fauna suggested a post-Pliocene, probably Pleistocene, age. They measured the particle size of the clastic framework as ranging from 0.05 to 1 mm. Noting the absence of coral fragments in the calcarenite, they suggested it was formed prior to the development of the fringing coral reef. Upton et al. (1967) found support for the Pleistocene age in the fact that the calcarenites (which they described as "false-bedded") lie unconformably on Plio-Pleistocene volcanics and have undergone considerable erosion since their formation.

While McDougall et al. (1965) considered the calcarenite may have been deposited on shallow banks subject to strong tidal currents, and afterwards uplifted, they alternatively suggested it may be an aeolian deposit, the clastic grains of which may either have been windswept from the top of the exposed submarine platform during a period of low sea-level, or were driven onto the coast by wave action, where wind became the main agency of transport. They also noted, significantly, that the calcarenite is found only on the present windward flanks of the island.

Braithwaite (1994) found that of all the widespread Pleistocene aeolian deposits of the western Indian Ocean, those of Rodrigues are unusual in that they consist predominantly of oolites with a relatively small bioclastic component. He noted:

> The oolites have a mean diameter of about 300 μm (medium sand), many nuclei are dark micritic peloids while others are bioclasts. ... The marked contrast in grain type between oolitic and bioclastic rocks are paralleled by differences in mineralogy and diagenetic history. The bioclastic limestones include both aragonite and calcite (identified by X-ray diffraction) but have only a sparse fine-grained blocky calcite cement. This is restricted to patches where it forms meniscus bridges between grains and occasional pendant drops, suggesting deposition in the vadose zone, although it lacks the fibrous or prismatic textures typical of such environments. By contrast, in oolites, the ooliths themselves consist of aragonite and they are commonly embedded in a coarse blocky calcite cement which may completely fill pores and locally extends inwards from grain surfaces as a neomorphic replacement of the original tangential structure.

Braithwaite further noted that an aeolian (rather than a submarine) origin for the Rodrigues calcarenites is confirmed by (a) the distinctive high-angle lamination which characterises the deposits, and (b) their occurrence over a wide range of elevations.

Braithwaite (1994) did not directly date the deposits but he asserted that they are not recent, and:

> The active phase of accretion was followed by a passive phase of colonization by trees, and the whole assemblage has been dissected by a mature karst system which, judging from the included fauna, is at least late Pleistocene (see Slater 1879).

Saddul (2002: 345), gave the timing of the significant lowering of sea level (130–240 m) which allowed aeolian forces to accumulate the inland calcarenite dunes as between 80,000 and 40,000 BP. Extensive studies in island calcarenite deposits elsewhere, however, including detailed dating and stratigraphic analyses, have led to the general abandonment of the notion that aeolian calcarenites formed during low sea stands. On the contrary, evidence from Bermuda and the Bahamas (Hearty and Kindler 1995), indicates that these deposits are associated with rapidly rising sea levels during interglacials, as reef and lagoon carbonates are reworked by littoral processes and placed on sand beaches where prevailing winds can move them into dunal deposits well above then-current sea level. On the island of Kauaì in the Hawaiian islands, Blay and Longman (2001) and Hearty et al. (2000) identified aeolian calcarenites they believed to be from three or more previous interglacials, and radiometric dating of a capping basalt flow showed that the most extensive of these was more than 350,000 years old, probably associated with a particularly high stand at Marine Isotope Stage 11. The last period of significant accumulation of calcarenites on stable platforms in the Bahamas occurred during the last interglacial before the Holocene, MIS 5, in which a series of reef deposits and calcarenite dune deposits were laid down ca. 120,000 years ago (Neumann and Hearty 1996).

Although much work on calcarenites still needs to be done in the Mascarenes, there is no reason to believe that eustatically-driven phenomena such as rapid sea-level rise during an interglacial would differ in the Indian Ocean. Indeed, at Androhomana Cave on the southeastern coast of Madagascar (Burney et al. 2008), aeolian calcarenites in which this large collapse feature formed show three distinct units separated by red paleosols, suggesting a complex history of accumulation similar to those described for islands of the Atlantic and Pacific. Thus whereas the age of the thick calcarenites of Rodrigues is not known, they are likely to be from one or more previous interglacials, perhaps ranging from 120,000 to 350,000 years ago. The late Pleistocene age of these deposits is also consistent with the extensive formation of caves and other karst features, some containing very large speleothems and thick clastic fills.

In terms of the 'Carbonate Island Karst Model' propounded by Jenson et al. (2006, as cited by Mylroie and Carew 2010; see also Chap. 4), Rodrigues fits the "Composite Island" category, with the eogenetic karst eroded, primarily, by allogenically recharged streams at, or slightly above, the carbonate/non-carbonate (basalt) rock contact.

12.4 The Karst

While karst was not their main concern, McDougall et al. (1965) noted that the limestone outcrops showed signs of rapid erosion (the gentle seaward slopes being fretted by solution) and that there was "notable development of swallow-holes and caverns." They also noted that while in one sector the limestone plain shelves gently into the lagoon and is overlain by recent 'beach-rock', elsewhere it terminates in low cliffs undercut at wave-level.

Saddul (2002: 340) described some of the karst features:

> On the surface, there is a multitude of pot holes, shallow depressions and water courses which remain dry most of the year. In some of them arable

Fig. 12.4 The calcarenite plain is dotted with karst pinnacles frequently standing over 2 m above the surface. The pinnacles, in turn, demonstrate surface karst features such as rillenkarren and solution tubes

> farming is practised where there are thin deposits of clay. River Anse Quitor flows in such a shallow depression. . . . Caverne Patate and Plaine Caverne are local names which remind us of the existence of subterranean galleries in the aeolianite plain of Plaine Corail. These are of great ecological value and touristic significance. In fact, there are more than eleven caverns varying in size. Underground streams do not seem to be a feature of these galleries except perhaps during storm periods, although an atmosphere of dampness prevails due to the seepage of water through the joints and crevasses of the cavern roof. . . . Inside Caverne Patate, "totem" pillars, stalactites and stalagmites have been formed by slow seepage of lime-charged water from the joints.

Saddul made no mention of the numerous karst features evident on the exposed surface of the aeolianite, such as solution flutes (rillenkarren), calcarenite pinnacles (spitzkarren, Fig. 12.4), solution pans (kamenitza) and the large solution/collapse features (Fig. 12.5) known locally as *canyons* or *cavernes* (though the latter normally refers to roofed caves). This suite of features were probably first recognized as a *subdued epikarst* by Halliday and Middleton (1996). Numerous surface examples of *rhizomorphs* or *vegemorphs* indicate that, at least during their formative stages, the calcarenite plains were heavily vegetated. In more recent times (post-settlement at least) the plains have been more akin to *bare karst,* carrying only extremely skeletal soils and sparse, mainly introduced hardy vegetation, primarily *Lantana*

Fig. 12.5 In the collapsed cave known as Canyon Tiyel; it is about 50 m wide, some 250 m long and up to 20 m deep. The silt floor is 25 m asl. This enclosure made an ideal initial holding pen for the giant tortoises 're'-introduced in 2007 and it remains the easiest place for visitors to meet them

camara. While the rock is poorly consolidated (it used to be cut into blocks with hand saws for use in building construction and is still used on a small scale to make tourist souvenirs), its surface is 'case-hardened' by repeated solution and redeposition of the carbonates.

The extensive calcarenite exposure between Anse Grande Var and Baie Topaze can be conveniently divided (by Rivière Anse Quitor) into Plaine Caverne (to the east) and Plain Corail (to the west – though it should be noted that the latter is a misnomer, since the calcarenite is not composed of coralline material). Both are underlain by numerous solution caves, the most extensive of which is Caverne Patate, at over 1 km. Grande Caverne reaches almost 0.5 km. Both are open for public viewing and display numerous large examples of secondary calcite deposits in the form of stalactites, stalagmites, columns, shawls (draperies), flowstone, gours (rimstones), and occasionally, helictites, cave pearls and dogtooth spar or aragonite.

Probably the first systematic surveying and documenting of caves in Mauritius was carried out by the Spéléo-Club Nivernibou, from Decize in France, in April 1991 (Billon et al. 1991). While making scant reference to Rodrigues in their report, they did claim to have explored and surveyed more than a kilometer of cave on the island. Subsequently two members of the team published the first maps of "Tamarin" (Grande Caverne) and "Cave aux Crabes" (Caverne de la Vierge) (Billon and Chojnacki 1993). Middleton visited the island in February and May 1995 (on the latter occasion accompanied by US speleologist Dr William Halliday), carrying out the first survey of Caverne Patate (Fig. 12.6), and surveys of the upstream portion of Caverne Safran and the larger part of Grande Caverne (more commonly known at that time as "Caverne Tamarin") (Fig. 12.7) (Middleton 1996).

The third documented visit by speleologists to Rodrigues was conducted by the Société d'Etudes Scientifiques des Cavernes de la Réunion, which had a team on the island for a week in December 1996 and produced a detailed report, including maps of six of the caves (Rivière et al. 1997). This report also covered the karst areas and their conservation, the island's history, and their finds of tortoise and solitaire bones and cave fauna.

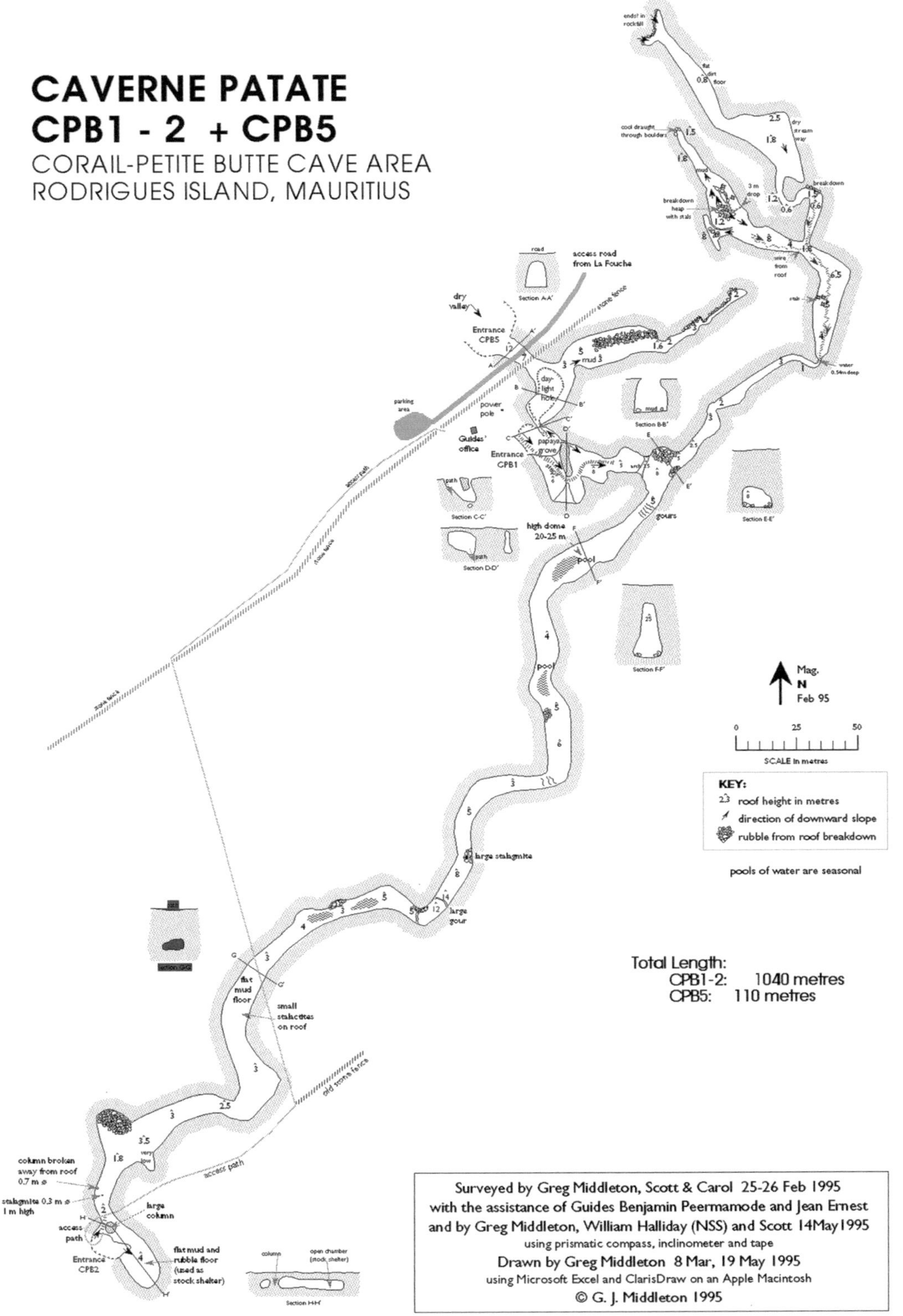

Fig. 12.6 Plan and some cross-sections of the longest cave on Rodrigues, Caverne Patate (1,040 m)

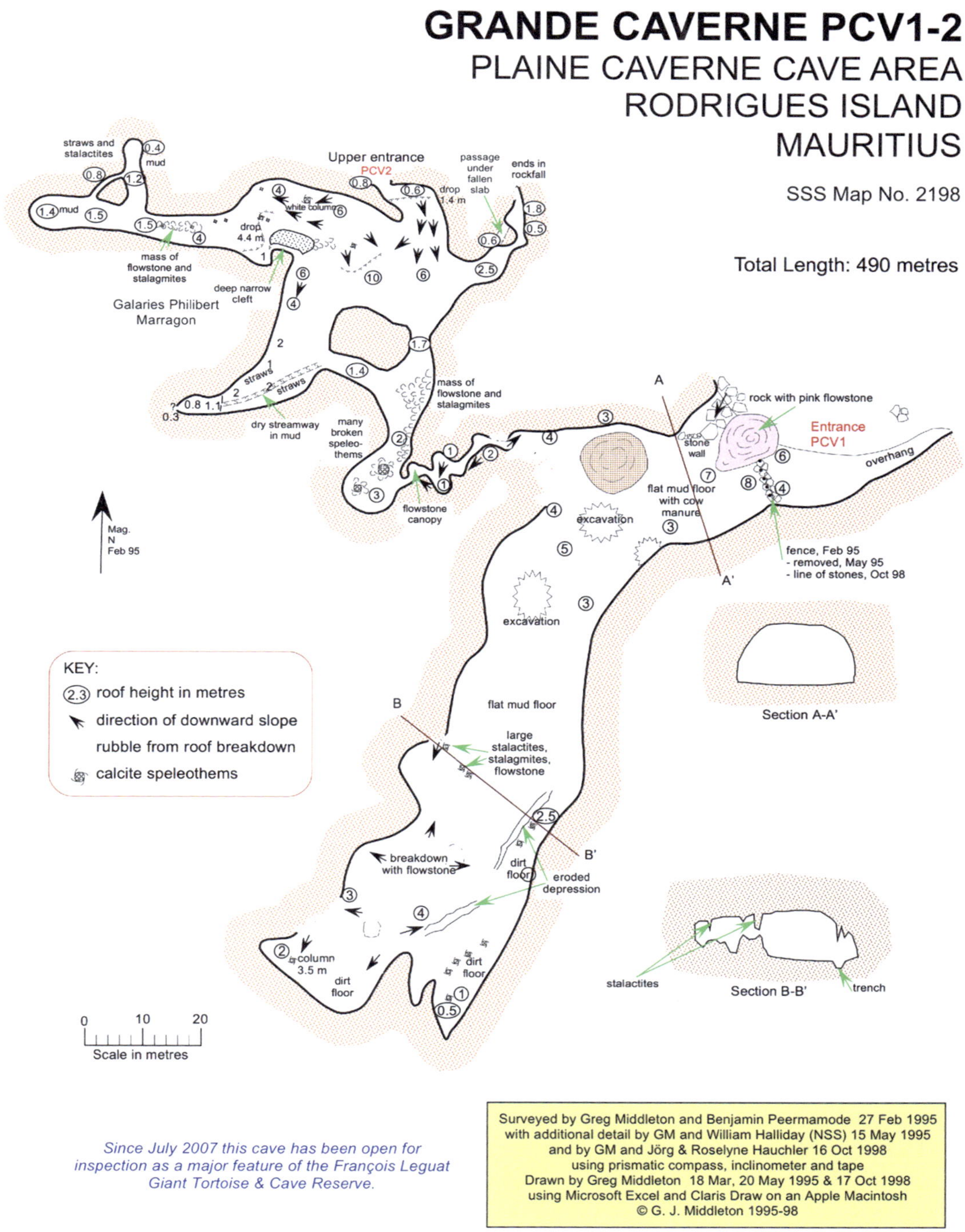

Fig. 12.7 Survey of Grande Caverne. At nearly 500 m in length it is the second longest cave on Rodrigues and has been a major source of fossil bones of giant tortoises and Solitaires

In 1998, Middleton was commissioned to carry out a survey of the caves of Mauritius, including Rodrigues, and produced a report making recommendations for their future conservation and management (Middleton 1998). Major recommendations included a karst national park for Rodrigues and the rehabilitation of Caverne Patate.

Of the known caves, Caverne Patate (1,040 m), Caverne Safran (425 m) and Caverne de La Vierge (280 m) are undoubtedly stream caves; Grande Caverne (490) and the Bambara series (40, 70, 80, 106 m) are at least in part stream caves; likely flank margin caves include Caverne Papaye (145 m) and Electricity Pole Cave. Other caves such as Caverne Bouteille and Caverne Monseigneur appear to be entirely phreatic in origin, having no obvious stream courses. The number of surveyed caves has now reached 30.

12.5 Environmental History

The (European) discovery of Rodrigues is credited to Diego Rodriguez in 1528, but no landing was effected (North-Coombes 1971: 19). The first recorded human visitors to Rodrigues were from a Dutch fleet under Admiral Wolfert Harmenszoon in September 1601. They collected birds (undoubtedly Solitaire *Pezophaps solitaria*, the closest relative of the Dodo *Raphus cucullatus* of Mauritius) and other "refreshments" but had difficulty finding water (North-Coombes 1971: 23).

The first human settlement on the island, in May 1691, was by a group of eight Huguenots under François Leguat, fleeing the Catholic persecution of Protestants in France. When further colonists failed to arrive within 2 years they abandoned the island, but during his time there Leguat had written a detailed account (Leguat 1708 but see Cheke and Hume 2008). There is much invaluable information in his *Voyage et Avantures*, although no mention of caves. Leguat provided the first description of a Solitaire. The bird was probably extinct by about 1761 and although its fossil remains are numerous in Rodrigues caves and received much scientific attention, the Solitaire never received the same scientific repute as the iconic Dodo of Mauritius. All known Solitaire bones have been recovered from caves in the calcarenite except for a single femur found recently in sand dunes at Mourouk (JP Hume, personal communication 2012).

The settlement of Rodrigues was a rather haphazard affair; a party was stranded there in 1725–26 and then a small post was established in 1735 to capture and ready tortoises for removal to Mauritius. Many passing vessels stopped to take tortoises as they were greatly prized as fresh meat on sailing ships. It has been calculated that between 1735 and 1771 a total of 280,000 animals were taken (Cheke and Hume 2008: 114).

More colonists settled on the island in the late 1790s. With the cessation of close grazing by tortoises, brush accumulated that lead to destructive fires that removed much of the remaining forest. Exotic animals and plants were steadily introduced and the native biota declined. As Cheke and Hume (2008:110) described the situation:

> During the 18th century Rodrigues regressed from desert-island paradise to ecological ruin. The administrators in [Réunion and Mauritius] bled it dry of its only 'useful' resource (tortoises and turtles) then abandoned it, burnt-over and desolate, to a handful of farming (and feuding) entrepreneurs.

The British used Rodrigues as a base where they assembled their troops for the takeover of Mauritius in 1809 but then left Rodrigues with its few settlers and their slaves much to their own devices until a surveyor, C.T. Hoart, was sent to map the island in 1825. He made only sparse comments about the vegetation but noted "there is no timber on the west and only a cover of squine grass, but on the limestone outcrop trees of *bois puant* and *benjoin*, though small, are hard and very plentiful" (quoted in Cheke and Hume 2008:148). Hoart also reported that the human population was still only 123 in 1826, consisting of three landowning white families and 100 slaves. In 1845 another surveyor, Thomas Corby, reported that Plaine Corail was already reduced to "a few dwarf trees in the interstices of the rocks and some scanty herbage".

Rodrigues was home to a number of endemic plants and animals, most notably the Solitaire and two giant tortoises; the smaller domed

Fig. 12.8 A bronze model of the extinct Rodrigues saddle-backed tortoise or Carosse, *Cylindraspis vosmaeri*, displayed at La Vanille Wildlife Park, Rivière des Anguilles, Mauritius. This was modeled on the only extant specimen in the Muséum National d'Histoire Naturelle, Paris

Cylindraspis peltastes and the larger 'Carosse' or saddleback *C. vosmaeri*, both of which are ecologically poorly known. There are only a few detailed illustrations of Rodrigues tortoises drawn from life. These are three views in pencil of a domed tortoise by the artist Paul Philippe Saugin de Jossigny, and a coloured rendition of the saddleback by Claude Aubriet. There is also a unique stuffed male specimen of the saddleback tortoise in Paris (Fig. 12.8). Yet Leguat had reported (in 1691): "sometimes you see two or three thousand of them in a flock: so that you may go above a hundred paces on their backs … without setting foot to ground" (North-Coombes 1971: 40).

According to Hume and Walters (2012), the Solitaire was probably extinct by about 1761 when Pingré, who was observing the Transit of Venus, failed to find any, although it was rumoured to still survive in remote places. The first calcite-encrusted bones were collected in 1786 and by 1830 had found their way to the great French natural historian, George Cuvier (North-Coombes 1971: 261). Subsequently this led to the arousing of interest in the 'bone caves' of Rodrigues and concerted efforts began to obtain further specimens.

The second Transit of Venus expedition (1874) brought a group of British scientists to the island where they made many observations beyond astronomical recordings. Balfour (1879), as mentioned, noted the limestone plain but erroneously assumed it was composed of coral. Of the caves he said:

> … some extend for a great distance through the rock, and are rich in stalactites and stalagmites; others again are mere small holes. The whole plain is riddled with these caves, and on walking over it one constantly passes small apertures and fissures, evidently "blow-holes" of some subterranean cavern. Wide and deep hollows are also met with, on the floor of which large fragments of limestone lie in confused heaps. These are apparently old caves of which the roofs have fallen in, and the continuation of the cavern may be found at either extremity. The floor of these hollows is composed of volcanic soil …

Balfour's colleague, naturalist Henry Slater, was especially tasked with searching for caves and within them for bones, particularly of Solitaire and tortoise. Slater (1879) made many observations on the caves and their contents, stating:

> The caves have been, if not originated, at least much enlarged by water, of which many bear abundant traces, and in the rainy season some are evidently the courses of subterranean streams.
>
> There exists near my camp at the caves a sort of ravine, terminated at each end by a cavern, and having others opening into it. The terminal caves and precipitous sides at once determined me that this has been a vast cavern, the roof of the greater part of which has fallen in; this fall has left a sort of ravine with a level bottom surrounded on all sides by precipices, nearly, if not quite, perpendicular, and having a height of from 30 to 90 feet; the bottom is now covered with earth and full of large trees, the tops of which rise to the level of the cliffs. Descent can only be effected with ease in two places, where two heaps of limestone blocks rear themselves against the precipices. There is no

reason to believe that water ever accumulates in the caverns opening into this 'gorge', and in these caves most of my specimens of any value were found ...

The depth of the bone-earth is very variable; in some caves we find it with a depth of from 6 inches to 3 feet; in others, however, it varies from 4 to 9 feet in depth. Below about 2 feet I never found many bones, which makes me believe that the agency which deposited the bones in the caverns, never operated until the latter days of the existence of the Solitaire. The bones might certainly have decayed, but yet I usually found that the bones which were well covered with earth were in much better preservation than those near or upon the surface, which were usually much decayed. This makes me think that the Solitaire resorted to the caverns in case of fire in the island, which has been known to have denuded it several times of its trees; more so, as in several cases I found nearly perfect skeletons, which lay evidently as they died; this precludes the idea that they were carried there by wild cats.

There were further occasional scientific raids on the caves for bones (with rapidly diminishing levels of success) and numerous tourist parties destroyed and removed large quantities of speleothems over the following century.

12.6 Restoration and Management

While the caves were a source of scientific interest and an attraction to souvenir-hunters, with the growth of tourism, their value as tourist attractions came to be realized, resulting in the development of a level of local management. This was not the case in 1970 when North-Coombs was writing (as he said that vandalism was continuing) but it was so by 1991 when the Australian Government donated electric torches to replace the destructive flaming kerosene-fuelled open torches that were still in use (Fig. 12.9) (Middleton 1996). At that time the local administration had appointed guides and was charging a fee to conduct tourists through Caverne Patate. Fortunately this cave required very little modification to make it accessible. Indeed, in 1995, the only improvement was a set of badly worn calcarenite steps leading up to the exit. By 2008 those steps had been entirely rebuilt and the guides had been given some rudimentary training in cave interpretation.

Fig. 12.9 As late as 1991, kerosene-fuelled flaming torches were in routine use by the guides at Caverne Patate. Soot staining from these has done incalculable long-term damage to the cave and its calcite speleothems. *(Photo courtesy Benjamin Peermamode.)*

Fig. 12.10 Aldabra Giant Tortoises, *Dipsochelys dussumieri*, grazing in front of the entrance to Grande Caverne. These tortoises are the closest surviving analogues of the Rodriguan saddle-backed tortoise. They are useful in controlling exotic vegetation as they graze this in preference to juvenile native plants

Meanwhile, Owen Griffiths, an Australian zoologist-entrepreneur who operates a wildlife park on Mauritius, took an interest in the plight of native Rodriguan biota. In 2001, he developed a concept for a giant tortoise and cave reserve to be situated on Plaine Caverne adjacent to an existing nature reserve at Anse Quitor (Griffiths 2001; Middleton 2001). The concept is essentially a "rewilding" project (*sensu* Donlan et al. 2005) involving the restoration of 19 ha of native vegetation on a section of calcarenite plain which included some of the 'canyons' or steep-sides gorges first recorded by Slater in 1874. Griffiths realized that these collapsed caves would make ideal enclosures to hold surrogates for the giant tortoises which once roamed the island. While primarily a conservation project, it includes significant educational, research, tourism and local involvement aspects. A long-term lease for the reserve was granted in 2005 and the labour-intensive task of restoring the native biota commenced. The *Lantana* which was the major vegetation on the karst has been largely removed and replaced with over 175,000 native plants that are now (April 2012) firmly established and self-seeding. Domestic cattle and goats have necessarily been excluded. The closest available analogues of the extinct tortoises, the Aldabra Giant Tortoise, *Dipsochelys dussumieri* (Fig. 12.10), and the Madagascar Radiated Tortoise, *Asterochelys radiata,* have been introduced (from captive populations on Mauritius), initially just to the floors of the canyons and more recently, in the case of *Dipsochelys,* to the wider reserve. Both species are breeding well and are helping to control many exotic plants (*Dipsochelys* does not browse juveniles of most native plants, preferring only older foliage or exotics – Jones 2008: 256.) The reserve, including Grande Caverne, which has been fitted with elevated pathways and lighting, was opened to the public on 14 July 2007 (Middleton 2007).

Researchers and post-graduate students are encouraged to undertake studies in the reserve. Of particular note in recent times are the investigations of:

H.B. Vanhof, Netherlands – analysis of the oxygen isotope record from a stalagmite from Caverne de la Vierge which indicated it grew between 4125 and 2260 calendar years BP with an alternation of humid periods lasting up to

Fig. 12.11 This hand auger was sunk to a depth of 10 m in the sediment on the floor of Canyon Tiyel. Radiocarbon dates to over 11,700 years BP were obtained, though there has evidently been some reworking of the basal sediments

~200 years with dry periods lasting decades. Apart from its local implications, this data has helped explain the mass mortalities of Dodos *(Raphus cucullatus)* and giant tortoises *(Cylindraspis* spp.*)* which occurred between 4,235 and 4,100 years BP at Mare aux Songes, Mauritius (Rijsdijk et al. 2011)

J.P. Hume, U.K. – excavation and studies of (mainly) extinct vertebrates, particularly birds, from the calcarenite caves (e.g. Rodrigues parakeet, *Psittacula exsul,* and Rodrigues parrot, *Necropsittacus rodericanus* – Hume 2007; Rodrigues turtle dove, *Nesoenas rodericana,* and Rodrigues blue pigeon, *Alectroenas payandeei* – Hume 2011). Ongoing work involves passerines (song birds) and seabirds.

D. Burney, Hawaiì, USA – palaeoecology studies to better understand the interrelationships between isolated endemic populations and human invaders and their companion species. Investigations in April–May 2011 revealed that the major gorge, Canyon Tiyel, contains sediment at least 10 m deep (Fig. 12.11). At a depth of 7.4 m the material yielded a calibrated age in excess of 11,000 years BP. These results indicate that if this feature is a former cave passage, the roof must have collapsed prior to the Holocene, further supporting the idea that the calcarenite and major cave passages formed prior to the present interglacial. Bones recovered from adjacent caves mostly contained insufficient collagen for dating, but one near the surface yielded an age of 2,850 ± 30 years BP, suggesting that some bone deposits may be much older. Further studies are contemplated (Burney 2011). As there has not been a general physical and biological survey of Rodrigues since 1874, the Reserve management hopes to organize such a survey in the coming decade.

12.7 Conclusions

Rodrigues is an extremely remote and unusual oceanic island, yet contains some elements such as its calcarenite deposits, its fauna and its reefs which enable comparisons to be made with related environments and biotas in other parts of the world. The island has considerable potential to provide unique insights into the processes of human-caused extinction and environmental change. In the François Leguat Reserve opportunities exist for continuing studies, for the development and testing of techniques for rewilding and conservation management, and for significantly improving the public understanding of the values of such areas.

Acknowledgements Assistance with the preparation of this text from Julian Hume, Lorna Steel, Aurele André and Owen Griffiths is gratefully acknowledged.

References

Balfour IB (1879) The physical features of Rodriguez. In: Hooker JD, Gunther A (eds) An account of the petrological, botanical and zoological collections made in Kerguelen's land and Rodriguez during the transit of Venus expeditions, carried out by order of Her Majesty's government in the years 1874–75, vol 168, Philosophical transactions of the Royal Society of London. G.E. Eyre and W. Spottiswoode, London, pp 289–290

Billon F, Billon C, Chojnacki P, Rousseau G (1991) Explorations souterraines à L'île Maurice. Compte rendu expédition Avril 1991. S.C. Nivernibou, Decize

Billon F, Chojnacki P (1993) Explorations souterraines à L'île Maurice. Sous le Plancher 1993(8):66–86

Blay CT, Longman MW (2001) Stratigraphy and sedimentology of Pleistocene and Holocene carbonate eolianites, Kauaì, Hawaiì, U.S.A. SEPM Spec Pub 71:93–115

Braithwaite CJR (1994) Quaternary oolites in the Indian Ocean. Atoll Res Bull 420:1–10

Burney DA (2011) Rodrigues Island: hope thrives at the François Leguat Giant Tortoise and Cave Reserve. Madagascar Conserv Dev 6(1):3–4

Burney DA, Vasey N, Godfrey LR, Ramilisonina JWL, Jungers WL, Ramarolahy M, Raharivony N (2008) New findings from Andrahomana Cave, southeastern Madagascar. J Cave Karst Stud 70(1):13–24

Cheke A, Hume J (2008) Lost land of the dodo. An ecological history of Mauritius, Réunion and Rodrigues. T & AD Poyser, London

Donlan CJ, Greene HW, Berger J, Bock CE, Bock JH, Burney DA, Estes JA, Foreman D, Martin PS, Roemer GW, Smith FA, Soulé ME (2005) Rewilding North America. Nature 436:913–914

Giorgi L, Borchiellini S (1998) Carte geologue de l'ile Rodrigues au 1: 25000. Le schema hydrogeologique. La notice explicative. Ministere Délégué à la Cooperation et de la Francophonie & Géolab, Paris, 28 pp + maps

Griffiths O (2001) 'Rodrigues giant tortoise reserve and cave project', submitted to Mauritian Ministry of Tourism, November 2001, 22 pp

Halliday WR, Middleton G (1996) A subdued karst on the island of Rodrigues, Mauritius. In: Fornós JJ, Ginés A (eds) Karren landforms. Universitat de les Illes Balears, Palma de Mallorca, pp 345–354

Hearty PJ, Kindler P (1995) Sea-level highstand chronology from stable carbonate platforms (Bermuda and the Bahamas). J Coastal Res 11(3):675–689

Hearty PJ, Kaufman DS, Olson SL, James HF (2000) Stratigraphy and whole-rock amino acid geochronology of key Holocene and last interglacial carbonate deposits in the Hawaiian Islands. Pac Sci 54:423–442

Hume JP (2007) Reappraisal of the parrots (Aves: Psittacidae) from the Mascarene Islands, with comments on their ecology, morphology, and affinities. Zootaxa 1513:1–76

Hume JP (2011) Systematics, morphology and ecology of pigeons and doves (Aves: Columbidae) of the Mascarene Islands, with three new species. Zootaxa 3124:1–62

Hume JP, Walters M (2012) Extinct birds. T & AD Poyser, London, 544 pp

Jenson JW, Keel TM, Mylroie JR, Mylroie JE, Stafford KW, Taborosi D, Wexel C (2006) Karst of the Mariana Islands: the interaction of tectonics, glacioeustasy and fresh-water/sea-water mixing in island carbonates. Geol Soc Am Spec Pap 404:129–138

Jones CG (2008) Practical conservation on Mauritius and Rodrigues. In: Cheke A, Hume J (eds) Lost land of the dodo. An ecological history of Mauritius, Réunion and Rodrigues. T & AD Poyser, London, pp 226–259

Leguat F (1708) Voyage et avantures de François Leguat, et de ses compagnons, en deux îles désertes des Indes Orientales .... de Lorme, Amsterdam (2 vols.)

McDougall I, Upton GJ, Wadsworth WJ (1965) A geological reconnaissance of Rodriguez Island, Indian Ocean. Nature 206:26–27

Middleton G (1996) Rambles under Mauritius #7 – Plaine Corail, Rodrigues. J Sydney Speleol Soc 40(6): 83–94

Middleton GJ (1998) 'The conservation and management of the caves of Mauritius (including Rodrigues).' Unpublished consultancy report to Department of Environment. 54 pp. + maps

Middleton GJ (2001) 'Rodrigues giant tortoise reserve and cave project: cave presentation concept plan', prepared for Owen Griffiths, Mauritius

Middleton G (2007) Grande Caverne – a new show cave opens in Rodrigues. Aust'asian Cave Karst Manage Assoc J 68:4–5

Mylroie JE, Carew JL (2010) Field guide to the geology and karst geomorphology of San Salvador Island [Bahamas]. Published by the Editors, USA

Neumann AC, Hearty PJ (1996) Rapid sea-level rise at the close of the last interglacial (substage 5e) recorded in Bahamian island geology. Geology 24(9):775–778

North-Coombes A (1971) The island of Rodrigues. The author, Mauritius (Reprinted 2002)

Rees SA, Opdyke BN, Wilson PA, Fifield LK (2005) Coral reef sedimentation on Rodrigues and the Western Indian Ocean and its impact on the carbon cycle. Philos Trans R Soc A 363:101–120

Rijsdijk KF, Zinke J, de Louw PGB, Hume JP, van der Plicht HJ, Hooghiemstra H, Meijer HJM, Vonhof HB, Porch N, Florens FBV, Baider C, van Geel B, Brinkkemper J, Vernimmen T, Janoo A (2011) Mid-Holocene (4200 kyr BP) mass mortalities in Mauritius (Mascarenes): insular vertebrates resilient to climatic extremes but vulnerable to human impact. Holocene 21(8):1179–1194

Rivière J, Brial P, Duquenois M (1997) Expédition Rodrigues 97. Annales de Volcanospéléologie de la Réunion, Bull 2(March)

Saddul P (2002) Mauritius: a geomorphological analysis. Mahatma Gandhi Institute, Moka (Revised edition)

Slater HS (1879) Observations on the bone caves of Rodrigues. In: Hooker JD, Gunther A (eds) An account of the petrological, botanical and zoological collections made in Kerguelen's land and Rodriguez during the transit of Venus expeditions, carried out by order of Her Majesty's government in the years 1874–75, vol 168, Philosophical transactions of the Royal Society of London. G.E. Eyre and W. Spottiswoode, London, pp 420–422

Strahm W (1989) Plant red data book for Rodrigues. Koeltz Scientific Books, Königstein

Upton BGJ, Wadsworth WJ, Newman TC (1967) The petrology of Rodriguez Island. Indian Ocean Geol Soc Am Bull 78(12):1495–1506

Caves and Karst Hydrogeology of the Mariana Islands

13

Kevin W. Stafford, Danko Taboroši, and John J. Jenson

Abstract

The Mariana Islands contain a complex and diverse assemblage of karst features associated with the eastern paleo-volcanic arc chain of the Mariana Ridge, including the islands of Guam, Rota, Aguijan, Tinian, Saipan and Farallon de Medinilla. Karst is dominated by flank margin cave development; however, fracture caves and contact caves are significant features throughout the region. Research in the Mariana Islands resulted in the development of the Complex Island component of the Carbonate Island Karst Model (CIKM) and was instrumental in the recognition of island aquifer compartmentalization as a result of differential tectonism and subsidence, as well as syndeposition of both carbonate and volcanic facies. Karst resources throughout the Marianas have been heavily utilized throughout history as water supplies and sites of habitation, refuge, defense and spirituality. Today, karst resources continue to play and important role in the lives of people throughout the Mariana Islands, both as groundwater and cultural resources.

K.W. Stafford (✉)
Department of Geology, Stephen F. Austin State University, Nacogdoches, TX 76962, USA
e-mail: staffordk@sfasu.edu

D. Taboroši
Island Research and Education Initiative, Palikir, Pohnpei, FM 96941, USA

Water and Environmental Research Institute of the Western Pacific, University of Guam, Mangilao, GU 96923, USA
e-mail: taborosi@gmail.com

J.J. Jenson
Water and Environmental Research Institute of the Western Pacific, University of Guam, Mangilao, GU 96923, USA
e-mail: jjenson@uguam.uog.edu

13.1 Introduction

The Mariana Islands are composed of 15 islands ('the Marianas') that make up the exposed portions of the Mariana Ridge (Fig. 13.1). They are located approximately 3,000 km east of the Asian landmass and approximately 160 km west of the Mariana Trench, between 13°and 21° north of the equator. The climate is a wet-dry tropical system with a distinct rainy season from July to September and a distinct dry season from February to March. Annual precipitation averages 200 cm with temperatures ranging from 20° to 32 °C (Doan et al. 1960). Dense woody vegetation dominates carbonate regions

M.J. Lace and J.E. Mylroie (eds.), *Coastal Karst Landforms*, Coastal Research Library 5,
DOI 10.1007/978-94-007-5016-6_13,

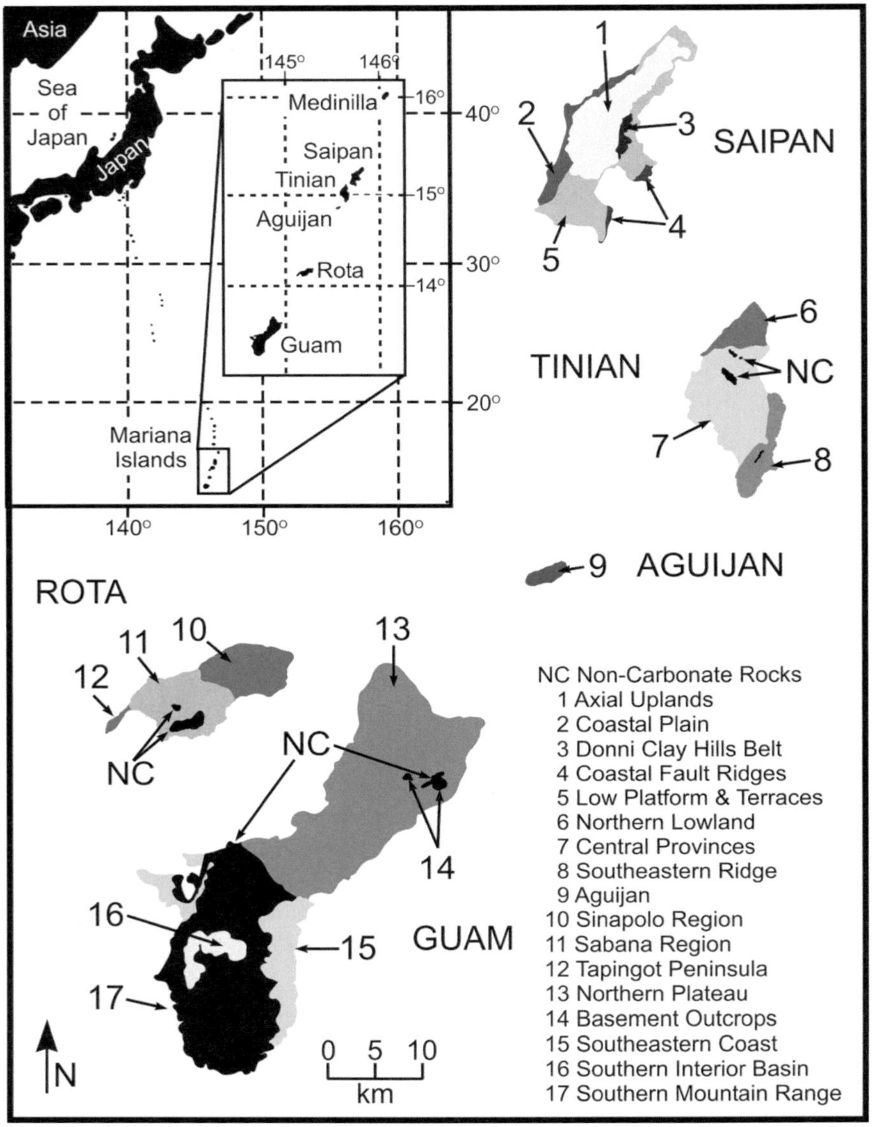

Fig. 13.1 Location of the Mariana Islands, with carbonate islands in expanded view

where soils are thin and groundwater recharge is rapid, whereas grass species dominate weathered volcanic terrains with thicker, clay-rich soils (Gingerich and Yeatts 2000). Inhabitants of the islands rely on groundwater resources extensively for water supplies, which are supplemented by spring capture and rain collection systems.

The Marianas have a long, rich history of habitation and utilization of water resources. The islands were first settled by the Chamorro people at least 4,300 years ago (Athens and Ward 2004). Western discovery started in 1521 by Spain with subsequent control by Germany, Japan and the United States throughout the late nineteenth and twentieth centuries. Today, Guam is a United States territory while the remainder of the Mariana Islands is associated with the United States as the Commonwealth of the Northern Mariana Islands (Rogers 1996). Ever since the initial settlement, the Marianas Islands have had their groundwater resources utilized by people. In prehistory, groundwater was extracted largely from shallow hand-dug vertical and Maui-type wells, which have been integral for agriculture and domestic use since that time, especially in karst areas with limited surface water resources.

In terms of hydrology and geomorphology, karst development in the Mariana Islands is highly complex. It shows the full range of island karst landforms (e.g. flank margin caves, banana holes, fissure caves, recharge caves, etc.) and evidence of associated hydrogeologic processes (e.g. allogenic and eogenic processes, freshwater-saltwater interaction, etc.). Being a larger island with much non-limestone land area, Guam also includes substantial karst development that would be more traditionally defined as continental-style. In that sense, some of its karst features are merely "karst on island" rather than island karst in true sense. Research conducted in the Marianas assisted in significantly expanding the concepts of the Carbonate Island Karst Model (CIKM) (see Chap. 4) that was developed throughout the Caribbean region, most significantly the recognition of compartmentalization of groundwater resources on carbonate islands and the development of the Complex Island karst model.

13.2 Geophysiography of the Mariana Islands

The Mariana Islands are exposed portions of two island arc complexes of the Mariana Ridge, which include an eastern paleo-volcanic chain and a western active-volcanic chain (Karig 1971). The islands of Guam, Rota, Aguijan, Tinian, Saipan and Farallon de Medinilla (Fig. 13.1) reflect the subaerial exposures of the eastern paleo-volcanic chain, while the remaining nine islands of the Mariana Island complex are associated with the western, active volcanic chain. The southern six islands of the Marianas exhibit similar geologic histories and general configurations consisting of Paleogene volcanic cores that are mantled by Neogene and Pleistocene carbonates and fringed by Holocene reef and beach deposits (Cloud et al. 1956; Doan et al. 1960; Tracey et al. 1964). The northern islands of the Mariana Ridge are generally younger and exhibit active volcanism to varying degrees with limited carbonate deposition. Volcanism that formed the base of southern Mariana Islands (i.e. Guam, Rota, Aguijan, Tinian, Saipan and Farallon de Medinilla) was primarily active during the late Eocene and early Oligocene with volcanism continuing on Guam and Saipan into the mid-Miocene (Dickinson 2000).

All of the islands associated with the eastern paleo-volcanic island arc exhibit distinct terrace levels associated with glacio-eustatic sea level change, which are compounded by differential rates of subsidence and uplift between and within individual islands, making it difficult at best to attempt any island-wide study of previous sea level fluctuations within the Mariana Islands (Keel 2005; Stafford et al. 2004a; Taboroši et al. 2004a). In the southern Mariana Islands, carbonates largely exhibit classic eogenetic characteristics, having never undergone deep-burial diagenesis. Littoral zones throughout the Mariana Islands exhibit biomechanical erosion, including erosion notches and littoral karren, associated with wave erosion, tidal fluctuations, salt spray and biological activity of intertidal organisms. Much of the general geologic studies conducted in the Marianas were associated with United States military operations during or shortly after World War II (e.g. Cloud et al. 1956; Doan et al. 1960; Tracey et al. 1964); however, most of the southern Mariana Islands have been studied significantly with respect to karst resources in the past two decades (e.g. Jenson et al. 2006; Mylroie et al. 2001; Stafford et al. 2004a, b; Taboroši et al. 2003).

Guam, as the largest of the Mariana Islands both in physical size and population, has been the most studied. The basis of all geologic studies on the island is the comprehensive work of Tracey et al. (1964), which has been supplemented by numerous investigations by the University of Guam and the Water and Environmental Research Institute of the Western Pacific. Guam covers 550 km^2, with a north-south length of 48 km and a width varying from 6 to 19 km (Taboroši et al. 2004a). Guam is effectively divided into two dominant geologic and hydrologic regions that are roughly equal in area as result of the Pago-Adelup Fault that bisects the island from west to east. Northern Guam is largely a low relief, carbonate plateau comprised primarily of the Mariana Limestone and bounded by steep coastal scarps. Southern Guam is a deeply dissected volcanic terrain with residual carbonates in upland and coastal areas (Tracey et al. 1964). Guam's highest elevation, 406 m, is in the highly eroded volcanics of southern Guam, where remaining carbonates in the interior exhibit karst development similar to that found in continental settings with strata that is more diagenetically mature than observed elsewhere in the Marianas (Taboroši et al. 2004a).

Rota is located approximately 60 km northeast of Guam, covers an area of 96 km^2, has an east-west maximum length of 20 km and ranges from 0.5 to 7 km wide (Keel 2005). No comprehensive geologic map or study of the island has been completed to date, with most research referencing Sugawara (1934), whose work focused on geomorphology and sedimentary facies. The island consists of six primary terrace levels that range from sea level to a maximum elevation of 491 m. The far eastern portion of Rota consists of a narrow, low elevation peninsula that connects an isolated, southwestern, topographic high to the majority of the island. The uplands region is dominated by highly eroded volcanics that have created thick residual soils that are also seen on steep scarps of south central Rota (Stafford et al. 2002). Carbonate units on Rota are similar to those observed on Guam with a dominance of Pleistocene-age Mariana Limestone strata. Neogene carbonates are observed adjacent to volcanics and have been documented as interfingering with carbonates in the extreme southeastern region of Rota (Keel 2005).

Aguijan lays approximately 75 km northeast of Rota, covers an area of 7.2 km^2, and has a maximum east-west length of 4.9 km and an average width of 1.6 km (Stafford et al. 2004b). Today Aguijan has no human occupation and primarily functions as a wildlife preserve governed by the municipality of Tinian with limited access. No detailed geologic studies have been conducted on Aguijan since the limited investigations by Tayama (1936). The island is characterized by three dominant terrace levels, which reach a maximum elevation of 157 m with no known volcanic rocks cropping out (Stafford et al. 2004a). Carbonate facies represent Plio-Pleistocene strata similar to those described as the Mariana Limestone elsewhere, with little Holocene deposits fringing the island. Unlike the rest of the Mariana Islands, Aguijan contains no beaches. Its coastlines are completely composed of steep scarps and cliffs (Stafford et al. 2004b). Approximately 1 km south of Aguijan, Naftan Rock is exposed subaerially covering less than 10,000 m^2 with an elevation of less than 20 m. Naftan Rock is connected to Aguijan by a shallow-water carbonate platform less than 10 m deep.

Tinian is located 9 km north of Aguijan, covers an area of 102 km^2, has a maximum north-south length of 20 km and varies in width from less than 1 km to almost 10 km (Stafford et al. 2005). Tinian is effectively divided into three physiographic regions, the Northern Lowland, the Central Plateau and the Southeastern Ridge. It reaches a maximum elevation of 187 m in the southeastern portion. Doan et al. (1960) provide the most comprehensive geologic study of Tinian, in which they describe most carbonate strata as Plio-Pleistocene Mariana Limestone with some Miocene carbonates proximal to Eocene volcanic units that crop out over less than 0.1 km^2 near the highest portion of the Central Plateau at 166 m elevation. Minor volcanics also crop out along the western edge of the Southeastern Ridge. Differential uplift and/or subsidence has split the three physiographic regions and obscured terrace levels; however, at least four distinct terrace levels can be detected throughout the island (Stafford et al. 2005). Modern beach and reef deposits are common throughout, but the majority of the coastline is composed of steep-walled scarps.

Saipan, the second largest island of the Marianas, is located 5 km northeast of Tinian, covers an area of 124 km^2, has a maximum north-south length of 23 km and varies in width from 2.5 to 10 km. Saipan was described by Cloud et al. (1956) as being composed of syndepositional volcanic and carbonate units with complex, contemporaneous structural modification. As a result, this is the most geologically complex, both structurally and lithologically, of the Mariana Islands. The southern portion of Saipan is dominated by a terrace that reaches 474 m elevation, at Mount Tagpochau, while the majority of the island is dominated by an axial upland region flanked by limestone terraces. The northern axial upland is separated by a volcanic ridge and the northern extension of the axial ridge culminates in a limestone plateau that drops precipitously for more than 180 m to the platform below at Bandadero Cliff. The axial upland is separated from coastal zones by 10–12 terrace levels attesting to the complex history of differential uplift, subsidence and interfingering of volcanic and carbonate strata (Jenson et al. 2006). Carbonate facies appear similar to those observed on Tinian; however, complex syndeposition complicates local stratigraphy and associated hydrology and has resulted in a much poorer understanding of the geologic history of Saipan than those islands south of it.

Farallon de Medinilla is located 83 km north of Saipan. It is a narrow uninhabited island, 2.8 km long with a maximum elevation of 81 m and an area of 0.9 km^2 (Lusk et al. 2000). Used as a military target range for many years, it is not accessible because of unexploded ordinance and little is known about its geology other than it appears to be Cenozoic limestone. Air photos indicate features that could be either flank margin or sea caves in the coastal cliffs, but otherwise the karst has not been described.

13.3 Karst Development of the Mariana Islands

Karst development throughout the Mariana Islands is complex and diverse, with the traditional island karst features such as flank margin caves and littoral eogenetic karren, to more unusual forms that are limited to tectonically active islands with complex carbonate and volcanic structure and even features reminiscent of continental settings. Certain common island karst features in other regions of the world, notably banana holes and pit caves, are poorly represented and limited throughout the Marianas. Littoral eogenetic karren is common throughout the Mariana Islands and forms rugged terrains proximal to coastlines where the interaction of salt spray and associated crystal wedging combine with the effects of freshwater/saltwater mixing and the activity of biologic organisms (Taboroši et al. 2004b, see also Chap. 2). Littoral eogenetic karren may locally extend uninterrupted along kilometers of scarped coastline in bands up to a 100 m wide. Such areas are either barren or poorly vegetated by a few halophytic and xerophytic scrub species tolerant of high salt content and desiccation. Such terrains contain extremely jagged rocks and crevices with meters of vertical relief. Cavernous porosity is well developed, with most caves being flank-margin caves, fissure caves, mixing zone fracture caves, and contact caves (Fig. 13.2). Sea caves and bioerosion notches occur along coastlines. They are predominantly biomechanical erosion features rather than karst *sensu stricto*.

Flank margin caves account for at least three fourths of all known cave development and have been documented extensively just above sea level along coastlines, up to several hundred meters elevation where they are exposed as discontinuous series of isolated chambers that have been breached by scarp retreat. Most individual flank margin caves are less than 4 m in height and cover less than 100 m^2 in area, but many were likely much larger in the past and today only represent the remaining innermost portions left behind as scarp edges have eroded and retreated (Stafford et al. 2005). Occasionally flank margin caves are encountered that have been breached by ceiling collapse and appear to otherwise be mostly intact. Some of these features can cover areas that exceed 5,000 m^2 and individual rooms that are more than 10 m tall, such as Liyang Dangkolo on Tinian (Fig. 13.3) (Stafford et al. 2004a). Complexes of flank margin caves often extend laterally over hundreds of meters as meter-thick horizons, similar to cave horizons on other islands such as Isla de Mona (Puerto Rico) and Curaçao. The most extensive series documented occur at

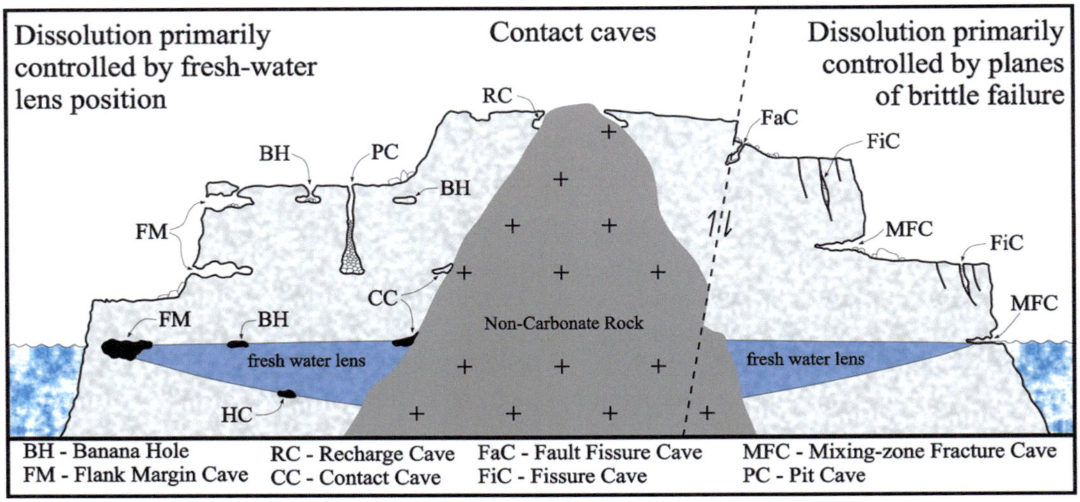

Fig. 13.2 Composite diagram showing primary cave types that occur throughout the Mariana Islands (vertically exaggerated)

Suicide Cliffs on the southeastern edge of Tinian, with almost 2 km of intermittent flank margin caves developed along an upper level terrace (Fig. 13.4) (Stafford et al. 2005) and at the Sagua Cave Complex along the southwestern coast of Rota with more than 600 m of semi-continuous flank margin cave development (Fig. 13.5) (Keel 2005). Other large semi-continuous and intermittent horizons of flank margin cave development are common throughout the Marianas, as well as numerous flank margin caves that appear isolated but likely represent exposure resulting from differential rates of erosion, scarp retreat and general denudation of geomorphic surfaces. Somewhat surprisingly, banana holes (or water-table caves) are exceedingly rare throughout the Mariana Islands (in contrast with classic eogenetic karst regions such as the Bahamas, where they are ubiquitous). The rarity of banana holes in the Marianas may be a result of the higher vertical relief, which does not promote the isolated collapse of relatively small water-table dissolution voids (Stafford et al. 2004b, 2005), or the lack of long-term sea level stillstands that would initiate voids within an easily collapsible vertical distance beneath the current land surface. Chapter 4 offers a new facies-restricted banana hole model which also explains their absence in the Mariana reef limestones.

Unlike traditional eogenetic karst regions studied in the Bahamas, the Mariana Islands have a complex tectonic history. This has resulted in the high elevation terrains with steep cliffs and fragmentation of islands into hydrogeologically isolated regions produced by differential rates of subsidence and uplift (Jenson et al. 2006). Associated with brittle deformation, high angle faults and bank-margin failures have acted as preferential flow paths for groundwater discharge. In general, these are referred to as fracture caves, but are more appropriately subdivided into three classes: fault caves, fissure caves and mixing zone fracture caves (Stafford et al. 2004a; Keel 2005). All three types of fracture caves were first described in the Mariana Islands with respect to eogenetic karst development and the CIKM.

Fault caves, as their name implies, have developed along high angle normal faults in the Mariana Islands (Stafford et al. 2004b, 2005). These are the least common of the fracture caves but show the effects of preferential dissolution related to fast flow routes created by the planar surface of fault displacement. Fault caves are developed parallel to the fault, vertically aligned along the fault plane, and show lateral widening at the intersection of the freshwater lens, suggesting that the features are formed both in

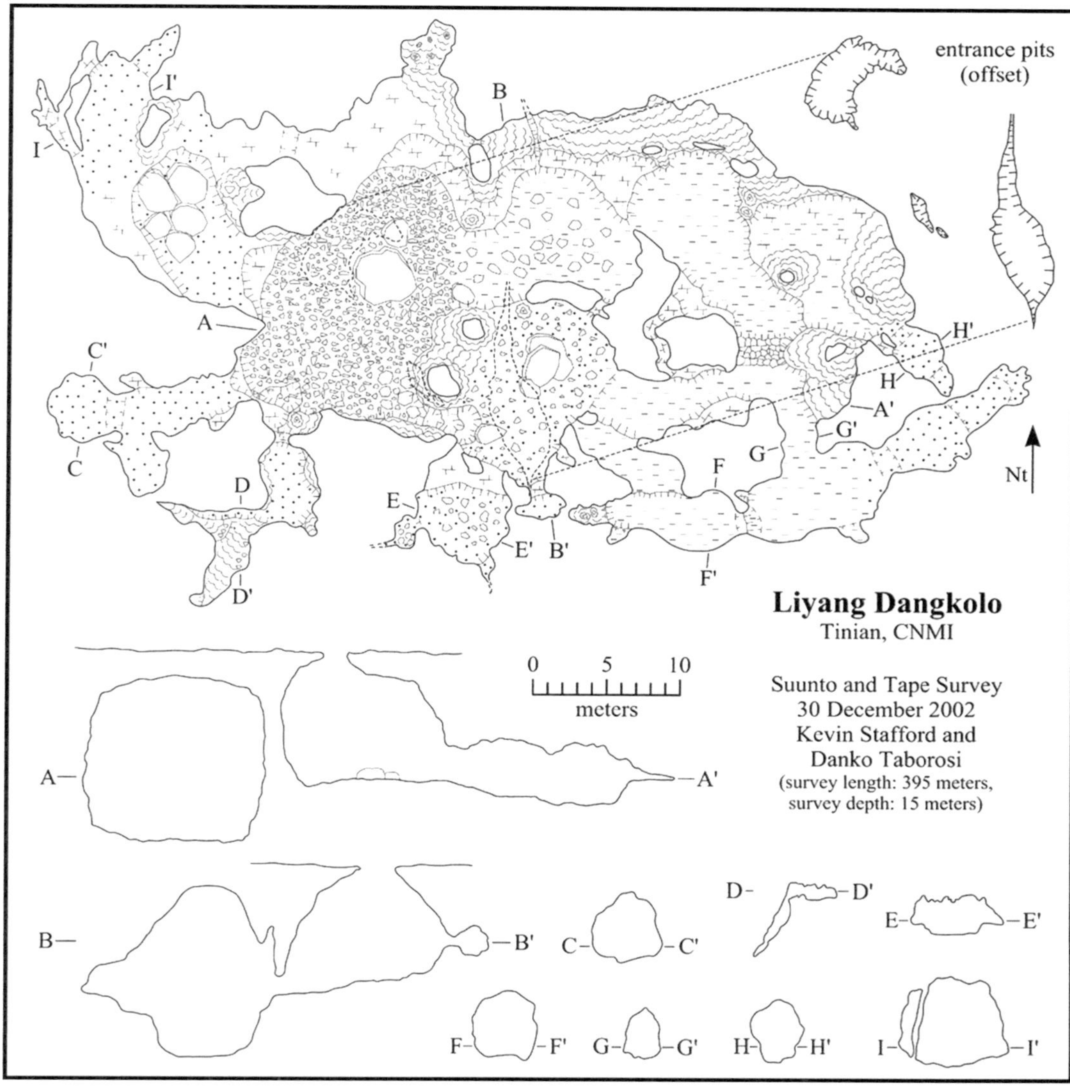

Fig. 13.3 Map of Liyang Dangkolo, a well-developed flank margin cave on Tinian. Note that standardized map symbology is based on Dasher (1994)

the vadose and phreatic zones (Stafford et al. 2004a). Although uncommon, they have been documented on both Tinian and Aguijan.

Fissure caves form as a result of the solutional enhancement of bank-margin failures resulting from coastal erosion and over steepening of scarp edges (Stafford et al. 2005). As scarps fail, fissures are created near-parallel to terrace and coastal edges that show little or no vertical displacement. The resulting fracture provides a vadose fast flow route for descending meteoric water and a preferential flow path for groundwater flow parallel to coastlines. These features tend to be horizontally limited from decimeters to meters in width, but can extend laterally perpendicular to scarps for hundreds of meters and tens of meters deep, often the deepest caves found in the Mariana Islands. Many of these features are unroofed or only have ceilings that are composed of breakdown blocks that have rotated into

Fig. 13.4 Typical segment of Suicide Cliffs on Tinian with extensive flank margin cave development. Note that the largest cave entrance on the right of the photo is approximately 4 m tall

Fig. 13.5 Typical segment of semi-continuous flank margin cave development that makes up part of the Sagua Cave Complex on Rota. Note that the flank margin caves are ~3 m above sea level and that a well-defined bioerosion notch has developed below them along the coastline

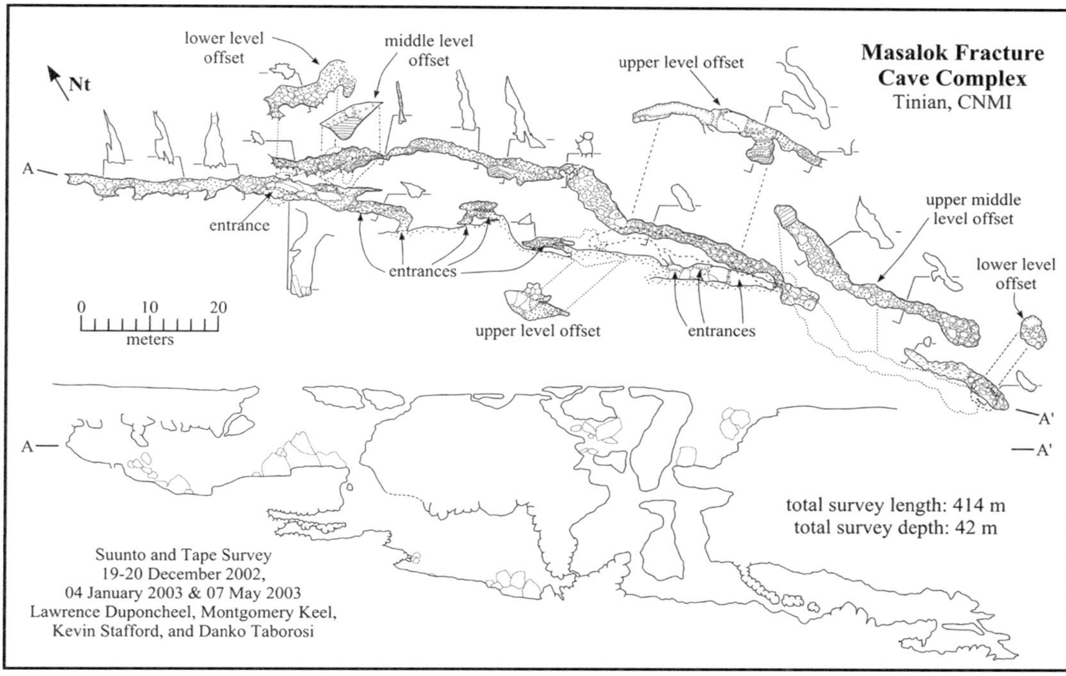

Fig. 13.6 Map of Masalok Fracture Cave Complex, a well-developed and laterally extensive fissure cave on Tinian. Note that standardized map symbology is based on Dasher (1994)

the fissure and become trapped. Similarly, most floors in fissure caves are composed of colluvium. The most extensive fissure cave thus far documented in the Marianas is Masalok Fracture Cave Complex (Fig. 13.6). It occurs along the eastern coast of Tinian and has a total depth of 42 m, length of 200 m and a maximum width of less than 5 m (Stafford et al. 2004a).

Mixing zone fracture caves occur as horizontally widened fractures perpendicular to scarps and coastlines where they represent active or former focused discharge features (Keel 2005). They have been documented throughout the Mariana Islands at elevations ranging from sea level to the highest terraces, but are most prevalent on Rota. In general, these features extend inland for tens of meters with narrow apertures along the fracture plane that is then horizontally widened up to several meters where they discharge at sea level or what is interpreted to represent past sea level elevations. However, Rota contains features that represent large-scale development of mixing zone fracture caves with individual features that extend up to 100 m inland with widths and heights of several meters as can be seen at Knuckle Bone Cave and Deer Cave (Fig. 13.7) (Keel 2005). Active mixing zone fracture caves occur throughout the Marianas, such as Cetacean Cave on Tinian and No Can Fracture on Guam (Fig. 13.8). The general model for mixing zone fracture caves incorporates the preferential flow of groundwater along fractures normal to coastlines that mixes with seawater to create a zone of mixing and increased solutional aggressiveness, which results in horizontal widening of the fracture (Keel 2005). Headward dissolution proceeds as the mixing zone moves inland, associated with greater circulation created by widening of the fracture proximal to the coastline. Some mixing zone fracture caves contain boneyard regions of complex, interconnected porosity at the most inland regions, suggesting these were the sites of most intense dissolution or maximum mixing at the speleogenetic termination of these features (Keel 2005). Vertical exaggeration of these caves is attributed by gradual change in relative sea

Fig. 13.7 Photo looking out of the entrance of Deer Cave on Rota, showing the general morphology of a typical mixing-zone fracture cave; bottom, map of Deer Cave

Fig. 13.8 Photo looking out the entrance of No Can Fissure Cave, an active mixing-zone fracture cave on Guam

level; however, some caves exhibit multiple levels of horizontal widening suggesting episodic uplift and re-equilibration of discharge horizons such as seen at Insect Bat Cave on Aguijan (Stafford et al. 2005)

The third major class of caves found in the Mariana Islands is that of caves formed at the contact between volcanic rocks and overlying carbonates (Jenson et al. 2006). Where volcanic rocks are exposed at the land surface, they rapidly weather into low permeability clays that promote overland flow. The result is the development of focused, allogenic groundwater recharge at the margin of upland volcanic terrains where overlying carbonate strata are in contact. This results in solutional porosity created by focused water unsaturated with respect to carbonate ions; however, these recharge caves (Fig. 13.2) are often limited in size not because of the lack of solutional potential by allogenic water but instead as a result of infilling by clays and clastic deposits that are mechanically weathered from the volcanic

Fig. 13.9 Photo of Alamagosa Cave, and active stream cave in Southern Guam

terrains supplying the overland flow. Recharge caves of varying size, but generally less than a few tens of meters in any maximum dimension, are found at all locations within the Mariana Islands where upland volcanics occur and overland flow is capable of developing (Jenson et al. 2006).

As with the focusing of overland flow in upland volcanic terrains, the subsurface focusing of groundwater flow along carbonate-noncarbonate boundaries can result in contact caves (Taboroši et al. 2004a). As volcanic rocks extend above the freshwater lens, autogenic recharge can be directed laterally along these surfaces as vadose flow is directed towards the freshwater lens. This focusing of waters along a low permeability boundary increases the potential for dissolution simply by the development of channelized flow in the vadose zone, a phenomenon that is common in diagenetically mature continental settings but often rare in eogenetic island karst. In southern Guam, in the more diagenetically mature rocks of the upland region, traversable stream caves have been documented where vadose flow is focused along this permeability boundary between volcanic basement rocks and limestone remnants above (Taboroši et al. 2004a). Alamagosa Spring, in southern Guam, issues from a stream cave that extends as a traversable passage for 250 m (Fig. 13.9). It is likely that contact caves are common throughout the Mariana Islands but remain undiscovered because they occur at contacts that are tens to hundreds of meters in the subsurface and do not have entrances at the land surface.

Pit caves are rare throughout the Mariana Islands. They form by near vertical migration of solutional vadose water, generally along fractures (Jenson et al. 2006). Rota and Tinian have only one documented pit cave each, while Saipan and Aguijan have none reported. On Guam, several pit caves have been identified, including one 45 m deep at Amantes Point (Fig. 13.10) and one 35 m deep within the Talafofo Caves complex (Mylroie et al. 2001).

Uniqueness of Mariana Islands karst is further demonstrated by two highly unusual caves unlike any so far documented in island karst environments elsewhere. Kalabera Cave on Saipan has been described as a phreatic-lift tube with more than 300 m of passages extending vertically 55 m (Fig. 13.11) (Jenson et al. 2006). It has formed as water traveled laterally downdip in confinement by volcaniclastic units before

Fig. 13.10 Photo of Amantes Pit on Guam, one of the few pit caves found in the Mariana Islands

intersecting a fracture that allowed upward migration of groundwater as artesian-style spring discharge creating a vertical, phreatic shaft where water flowed upwards in the past.

Liyang Atkiya, the largest cave documented on Aguijan, has a length greater than 200 m and a depth over 50 m (Fig. 13.12) (Stafford et al. 2005). Its entrance is at the base of the mid-level terrace and connects to a large room up to 25 m wide and 10 m tall, reminiscent of a large flank margin cave. However, the inland portion of the chamber is connected to a long, near linear passage that slopes gently into the subsurface before eventually branching into two passages that subsequently divide after several tens of meters into maze-like passages and boneyard. These lower passages are similar to those observed in hypogene continental settings where sluggish groundwater mixing dominates dissolution, but centimeter-scale scallops on walls of the passage that connects the lower maze area to the large upper chamber suggest upward, rapid

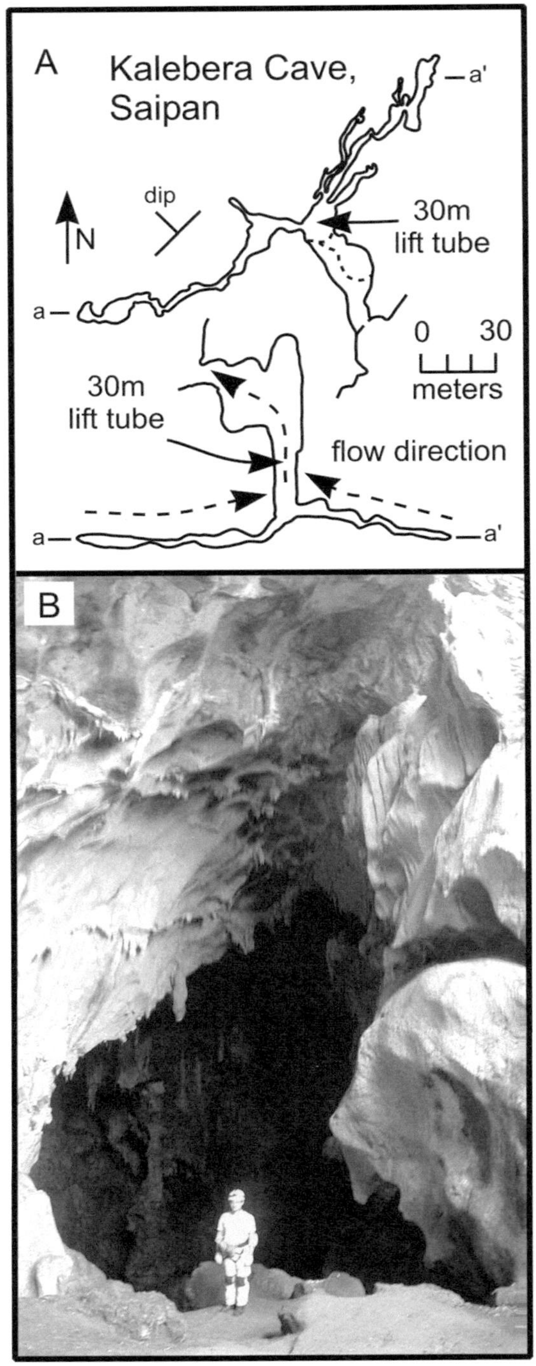

Fig. 13.11 Photo and map of Kalabera Cave

conduit flow (Stafford et al. 2004a). The speleogenetic origin of Liyang Atkiya is problematic, with theories ranging from thermal convection created by an interior island core that was still hot from volcanic activity to a complex version of the phreatic lift tube seen at Kalabera Cave.

While closed depressions are often used as indicators of karst development in continental

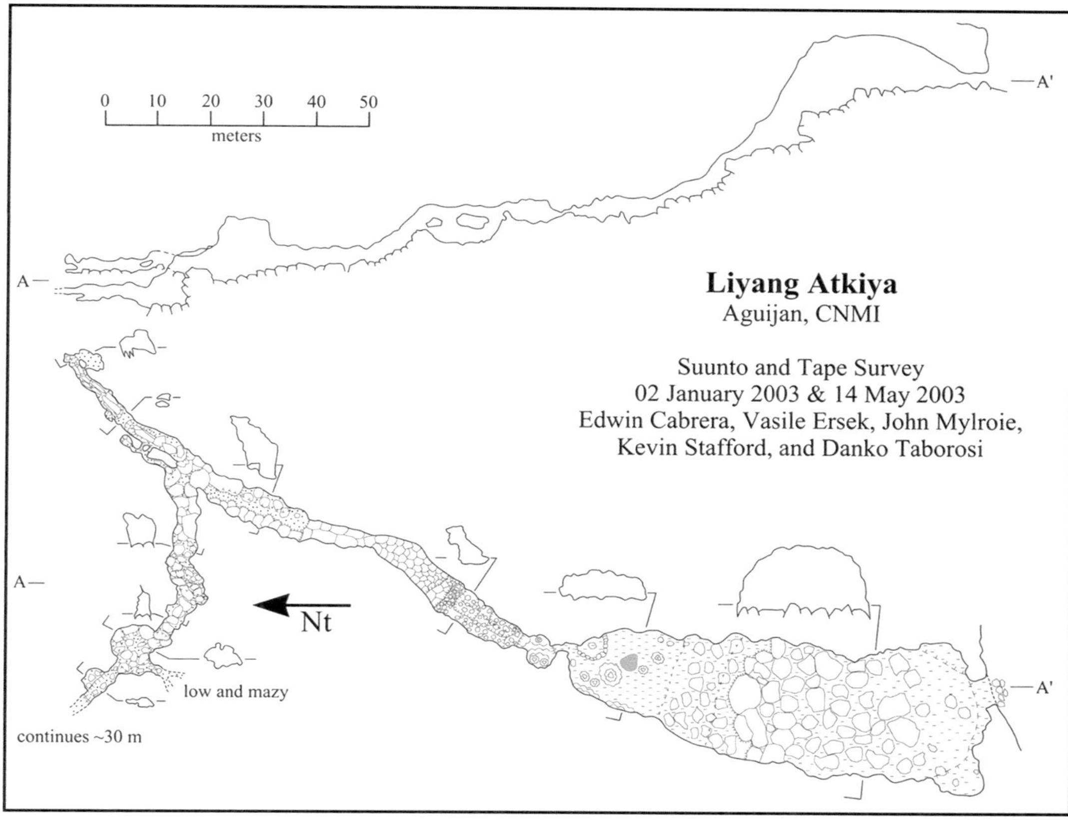

Fig. 13.12 Map of Liyang Atkiya on Aguijan, illustrating the complex speleogenetic origin of this unique cave. Note that standardized map symbology is based on Dasher (1994)

settings, they have proven to be poor indicators throughout the Mariana Islands. Most closed depressions in the Marianas appear to be constructional features associated with original depositional environments or are anthropogenic in origin. In upland areas where allogenic waters focus groundwater recharge, closed depressions can be useful indicators of the presence of recharge caves but otherwise should be viewed questioningly throughout the Mariana Islands. Because eogenetic rocks retain high primary porosities and permeability, most groundwater recharge on carbonate terrains is autogenic and thus does not promote the development of solutional sinkholes.

Karst springs are common throughout the region, with focused discharge through mixing zone fracture caves and perched stream caves described above. Coastal discharge is observed along beaches throughout the Mariana Islands at low tide, attesting to intense, diffuse discharge along the margin of the freshwater lens where modern flank margin caves are likely forming in the subsurface. Significant springs and other groundwater sources are discussed in more detail below.

13.4 Groundwater Resources of the Mariana Islands

Fluvial systems and surface water are sparse throughout the Mariana Islands, with most occurrences limited to volcanic terrains where allogenic water is directed to carbonate boundaries, regions where the land surface dips below the water table, and coastal discharge.

Guam provides the exception to this general rule, where southern Guam has a well-developed surface drainage network over volcanic terrain (Taboroši et al. 2004a). Most groundwater resources are directly associated with karst development with general behavior similar to that proposed by the conceptual carbonate island karst model of Mylroie et al. (2001), which was later expanded to include the complexities of carbonate-noncarbonate facies interfingering and syndepositional tectonics (Jenson et al. 2006). Research throughout the Marianas has led to the recognition of aquifer compartmentalization and the associated development of preferential fast flow paths along fracture surfaces. Individual carbonate islands of the Marianas may exhibit all four classes of island karst (i.e. simple, carbonate-cover, composite and complex) as predicted by the carbonate island karst model. No single class can be applied to an entire island because of tectonic overprinting and differential rates of uplift and subsidence within individual islands. Groundwater management throughout the Mariana Islands therefore relies on extraction techniques that must be adapted to the local situation, whether it is the use of traditional Maui-type wells that skim water off the top of the freshwater lens or the extraction of groundwater from parabasal regions where underlying, low permeability volcanics prevent or limit the potential for upconing and saltwater intrusion.

Guam hosts a population of greater than 175,000 people and relies heavily of the karst aquifer of northern Guam for its water resources, which includes more than 1,600 Ls^{-1} demand. Hydrogeologic modeling indicates that sustainable yields of greater than 2,200 Ls^{-1} are possible if managed properly; however, overpumping has resulted in saline intrusion at some well sites. Water quality is maintained in basal regions if pumping does not exceed 17 Ls^{-1}, and 31–47 Ls^{-1} in parabasal regions (Mink and Vacher 1997), but constant monitoring is necessary to preserve sustainable resources with more than 130 wells producing from the North Guam Lens Aquifer, which provides more than 80 % of the total water resources for the island. Other water resources are derived from isolated wells and throughout southern Guam, primarily the limestone fringe. The U.S. Navy has a traditional surface reservoir built on volcanics in southern Guam. Rainwater capture and water conservation practices assist in reducing groundwater demand; however, increasing populations and significant tourism place high demands on the limited water resources available throughout Guam.

Rota provides a unique example in the Marianas of groundwater resources that can be exploited by capturing spring discharge. While most springs in the Marianas are associated with sea level and coastal discharge, several springs occur on a mid-level scarp on the south slope of the Sabana region of Rota. At this location, underlying volcanic rocks are exposed at the slope surface and allogenic waters that recharge in upland regions, and autogenic recharge to adjacent carbonates discharge to the surface from limestone springs. Water Cave has been modified with a retention wall and connected to a plumbing network that supplies the majority of the island water needs (Fig. 13.13). Because the spring is located at an elevation of 350 m, it is not only an excellent source but also allows gravity-driven distribution to the vast majority of the population. While currently these discharging contact caves satisfy most of the islands potable water needs, this may change in the future if population increases and natural discharge volume becomes insufficient. In that case, groundwater extraction by wells will become important.

Aguijan has no known natural water sources at the surface and its small area-to-height ratio does not promote shallow groundwater resources. This, in conjunction with difficulty of access to the island because of the lack of natural beaches, has effectively made the island uninhabitable in modern times. However, during the Japanese period, the problem of limited water resources was addressed through the construction of large cisterns and storage tanks that captured rainwater. Remnants of these rain collection systems are the only known sources of surface water on Aguijan today, providing water resources to island visitors and local fauna alike.

Fig. 13.13 Water cave on Rota provides the primary source for potable water for the entire island. The cave entrance has been modified to impound water and collect it for distribution throughout the island

Tinian had an extensive network of wells installed during and shortly after World War II. Freshwater naturally occurs at the land surface only in the Northern Lowland, where the water table is above the land surface, and along coastlines where it discharges through discrete fractures or diffuses through Holocene beach sands. Today, the majority of Tinian's extracted water comes from a Maui-type well system situated in Median Valley, a constructional topographic depression between the Central Plateau and Southeastern Ridge. While groundwater extraction on Tinian utilizes traditional techniques that limit the potential for significant upconing and saltwater intrusion, intense agriculture, including farming and livestock operations near municipal wells, increase the chance of water resource contamination.

Saipan, with its complex syndepositional carbonate and volcanic rocks coupled to complex syntectonic deformation, is the most hydrogeologically complex island of the Marianas (Jenson et al. 2006). Extensive compartmentalization of the aquifer system has resulted in numerous small aquifers with limited connectivity, which greatly complicates water resource management. Water resources on Saipan rely heavily on a combination of Maui-type and conventional extraction wells, with more than 90 % of water resources extracted from 140 shallow wells. With the highest and most dense population within the Northern Mariana Islands, Saipan currently does not have adequate water resources, both in quantity and quality. Carruth (2003) reported that chloride concentrations range from less than 100 mg L^{-1} to more than 2,000 mg L^{-1}, with average daily total groundwater extractions exceeding 11 Mgal day^{-1}. Saipan, like Guam, is struggling to meet the water resource demands of rapidly increasing population.

Throughout the Mariana Islands groundwater contamination is an ever present threat because of the high permeability of the eogenetic carbonate facies. Fractures that modify ground water flow and compartmentalize island aquifers work both advantageously and disadvantageously in controlling contaminant transport. While these fast flow paths have the potential to rapidly transport contaminants into the subsurface, they also rapidly move groundwater towards coastal

Fig. 13.14 Section of cave wall in Pictograph Cave on Rota showing early Chamorro utilization of cave resources. Image height is approximately 1 m

discharge points. This rapid lateral movement may limit the potential for groundwater contamination locally, but may enhance the possibility of contaminants affecting nearshore ecosystems, especially coral reefs. The future of groundwater resources throughout the Mariana Islands will require the development of strategic, sustainable, groundwater management plans that are supplemented by rainwater collection. Research at the Water and Environmental Research Institute of the Western Pacific at the University of Guam is actively developing and expanding models and groundwater monitoring, in conjunction with public education and outreach programs to ensure the future sustainability of the water resources throughout the Mariana Islands.

13.5 Utilization of Karst Resources in the Mariana Islands

The Mariana Islands have a rich cultural history, which included extensive use of caves and karst resources. The indigenous Chamorro people first settled the Mariana Islands around 2300 BCE from Southeast Asia (Athens and Ward 2004). Their prehistoric material culture is still evident throughout the Marianas, both on the surface and within caves. Shallow trenches used for extraction of groundwater indicate that the Chamorro developed skimming-type groundwater extraction to utilize freshwater resources while limiting the potential for saltwater intrusion. Caves throughout the Marianas show evidence of continued use by prehistoric Chamorro populations, including archeological remains and extensive rock art, primarily pictographs (Fig. 13.14), suggesting that natural karst features were utilized for both habitation and ceremonial purposes.

After Ferdinand Magellan discovered the Marianas in 1521, Spain occupied the islands for nearly four centuries, but sold them to Germany in 1899. Germany retained control of the Mariana Islands until 1914, when Japan took control of the islands at the start of World War I, with the exception of Guam, which was taken in 1898 by the United States. Japan controlled the Marianas throughout the succeeding three decades, including two world wars and the

Fig. 13.15 Large, 16 cm Japanese cannon emplace in the entrance of Canon Cave, a modified flank margin cave on Tinian. Cave height is 6 m

intervening years, which included the capture of Guam from the United States in 1941. During the Japanese era, karst resources were again highly utilized, but this time they were not only used for habitation but also from a strategic military advantage. Caves were utilized as fortifications, water sources, and refuges (Taboroši et al. 2007). Many caves were extensively modified with reinforcements by the Japanese military at this time, while pits and tunnels were also excavated to create man-made caves that supplemented the natural ones. Many of the coastal caves were modified as defensive fortifications with armored positions ranging from-large caliber cannons that completely filled cave entrances (Fig. 13.15), to small-caliber weaponry that utilized modification of natural features to form pillboxes. Because of the effectiveness of these natural fortifications, the Mariana Islands are remembered as sites of some of the bloodiest battles of World War II during the United States' campaign to secure this region. The extensive use of karst resources during World War II continued the long history of utilization of karst resources in the region (Taboroši et al. 2007).

The United States regained control of Guam in 1944 and subsequently captured the Northern Mariana Islands, including the strategic capture of Saipan and Tinian which led to the end of World War II, while the northernmost islands of the Marianas and Rota were never physically captured by the United States, but instead were isolated from Japan by United States' control of maritime activities in the region (Rogers 1996). When World War II ended in 1945, the caves of the Marianas continued to serve as habitation and refuge for stragglers that remained in hiding (Trefalt 2003). In 1972, the last Japanese soldier was found hiding in a cave in southern Guam, nearly three decades after the end of the war.

After World War II, Guam remained a United States territory while the remaining Mariana Islands were governed as the Commonwealth of the Northern Mariana Islands, which was established in 1975 through a covenant with United States. Today, karst resources are still heavily utilized throughout the Marianas. Many sites have been developed into memorials to remember those lost during World War II, including the development of shrines in some caves. Several caves have been developed for tourism, both by the local governments and private owners. Water Cave on Rota continues to be the primary water resource for the island, while other coastal features throughout the islands are popular recreation sites by both locals and tourists. Many caves contain rich archeological records that attest to the long use of karst resources in the Mariana Islands, primarily from the prehistoric Chamorro people and the intense military-related utilization by the Japanese. Today, as population growth continues, ever greater pressures are being placed on these karst resources, attesting to the need for increased management of both the natural and cultural resources of karst in the Marianas, a similar problem faced in karst regions worldwide.

13.6 Significance of the Mariana Islands in Karst Science

The interaction of tectonics, sea level change and karst solution processes have created complex geomorphic and hydrogeologic terrains throughout the Mariana Islands. This complexity led to the creation of the fourth conceptual category of the Carbonate Island Karst Model, the Complex Island, which addresses the complications of interfingering between soluble and non-soluble eogenetic facies as well as significant, syndepositional tectonism (Jenson et al. 2006). In addition to the Complex Island model, the Mariana Islands advanced understandings of island karst in tectonically active settings, including the introduction of the three genetic types of fracture caves: fault caves, fissure caves and mixing zone fracture caves (Keel 2005; Stafford et al. 2004a). Brittle deformation associated with differential uplift and subsidence has resulted in aquifer compartmentalization in the Mariana Islands, which shows that the conceptual CIKM must be used cautiously and cannot always be applied to entire islands. As brittle deformation produces planar fast flow surfaces, freshwater lens morphologies become altered such that groundwater flow is focused towards these higher permeability flow regimes, thus effectively isolating portions of island aquifers from each other.

Since the beginning of human occupation and continuing until today, karst resources of the Marianas have been an integral part of the daily lives of people living throughout the Mariana Islands. Modern use of caves as recreation and spiritual sites is continuing the long heritage of incorporating karst resources into the local community and culture. Today, the Mariana Islands face the same problems they have in the past with groundwater management. As populations increase throughout the islands, ever greater pressures are being placed on groundwater resources making it crucial to manage these resources to prevent saltwater intrusion from overpumping. Resource management is integral in preventing and reducing contamination of groundwater from surface and near-surface contaminants that can potentially enter the groundwater system with little or no filtration because of the high permeabilities exhibited by these eogenetic carbonate rocks. As with all karst terrains, the key to the future sustainability of natural resource management is through public education on the susceptible nature of their groundwater resources. In the Mariana Islands this includes both modern and ancient techniques, whether it is the use of models that demonstrate groundwater flow and speleogenesis or the implementation of traditional Maui-type wells that extract groundwater by skimming it from the surface of the freshwater lens.

References

Athens JS, Ward JV (2004) Holocene vegetation, savanna origins and human settlement of Guam. Records of the Australian Museum, Supplement 29. Australian Museum, Sydney

Carruth RL (2003) Ground-water resources of Saipan, Commonwealth of the Northern Mariana Islands: water-resources investigations report 03–4178. U.S. Geological Survey, Honolulu

Cloud PE Jr, Schmidt RG, Burke HW (1956) Geology of Saipan, Marian Islands, Part 1, General geology, 280-A. U.S. Geological Survey professional paper. Government Printing Office, Washington, DC

Dasher GR (1994) On station. National Speleological Society Inc, Huntsville

Dickinson WR (2000) Hydro-isostatic and tectonic influences on emergent Holocene paleoshorelines in the Mariana Islands, Western Pacific Ocean. J Coast Res 16–3:735–746

Doan DB, Burke HW, May HG, Stensland CH, Blumenstock DI (1960) Military geology of Tinian, Mariana Islands. Chief of Engineers, U.S. Army

Gingerich SB, Yeatts DS (2000) Ground-water resources of Tinian, Commonwealth of the Northern Mariana Islands, Water-resources investigations report 00–4068. U.S. Department of the Interior

Jenson JW, Keel TM, Mylroie JR, Mylroie JE, Stafford KW, Taboroši D, Wexel C (2006) Karst of the Mariana Islands: the interaction of tectonics, glacio-eustasy, and freshwater/saltwater mixing in island carbonates. In: Harmon RS, Wicks C (eds) Perspectives on karst geomorphology, hydrology, and geochemistry – a tribute volume to Derek C. Ford and William B. White, Geological Society of America special paper 404. Geological Society of America, Boulder

Karig DE (1971) Structural history of the Marian Island arc system. Geol Soc Am Bull 82:323–344

Keel TM (2005) Caves and karst of Rota (Luta), CNMI – technical report 107. Water and Environmental Research Institute of the Western Pacific, University of Guam, Mangilao

Lusk MR, Bruner P, Kessler C (2000) The avifauna of Farallon de Medinilla, Mariana Islands. J Field Ornithol 71(1):22–33

Mink JG, Vacher HL (1997) Chapter 25: Hydrogeology of northern Guam. In: Vacher HL, Quinn T (eds) Geology and hydrogeology of carbonate island: developments in sedimentology 54. Elsevier Science, New York

Mylroie JE, Jenson JW, Taboroši D, Jocson JMU, Vann D, Wexel C (2001) Karst features of Guam in terms of a general model of carbonate island karst. J Cave Karst Stud 63–1:9–22

Rogers RG (1996) Destiny's landfall: a history of Guam. University of Hawaii Press, Honolulu

Stafford KW, Mylroie JE, Jenson JW (2002) Karst geology and hydrology of Tinian and Rota (Luta), CNMI: technical report no. 96. Water and Environmental Research Institute of the Western Pacific, University of Guam, Mangilao

Stafford KW, Mylroie JE, Jenson JW (2004a) Karst geology of Aguijan and Tinian, CNMI: cave inventory and structural analysis of development: technical report 106. Water and Environmental Research Institute of the Western Pacific, University of Guam, Mangilao

Stafford KW, Mylroie JE, Taboroši D, Jenson JW (2004b) Eogenetic karst development on a small, tectonically active, carbonate island: Aguijan, Mariana Islands. Cave Karst Sci 31(3):101–108

Stafford KW, Mylroie JE, Taboroši D, Jenson JW, Mylroie JR (2005) Karst development on Tinian, Commonwealth of the Northern Marian Islands: controls on dissolution in relation to the carbonate island karst model. J Cave Karst Stud 67(1):14–27

Sugawara S (1934) Topography, geology and coral reefs of Rota Island. Tokyo, Pacific Geologic Studies, Military Geology Branch, United States Geological Survey

Taboroši D, Jenson JW, Mylroie JE (2003) Zones of enhanced dissolution and associated cave morphology in an uplifted carbonate island aquifer, northern Guam, Mariana Islands. Speleogenesis Evol Karst Aquifers 1:1–16

Taboroši D, Jenson JW, Mylroie JE (2004a) Karst features of Guam, Mariana Islands. Water and Environmental Research Institute of the Western Pacific, University of Guam, Mangilao

Taboroši D, Jenson JW, Mylroie JE (2004b) Karren features in island karst: Guam, Mariana Islands. Z Geomorphol 48:369–389

Taboroši D, Jenson JW, Stafford KW, Matsuzawa R (2007) World War II artifact and wartime use of caves in the Mariana Islands, western Pacific. J Speleol Soc Jpn 32:42–53

Tayama R (1936) Geomorphology, geology, and coral reefs of Tinian Island together with Aguijan and Naftan Islands. Institute of Geology and Paleontology, Tohoku University, Sendai

Tracey JI Jr, Schlanger SO, Stark JT, Coan DB, May HG (1964) General geology of Guam, 403-A. U.S. Geological Survey professional paper. Government Printing Office, Washington, DC

Trefalt B (2003) Japanese army stragglers and memories of the war in Japan, 1950–1975. RoutledgeCurzon, London

Littoral Cave Development on the Western U.S. Coast

14

David E. Bunnell and Johanna L. Kovarik

Abstract

Sea caves have been documented in all four states on the Pacific coast of the United States. By far the most extensive written documentation, along with cave surveys, is from California due to the efforts of the California Sea Cave Survey since the early 1980s. The largest concentration of sea caves there are those in the Channel Islands of southern California, where over 15 km has been surveyed in 380 caves. Painted Cave on Santa Cruz Island is one of the world's largest at 375 m in length. But large caves are also known from the central California counties of Santa Cruz, San Luis Obispo, and Mendocino. Currently Sea Lion Caves in Oregon is the longest known on the Pacific coast of the U.S. at 401 m, as well as one of the world's longest, and among the few sea caves in Oregon with survey data available. Sea caves are known from several other regions in the state. Washington State has scattered areas with sea caves and few surveyed, with perhaps the largest concentration at Cape Flattery on its northwestern tip. Alaska has seen some significant survey activity resulting in documentation of a few dozen caves. Potential is great there for some large sea caves to be explored in the future, especially in the Aleutian Islands. Southeast Alaska contains the best documented caves to date, some within carbonate rocks with speleothems and some now found well above the current littoral zone due to uplift following retreat of coastal glaciers. The latter has resulted in more caves with archaeological significance since their drier environment was more hospitable as shelter than is typical of littoral caves in the region. Few such uplifted caves are known in the other Pacific states other than some well-documented Channel Islands caves.

D.E. Bunnell (✉)
California Sea Cave Survey, PO Box 879,
Angels Camp, CA 95222, USA
e-mail: davebunnell@underearth.us

J.L. Kovarik
Department of Geology, University of South Florida,
Tampa, FL 33620, USA
e-mail: jkovarik@mail.usf.edu

14.1 Geologic Overview of the West Coast

Littoral cave development is strongly influenced by geology, including rock type, structure, and movement. Faults, joints, and bedding planes

M.J. Lace and J.E. Mylroie (eds.), *Coastal Karst Landforms*, Coastal Research Library 5,
DOI 10.1007/978-94-007-5016-6_14,

form zones of weakness in bedrock allowing focused erosion. This mechanical weathering, at times combined with dissolution, eventually widens and lengthens these zones of weakness into caves, as opposed to uniformly weathering an outcrop. Contacts between rock units with differential resistance to erosion and solution also play a role in many regions. The west coast contains a wide variety of rock deformed in places by the contact of the Pacific and North American plates (Sherman 2005; Johnson and Ledesma-Vasquez 2009).

In terms of sea cave formation itself, most activity is confined to the time since the last ice age (about 20,000 ybp), since sea levels have been rising continually since then and the zone of active sea cave formation along with it. As mentioned in Chap. 1, sea levels have only been relatively constant over the last 3,000 years; thus, most west coast sea caves where the coastal sea cliffs have not changed in relation to sea levels will be in that age range.

14.2 Alaska

The coastline of Alaska is approximately 70,800 km of diverse geology from southeast Alaska and the Alexander Archipelago to the shores of the Chukchi and Beaufort Seas along the northern coast. Currently, the greatest number of known sea caves is located in southeast Alaska, where geologists and cavers have conducted organized cave mapping since the 1990s. Additionally, archaeologists and paleontologists have documented sea caves in the Shumagin Islands of the Aleutian chain, and lava tubes have been noted on the Pribilof Islands in far western Alaska. Supplementary knowledge of sea caves on the Alaskan coast is not well documented; however travelers have reported sea caves in Prince William Sound on Glacier Island as well as to the west and further along the Aleutian chain. Researchers have long studied southeast Alaskan sea caves as important repositories of archaeological and paleontological information which bear directly on the history of the populating of North America.

14.2.1 Southeast Alaska

The geology of southeast Alaska consists primarily of accreted terrains including the Alexander terrain into which most mapped sea caves in southeast Alaska have formed. This terrain collided obliquely with the landmass of North America and accreted onto the continental margin during the middle Jurassic to Late Cretaceous time (Coney et al. 1980). The process of accretion resulted in fragmentation and smearing of sections of the terrain northward, while other portions remained in place. Faults and fractures resulting from the northward movements of the Alexander Terrain are dominated by northwesterly trending strike-slip faults and second order intersecting north-trending strike-slip faults, which often define sea cave formation along surge channels. In southeast Alaska, volunteer cavers, the US Forest Service, and the Tongass Cave Project have mapped over 70 sea caves. These caves are invaluable to the local native groups in southeast Alaska, yielding archaeological finds such as tools, fibers, and other evidence of early usage. Additionally, in recent focused surveys, Alaskan sea caves have gained importance as a data for development of a paleoshoreline predictive model. Seventeen caves mapped during the summer of 2010 demonstrated by measured elevation above mean low lowest water (MLLW) and average ceiling height that these caves began forming approximately 10,000 radiocarbon years before present (RYBP), and were uplifted from marine influence approximately 4,500–3,200 (RYBP) (Baichtal 2011).

14.2.1.1 On Your Knees Cave (67.9 m)

On Your Knees Cave is an uplifted sea cave formed in Silurian limestone located on the outer coast of southern southeast Alaska. It was discovered during contract karst evaluation work in 1992 by the Tongass Cave Project. During initial surveys, cavers discovered bear bones in the cave, and the short cave was recommended for protection. Further work in 1994 by Dr. Tim Heaton uncovered human remains and additional archaeological artifacts within the cave. These finds within On Your Knees Cave resulted in 16 years' worth of collaboration between the US Forest Service and the Alaska Native Tribes. With funding from the National Science Foundation, Sealaska Corporation, and the US Forest Service, researchers found the human remains to be those of a young man, and dated them to 10,300 ybp (years before present). In 2008 all collaborators on the project came together to bury the bones of the young man, named in the local Tlingit language Shuká Ḵaa ("Man Ahead of Us").

14.2.1.2 Swooping Owl Cave (120 m)

Swooping Owl Cave is located in southern southeast Alaska. Tongass National Forest Geologist Jim Baichtal discovered the cave initially, during which time he spooked a large owl roosting in the cave, resulting in the name. A group of volunteers working with the Forest Service returned to the cave and mapped it in 2010. The cave is located approximately 12 m above sea level, and both the cave and the tidal channel leading to it are formed in granodiorite bedrock along distinct joints trending east-west. The entrance at the dripline is over 15 m in height. The cave contains speleothems, as well as wood, debris, and bones in the back of the cave. This cave is unique due to its length and the rock in which it has formed—most known sea caves in southeast Alaska are found in carbonate rock.

14.2.1.3 Puffin Grotto (247.7 m)

The Tongass Cave Project discovered and mapped Puffin Grotto on an outer island in southeast Alaska in 1994 (Fig. 14.1). The cave formed in Silurian marble, and the Tongass Cave Project report suggests that, due to passage morphology, mechanical as well as chemical processes might play a role in speleogenesis. Multiple entrances lead eventually to a chamber in the rear of the cave named, "Jonah's Room". The cave is a treasure trove of resources, including whale bones and moonmilk, and is one of the longest mapped sea caves in southeast Alaska.

14.3 Washington State

Washington coast geology is a result of tectonic forces caused by the combined movements of the large Pacific and North America Plates and the smaller Juan de Fuca Plate. There appear to be no formal surveys of any sea caves in Washington State, but there are several areas where sea caves are known. The most significant of these areas is Cape Flattery, which is at the northwest tip of the Olympic peninsula. Cave entrances are visible from a trail established by the Makah Indians, which circumnavigates the tip of the Cape, as well as on nearby Tatoosh Island.

Along the coast between Cape Flattery and Point Grenville, cliffs rise 15–90 m above a wave-cut terrace. This wave-cut terrace, which normally extends about 800 m from shore, is nearly 3 km wide west of Ozette Lake (Bird 2000). Small islands, sea stacks, and rocks dot the surface of the terrace.

Another region of interest on the Olympic Peninsula is Shi Shi Beach (within Olympic National Park) and the Point of Arches, where there are numerous sea stacks and arches. A 1907 postcard shows a dramatic double arch visible from the Point of Arches. There is at least one large sea cave known in this area, and likely others in the cliffs, which reach 45 m high. Other areas further south with sea caves mentioned in online narratives include Elephant Rock (a double sea arch) and Ruby Point.

14.4 Oregon

There are numerous regions in coastal Oregon where sea caves can be found, but similar to Washington there is a dearth of published material or formal surveys of caves. The most significant documented cave is the commercial Sea Lion Caves near Florence. It is formed in Eocene basalt in Sea Lion Point (Lund 1971). Narrow benches near the base of the cliffs here

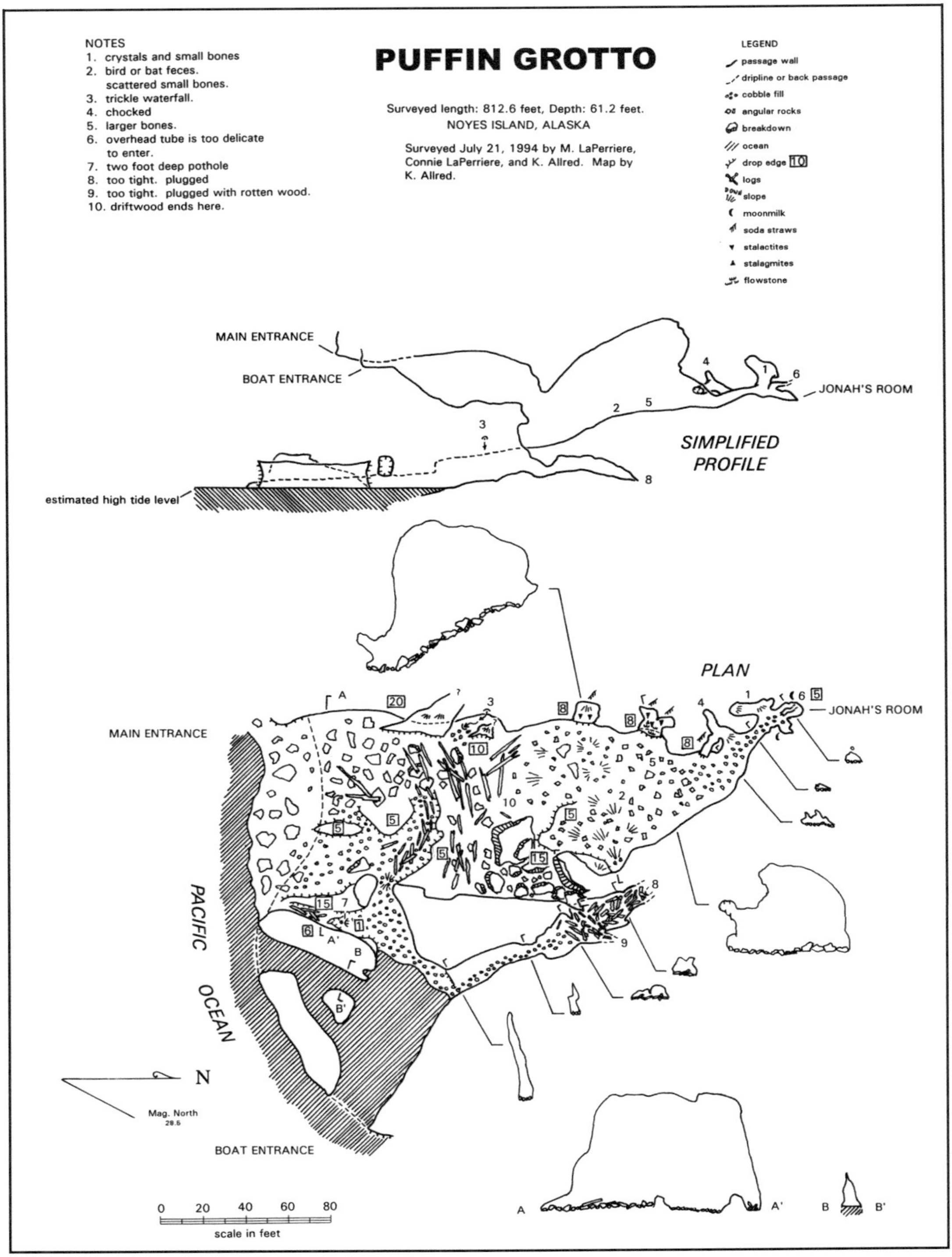

Fig. 14.1 Puffin Grotto, Alaska

Fig. 14.2 Sea Lion Caves, Oregon

formed on less resistant fragmented basalt. The cave formed along two intersecting fractures, one trending north-south and a second one east-west. The cave was developed for tourism and opened in 1932; tourists enter along a tunnel in the north-south fracture zone. This portion is well above present sea level, probably reflecting regional uplift relative to sea level. The basalt here is stratified with layers of ash visible in the cave walls (Fig. 14.2), and it is likely that this afforded an additional zone of weakness combined with intersecting faults.

Sea Lion has long called itself the "world's largest sea cave." A 1970 map shows about 183 m of passage, but the longer southern passage beyond the 60 m-wide inner chamber was not surveyed. The management contracted with a professional survey crew in 1996 and gave a figure of 401 m along the length of the north-south tunnel, which was used as the basis for the claim as largest (verified by unpublished documents provided to the first author by the management). Based on this survey, their claim appears correct regarding length (http://www.caverbob.com/seacave.html). However, if "largest" is viewed in terms of volume, the floor area of 1.8 acres (7,284 m^2) is a good bit less than New Zealand's Riko Riko Cave's laser-surveyed area of almost double that value (Elrick 2003).

Another well-known littoral cave feature is a large littoral sinkhole, which is featured in Devil's Punchbowl State Park, 12 km north of Newport. Formed in sandstone, there are at least three short, separate seaward openings into the sink, whose floor is filled with water except at low tide (Fig. 14.3). It measures about 27 m across and is about 15 m deep, with sheer walls.

A prominent littoral sink is also found in the Samuel Boardman State Scenic Corridor (Bunnell 2008a), 4.5 km north of Brookings. Park information suggests that this feature and a series of some half dozen arches that extend from it are remnants of what was once a large sea cave.

Other areas with sea caves include Hug Point (Niem 1975) where caves are formed in sandstone, Cape Kiwanda, and Cape Lookout. A particularly large cave is in a point about 1.5 km north of Oceanside and known as Lost Boy Cave (personal communications between the first author and Oregon Grotto members).

14.5 California

There are almost certainly many sea caves in the far northern counties of California, but little has been published about them. Our focus here

Fig. 14.3 Devils Punchbowl, Oregon. (**a**) Seaward openings viewed from inside the Punchbowl, during a negative tide. (**b**) Punchbowl viewed from above during a high tide

will be regions where formal surveys have been undertaken in several counties, and the Channel Islands of southern California where the largest concentration of documented sea caves is to be found. Summaries of the coastal survey work appear in Bunnell (1998) which addresses each county with sea caves. Here we present only those areas with the most significant caves in terms of length or abundance.

14.5.1 Mendocino County

Bill Halliday, in his 1962 description of the caves of Mendocino County in *Caves of California*, described the county as being "the most extensive locality of littoral pseudokarst on the Pacific coast." Based on surveys, he is quite right in that assessment, though Santa Rosa Island, discussed below, comes close. There are six, possibly seven

Fig. 14.4 Mendocino littoral sinks. (**a**) Fully collapsed littoral sink on the Stornetta Pulbic Lands, Mendocino County, CA. (**b**) Second littoral sink forming within 50m of the sink shown in (**a**), not yet fully collapsed

major littoral sinks or "punchbowls" associated with caves in Mendocino. Two are easily accessible. There is a 30 m-diameter, circular collapse pit, 20 m deep, into a sea cave at Russian Gulch State Park. The cave passage leading into it covers about 75 m, based on a surface survey by Bruce Rogers (Rogers 1984), who noted that the predominant rock type in this area is a "Cretaceous-age Franciscan terrain greywacke and the cave is formed along a series of prominent near-vertical joints." A punchbowl at Mendocino Headlands State Park is an impressive feature with a 12 m freefall drop to the water below. Another monster sink, similar in size and depth to the one at Russian Gulch, is found on the south side of Little River. Based on surveys it is 40 × 37 m around the rim and 20 m deep (see map in Bunnell 2008a). At the bottom, an opening admits to a chamber where two large tunnels intersect to form the main cave. A sunbeam enters from the right hand tunnel and lights up the interior of the cave in clear weather.

On Point Arena, a large punchbowl can be seen from the lighthouse. While it certainly had cave passage entering in the past, judging from a 70-year old photo at the museum, there is no longer any roofed-over passage entering into it. South of Point Arena are the Stornetta Public Lands, recently opened to the public. Here there are a pair of littoral sinks, one fully collapsed and a nearby one in partial collapse, much like a solutional doline (Fig. 14.4) This is the only such littoral sink that the first author has observed during the pre-collapse phase out of dozens he

Fig. 14.5 Cave of the Lost Soles in Mendocino County, CA

has documented (Bunnell 2008a). The Stornetta lands contain many large cave openings and a large cave in the point just south of Point Arena. This 143 m-long cave has formed from the intersection of two fault-controlled passages at right angles. Because one of them was filled with tennis shoes—apparently flotsam from a cargo ship—it was named "Cave of the Lost Soles" (see Bunnell 1998 for a map) (Fig. 14.5).

Other large sea caves are known on the Mendocino Headlands and in Peters Creek Cove, although little survey has been done on the Headlands. There are numerous circular coves along the Mendocino coast, which may well have resulted from the collapse of other littoral sinks over time. A determination of why this section of coast is so prone to formation of such features would make an interesting geological study.

14.5.2 Marin County

Most of the caves in Marin County are found in two National Parks. The scenic Point Reyes National Seashore, at the northern edge of the San Andreas fault, marks the western edge of the Salinian Block, a huge slab of granite and metamorphosed sedimentary rocks that extends south through Santa Cruz County and down into Monterey County and the northern Big Sur coast. Rogers claims there to be over 220 sea caves in the park, including 10 caves over 30 m long (Rogers 2010). Many of these are on the hammerhead-like peninsula of granite which is Point Reyes. The longest of these is Point Resistance Cave (118 m). According to an account by Rogers (1986), California's first sea cave map was likely produced by two lighthouse keepers at Point Reyes in 1927. These caves were not walk-in caves—in order to access them, the lighthouse keepers might have descended the 30 m cliff beneath the lighthouse hand-over-hand on thick manila ropes, possibly sporting kerosene lanterns. Their map depicts two caves, one about 85 m long, with an upper dry level above a very actively forming lower level. "Stalactites found here" is shown on one end of a cross-faulting passage.

Other significant caves are El Reyes or Elephant Cave north of the peninsula, mapped to 96 m, formed along three parallel faults, and El Sobrante Pit, a littoral sink that is 51 m feet deep on the landward side and perhaps the deepest known in California (Rogers 2011). Further south, Rogers reports 119 sea caves in the Marin Headlands and other portions of the Golden Gate National Recreation Area.

14.5.3 Santa Cruz County

Much of the Santa Cruz coast is dominated by sedimentary cliffs up to 30 m high, formed in the Santa Cruz mudstone or Monterey Shale. Santa Cruz County has one of the largest concentrations of coastal sea caves in California, with 120 caves mapped, five of them over 90 m long. This includes the coast's second largest sea cave, Forbidden Fissures (213 m). The county also contains what may be California's largest sea cave chamber at 47 m wide, in Sarawak Cave. One unusual cave is Basketball Cave (67 m) which has two separate pit entrances in addition to its main seaward entrance. A drop of 12 m is afforded by the highest of these, into the back end of the cave.

14.5.4 Monterey County

This county is dominated by miles of rugged Big Sur cliffs, much of them composed of volcanic rocks battered by frequently rough surf. The geology is complex, with the northern coast part of the Salinian complex and the southern part of the Franciscan complex. Much of this rugged coast is inaccessible without a boat, and only four caves have been mapped. One of them proved to be quite large: Bixby Landing Cave, incompletely mapped at 155 m. Fifteen miles south of Monterey on twisty Highway 1, the Bixby Creek Bridge is a major landmark of the Big Sur coast. At 213 m long and 79 m high, it is one of the largest concrete span bridges in the world. The cave is found in the basalt peninsula jutting out on its north side. The cave has a single entrance, though surge activity in one of the unexplored side passages suggests a back entrance. Several passages radiate off this room, all of them water-floored even at low tide.

14.5.5 San Luis Obispo County

Caves are found in four areas of the county: at Montana d'Oro State Park (just south of Morro Bay), in and around Point Buchon, at Cave Landing (just south of Avila State Beach), and along the cliffs between Shell Beach and Pismo State Beach.

14.5.5.1 Montano D'Oro State Park

Located just south of Morro Bay, the park contains about 1.5 km of rugged coastline. Sea stacks, arches, and a few small caves are found. Much of the rock has a near vertical dip, which has resulted in the formation of long finger coves. Significant sea caves are confined to a few areas with horizontal bedding, at the park's southern end and at Spooner's Cove. The caves are formed in Pleistocene marine terrace deposits: loosely consolidated white and orange-brown sandstone and conglomerate.

14.5.5.2 Point Buchon Region

This includes the coast beginning just north of Montano d' Oro and continuing north. Much of the rock here dips steeply (50° at Point Buchon) and lateral passages have been cut along the dip in many caves. The rock here is a continuation of the Pleistocene deposit mentioned above. Here are found some of the largest coastal caves in California, including the largest, Sea Maze Cave, with several parallel passages, and a survey length of 229 m (Fig. 14.6). Point Buchon itself is riven through with an arch and a maze cave enterable on both sides of the point, with 141 m of surveyed cave. Nearby Double Door Cave was surveyed to 123 m.

14.5.5.3 Cave Landing

The Cave Landing area is just east of Avila State Beach; it is bordered by Fossil Pt. on the west and Mallagh Landing on the east. The area is also known as Pirate's Cove, and there are tales of pirates hiding their treasure in some of these caves. It contains numerous sea arches (in the center of the cove is a triple arch). Caves here are formed in orange Pleistocene-age sandstones,

Fig. 14.6 Sea Maze Cave map, California

often with colorful bands of red or yellow. There are several caves here with two distinct levels, probably as a result of local coastal uplift. These upper levels remain above the typical surf zone. Ten caves were surveyed in this area, the longest of them 88 m.

14.5.5.4 Shell Beach

The caves here are formed in Pleistocene sandstones. This sandstone is orange on the mainland but a chalky white in the large offshore stack in which the Cave-in-Rock caves are found. Some of the caves show evidence of local faulting, though none is indicated for the cave study area on the geological map of the Arroyo Grande Quad (C.A. Hall 1973). Ten caves have been surveyed here, ranging up to 70 m in length. One notable cave found at the end of Cliff Drive is expressed as a 15 m-deep littoral sinkhole. It was formerly developed as a tourist attraction under the name Dinosaur Cave and is also known as the Caverns of Mystery. It was the first sea cave surveyed by the California Sea Cave Survey (CSCS), formed in 1982 as a project of the National Speleological Society.

14.5.6 San Diego Region

14.5.6.1 Sunset Cliffs/Point Loma

Sunset cliffs is a scenic residential area in the city of San Diego, bordering the northernmost section of the sea cliffs which are continuous to the tip of Point Loma. Ladera Street forms a rough boundary between Sunset Cliffs and Point Loma (see description below). Although the cliffs extend northward as far as Ocean Beach, the major caves of Sunset Cliffs lie between Osprey and Ladera Streets. The cliffs are composed of Upper Cretaceous and Pleistocene sedimentary rocks, predominantly olive-gray, thin bedded marine sandstone and siltstone known as the Point Loma formation (Kennedy 1973). Two prominent vertical joint sets cut this formation at intersecting angles that strike N 30–40 W and N 40–50 W. This has resulted in formation of large rectangular blocks such as the prominent offshore stack near Froude Street.

Point Loma is a 2 km long peninsula extending south from Ladera Street. Cabrillo National Monument occupies its southern end. Numerous caves are found along the ocean side of the point. The geological setting is the same as described above for Sunset Cliffs.

Kennedy (1973) concluded that the average rate of erosion of Sunset Cliffs is about one-half inch (1.27 cm) per year. By far the major contributor to erosion of these cliffs in general has been the collapse of sea caves. The large blowhole above Big Blowhole Cave is an example of such a collapse. A much larger blowhole existed at the foot of Osprey Street as recently as 1973, but its seaward wall has since collapsed; it was illustrated on old postcards and called the Devil's Pot. More recently, a large portion of the entrance chamber to Mayan Temple Cave was removed by the extensive storms in 1983, shortening the cave by a good 12 m or so (fortunately, we hadn't mapped it yet). Since some of these caves extend under the road and close to residences, the Army Corps of Engineers piled large rocks in front of many of the cave entrances in 1971 to prevent further erosion. Interestingly there is little if any rock piled in front of two of the largest cave entrances, Big Blowhole and Seaweed. On the other hand, the piled rock almost completely blocks one of the entrances to Rat Cave. All these caves were surveyed in the early 1980s by the CSCS and range up to 100 m in length.

14.5.6.2 La Jolla

Perhaps the most famous of the sea caves on the California coast are the "seven sea caves of La Jolla." The caves are located in a 60 m-high, north-facing cliff composed predominately of sandstone of the Cretaceous Chico formation (Halliday 1962); horizontal layers of shale, mudstone and conglomerate are also evident. Fragments of conglomerate jut out from the cliff face and some segments have been carried away by the waves, leaving potholes of varying diameters. Some fossils may also be found in the shale and mudstones (Winsted 1913).

Postcards of the cliffs dating back to 1899 show they have changed little over the years, indicating that the rate of erosion here is probably not great. The only major recent alteration seems to have been the collapse of an arch, which occurred sometime around 1927 (Shepard and Grant 1947). By studying the graffiti carved in the caves and on the cliffs between 1930 and 1940, K. O. Emery estimated the rate of erosion to be about one foot (30.5 cm) every 600 years (Emery 1941). In addition to focused wave energy (Waterstrat et al. 2010), marine animals such as chitons contribute to sea cave development by drilling holes in the rock (Moore 1954).

The caves are accessible by climbing down to the ocean at either the east or west end or via the tunnel leading from inside the gift shop to the back of commercially developed Sunny Jim's Cave. They were mapped in 1982 by the CSCS (Fig. 14.7). The three most notable caves.

14.5.6.3 Sunny Jim's Cave (99 m)

This is the most well-known and heavily visited of the La Jolla caves. A German professor Gustav Schultz, hired laborers who worked 2 years with a pick and shovel building a tunnel from the surface to the back of the cave. A few years after the tunnel's completion in 1903, electric lights and 133 steps were added.

The name for this cave is derived from the obvious silhouette of the profile of a man outlined by the rocks at the entrance. As seen from the wooden platform in the back of the cave, "Sunny Jim" a British cartoon character from the 1920s, appears to be facing left and wearing a pointed stocking cap. Numerous versions of this classic entrance shot are featured on postcards dating back to the early 1900s. The main portion of the cave is a 30 m-long breakdown-floored room. The breakdown has a characteristic purple color from red algae. Two long side tunnels formed along joints (Moore 1954) extend from the main room.

14.5.6.4 Arch Cave (185 m)

The largest and most complex of the La Jolla caves, an impressive 12 m-high arch separates the two main sections. It is likely that this single cave is actually considered to be two of the traditional "seven caves" since, when viewed from the ocean, the arch appears to separate the two large entrances. Several passages up to 7 m high and 30 m long radiate back from the multiple entrances.

14.5.6.5 White Lady Cave (81 m)

After Sunny Jim Cave this cave is the most well-known and apparently the only other cave with a historic name. The legend behind the name is that a young couple were honeymooning in the area and visited the cave at low tide. They were preoccupied and failed to notice the rising tide, and were never seen again. The entrance is said to show a profile of the young woman in her bridal gown, and an old 1920s-era postcard shows her silhouette in the entrance and names the cave.

The cave has two large entrances and is easily accessible from the eastern side of the cliffs, making it one of the more highly visited caves in the area.

14.6 The Channel Islands

The Channel Islands are a chain of 8 islands lying off the coast of southern California, ranging from 22 to 60 km out to sea. Sea caves are known to occur on all of them but survey efforts have been confined to the five that comprise Channel Islands National Park (Anacapa, San Miguel, Santa Barbara, Santa Cruz, and Santa Rosa). With over 15 km of passage mapped in over 380 caves, the park has more caves than any western park other than perhaps the Grand Canyon. Published accounts of the Island caves are known from 1891, when Lorenzo Yates describes several in the *American Geologist*. Painted Cave and Cueva Valdez, two large caves on Santa Cruz Island, are well known and marked on the USGS topographic maps.

Santa Cruz and Anacapa formed from basalt, extruded as the Farallon plate was subducted under the North American plate, during the Miocene. Two to five million years ago they were uplifted, resulting in hundreds of faults and fractures which gave the surrounding seas an entry point to erode into the cliffs. Almost every cave on these islands has formed along an obvious fault, whereas others have formed along a contact zone between basalt and less resistant agglomerate. Complex caves and large internal chambers have formed where one of more faults intersect.

Pleistocene times (beginning two million years ago) also marked the onset of a continuing pendulum effect within global climatic regimes. A series of recurrent ice ages have influenced sea level stands within a vertical range of >100 m. The most recent melting of ice age glaciers, commencing some 17,000 years ago, launched the latest period of rapid sea level rise. Well-formed marine terraces on Santa

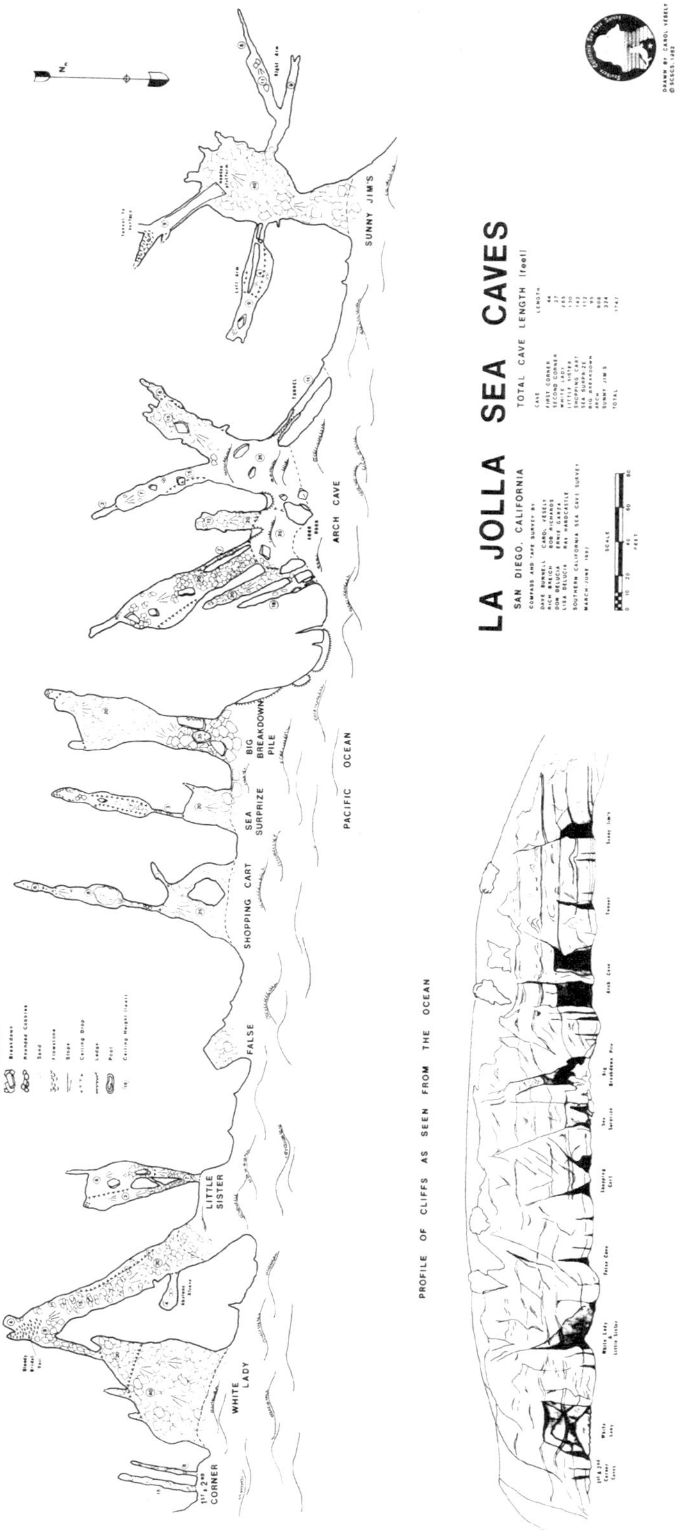

Fig. 14.7 La Jolla Caves map, California

Cruz Island mark the interplay of gradual uplift with episodic encroachment of the sea. These surf-cut platforms may be most readily observed in the alluvial shale sediments atop the northwest face of the island, above a large concentration of caves that penetrate especially deeply into the cliffs below. Draining along the platforms sometimes follows gullies formed along faults which sometimes correspond to caves below, such as above Painted Cave. This may have facilitated influx of meteoric water, which in turn has likely facilitated erosion from the surf below.

The interplay of regional variations in uplift and more global sea level rise may help to explain the extreme vertical development of some Channel Island sea caves. Littoral erosion has taken its course within a vertical range that has remained more or less constant, though island elevation, measured relative to sea level, has undergone continual change. In light of this change, one may speculate that extensive, presently submerged caves could be found where little uplift has occurred, in passage carved out at lower sea level stands. Although few have yet been described, at least one of these caves (Seals' Cove Submerged Cave on Santa Cruz) appears to be quite large and complex-enough so that a team of divers became lost inside and were fortunately rescued by a search and rescue (SAR) team from Santa Barbara, having survived by finding trapped air pockets. Summaries of sea cave work in the Channel Islands can be found in Bunnell (1983, 1993b, 2008a, b, 2009) and in two books: Bunnell (1988) and Bunnell (1993a).

The larger Channel Islands were home to populations of Native Americans, most recently the Chumash Indians, since at least 11,700 ybp (Schoenherr et al. 1999). Anacapa and Santa Barabara islands lack permanent water and appear to have been visited but not inhabited. Relict sea caves with pictographs have been found on Santa Cruz and San Nicholas Islands, and kitchen midden sites were observed in some of the Santa Cruz Island caves (Bunnell 1988). One of these was reported to the Santa Barbara Museum of Natural History, who later excavated an intact Chumash harpoon (Timbrook and Johnson 1988).

14.6.1 Anacapa Island

This group of three small islets form a shoreline composed almost entirely of 46–60 m high volcanic cliffs, with a few beaches appearing at low tide. Some geology texts suggest that the islets were actually split by collapse of sea caves, and at least one intact cave penetrates through one of the islands. Concentrated survey efforts here from 1986 to 1993 yielded 135 caves with an aggregate length of 6.26 km (Bunnell 1993a). Longest of these is Catacombs Cave, 246 m. This unusually maze-like sea cave has formed along at least 6 intersecting faults. Cathedral Cave is featured on park maps and is popular with sea kayakers who enjoy its spacious 241 m length that affords a through-trip. Figure 14.8 shows the large volume of another of these caves (Half-Gone) and the chunky weathering that littoral erosion of basalt produces.

14.6.2 Santa Barbara Island

Smallest of the five islands in the park, the island is barely 1 mile across. In one 4-day excursion the first author's team surveyed a dozen caves ranging up to 175 m in length. These caves are formed in the basaltic cliffs and are concentrated near Arch and Webster Points. The latter has an unusual concentration of deep parallel cave systems which don't intersect for the most part. Work here has yet to continue, with at least a dozen other known caves to survey.

14.6.3 Santa Cruz Island

Santa Cruz is the largest and most mountainous of the Channel Islands, 22 miles long, 6 miles at its widest, and over 77 miles of shoreline. Over 120 caves have been mapped, and after two decades

Fig. 14.8 Half Gone Cave on Anacapa Island

of survey there are 26 caves over 90 m long and 46 over 60 m long, with a combined length of over 7.5 km. Many caves remain to be surveyed on this island (Bunnell 2009).

Many of these large caves are found among the sheer cliffs on the west end of the island, typically single passages penetrating 60–100 m or more into the cliffs. Painted Cave is the largest and most awe-inspiring of all the caves on the island (Fig. 14.9-Painted Cave map). It is presently the world's second largest surveyed sea cave. The 39 m-high entrance leads into a wide passage which heads straight back for 182 m, the ceiling lowering steadily to a point where it is only about 5 m above water level. Beyond the low point, the cave opens into a dark inner chamber, some 60 m long, 38 m wide, and up to 12 m high (Fig. 14.10). Two ledges at one end of the chamber often house a large population of California sea lions. The seal population varies seasonally, with the largest population in winter and spring and relatively few seals in the summer months. Two passages lead from the main chamber to cobble beaches. The large size of the inner chamber is attributable to the presence of a second fault which intersects the fault forming the cave's major axis at a 60° angle. The high internal humidity of the chamber, due largely to its restricted entrance, probably accelerates weathering of the rock within. Other impressive caves include the Blimp Hangar, with its 60 m-high entrance, one of the larger cave entrances of any type in the western USA.

14.6.4 San Miguel Island

San Miguel is the third largest of the Channel Islands but only has a relatively short section of shoreline on its eastern side composed of the cave-forming Miocene-age volcanics. Orr (1951) reports only six caves. However, San Miguel is unique in having the only sea caves known in the Channel Islands with burials of pre-contact Native Americans in some. One such relict sea cave, Daisy Cave, has been extensively studied (Erlandson et al. 1996) and midden deposits there have been radiocarbon dated to 11,700 ybp. It is the oldest coastal shell midden known in North America (Schoenherr et al. 1999).

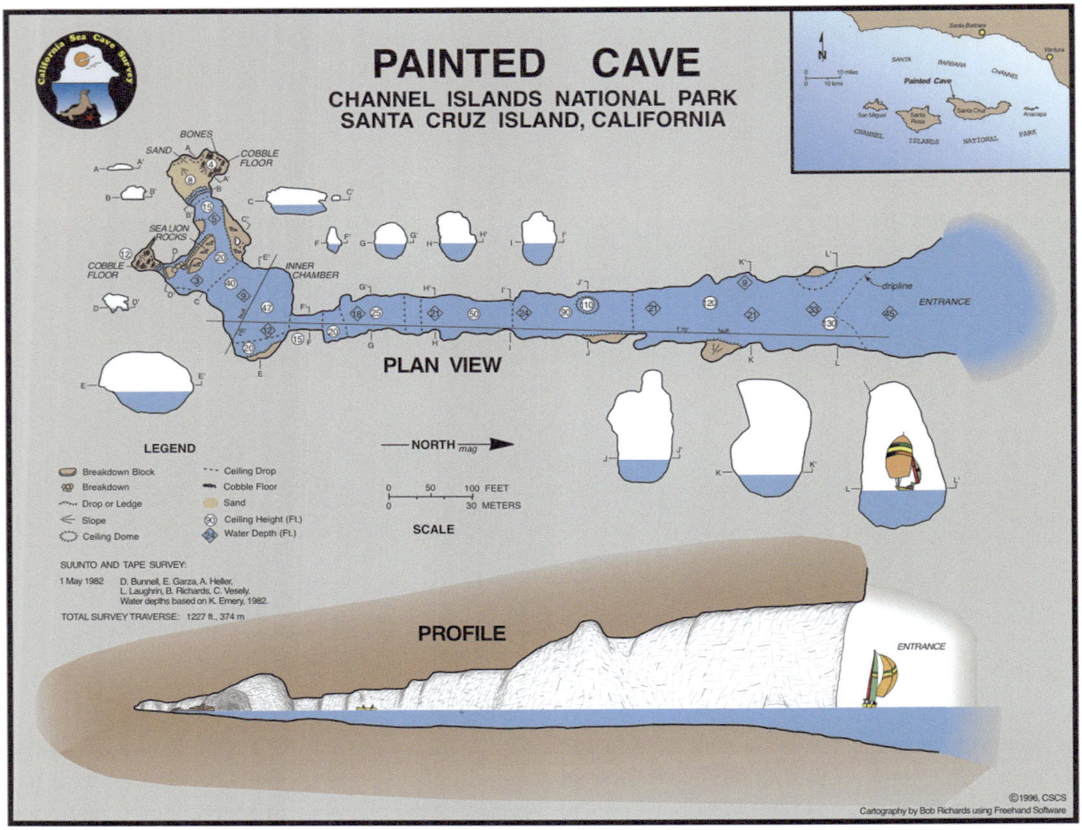

Fig. 14.9 Painted Cave map, Santa Cruz Island

Fig. 14.10 Painted Cave inner chamber, Santa Cruz Island

14.6.5 Santa Rosa Island

Santa Rosa is the second largest of the Channel Islands and offers a wider variety of additional host rocks for sea cave development beyond the basalt that predominates on the other islands: sandstone, conglomerate, mudstone. Survey work here is also incomplete but to date have 45 caves have been surveyed by the CSCS. The largest so far is Jumbo-Gumbo Cave, an impressive borehole passage up to 30 m wide and 204 m long, littered with huge breakdown blocks, and formed in basalt. Three caves have large littoral sinkholes, including the 138 m Witch's Cauldron, with a skylight over 30 m in width. This formed at the intersection of two cross-faults in marine mudstone and conglomerate.

14.7 Conclusions

Despite years of survey work on the west coast of the United States, there are likely hundreds of caves remaining to be explored and surveyed, especially in Washington and Alaska. By far the largest and most enduring caves are formed within coastal volcanics, and typically longer caves, larger chambers, and littoral collapses tend to form on intersecting faults and on the windward coasts of islands and peninsulas. Sea caves lifted above the active tidal zone by regional uplift have proved important in archaeological studies in Alaska and the California Channel Islands.

References

Baichtal JF (2011) Timing and development of sea caves along the coastline of southern Southeast Alaska. Beneath the Forest. National cave and karst newsletter published by the U.S. Forest Service 4:2

Bird E (2000) Shore platforms at cape flattery, Washington. Wash Geol 28(1/2):10–15

Bunnell D (1983) The amazing caves of Santa Cruz Island. NSS News 41(2):86–91

Bunnell D (1988) Sea caves of Santa Cruz Island. McNally and Loftin, Santa Barbara, 123 p

Bunnell D (1993a) Sea caves of Anacapa Island. McNally and Loftin, Santa Barbara, 207 p

Bunnell D (1993b) Sea caving in the Channel Islands. NSS News 51(6):150–159

Bunnell D (1998) California's coastal sea caves. NSS News 56(10):292–300

Bunnell D (2008a) Vertical sea caving. NSS News 66(10):11–18

Bunnell D (2008b) Return to painted cave. NSS News 66(1):19–21

Bunnell D (2009) Return to Santa Cruz Island. NSS News 67(1):12–19

Coney PJ, Jones DL, Monger JWH (1980) Cordilleran suspect terranes. Nature 288:329–333

Elrick J (2003) Three dimensional sea cave volume determination and accuracy modelling. Surv 480 research project for bachelor of surveying degree, National School of Surveying, University of Otago, Dunedin, New Zealand

Emery KO (1941) Rate of surface retreat of sea cliffs based on dated inscriptions. Science 93(2426): 617–618

Erlandson JM, Kennet DJ, Ingram BL, Guthrie DA, Morris DP, Tveskov MA, West GJ, Walker PL (1996) An archaeological and paleontological chronology for daisy cave (CA-SMI-261), San Miguel Island, California. Radiocarbon 38(2):355–373

Hall CA (1973) Geology of the Arroyo Grande 15′ quadrangle, San Luis Obispo County, CA. Published by the US Geological Survey, California

Halliday WR (1962) Caves of California. Special report of the Western Speleological Survey, National Speleological Society

Johnson ME, Ledesma-Vasquez J (2009) Atlas of coastal ecosystems in the western gulf of California: tracking limestone deposits on the margin of a young sea. University of Arizona Press, Tuscaloosa

Kennedy MP (1973) Sea-cliff erosion at sunset cliffs, San Diego. Calif Geol 26:27–31

Lund E (1971) Coastal landforms between Florence and Yachats, Oregon. The Ore Bin 33(2):21–44

Meighan CW (2000) Overview of the archaeology of San Clemente Island, California. Pac Coast Archaeological Soc Q 36(1):1–17

Moore DG (1954) Origin and development of Sea caves. Natl Speleological Soc Bull 16:71–76

Niem A (1975) Geology of Hug Point State Park Northern Oregon Coast. The Ore Bin 37(2):17–36

Orr PE (1951) The Orca goes underground. Published by the Santa Barbara Museum of Natural History Museum Talk 26:2

Rogers B (1984) Mendocino sea caves, in Short Rounds, a newsletter of the Golden Gate Grotto of the National Speleological Society, June issue

Rogers B (1986) Caves at Point Reyes. Short Rounds, a newsletter of the Golden Gate Grotto of the National Speleological Society, Feb 1986

Rogers B (2010) Point Reyes, landscape in the past, Part 1. California Caver, Western Region of the National Speleological Society, 245(Spring issue)

Rogers B (2011) Point Reyes, landscape in the past, Part 3. California Caver 248(Winter issue)
Schoenherr AC, Feldmeth R, Emerson MJ (1999) Natural history of the Islands of California. University of California Press, Berkeley
Shepard FP, Grant US (1947) Wave erosion along the southern California coast. Bull Geol Soc Am 68: 919–926
Sherman DJ (2005) North American coastal geomorphology. In: Schwartz ML (ed) Encyclopedia of coastal science. Springer, Dordrecht, pp 721–728
Timbrook J, Johnson J (1988) Rare Chumash harpoon discovered. Bulletin of the Santa Barbara Museum of Natural History (Number 121), October
Waterstrat WJ, Mylroie JE, Owen AM, Mylroie JR (2010) Coastal caves in Bahamian eolian calcarenites: differentiating between sea caves and flank margin caves using quantitative morphology. J Cave Karst Stud 72:61–74
Winsted HL (1913) The sea caves at La Jolla, California. J Geogr 125–126, Dec 1913

Climatic Influences on Coastal Cave and Karst Development in Florida

15

Jason S. Polk and Robert Brinkmann

Abstract

Coastal karst development in Florida is a complex, temporally variable phenomenon that was influenced by eustatic processes and long-term climatic variability through the Cenozoic up through the present. Much of the past history involving coastal and climatic influences on cave and karst evolution in Florida is still not well understood. Currently, Florida is home to almost 20 million people that reside one of the world's largest and most productive karst aquifer systems, featuring hundreds of springs, caves, and sinkholes as part of its karst geomorphology. The large carbonate platform is exposed to hurricanes, sea level rise, and continued groundwater withdrawal. Therefore, it is important to consider Florida's karst landscape as an evolving geomorphological system under both past and current environmental conditions. It is important to understand the rock, structure, gradient, fluid, and time elements of Florida's karst development, with a particular emphasis placed on the climatic influences that shaped the landscape. Focusing on the central and northern regions of the peninsula where the ridgelands and coastal lowlands contain the most exposed karst features, the elements of gradient, fluid, and time interact with climate in such a way that the karst landscape development can be partially explained by changes in the hydrology related to water table fluctuations, sea-level changes, and variability in precipitation patterns over time. Aquifer development, speleogenesis, and paleoclimatic changes on these processes are discussed, while noting that the primary mechanisms controlling cave development,

J.S. Polk (✉)
Department of Geography and Geology, Western Kentucky University, Bowling Green, KY 42101, USA
e-mail: jason.polk@wku.edu

R. Brinkmann
Department of Geology, Environment, and Sustainability, Hofstra University, Hempstead, NY 11549, USA
e-mail: Robert.Brinkmann@hofstra.edu

M.J. Lace and J.E. Mylroie (eds.), *Coastal Karst Landforms*, Coastal Research Library 5,
DOI 10.1007/978-94-007-5016-6_15, © Springer Science+Business Media Dordrecht 2013

sinkhole occurrence, and karst evolution are past changes in sea level and water table movement. The state's karst regions are described within this context to organize the discussion of the coastal karst development of peninsular Florida under variable climatic conditions.

15.1 Introduction

Coastal karst landscapes occur in many areas of the world, particularly on islands in tropical regions. Subtropical peninsular Florida, almost completely underlain by carbonate bedrock, provides a geographically different case study of coastal karstification heavily influenced by eustatic processes and climatic variability in a region where continental processes are more often associated with karst development. Boasting one of the world's largest and most productive karst aquifer systems, along with hundreds of water- and air- filled caves, Florida is home to almost 20 million people who depend on its precious karst groundwater. Yet, it is still quite possibly one of the least understood coastal karst systems. Part of this is because of Florida's complex coastal karst geomorphology, which evolved under highly variable climatic and geological conditions throughout the Cenozoic, with periods of rapid karstification through the Neogene and Quaternary. With new attention being given to global climate change and sea level rise, only recently has research focused more on understanding the paleoclimate of Florida and the surrounding region (Grimm et al. 1993, 2006; Watts et al. 1992; Watts and Hansen 1994; Alvarez-Zarikian et al. 2005; van Beynen et al. 2007a, b, 2008; Polk 2009, Polk et al. 2012; Morrissey et al. 2010). It is from this perspective that the influence of climatic changes on karst development and water resources in the state needs advancing. As climate change uncertainties persist, only time will provide clues to the potential impacts on Florida's karst landscape. The possible implications on the vulnerability of the peninsula's natural resources, including springs, lakes, and its massive, yet finite, aquifers, provide impetus for further efforts to be undertaken to better understand this unique karst landscape.

From a geomorphological viewpoint, changes in equilibrium result from a variety of causes that influence landscape evolution. In particular, changes in climate often are a driving factor for geomorphological events, thus it is from this perspective that Florida's karst landscape evolution will be discussed. While the examples presented herein will be primarily qualitative and descriptive in nature, research in the past few decades has provided more quantitative measurements of coastal karst development in Florida, particularly with respect to air-filled and underwater caves, and phreatically formed conduits. This chapter will focus on a more broad-based explanation of certain major karst regions and features in the state, with an emphasis on the west-central and northern areas to outline coastal karst geomorphology and cave development under varying climatic conditions in Florida.

Moving north to south, Florida's geology and geomorphology is more heterogeneous than expected, particularly in reference to cave development and karstification (White 1970). Over the past several million years, the Florida platform was geologically stable, without the influence of tectonics to cause significant or abrupt changes in the topography or elevation of the landscape (Beck et al. 1985). Thus, climatically induced sea-level variability, and concomitant deposition and erosion, is the primary driver of changes in the both the size and relief of the Florida platform, and also a major cause of water table fluctuations and subsequent karstification at varying elevations within the state. In considering the evolution of karst development in subtropical regions with fairly young, eogenetic carbonate rock, often the focus is on the geological influences and their interaction with hydrological variations over time; however, the influence of climate variability, including glacioeustatic changes, receives less attention in the overall discussion of its role in the geomorphic evolution of karst landscapes.

In the case of Florida, this influence, combined with the state's coastal setting, is a primary factor in its karst development (Brinkmann and Reeder 1994; Florea 2006; Gulley et al. 2012a, b).

In any karst setting, there are five primary factors that contribute to the initial ability for karstification and speleogenesis to occur, along with continued evolution of the landscape's karst features (Groves 1993). These include: the type of rock, the geologic structure (geometry), gradient (induces flow paths and drainage), fluid (aggressive fluid capable of rock dissolution), and time. Two other factors should be added to this list: climate and geologic setting—in this case that interests us in this paper, the geologic setting is the proximity to the marine coast.

15.2 Geologic Setting

In Florida, the geologic setting is fairly straightforward upon first analysis. The peninsula consists mostly of carbonate rock (limestone and dolostone) that extends for depths up to several thousand meters in some areas (Lane 1994). Gypsum is also present at depth. These carbonate and evaporitic rocks overlie igneous and sedimentary (sandstone, shale) basement rocks formed in the Precambrian and Cambrian, with carbonate rock deposition beginning in the Cretaceous period after the platform was flooded by the Northern Atlantic's shallow waters (Lane 1994; Florea 2008) (Figs. 15.1 and 15.2). Extreme sea level fluctuations during the Cenozoic helped to create the topographical nuances within the peninsula still present today, with unconformities occurring from variable periods of carbonate rock deposition, and sedimentation from the eroding Appalachian Mountains during the Paleogene. Continued sedimentation occurred throughout the Neogene and Quaternary, primarily consisting of thin beds of quartz sands, carbonates, and clays (Lane 1994). Progressing upward through the sequence, the bedrock and overlying sediments become increasingly complex due to variations in sea level and sediment deposition causing irregular and disjointed features, unconformities, and missing units that were eroded away during past sea-level lowstands.

Most of the exposed carbonate rocks are in the northern, central, and western coastal lowland areas of the state, and are Middle to Late Eocene in age, with some outcrops of mid-Oligocene rocks (Figs. 15.1 and 15.2). Many of these formations are overlain by siliclastic sediments that form the Miocene Hawthorn Group and Pleistocene and Quaternary siliclastics and marine sands to create a covered, or mantled karst system (Kuniansky 2001). In the central and northern parts of the state, much of the Oligocene Suwannee Limestone has eroded away, leaving the Eocene Ocala Limestone exposed at the surface. The Ocala is formed from multiple depositional sequences up to 35 m thick, with a total thickness up to 120 m or more in some areas (Miller 1986), and also has dolomitic layers within it at various locations throughout the state (see Florea 2008 for a detailed description). These can cause some semi-confining areas where indurated dolomite exists (the Ocala also contains sucrosic dolomites with higher porosity) (Gaswirth et al. 2006). The Upper Floridan Aquifer also flows within these stratigraphic units (Fig. 15.2). Below the Ocala, the Avon Park, Oldsmar, and Cedar Key Formations are found, which are also Eocene in age and form part of the Lower Floridan Aquifer (a middle confining unit exists within the Avon Park formation). Significant karst features exist in these formations as well, primarily as water-filled conduits and paleokarst features, and are discussed in more detail below.

15.2.1 Structure

Florida's major karst forming rocks are young, eogenetic carbonates, ranging from 48 to 25 million years old, and thus contain high porosity and permeability characteristic of recent carbonate deposition that has not undergone significant induration or compaction. Often, the porosity is 30–40 % and the matrix permeability can be as high as $10^{-14.5}$ m^2 (Budd and Vacher 2004; Florea and Vacher 2007) (Fig. 15.3a). Most of Florida's stratigraphy is horizontal, with some

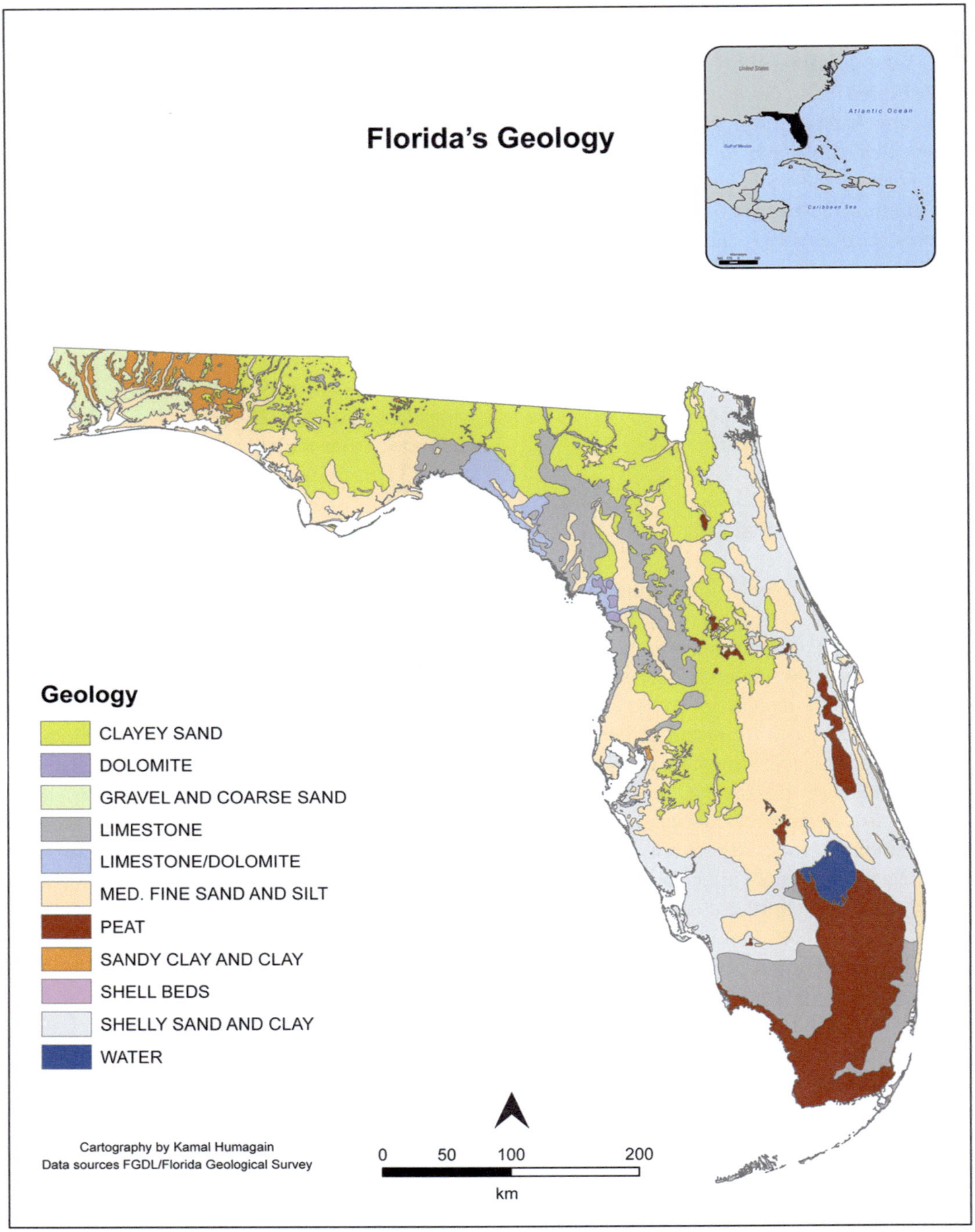

Fig. 15.1 Map of Florida's surface geology indicating areas of carbonate rock outcrops and exposures (Adapted from Florea 2008)

Period	Epoch	Lithostratigraphic Unit	Hydrostratigraphic Unit
Quaternary	Holocene	Pleistocene-Holocene Undifferentiated sediments	Surficial Aquifer System
	Pleistocene		
	Pliocene		
	Miocene	Coarse clastics Miccosukee Formation Cypresshead Formation	Intermediate Confining Unit
Tertiary		Hawthorn Group	
	Oligocene	Suwannee Limestone	Floridan Aquifer System
	Eocene	Ocala Limestone	
		Avon Park Formation	
		Oldsmar Formation	
	Paleocene	Cedar Keys Formation	Lower Confining Unit

Fig. 15.2 Northern and central Florida's stratigraphy, including lithology and hydrologic sequences. Units not to scale (Modified from Scott (1988) and Moore et al. (2009))

cross-bedding and gentle dipping that occurs in several areas of the state (Fig. 15.3b). The primary topographic highs occur along the Peninsular Arch, which runs diagonally almost parallel to the eastern coastline through the center of the state (Beck et al. 1985). Typically, bedding planes and a regional joint system trending northeast or northwest in direction dominate the stratigraphy, and influence the development of caves and sinkholes throughout the northern and central regions of the state (Florea 2006). Fracture systems and vadose matrix flow are the primary ways by which water enters the karst system, with some sink and rise systems existing in north Florida such as the Santa Fe River sink and rise system (Martin and Dean 2001; Gulley et al. 2012a, 2012b). The flow elevation and direction are dictated by confinement layers and gradient. Overall, the structure of the bedrock, including its high matrix porosity and permeability, prominent regional joint and fracture systems, and continuous bedding plane contacts allow for diffuse recharge and homogenous groundwater flow. This combination of characteristics also contributes to the ability of precipitation and water table fluctuations to be driving factors in karst development by cutting across the structural features to follow the gradient to sea level, which serves as the ultimate base level for the region.

15.2.2 Gradient

The Florida peninsula ranges in elevation from near sea level at the coast to its highest point, 345 ft (105 m) in northern Florida at Britton Hill. Along the major topographical highs, including the Brooksville Ridge, Ocala Uplift, Cody Scarp, Lake Wales Ridge, and others, elevations range from 15 to 295 ft (5–90 m), and thus provide a slight gradient to influence flow regimes, with most of the Upper Floridan Aquifer's bedrock having a regional dip of less than one degree (Scott et al. 2004; Florea 2006) (Fig. 15.4). The hydraulic gradient can increase movement toward the coastline due to deeper conduits that exist in

Fig. 15.3 (**a**) Close-up of the eogenetic carbonate bedrock showing its highly porous and permeable character in Finch's Cave, Marion County. (**b**) Cross-bedding in Maynard's Cave in west-central Florida, indicating past evidence of coastal processes at work during deposition of the Ocala Limestone when the shoreline was further inland than the present

the lower formations, with freshwater found at depths in places such as Weeki Wachee Spring, which has flow rising from a depth of over 400 ft (120 m) before reaching the surface at the spring vent. The entire platform, including what is now under ocean water, is over 300 miles (480 km) wide and extends over 150 miles (240 km) into the Gulf of Mexico and over 70 miles (113 km) into the Atlantic Ocean (Randazzo and Jones 1997; Scott et al. 2004). Only a small portion (less than 50 %) of the peninsula is exposed at the present sea level. During the past peak interglacial periods, it has been argued that sea level was more than 100 ft (30 m) above the current level, leaving much the peninsula as islands (Lane 1986). In addition, during low sea-level

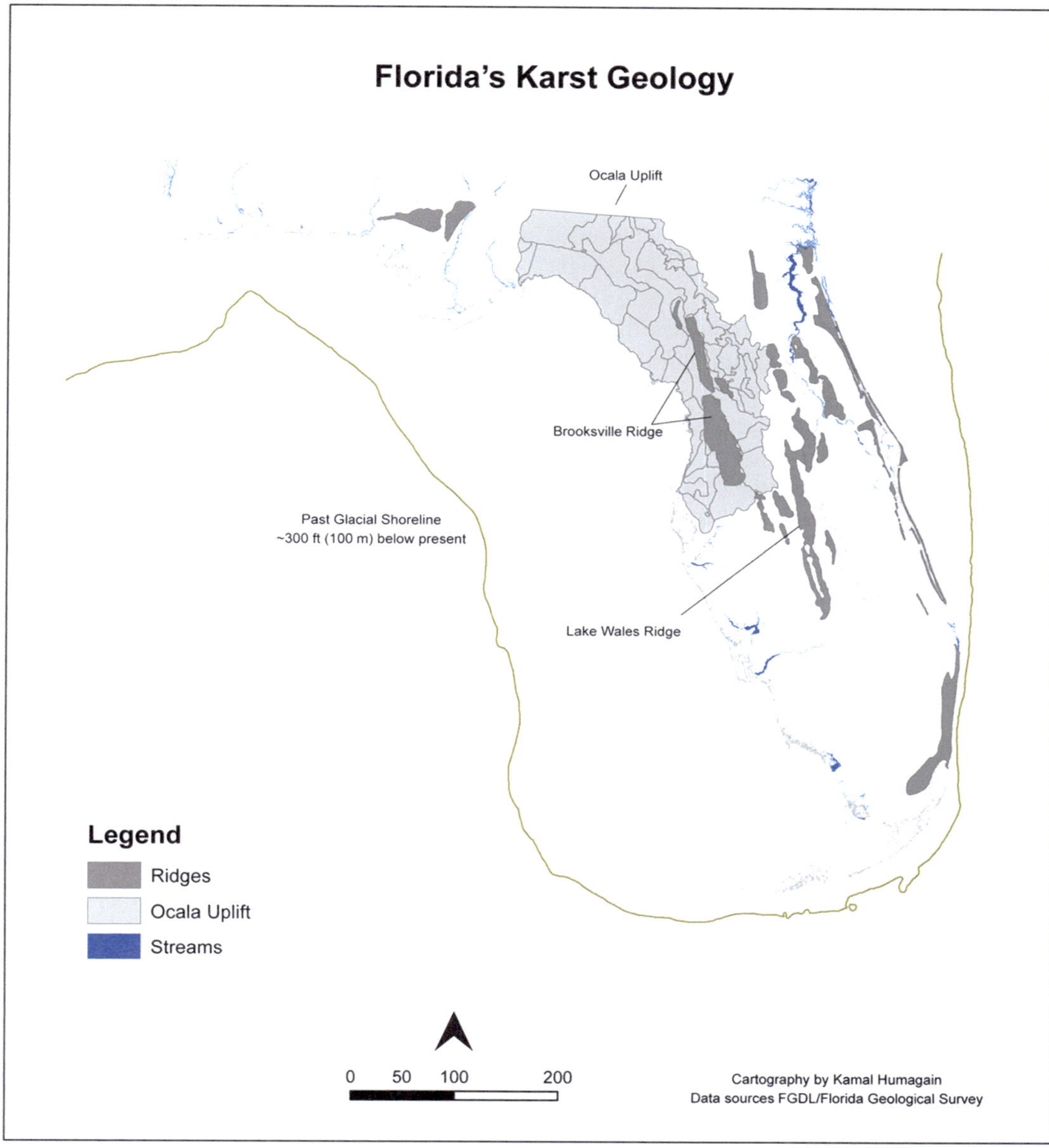

Fig. 15.4 Ridges and prominent karst areas described in this chapter. Florida's major ridges are shown here, including the *yellow line*, which indicates the continental shelf and edge of the platform at its widest during past glacial minimums (~100 m lower). Coastal marine terraces often run parallel to this line moving inward toward the modern shoreline of Florida

stands, the peninsula was significantly larger as evidenced by offshore karst features. Despite the low gradient, there is enough recharge in the aquifer systems to produce significant volumes of water flowing toward the coastline, and thus the formation of many large, first magnitude springs (Scott et al. 2004). The low gradient and relief may make the existing karst features subtler than in other karst settings, but they are still indicative of common karst processes at work, including the movement of fluid necessary for karst development to occur. The influence of sea level fluctuations is significant on determining the size, shape, and gradient of the peninsula and in turn the water table as well.

15.2.3 Fluid

More than 700 springs flow from underground in Florida, with a total discharge of over eight billion gallons per day (30 billion liters) (Lane 1986; Scott et al. 2004). It is estimated that over 2 quadrillion gallons (75 quadrillion liters) of water flows within the Floridan aquifer system (Scott et al. 2004). Some surface drainage occurs in the major rivers, such as the Santa Fe, Withlacoochee, and Hillsborough, to name a few, while many karst springs also flow into these river systems. These rivers flow at base level and intersect the water table, with certain features, such as sinkhole complexes, karst windows, and some air-filled caves also comprising part of the recharge system. Most springs occur in the central and northern karst regions (Fig. 15.5a), and flow from the Upper Floridan Aquifer (Scott et al. 2004). The interaction of these hydrologic features within the peninsula directly contributes to karst development by creating drainage basins, removing dissolved bedrock, and organizing flow paths toward the coast. Hence, changes in precipitation, water table elevations, and sea level all influence the existence, flow direction, and discharge of these features throughout changing climate regimes.

Along the coast of Florida, there also exist several submarine springs, primarily on the Gulf of Mexico side of the peninsula (Scott et al. 2004). Many of these springs are poorly documented and could potentially provide insight into the past hydrologic regime during lower sea-level stands. These springs are likely older features that formed due to the past topography during sea level regressions, and lends merit to the idea that most of the subsurface flow in the geologic past was toward the western side of the state off the topographic highs found along the central axis of the current peninsula, and followed the increased gradient where the peninsula had extended several kilometers into the Gulf (Beck et al. 1985). The most well known is the Spring Creek Group, which is estimated to produce over one billion gallons (~3.7 billion liters) of water per day (Scott et al. 2004).

Precipitation provides the primary form of recharge to the Floridan Aquifer System, with the state receiving an average of 50 in. (127 cm) of rainfall per year in the regions described herein (Winsberg 2003). Of this, over 70 % occurs during the summer, although effective recharge rates vary between 1 and 10 in. (2.54–25.4 cm) per year over about 55 % of the state, primarily due to high evapotranspiration rates (Winsberg 2003; Scott et al. 2004; Alvarez-Zarikian et al. 2005; Polk et al. 2012). Warm, humid air from the Gulf of Mexico drives many of the rainfall events, along with large frontal systems from the west, both of which can produce frontal and large convective storms with substantial amounts of precipitation (Alvarez-Zarikian et al. 2005; van Polk et al. 2012). Intense hurricanes and tropical storms also impact the peninsula on a regular basis, primarily in the late summer, and can contribute significant amounts of precipitation to recharge and water table fluctuation. These storms vary in frequency and severity, and thus can impact cave and karst development through influencing recharge (Florea and Vacher 2007; van Polk et al. 2012).

The ten largest freshwater springs in the state discharge an average of almost 7,000 f^3/s (~200 m^3/s) of water, which then removes over 2,000 t (1,814,369 kg) of dissolved solids per day (Table 15.1). In turn, each respective spring basin's discharge equates to overall landscape lowering from carbonate rock dissolution of approximately 1–2 in./year (2.5–5 cm/year), which makes its way to the Gulf of Mexico or Atlantic Ocean (Lane 1986). The surface is actually lowering at a much slower rate, indicating that the bulk of the dissolutional removal is coming from within the aquifer itself. If these values remained constant throughout millions of years, then the karst landscape of Florida would be very different today; hence, changes in precipitation amounts,

Fig. 15.5 (**a**) Rainbow Springs in Marion County, a first-magnitude spring and the fourth largest in the state. It is a state park, and serves as the headwaters for the Rainbow River. (**b**) Peck's Sink in Brooksville, FL, which drains the entire city. Note the person in white for scale (Photo by Tom Turner)

discharge, and drainage basins influence the amount of landscape denudation that occurs. Understanding climatic variability over geologic timescales is important for this reason, as making assumptions about future landscape lowering rates and karst dissolution processes from modern climate and hydrologic data may be inaccurate.

Table 15.1 List of Florida's top ten largest springs by magnitude of average discharge per year

Rank	Spring	Cave system mapped	County	Average discharge (cfs)
1	Spring creek	No	Wakulla	2,003
2	Crystal river (King's Bay)	No	Citrus	916
3	Silver	Yes	Marion	799
4	Rainbow	No	Marion	711
4	Alpaha Rise	No	Hamilton	608
6	St. Marks Rise	No	Leon	519
7	Wakulla	Yes	Wakulla	391
8	Wacissa	No	Jefferson	388
9	Ichetucknee	Partial	Columbia	360
10	Holton	No	Hamilton	243

Adapted from Scott (1997) and Spechler and Schiffer (1995)

15.3 Florida Paleoclimate (Time and Climate)

Relatively little is known about the paleoclimate of Florida because it is climatologically complex. In the case of Florida, gradient, fluid, and time all interact together and link inherently to climate as a main driver of coastal karst development in Florida. Understanding climate change in Florida throughout the Quaternary and beyond is important to determine the driving forces that influence the geomorphology of karst development, including precipitation and sea level changes. As touched on in the above sections, several themes emerge regarding sea level fluctuations, water table movement, and precipitation variability as primary influences on karst development in the state, all of which are predominantly controlled by its coastal setting. The peninsula may not seem to be entirely coastal in nature, as there are several areas of high elevation and certainly the coast is not within sight from every physiographic region. At various times in the past, Florida's size and shape changed often in response to sea-level transgression and regression. Thus, it is important to discuss some of the past changes in climate that would have influenced these events and contributed to the nature by which the landscape's geography was different than in modern time.

Large scale studies, such as the one by Stahle and Cleaveland (1992) reconstructing spring rainfall over the southeast U.S. for the past 1 kyr using tree rings, provide some paleoclimate data for Florida. While this study is helpful in understanding broader influences of climate change in Florida, more localized aspects are not addressed, such as regional variations and anthropogenic influences on climate and the environment. More localized studies, such as that of Kelly (2004), show that the Atlantic Multidecadal Oscillation (AMO), a long-term, cyclic variation in the sea surface temperatures of the Atlantic Ocean, caused periodic changes in the flows of major rivers in Florida over the past few decades. Interestingly, they also discovered that the trends in flow were reversed when comparing rivers in the northern part of the state to those in the south. This has possible implications for how recharge and the resultant dissolution rates of the carbonate bedrock could vary throughout the state based on climatic differences.

Most paleoclimate records indicate continuous variability over the past few millennia in precipitation and recharge, including substantial changes in water table elevations from lake and spring records. One example is that of Wakulla Springs in northern Florida, where prehistoric fossils found deep within the spring passage indicate it was likely dry during certain periods of lower sea level in the Pleistocene, when it ceased functioning as a spring (Lane 1986). Other examples like this exist, providing an explanation as to how the subsurface drainage system and large conduits like those found in Wakulla, Weeki Wachee, and other major springs (some measuring more than 100 ft (30 m) wide by 150 ft (45 m) high), existed prior to the last ice age, and are likely even older, having evolved through multiple periods of water table fluctuation caused by sea level changes. Other studies using coastal sediments and changes in storm phases further contribute to the regional study of Florida climate change (Otvos 2005), and address mainly coastal processes. Studies involving paleoclimate reconstruction from Lake Tulane (Grimm et al. 1993, Grimm et al. 2006; Watts and Hansen 1994) and Camel Lake (Watts et al. 1992) provide low-resolution data for long-term paleoclimate reconstruction, due to many lakes in Florida having undergone periods of desiccation during sea-level lowering. However, these data are important in providing some of the longest records of precipitation and water table variability in the state.

Studies of paleosink and lake sediments provide insight regarding changes in the water table in response the sea level fluctuation and precipitation variability. Little Salt Spring and Lake Tulane are two examples where paleoclimate records of human occupation and changes in the water levels give clues to past climatic conditions over the past few thousand years. Occupation of the Little Salt Spring site reveals human occupation during more arid conditions over the past 5 kyr (Alvarez-Zarikian et al. 2005), while the lake sediment record itself extends to 12 kyr and the beginning of the lake's existence. The record indicates slow sea level rise from 12 to about 6 kyr, with more abrupt variations occurring during the Late Holocene. A longer lake record produced for central Florida by Huang et al. (2006) for Lake Tulane covers the past 62 kyr and shows that Florida was drier during the last glacial period than the past few millennia based on vegetation reconstruction. Grimm et al. (2006) also provide a 60 kyr sediment core record from Lake Tulane that indicates periods of abrupt warm, wet periods during the Pleistocene, yet it was overall cooler, and that the Holocene was warm and wet as well. These climatic changes are attributed to changes in the North Atlantic Thermohaline Circulation, influences of El Nino that may contribute to changes in winter precipitation, and shifts in other teleconnections in the region, such as the ITCZ and the Gulf Stream. Studies by Haug et al. (2001) and Cane (2005) also recognized the influences of the Intertropical Convergence Zone (ITCZ), the North Atlantic Oscillation (NAO), and El Niño on Florida's climate.

To date, several terrestrial studies examined regional paleoclimatic data from Florida caves (van Beynen et al. 2007a, b, 2008; Polk 2009). These studies examined several speleothems from caves along the Brooksville Ridge and Ocala area in west-central Florida to create a record of seasonal changes precipitation and vegetation over the past 30 kyr using stable isotopes and trace elements in speleothem calcite. The fact that stalagmites were growing in an air-filled cave during the last glacial maximum provides insight to the wetness of Florida during this period, and the location of the water table, which would have been below the caves with active formations. Part of this body of work found that the North Atlantic Oscillation (NAO) and Intertropical Convergence Zone were influential in changing Florida's climate on decadal to centennial scales through the Late Holocene (Fig. 15.6). These atmospheric-oceanic teleconnections influence precipitation patterns, sea surface temperatures, and wind patterns on cycles varying from years to centuries, and over time contribute to the climatic shifts that drive sea-level changes and influence coastal karst development in Florida (Polk 2009, Polk et al. 2012).

Stahle and Cleaveland (1992) found no synchronicity between the NAO, ENSO, or the Pacific/North America (PNA) circulation (reflective of PDO) during periods of precipitation change in the southeastern U.S. using tree ring records. However, Hagemeyer (2006) attributes modern storm frequency and severity to interplay between the PNA, NAO, and ENSO. He found that El Niño conditions coupled with negative NAO and positive PNA indices provide optimal conditions for increased winter precipitation in Florida. Conversely, the opposite occurs during periods of La Niña conditions and positive NAO and negative PNA conditions, with reduced winter storm frequency and rainfall amount. Cronin et al. (2002) also found a connection between the NAO, ENSO, and the PNA with

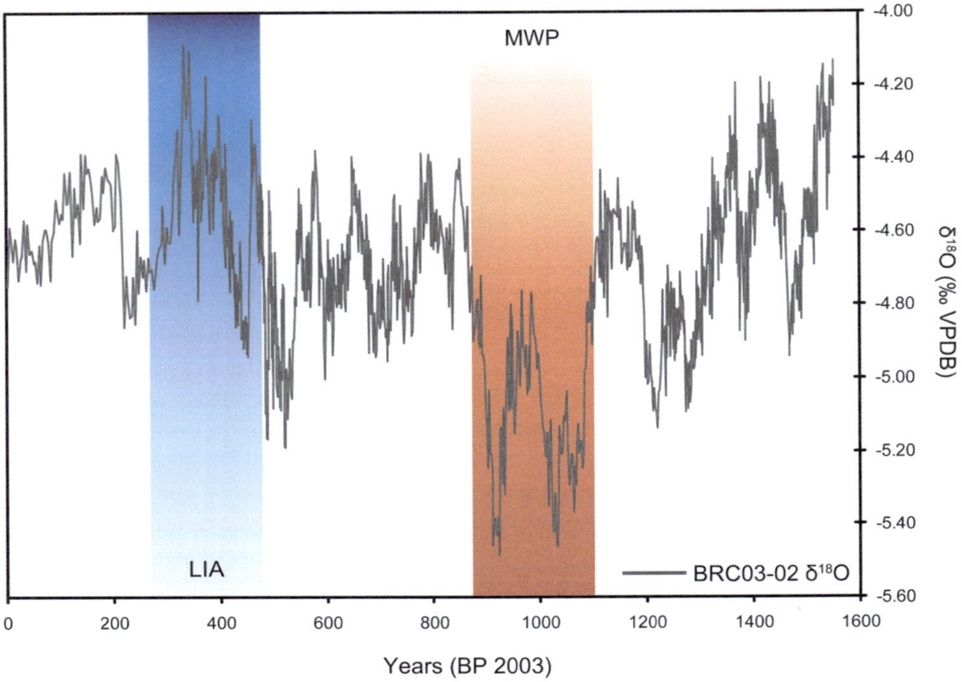

Fig. 15.6 Paleoclimate record from BRC Cave in west-central Florida derived from oxygen isotopes from a stalagmite. The record clearly shows the Little Ice Age (LIA) and Medieval Warm Period (MWP), which were two anomalous periods of temperature and precipitation extremes over the past 1,000 years. More negative values indicate increased precipitation (Modified from Polk 2009)

regard to winter precipitation since the late 1800s in Florida. These teleconnections also affect other areas of the North Atlantic, but long-term, high-resolution data are lacking to provide the information about the persistence and robustness of changes in these atmospheric-oceanic systems, which are drivers of climatic variability over multiple timescales that influence coastal karst development in Florida.

15.4 Coastal Influences on Cave Development

The previous sections outlined the major elements that combine together in generating karst development in Florida, while also discussing some of the driving forces that control the complexities of the landscape, including how climate change interacts with the peninsula. Concurrently, it is important to focus on some of the specifics of how Florida's coastal setting influences cave and karst feature development in the northern two-thirds of the state. As mentioned, the influences of sea level changes, water table fluctuations, and precipitation variability all influence karst development through time. Florida's coastal setting is tied to these three parameters, because it directly impacts where the shoreline existed in the past and the shape and size of the fresh-water lens, and because it provides a moisture and convection source for precipitation. Denizman and Randazzo (2000) likely provide the best overview of the other complexities governing the formation of Florida's karst landscape, which includes the influence of climate change on sea-level and paleowater tables, the highly permeable eogenetic limestone, the influence of siliclastic sediments, and the evolution of a both a surface and subsurface drainage system. Each of the above factors would vary in impact at different phases of karstification through time. Denizman and Randazzo (2000) also propose the probable causes of dissolution to include coastal mixing zone, water table, and hypogenic origins. Some of these dissolution mechanisms are better understood in the state than others (Martin and Dean 2001; Florea 2008, Moore et al. 2010; Gulley et al. 2012a, b), with hypogenic processes being the least studied thus far. Therefore, in the following sections some of the associated influences on cave and karst development described above are discussed within the context of these influences.

15.4.1 Sea-Level Influences

Sea level has fluctuated greatly throughout the Cenozoic period, with the rises and falls in the Quaternary being most consistent in their general timing and magnitude throughout several glacial cycles (Fig. 15.6; see also Fig. 1.5a). As these occurred, changes in Florida's landscape, including the size of the peninsula (Fig. 15.4), sediment deposition, water table fluctuations, and shifts in the hydrologic regime followed suit, leaving behind a paleoenvironmental record of the evolution of the karst landscape in several forms. One such indicator is that of sinkholes, which in Florida are widespread and numerous, with many resulting from the coalescence of multiple sinkholes. Often, these closed depressions are water-filled, and form the majority of Florida's over 7,000 freshwater lakes (Lane 1986), also providing insight to how groundwater has responded to changes in sea level and climate over time (Fig. 15.5b). Many of these karst features formed from changes in sea level during the Miocene and Pleistocene, perhaps even resulting from depressions on the ocean floor during lower sea-level stands, with most forming as a result of continued dissolution and subsequent cover-collapse (Lane 1986; Denizman and Randazzo 2000). The formation of karstic depressions that dot the Florida platform by the thousands likely formed in response to changing sea levels since the Miocene in the Suwannee River basin, specifically along paleo marine terraces (Denizman and Randazzo 2000). Changes in the water table over time, driven by climatic shifts, would also have contributed significantly to the formation of many of these sinkholes (Beck et al. 1985; Kindinger et al. 1999; Tihansky 1999). As mentioned in Sect. 15.3, lake levels and periods of dessication identified in paleoclimate studies provide a record of how the water table and climate interacted in the past. Precipitation is a primary hydrologic factor in the continued existence of these surficial water-filled features by providing recharge to the aquifers to which many of them are connected, as seen in their water levels fluctuating in conjunction with flood or drought events. Thus, changes in sea level and precipitation over geologic time in Florida both contributed to its karst development through rock-fluid interactions that were continuous throughout the Quaternary, and likely prior to that, through complex climate interactions and phasing require further study.

Another indicator of climate-driven sea-level variability is that the peninsula is characterized by several extensive marine terraces oriented mostly north–south moving inland from the Gulf Coast, which represent past sea-level stands that lasted for varying periods of time (Cooke 1945; Opdyke et al. 1984; see Figure 1 in Florea et al. 2007). Sea level would have remained at constant elevations the longest during the glacial periods, when the world's oceans were trapped as ice. The most recent glacial period, around 20 kyr ago, would have placed sea level over 100 m below the current level, and Florida would have doubled in size (Fig. 15.4), primarily extending westward toward the more distant Gulf of Mexico (Florea 2006). This influence on the climate and hydrology of the peninsula was severe, and its impact on karst development is discussed more below.

During recurring changes in sea level, the terraces became more prominent and provide insight to the landscape's response to changing climatic conditions, including their influence on cave development. Stringfield and LeGrand (1966) first proposed sea level controls on cave development in Florida (in Florea 2006). Brinkmann and Reeder (1994) further proposed that caves forming along the Brooksville Ridge and Ocala Arch, which is where many of the modern air-filled caves are located, corresponded with sea-level stages, and formed from mixing of recharge water, groundwater, and seawater at various stages.

Fig. 15.7 Water table fluctuations indicated on the walls of Morris Cave in Citrus County provide insight to both the seasonal and paleowater table and their changes over time (Photo by Tom Turner)

Florea et al. (2007) suggest that sea-level fluctuations had direct influence on the paleo water table and the development of caves in north and west-central Florida at multiple elevations corresponding to these past levels (Fig. 15.7).

Florea et al. (2007) showed that cave levels, especially for air-filled caves, in west-central Florida follow distinct elevations related to past sea level-induced coastal terraces, including the Wicomico, Penholoway, Talbott, and Pamlico and Silver Bluff terraces, which range in elevation from 2 to 30 m above sea level (see Florea et al. 2007, Figure 11 for a conceptual model). Some of the most extensive air-filled caves in the area are proposed to have formed in the Penholoway terrace (~21 m elevation), which would indicate they originated during an extended time of sea-level stability. Since glacioeustatic fluctuations are cyclic, with the ice ages lasting much longer than the interglacial periods for much of the Quaternary, it is likely that only during some of the longer periods of submersion during high sea-level stands were these caves able to form along the elevated water table.

Timing of the formation of these coastal marine terraces is constrained by various data, yet there are still some unknowns with regard to their ages. It is likely that the older terraces are at higher elevations, and this theory fits with the existing data, although some discrepancies exist (Florea et al. 2007; Adams et al. 2010). In northern Florida, some studies have proposed that isostatic rebound from the removal of carbonate rock over timescales as short as thousands of years may have altered the elevation of some coastal marine terraces (Opdyke et al. 1984; Adams et al. 2010). These estimates are as high as 1 m of rock removed per 11.2 kyr, which could cause uplift on the order of 0.047 mm/year through karstification. This subsequently would impact the significance and location of the major ridges and terraces in Florida, associating them with both past sea level and isostatic uplift to dictate their varying position and age in the landscape through geologic time. The continued karstification and formation of caves at different terrace levels would not be impacted by this process (Florea et al. 2007), although it could alter the

Fig. 15.8 The entrance to Blue Grotto, a spring over 30 m deep in Levy County and a tourist attraction for divers

elevation at which current cave levels appear in the modern landscape compared to their original position.

The larger, more extensive underwater conduits found in the Avon Park, Oldsmar, and Cedar Keys formations likely formed during the longer periods of low sea-level stands that lasted for tens of thousands of years during glacial periods, when the water table and the platform's gradient and hydrologic regime was significantly altered compared to the current setup. Terraces present at these elevations are less distinct in relation to cave formation likely because of their more continuous subjection to sea-level variability. Florea et al. (2007) also recognize this possibility and the potential influence of a sloping water table during the various sea-level stands that would have impacted caves at multiple levels once the primary dissolution of bedrock at previous water table levels had occurred. Thus, the influence of sea level would have been substantial in controlling both vadose and phreatic conduit development, with the lower elevation passages being inundated more often during the evolution of the platform through the Quaternary. Hence, understanding cave development in Florida is often complex at different locations throughout the State due to the subtle, yet important, variability in elevation of the major structural components, along with the overprinting of paleowater tables. The eogenetic carbonate bedrock with its high matrix porosity and permeability also is important in providing pathways for dissolution and cave development that do not necessarily follow the stratigraphic or water table controls that would limit passage formation to distinct elevations (Florea et al. 2007).

Florida has many extensive underwater cave systems, which are often associated with major spring features and provide direct connections within the aquifer system (Fig. 15.8). Moore et al. (2010) and Gulley et al. (2012a, b) provide some insight to the phreatic cave development processes that occur in Florida, supporting the concept of sea level fluctuations driving water table changes as primary causes of conduit development, with the added complexity of structural and lithological influences of the bedrock. In eogenetic karst settings such as Florida, while movement of water within conduits in the bedrock is a primary component of aquifer flow, additional flow occurs between matrix storage and the conduits due to the rock's high porosity and permeability. This contributes to dissolution occurring through conduit enlargement and within the matrix as well (Moore et al. 2010). Geochemical analysis of groundwater through a mixing-model approach in northern Florida indicates that during allogenic recharge events, conduit water permeates the surrounding matrix through the walls and creates a friable halo of easily weathered rock surrounding the conduit (Florea and Vacher 2007; Moore et al. 2009, 2010). Unlike the air-filled caves described previously, these phreatic conduits form continuously from both water already existing within the system and from inputs of undersaturated water through allogenic recharge, or from deeper sources due to gradient or water table fluctuation over time (Moore et al. 2009; Gulley et al. 2012a).

15.4.2 Fluid and Climate Interactions

The development of phreatic conduits, as described in the previous section, is a complex process due to variations in recharge, sea level, water table, and the eogenetic rock matrix and conduit flow patterns. In the various confined and unconfined components of the Floridan aquifer, flow regimes differed throughout the past, with discharge likely occurring at diffuse points along the coast or through submarine springs. The formation of rivers in unconfined areas, such as the Santa Fe, may have evolved from runoff on confining layers (Gulley et al. 2012a, b). These river systems provide direct interaction between allogenic surface water and the phreatic zone, thus providing a potential change in fluid dynamics that could cause additional cave development. Variations in fluid interactions from groundwater and meteoric recharge within the system in response to changing sea levels and precipitation patterns include geochemical differences through time that would have impacted karst development. This process is more straightforward in air-filled voids as previously described, which is mainly from percolating meteoric water and water table influences, but the more complex interactions within the phreatic zone from changes in groundwater flow paths, saturation indices of phreatic water, and changes in CO_2 availability.

Gulley et al. (2012b) proposed that CO_2 moving through the vadose zone to the water table could have played a primary role in the development of phreatic conduits within the Florida peninsula. As changes in sea level occurred, flooding of relict vadose passages and the formations of rivers by a rising water table would have connected the surface and subsurface systems, introducing undersaturated surface runoff, and increasing dissolution and conduit development. Rather than mixing zone formation from saltwater inundation at higher shorelines, CO_2 diffusion into the lower fresh water table within the platform could cause laterally expansive cave systems without surface entrance connections (spring systems) to form (Gulley et al. 2012b).

While some coastal mixing zone dissolution does likely occur, this does not appear to have been highly influential in the development of coastal karst landforms in Florida, nor is there much cave development from tidal or wave action due to the low energy environment of the Gulf of Mexico shoreline (Gulley et al. 2012a, b). There is a lack of much exposed limestone on the Atlantic side of the peninsula, although there are some caves in south Florida in exposed outcrops that are accessible. However, in support of the previous explanation of phreatic conduit development, seawater-freshwater interactions at deeper levels within the Florida peninsula likely play a role in the freshwater aquifer's development over glacial scales, thus having some impact on karst aquifer formation and conduit development. Morrissey et al. (2010) found that during the last glacial maximum, in south Florida the upper aquifer was mainly recharged from precipitation, while the lower aquifer was inundated with saltwater. As sea level rose, the freshwater circulation overlying it was slowed and trapped at higher elevations, leaving the saline water below, which would have an impact on the water table elevation for extended periods of time. This also indicates that meteoric water recharge, which would be significant for bedrock dissolution, would have contributed to continued karstification throughout the past (and possibly several past) glacial periods (Morrissey et al. 2010). This explanation supports the idea of the fresh water table being the most influential area of conduit development, including that of surrounding matrix dissolution, which over long periods of time would contribute significantly to karstification of the peninsula at multiple levels within the platform and under all climate conditions.

15.5 Cave Characteristics and Development

Many caves form along joints, fractures, and bedding planes, and often have maze-like patterns, although more extensive, linear caves do exist (e.g. Warrens Cave) (Brinkmann and Reeder 1994; Florea 2006). Previous phreatic caves during higher sea-level stands are now abandoned and act as vadose zone recharge points where meteoric water is pirated underground to recharge the aquifer system, causing further dissolution and collapse at joint intersections (Brinkmann and Reeder 1994). Evidence of recent Quaternary overprinting occurs where vertical solution features exist along upland areas, such as the Brooksville Ridge, where vertical pits intersect along enlarged joints or bedding plane controlled conduit systems that may not have previously had an entrance at the surface. Often, the caves have a plus sign shaped passage from these intersections (Fig. 15.9a), and are usually shorter than they are wide, although much variation does exist from breakdown and complex morphologies (Fig. 15.9b) related to lithology and structure (Florea 2006).

Fossils are commonly found in the fossiliferous limestone, and passages are usually small and relatively short in air-filled caves, with speleothems uncommon in most of the region (Polk 2009; Harley et al. 2010). The wall rock is often unstable, and there are few vertically extensive caves, although some do exist (Fig. 15.10). Those at lower elevations often contain passages that contain water when intersecting the water table, and these occasionally connect to the deeper aquifer and are able to be further explored through diving. Fluctuations in the water table are visible in some caves, and sinking streams, flooding, and aquatic biota occur in some caves connected to the aquifer system.

Several mechanisms of cave development are possible within the peninsula for both air- and water-filled caves. Overall, the controlling mechanism for changes in water table elevation is sea-level fluctuation, with air-filled caves likely developing in correspondence to elevations of past marine terraces and water table elevations. Phreatic caves are more complex, yet follow a similar pattern of development influenced by both water chemistry and gradient, and corresponding to changes in sea level and the resultant water table (Gulley et al. 2012a, b). There are exceptions to this, including the possibility of hypogenic conduit formation from previous changes in the hydrologic regime from rising sulfur-rich fluids during sea-level fluctuation (Florea et al. 2007). Also, some occurrences of back-flooding, anastomotic-like maze cave passages forming along rivers do exist, such as the Ovens found along the Chipola River (Fig. 15.11a) and Thornton's Cave and several large remnant passages found along the Withlacoochee River, which act as estavelles dependent upon water table and river level differences (Fig. 15.11c) though changes in base-level influencing the location of rivers could also cause similar features during normal flow.

Therefore, several different types of caves exist within the Florida platform, each caused by interactions of young eogenetic carbonate rock, sea level and water table fluctuations, and changing precipitation patterns over millions of years. Together, these parameters create a complex karst topography that is driven by hydrologic changes caused by gradient, geochemistry, and climatic variations in a coastal setting that provides a dynamic environment despite the appearance of the peninsula being a flat, featureless plain on the surface. The following sections describe some of the major karst regions in central and northern Florida and their corresponding features, which result directly from the aforementioned elements acting upon the peninsula through the Cenozoic.

Fig. 15.9 (**a**) Characteristic "plus sign" passage shape in a cave in central Florida. (**b**) Werner Cave in west-central Florida, which exhibits a water table connection and illustrates the common passage shape (wider than tall) of many caves in the area (Photos by Tom Turner)

15.6 Karst Regions of Florida

As noted in the previous sections, Florida's karst developed under a complex, temporally changing system throughout the Cenozoic. The landforms that evolved over millennia formed as part of an interaction of rainfall, rapid sea-level change as a result of shifting climate, and local topography. Today, we have a suite of eogenic karst landforms that help us interpret the puzzling Cenozoic geologic history of Florida. Nowhere else in the

Fig. 15.10 Maynard's Cave in Citrus County, which boasts one of the largest rooms in the region, and has two vertical entrances over 20 m high (Photo by Tom Turner)

world is there a rock assemblage with such a significant karst history. In the following sections, the major landform regions will be reviewed. Within each section, important landforms or interpretations are highlighted. The major landform regions that are discussed are the Peninsular Ridgelands, the Panhandle Uplands, the North Peninsular Lowlands, the Everglades, and the Gulf of Mexico and Peninsular Lowlands. While all areas of Florida are in some way impacted by karst, it must be noted that some areas of Florida have limited significance from a karst geomorphology perspective. Thus some areas, like most barrier islands, the Keys, and the Apalachicola River delta, are not part of this discussion.

It must also be noted that much of the karst in Florida can be classified as covered (mantled) or non-covered karst. The non-covered karst environments are found in some of the ridgelands (some of the ridges are covered with marine sands or dunes), and some coastal platforms in the Gulf of Mexico and Peninsular Lowlands, particularly near Ozello and the Big Bend area of Florida in the extreme northeastern portion of the Gulf of Mexico. Throughout most of the Everglades, limestone is also very near the surface. In other areas of the state, the karst is covered with variable thicknesses of marine sediments—largely fine to medium sand at the surface with highly variable sediments at depth. This cover

Fig. 15.11 (**a**) The Ovens cave complex located along the Chipola River. The caves are maze-like passages along the river and likely acted as phreatic inputs (springs) in the past during higher water levels, and now serve as overflow areas that backflood during major flooding events. (**b**) Typical passage shape and decorations in Florida Caverns State Park. (**c**) Water flowing out from Thornton's Cave along the Withlacoochee River. The cave has many entrances and acts as an estavelle, wherein water flows out or in depending upon the hydrologic conditions. After a wet summer and several large rain events, the water table had risen and was flowing out of the cave and into the river (Photo by Tom Turner) (**d**) The rolling, undulating topography of west-central Florida's karst landscape, which is more hilly and has higher elevations than expected for Florida (Photo by Grant Harley)

is particularly important in some areas because it masks karst formation (particularly south of Tampa and Orlando), or leads to raveling of sand into subsurface voids to create distinctly subtle karst features in the central and northern Gulf and Peninsular Lowlands.

15.6.1 The Peninsular Ridgelands

The Peninsular Ridgelands consist of over a dozen ridges or fragments of ridges that extend roughly north–south across the Florida Peninsular from near Georgia south to the northern edge of the Lake Okeechobee (Fig. 15.4). The ridges have been described by others in great detail (Cooke 1945; Lane 1986; Scott 1988; Florea 2006). They are erosional remnants of some of the oldest rocks that are near the surface in Florida. The rocks underlying the ridges consist of limestone that formed in the early to mid-Cenozoic in the Oligocene, Miocene, and Pliocene. Surrounding the ridges are the Gulf of Mexico and Peninsular Lowlands that will be described later. Some of the better-known ridges are the Lake Wales Ridge, the Brooksville Ridge, and the Crescent City Ridge (Fig. 15.4). The highest point on the ridge system is found at

312 ft (95 m) above sea level on the Lake Wales Ridge. The topography of the ridges is highly irregular due to the numerous complex sinkholes that punctuate the landscape (Fig. 15.11d). The local relief can range from a few feet (~1 m) to over 200 ft (60 m) in some locations.

The Peninsular Ridgelands are important because they contain the oldest assemblages of karst landforms in Florida. They are erosional coastal remnants. During high sea-level stands, these ridges were islands in the surrounding shallow sea across the Florida platform. In many ways, the ridges were islands similar to the Bahamas Archipelago (Chap. 7). They were small, discontinuous remnants of rock subject to marine erosion and deposition, wind processes, and most importantly, karst processes. During high sea-level stands, the water table was much higher. Today, the rocks in the highest portions of the ridges are well drained. Because of this topographic position, karst processes in many portions of the ridges were much more active in the past than today. Thus, most of the karst features that are found in the ridge are paleokarst features. While karst processes are still occurring, the processes that created the complex cave and sinkhole features on the ridges acted in the distant past.

Some of the most notable features on the ridges are complex sinkholes and caves. Most of the sinkholes found on the ridges are dry or seasonally filled. Their bottoms are usually much higher than the surrounding lowlands, so water drains quickly from them into the deeper Floridan Aquifer system. Indeed, the entire ridge system is very well drained with few lakes or streams, although some large sinkholes and solution valleys would have been flooded to form lakes and streams during high sea-level stands. Interestingly, the sinkholes on the ridges are much more complex than most others found in the surrounding lowland. They are larger and much less circular. Some of them clearly formed from the coalescence of multiple sinkholes.

The caves in the ridges are largely air filled or seasonally flooded. They typically consist of short (less than 100 m) of passageways that are often interrupted with breakdown (Fig. 15.12). It is evident that some of the caves were much larger in the past but have been truncated by collapse features or sinkhole formation. There is evidence of past speleothem formation in many of the caves, but in most caves, there is limited decoration. Many of the caves have considerable sedimentation due to the overland flow of water and concomitant sediment transport. There is also evidence of sediment flow into some caves from voids within the rock that channel overlying sediment into them. The sedimentary structures indicate that there are seasonal contributions similar to lake varves. In addition, in some instances there is evidence of distinct flooding due to extreme rainfall event brought about by hurricanes, tropical storms, or other extreme events. Some more complex caves, including some maze caves, indicate a groundwater origin for them. However, many of the caves contain clear phreatic features that suggest that the caves formed in flowing groundwater conditions (Gulley et al. 2012a, b).

Because the ridges are so narrow, the karst formation processes acting on them during high water stands were likely accentuated by saltwater and freshwater mixing adjacent to the coastal edge of the islands. When freshwater and salt water mix in the subsurface it becomes more corrosive to limestone. Dissolution of the rocks in such zones is accentuated and voids and caverns expand (Mylroie and Mylroie 2007). Because of the narrow, finger-like form of the ridges, most of the areas of the paleo-islands would have been impacted by this aggressive water, along with water table driven dissolution from fresh waters of different saturation when meteoric recharge entered the system. Therefore, it is not surprising that one finds some of the most extensive karst features found at the surface in Florida on these ancient ridges. While these forms are often attributed to age, the aggressive waters contribute to the relatively advanced karstification evident on the ridges. Certainly, they have been subject to karst processes longer than other portions of the state. But, the presence of mixing zone environments surrounding these narrow paleo-islands certainly impacted the form and extent of karst features present on the ridges.

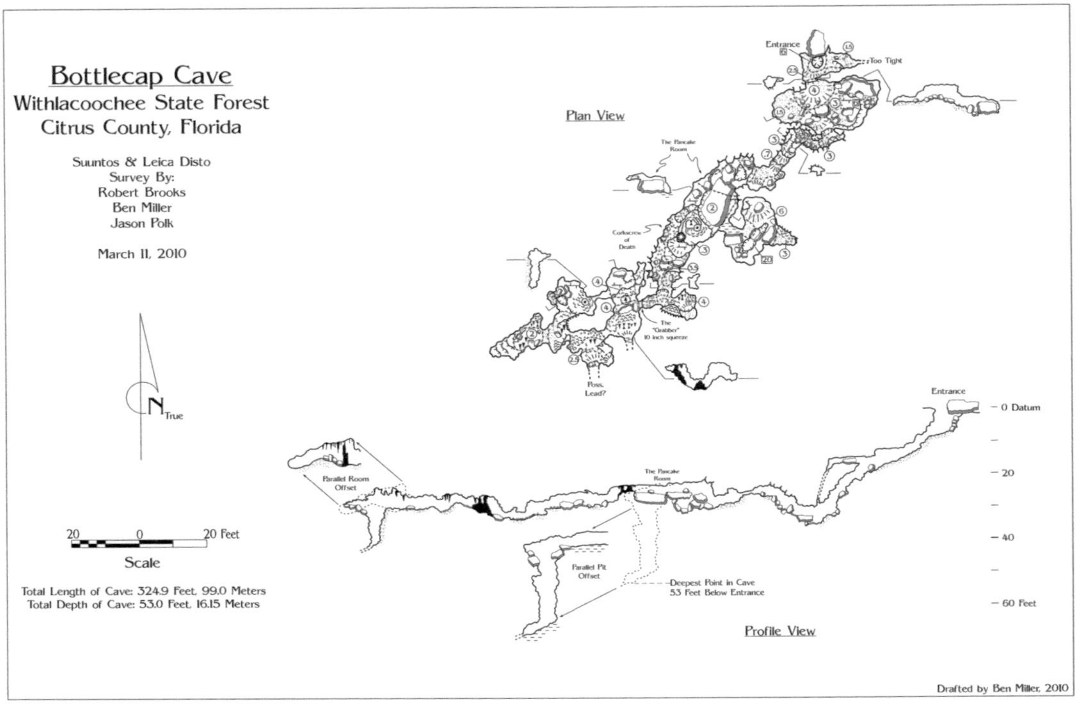

Fig. 15.12 Map of Bottlecap Cave in Citrus County. This cave is linear in nature, serving as an input to the aquifer, and it should be noted the multiple vertical pits found within the passage that indicate the extent of development along both the horizontal and vertical planes. The cave have extensive breakdown throughout, another common characteristic of the crumbly and soft eogenetic carbonate rock (Eocene Ocala Limestone here) found in Florida

Historically, agriculture has been problematic on the ridges due to the lack of water caused by the rapid infiltration and drainage. Indeed, the ridges are home to unique plant and animal assemblages that evolved on this unique dry environment, including some native cacti. Orange groves have survived on some of the southernmost ridges when they are irrigated. However, many of the ridges have been developed for real estate. The ridges are popular because they provide rare elevated views of the Florida landscape. Portions of the ridges have been preserved in forests and parks. Yet, the Peninsular Ridgelands contain some of the most threatened karst landscapes in the United States. Many of the caves have been vandalized and some of the sinkholes have been modified for agriculture or real estate development. Plus, due to modification of hydrologic systems, some of the caves and sinkholes have been altered by storm water schemes.

15.6.2 The Panhandle Uplands

The Panhandle Uplands of Florida contain the highest point in Florida at 345 ft (105 m) at Britton Hill near the Alabama State Line (Fig. 15.4). This region is an extension of the karst uplands of south Georgia and Alabama and is highly weathered Miocene, Oligocene, and Eocene limestone similar to that found in the Peninsular Ridgelands. However, here, instead of existing as islands in the higher sea level stands, these rocks served as the coastal edge of the continent. Thus, their southern edges represent coastal sea-level stands that are highly modified by paleo wave action. Valleys within these uplands served as estuarine environments. Today, vast extents of these uplands are preserved as forests or preserves and the wood is harvested for timber and paper. This upland region is known as the Marianna Lowlands in the western portion of the uplands and the Tallahassee Hills to the east.

Some of the better-known features in the Marianna Lowlands are the Florida Caverns—Florida's only commercial cave system and DeFuniak Springs on the southern portion of the uplands where water exits underground karst systems into the lowlands (Fig. 15.11b). Florida Caverns is specifically located within the Marianna Lowlands physiographic province. The park was designated in the early twentieth century as Florida's 7th state park prior to the discovery of the main tourist cavern. Other smaller caverns were known in the area. However, the caverns were discovered during the construction of facilities by the Civilian Conservation Corps. The Florida Caverns are Florida's only fee-based cave system. They contain a variety of interesting speleothems including soda straws, stalagmites, stalactites, and cave drapery (Florida 2006). The host limestone has striking echinoid fossils. The Chipola River Flows through the park, in part under a natural bridge.

East of the Marianna Lowlands are the Tallahassee Hills, which are part of a broader Red Hills physiographic region that extends into southern Georgia. The Tallahassee Hills are a karstic system characterized by relatively high relief, sinkholes, and karst valleys (Cooke 1945). To the south, the hills are separated from the Gulf Coastal Lowlands by a prominent east–west trending scarp called the Cody Scarp. This scarp is a paleoshoreline delineating the edge of the continent during the Pleistocene. Behind the scarp, the rocks have been subject to karst weathering for millions of years with persistent mixing zones occurring near the edge of the scarp. The presence of the mixing environment led to the formation of extensive voids and concomitant depression features.

15.6.3 The Everglades

The Florida Everglades are not normally considered a karst landscape. They are a flat wet plain very near sea level. Water draining from south of Orlando and Lake Okeechobee flows across the Everglades in a south-southwest direction toward the 10,000 islands region near Fort Myers. However, since the late nineteenth century, a great number of engineering projects have modified the Everglades for flood control and to expose more agricultural land for development. Many of these projects are being re-engineered to try to bring back some of the natural elements of the region.

While the Everglades are recently exposed (approximately 5,000 years ago), their formation has been taking place for millions of years. Sediment brought south by paleo-rivers and by ocean currents filled the waters in what is now southern Florida. Thick layers of sand, silt, clay, and lime-rich sediment formed to create a complex sedimentary assemblage. Quaternary, Pliocene, and Miocene sediments and rocks extend for hundreds of meters beneath the southern portion of the Florida peninsula. Up until recently, this area was an area of net deposition. Since exposure, the landscape does receive some sediment from overland flow and storm surges, but the great input of sediments and limey material is far less than in the past. Thus, the landscape can be seen as a relatively stable plain.

What dominates the Everglades today is its low-nutrient ecosystem known as "The River of Grass"—a phrase made famous by Marjorie Stoneman Douglas' important book on the Everglades (Douglas 1997). The grasses make up a unique marsh punctuated by tree islands. The system evolved with Florida's unique wet summer, dry winter climate that is known for frequent summer thunderstorms. Lightening frequently starts fires in the Everglades, particularly during the early portion of the rainy season when the glades are relatively dry. Thus, while there are some areas where peat has accumulated, most of the Everglades are underlain near the surface by Pleistocene limestone.

Because the Everglades are flooded for much of the year and because the limestone is young, there are very few karst features present in the Florida Everglades. However, there are some. There are a number of small subtle sinkholes and there are also small caves on some of the small ridges that can be found on the eastern fringe of the glades. However, none of these features match the size and extent of karst features found in the rest of the state. One notable feature, Lake

Okeechobee, is often mistakenly interpreted as a karst feature when in fact it is a depressional remnant left over from the paleosea that once covered much of south Florida.

Similarly, tree islands are quite common in the Everglades. Tree islands are small (1–10 acre) oval areas that are vegetated by trees. They stand out in the sea of grass as geographic markers and are ecologically significant. Some have interpreted these features as subtle karst depressions, similar to cypress domes found in other flat plains of the state. However, evidence suggests that many of these are middens and settlement sites that formed from prehistoric waste deposited at the surface by humans. What is fascinating about the tree islands of the Everglades is that they seem to slowly move with the flow of the water through the Everglades system. However, they are decidedly not karst features. Although most of the Everglades can be considered a flooded karst system, the environment is too young for the development of distinctive karst landforms. Instead, overland flow, human modification, and vegetative processes dominate the region's geomorphology.

15.6.4 The Gulf of Mexico and Peninsular Lowlands

The Gulf of Mexico and Peninsular Lowlands are found throughout the entire State of Florida in areas not discussed in previous sections. This lowland is a relatively flat plain, with a great deal of variety from north to south. In many areas of the north, bedrock is at or near the surface. In the south, the surface cover thickens. This makes karst landforms in places like Sarasota and De Soto Counties relatively rare. Places where the limestone is near the surface, particularly in the Big Bend Region (near the northeastern Gulf of Mexico) and Ozello, are heavily modified by wave, tide, and current activity due to their relatively low elevation and position as a marine platform (Fig. 15.13). However, they also display extensive surface and subsurface features such as vugs, small solution valleys, and springs (Knochenmus and Yobbi 2001). There is also evidence of biokarst activity along some solution zones.

Upland from the platform, one finds that until one reaches the ridges or peninsular highlands, the karst is covered with marine sediments, particularly fine to medium sand, varying in thickness up to 20 m or more (Beck et al. 1985). Sinkholes are common here and many cover collapse and cover subsidence sinkholes occur each year. Notable areas where sinkhole hazards are problematic are in Orlando, Ocala, and Tampa where hundreds of property insurance claims are made every year as a result of damage due to subsidence (Fig. 15.14a). Sinkhole law has evolved in Florida in recent years and due to sizable payouts by insurance companies, homeowners must pay for any extra sinkhole insurance and the property must be inspected prior to being granted a policy. Claims can occur during extreme dry conditions when the water table is low or during the height of the rainy season with the ground is saturated.

There are also many sinkhole lake districts in this area. Indeed, the City of Orlando is built around karst lakes. The larger lakes in Florida typically form from the coalescence of multiple sinkhole depressions. Some clearly have complex circular forms. Some of the smaller, circular lakes formed from single occurrences. What is important to note is that these lakes formed since the last drop in sea level (Denizman and Randzzo 2000). If they existed during high sea-level stands, they would have been filled with sediment. Indeed, there is evidence of paleosinkholes that formed in Florida prior high sea-level stands. They appear as flat surfaces on the landscape. However, when excavations are made for development, the depression form, filled with marine sediment, is exposed. There is also evidence of sinkhole formation within the offshore platform during low sea-level stands. Seawater is not conducive to the solution of limestone. Thus, any karst landform located offshore most likely formed during extreme low sea-level stands when the Florida platform

Fig. 15.13 Karst and pit development, as well as exposure of a NW trending enlarged joint along the coastline of Ozello near the coast in Citrus County. This type of surface is typical where the epikarst has direct connection to the water table

was exposed. There are dozens of known paleosinkholes in the Gulf of Mexico, some of which are sites of saltwater springs.

The bedrock in the Gulf and Peninsular Lowlands is highly porous and permeable. There are few air-filled caves that can be entered because the water table is very close to the surface. However, there are many known sites where cave divers have mapped extensive cavernous passageways. Some are simple phreatic tubes and others are more complex, even maze-like in form. It is clear from the evidence that there are multiple formation paths for subsurface voids in Florida. Some form from flowing water and others form from water table interactions.

Springs are also found throughout this lowland landscape. Many of them have been developed for tourism or parks (Weeki Wachee, Juniper Springs, and Rainbow Springs for example) (Fig. 15.8). They discharge water that flows to the springs underground from the upland ridges or from other areas within the lowland. On occasion, the springs lead to spring runs (e.g. Rainbow

Fig. 15.14 (**a**) Sinkhole that opened in the middle of the road during a rain event near Tampa. (**b**) Extensive amounts of trash in Dames Caves in west-central Florida. The caves are often visited by locals, as they are well known to the public and have been vandalized and abused extensively over several decades. Recent efforts by the Withlacoochee State Forest, Southwest Florida Water Management District, Hoffman Environmental Research Institute at Western Kentucky University, and the Karst Conservancy have led to the cleaning and preservation of the site, along with interpretive sign installation to help visitors understand the importance of Florida's karst landscape and groundwater resources (Photo by Tom Turner)

River) that flow to the ocean or other water body. Many lakes and rivers are also spring fed. Some of the springs have been exploited for water production. For example, Crystal Springs, which drains into the Hillsborough River, is a primary source for Zephyrhills brand bottled water.

15.7 Impacts to Florida Karst

There is extensive evidence for the destruction and damage of karst systems in Florida (Harley et al. 2011). For example, in recent years, it has been noted that the quality of the water in many of the springs in Florida has been deteriorating, in part, due to the impact of nitrate pollution that has seeped into the groundwater. Some areas have seen land subsidence due to pumping of groundwater to supply growing metro regions. In addition, many karst depressions (some of which were wetlands) have been significantly altered through the development process. Some have been filled in to make way for housing projects, roads, or commercial developments. Others have been modified due to changes in hydrology brought about due to changing surface conditions. Caves, too, have been damaged (Turner 2003; Fig. 14b). Steps have been taken to try to protect these karst systems through laws aimed at preserving wetlands, protecting caves and groundwater systems, and reducing pollution.

15.8 Summary

The karst platform of Florida has more of a complex geologic history than one would expect. The five main elements of rock, structure, gradient, fluid, and time with the added influence of climate change and the coastal setting are responsible for the geographic variability of karst landforms present in the state. Throughout the Cenozoic, geologic and climatic processes acted upon the peninsula, with sea-level fluctuations causing substantial differences in its size and gradient, while also forming extensive marine terraces and fluctuating water-table elevations that influenced carbonate dissolution in conduits and the rock matrix. The platform's hydrology varied greatly over the past few million years with the influence of glacial cycles and changes in precipitation. This impacted the evolution of Florida's large aquifer system, which influenced on sinkhole development, flow regimes, and the geochemistry of the aggressiveness of water acting upon the platform. Caves exist at elevations that correspond to coastal marine terrace and paleowater table elevations, with modern underwater caves forming at more complex elevations from phreatic processes during water-table fluctuations that inundated them many times throughout the Quaternary. Overall, Florida's coastal karst processes are complex, but also reveal the importance of climate change's influence on sea level and recharge, which in turn directly influence the development of karst features on the peninsula. Additional research on Florida's coastal karst geomorphology and increased attention to the importance and vulnerability of its karst features and groundwater are necessary in order to continue to improve understanding and awareness.

References

Adams PN, Opdyke ND, Jaeger JM (2010) Isostatic uplift driven by karstification and sea-level oscillation: modeling landscape evolution in north Florida. Geology 38(6):531–534

Alvarez-Zarikian CA, Swart PK, Gifford JA, Blackwelder PL (2005) Holocene paleohydrology of little slat spring, Florida, based on ostracod assemblages and stable isotopes. Palaeogeogr Palaeoecol Palaeocl 225:134–156

Beck BF, Ceryak R, Jenkins DT, Scott TM, Spangler DP (1985) Karst hydrogeology of Central and Northern Florida: The Florida Sinkhole Research Institute, University of Central Florida, Orlando

Brinkmann R, Reeder P (1994) The influence of sea level change and geologic structure on cave development in west-central Florida. Phys Geography 15(1):52–61

Budd D, Vacher HL (2004) Matrix permeability of the confined Floridan aquifer, Florida, USA. Hydrol J 12:531–549

Cane MA (2005) The evolution of El Nino, past and future. Earth Planetary Sci Lett 230:227–240

Cooke CW (1945) Geology of Florida. Florida Geol Surv Bull 29:339 p

Cronin TM, Dwyer GS, Schwede SB, Vann CD, Dowsett H (2002) Climate variability from the Florida Bay sedimentary record: possible teleconnections to ENSO, PNA, and CNP. Clim Res 19:233–245

Denizman C, Randazzo AF (2000) Post-Miocene subtropical karst evolution, lower Suwannee river basin, Florida. GSA Bull 112(12):1804–1813

Douglas MS (1997) The everglades: river of grass, 50th anniversary edition. Pineapple Press Inc, Sarasota, 478 p

Florea LJ (2006) Architecture of air-filled caves within the karst of the Brooksville ridge, west-central Florida. J Cave Karst Stud 68(2):64–75

Florea LJ (2008) Geology and hydrology of karst in west-central and north-central Florida. Caves Karst Fl 225–239

Florea LJ, Vacher HL (2007) Eogenetic karst hydrology: insights from the 2004 hurricanes, peninsular Florida. Groundwater 45(4):439–446

Florea LJ, Vacher HL, Donahue B, Naar D (2007) Quaternary cave levels in peninsular Florida. Quaternary Sci Rev 26:1344–1361

Florida DEP (2006) Florida Caverns State Park unit management plan. http://www.dep.state.fl.us/parks/planning/parkplans/FloridaCavernsStatePark.pdf. Accessed 3 Dec 2012

Gaswirth SB, Budd DA, Crawford BR (2006) Textural and stratigraphic controls on fractured dolomite in a carbonate aquifer system, Ocala limestone, west-central Florida. Sediment Geol 184:241–254

Grimm EC, Jacobson GL, Watts WA, Hansen BCS, Maasch KA (1993) A 50,000 year record of climate oscillations from Florida and its temporal correlation with the Heinrich events. Science 261(5118):198–200

Grimm EC, Watts WA, Jacobson GL, Hansen BCS, Almquist HR, Dieffenbacher-Krall AC (2006) Evidence for warm wet Heinrich events in Florida. Q Res 25:2197–2211

Groves CG (1993) Early development of karst systems. Dissertation published by the University of Virginia

Gulley JD, Martin JB, Spellman P, Moore PJ, Screaton EJ (2012a) Influence of partial confinement and Holocene river formation on groundwater flow and dissolution in the Florida carbonate platform. Hydrol Process. doi:10.1002/hyp.9601

Gulley JD, Martin JB, Moore PJ, Murphy J (2012b) Formation of phreatic caves in an eogenetic karst aquifer by CO_2 enrichment at lower water tables and subsequent flooding by sea level rise. Earth Surface Processes and Landforms. DOI: 10.1002/esp.3358

Hagemeyer BC (2006) ENSO, PNA and NAO scenarios for extreme storminess, rainfall and temperature variability during the Florida dry season. In: Preprints, 18th conference on climate variability and change, American Meteorological Society, Atlanta, CD-ROM P2.4

Harley GL, Reeder P, Polk JS, van Beynen P (2010) Developing a GIS-based inventory for the implementation of cave management protocols in Withlacoochee state forest, Florida. J Cave Karst Stud 72(1):35–42

Harley GL, Polk JS, North LA, Reeder P (2011) The application of a cave inventory system to stimulate development of management strategies: an example from west-central Florida, USA. J Environ Manag 92: 2547–2557. http://dx.doi.org/10.1016/j.jenvman.2011.05.020

Haug GH, Hughen KA, Sigman DM, Peterson LC, Rohl U (2001) Southward migration of the intertropical convergence zone through the Holocene. Science 293:1304–1308

Huang Y, Shuman B, Wang Y, Webb T, Grimm EC, Jacobson GL (2006) Climatic and environmental controls on the variation of C_3 and C_4 plant abundance in central Florida for the past 62,000 years. Palaeogeogr Palaeocl Palaeoecol 237:428–435

Kelly M (2004) Florida river flow patterns and the Atlantic multidecadal oscillation. Draft report, Southwest Florida Water Management District, 80 p

Kindinger JL, Davis JB, Flocks JG (1999) Geology and evolution of lakes in north-central Florida. Environ Geol 38(4):301–321

Knochenmus LA, Yobbi DK (2001) Hydrology of the coastal springs ground-water basin and adjacent parts of Pasco, Hernando, and Citrus Counties, Florida. Water-resources investigations report 01–4230. United States Geological Survey, 88 p

Kuniansky EL (2001) US Geological survey karst interest group proceedings, St. Petersburg, Florida, 13–16 Feb 2001, USGS water-resources investigations report 01–4011, 211 p

Lane E (1986) Karst in Florida, Florida Geological Survey special publication no. 29. Florida Geological Survey, Gainesville, 100 p

Lane E (1994) Florida's geological history and geological resources, Florida Geological Survey special publication no. 35, Florida Geological Survey, Gainesville, 46 p

Martin JB, Dean RW (2001) Exchange of water between conduits and matrix in the Floridan aquifer. Chem Geol 179:145–165

Miller JA (1986) Hydrogeologic framework of the Floridan aquifer system in Florida and in parts of Georgia, Alabama, and South Carolina. U S Geol Surv Prof Pap 1403-B:91p

Moore PJ, Martin JB, Screaton EJ (2009) Geochemical and statistical evidence of recharge, mixing, and controls on spring discharge in an eogenetic karst aquifer. J Hydrol 376:443–455

Moore PJ, Martin JB, Screaton EJ, Neuhoff PS (2010) Conduit enlargement in an eogenetic karst aquifer. J Hydrol 393:143–155

Morrissey SK, Clark JF, Bennett M, Richardson E, Stute M (2010) Groundwater reorganization in the Floridan aquifer following Holocene sea-level rise. Nat Geosci 3:683–698. doi:10.1038/NGEO956

Mylroie JE, Mylroie JR (2007) Development of the carbonate island karst model. J Cave Karst Stud 69:59–75

Opdyke ND, Spangler DP, Smith DL, Jones DS, Lindquist RC (1984) Origin of the epeirogenic uplift of Plio-Pleistocene beach ridges in Florida and development of the Florida karst. Geology 12:226–228

Otvos EG (2005) Holocene aridity and storm phases, gulf and Atlantic coasts, USA. Q Res 63:368–373

Polk JS (2009) Proxy records of climate change in tropical and subtropical karst environments. PhD dissertation, University of Florida, 658 p

Polk JS, van Beynen P, Wynn J (2012) An isotopic calibration study of precipitation, cave dripwater, and climate in west-central Florida. Hydro Proc 26(5):652–662

Randazzo AF, Jones DS (eds) (1997) The geology of Florida. Florida University Press, Gainesville, 327 p

Scott TM (1988) The lithostratigraphy of the Hawthorn Group (Miocene) of Florida, Florida Geological Survey bulletin 59. Florida Geological Survey, Tallahassee, 148 p

Scott TM (1997) Miocene to Holocene history of Florida. In: Jones DS, Randazzo AF (eds) The geology of Florida. University Press of Florida, Gainesville, pp 1–12

Scott TM, Means GH, Meegan RP, Means RC, Upchurch S, Copeland RE, Jones J, Roberts T, Willet A (2004) Springs of Florida, Florida Geological Survey bulletin no. 66. Florida Geological Survey, Tallahassee

Spechler RM, Schiffer DM (1995) Springs in Florida, U.S. Geological Survey fact sheet FS-151-95, U.S. Geological Survey, Tallahassee, 2 p

Stahle DW, Cleaveland MK (1992) Reconstruction and analysis of spring rainfall over the southeastern U.S. For the past 1000 years. Bull Am Meteorol Soc 73(12):1947–1961

Stringfield VT, LeGrand HE (1966) Hydrology of limestone terrains in the coastal plain of the Southeastern United States, Geological Society of America special paper 93. Geological Society of America, Denver, 41p

Tihansky AB (1999) Sinkholes, west-central Florida. In: Galloway D, Jones DR, Ingebritsen SE (eds) Land subsidence in the United States, United States Geological Survey circular 1182. United States Geological Survey, Tampa, 177 p

Turner T (2003) Brooksville ridge cave: Florida's hidden treasure. NSS News, May 2003:125–131, 143

van Beynen PE, Soto L, Pace-Graczyk K (2007a) Paleoclimate reconstruction derived from speleothem strontium and $\delta^{13}C$ in Central Florida. Quatern Int. DOI:10:1016/j.quaint.2007.03.019

van Beynen PE, Asmerom Y, Polyak V, Soto L, Polk JS (2007b) Variable intensity of teleconnections during the Late Holocene in Subtropical North America from an isotopic study of speleothem from Florida. Geophys Res Lett 34: L18703. DOI:10:1029/2007GL031046

van Beynen PE, Soto L, Polk J (2008) Variable calcite deposition rates as proxy for paleo-precipitation determination as derived from speleothems in central Florida. J Cave Karst Stud 70(1):1–19

van Polk JS, Beynen PE, Wynn JG (2012) An isotopic calibration study of precipitation, cave dripwater, and climate in west-central Florida. Hydrol Process. doi:10.1002/hyp.8169

Watts WA, Hansen BCS, Grimm EC (1992) Camel Lake: A 40,000-yr record of vegetational and forest history from northwest Florida. Ecology 73(3):1056–1066

Watts WA, Hansen BCS (1994) Pre-Holocene and Holocene pollen records of vegetation history from the Florida peninsula and their climatic implications. Palaeogeogr Palaeocl Palaeoecol 109:163–176

White WA (1970) The geomorphology of the Florida Peninsula, Bulletin no. 51. Florida Bureau of Geology, Tallahassee, 164 p

Winsberg MD (2003) Florida weather. University of Florida Press, Gainesville

Overview of the Controls on Eogenetic Cave and Karst Development in Quintana Roo, Mexico

16

Patricia N. Kambesis and James G. Coke IV

Abstract

The northeast coast of the state of Quintana Roo, Mexico is known for its vast underwater cave systems, numerous cenotes, and unusual coast line features that are all expressions of the coastal hydrology of an eogenetic karst region. Variations in geological and hydrological controls, and boundary conditions, have resulted in the formation of extensive linear phreatic caves located in the phreatic and vadose zone, and relict flank margin caves in the vadose zone. A significant number of the region's cenotes are portals into underwater cave systems. Dry sinkholes provide access to caves currently located in the vadose zone. A combination of karstification and littoral processes have affected the northeast coastline of Quintana Roo resulting in the formation of features such as coastal and off-shore springs, caletas, and crescent-shaped beaches. The juxtaposition of extensive inland recharge, diagenetically immature carbonates, intersection of regional fault, fracture and lineament trends, and mixing zone dynamics, all within glacioeustatic-driven fresh/saline water regime have resulted in a density-stratified coastal karst aquifer drained by an extensive network of conduits. Sea-level change has stranded the higher-elevation conduit networks and the flank margin caves in the vadose zone. The conduits, indicative of turbulent flow, and flank margin caves formed from laminar flow, are both the result mixing-zone dissolution. Surface and littoral erosion facilitate the continued karstification of the inland and coastal areas of Quintana Roo.

P.N. Kambesis (✉)
Department of Geosciences, Mississippi State University, Mississippi State, MS 39762, USA
e-mail: pk249@msstate.edu

J.G. Coke IV
Quintana Roo Speleological Survey, 7353 Myrna Blvd, Kent, OH 44230, USA
e-mail: chac@consolidated.net

16.1 Introduction

Carbonate platforms, and islands and continents with carbonate coastlines, share a unique hydrology that involves the interaction of a fresh-water lens and saline waters and are classified as eogenetic karst. Vacher and Mylroie (2002)

M.J. Lace and J.E. Mylroie (eds.), *Coastal Karst Landforms*, Coastal Research Library 5,
DOI 10.1007/978-94-007-5016-6_16, © Springer Science+Business Media Dordrecht 2013

defined eogenetic karst, which they initially applied to small carbonate islands, as a dual porosity system with permeability that increases with the development of small-scale secondary porosity features of touching vug channels and preferred passageways within a high porosity intergranular matrix. Cavernous dissolution features that form in eogenetic karst include epikarstic secondary porosity, flank margin caves, allogenically recharged vadose stream caves, bank margin fracture-guided systems, and linear phreatic caves. Vacher and Mylroie (2002) also suggested that island size and thus catchment size control the nature of coastal cave development. Small islands have a very large perimeter to area ratio, and meteoric catchment can easily be discharged to the sea as laminar flow, creating classic flank margin cave conditions. However, as islands (and carbonate platforms) grow larger, the area increases by the square, but the perimeter only linearly, so that the meteoric catchment increases faster than the available discharge perimeter, and diffuse flow paths become longer. Under these conditions, Vacher and Mylroie (2002) hypothesized, conduit flow becomes sustainable and cave development makes a switch to integrated turbulent flow systems.

Coastal carbonate strata of eogenetic karst are typically diagenetically immature, tend to be near horizontal in attitude, and in close proximity to marine water such that small changes in global sea level result in large changes in sub-aerially exposed zones (Vacher and Mylroie 2002). The incursion of saline water into the coastal carbonate rock, and decoupled meteoric recharge toward the coast, form a density-stratified aquifer with a fresh-water lens atop saline water. Both waters may be equally saturated with respect to carbonate but differ chemically in terms of CO_2 partial pressure, salinity or other chemical property. At their interface the waters mix and form a zone that is significantly undersaturated with respect to carbonate (Plummer 1975; Wigley and Plummer 1976). Depending on hydrogeologic boundary conditions, the mixing zone (or halocline) can produce extensive dissolutional cave systems indicative of turbulent flow and/or flank margin caves that result from laminar flow (see also Chap. 4).

In karst landscapes, regional hydrology and resultant cave type vary with mode of recharge, catchment size, rock/water interaction, and hydrodynamics. Cave passage distribution, morphology, and density are controlled by lithology and structure. In the eogenetic coastal setting, the overall hydrologic regime is affected by eustatic sea-level fluctuations and/or tectonics and can result in extensive polygenetic caves that have developed in different tiers and may be overprinted with features associated with turbulent and/or diffuse flow. The dissolutional processes that form coastal caves can also impact the morphology of the associated coastline.

The Yucatan peninsula epitomizes an eogenetic, coastal carbonate region that has been extensively karstified by the interaction of mixing-zone hydrology, littoral processes, and glacioeustacy. The Yucatan Peninsula (Fig. 16.1) is the aerially emergent part of the greater Yucatan Platform – a carbonate platform with a surface area of 300,000 km^2 (Bauer-Gottwein et al. 2011). The low-elevation, heavily karstified peninsula encompasses over half of the total platform surface area, and separates the Gulf of Mexico from the Caribbean Sea. The western submerged part of the platform is called the Campeche Bank and extends 200 km northwest into the Gulf of Mexico at depths of less than 200 m. The eastern submerged bank extends up to 10 km from the Caribbean shoreline with a 400-m loss of elevation into the Yucatan Basin east of Cozumel (Beddows 2003). Platform asymmetry is due to down-faulting that has lead to the development of fracture zones parallel to the Caribbean coast (Beddows 2004). The peninsula has been tectonically quiescent since the late Pleistocene (Weidie 1985) so major variations in sea level during that time are solely attributed to glacioeustatics. Significant physical features of the Yucatan peninsula include the Chicxulub impact zone and its associated ring of cenotes, an extensive, density-stratified karst aquifer that holds many caves and cave systems

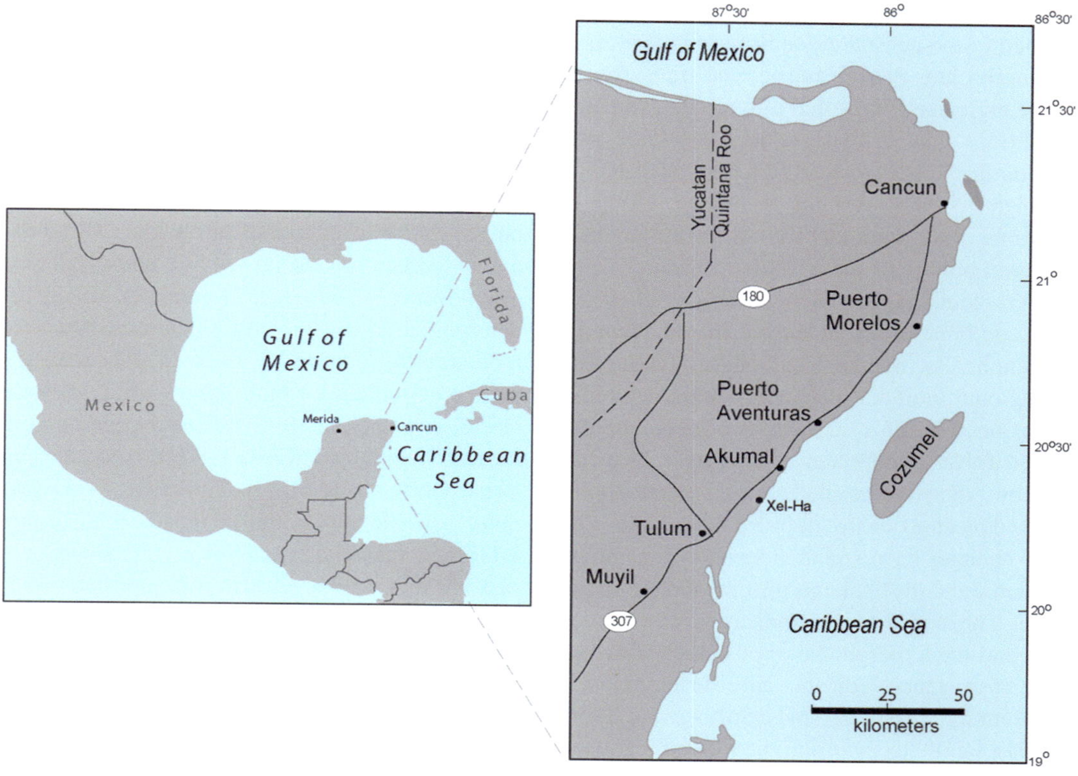

Fig. 16.1 Location map of Quintana Roo

both submerged and within the vadose zone, and a vast number of sinkholes, both water-filled (cenotes) and dry that are associated with the groundwater hydrology of the region (Bauer-Gottwein et al. 2011).

The state of Quintana Roo, Mexico, located along the northeast coast of the Yucatan peninsula (Fig. 16.1), holds a prime example of a carbonate coastline whose complex regional hydrology has resulted in the formation of an extensive conduit-drained aquifer. From Miyul to Puerto Morelos over 1,000 km of submerged cave passages have been documented (QRSS 2013). The world's longest underwater cave systems are, in ascending order, Sistemas Ox Bel Ha (242 km), Sac Actun (220 km) and Dos Ojos (82 km) (QRSS 2013). A dry passage connection between Sistemas Sac Actun and Dos Ojos gives Sac Actun the status of world's second longest cave at 308 km (Heyer and Sprouse 2012). In addition to the underwater systems, over 100 km of dry phreatic conduits have been documented in the vadose zone (Sprouse, 2013, personal communication), as well as relict flank margin caves located in eolianites along the coast (Kelly et al. 2006).

North, south and west of this extensive block of cave development, the density of caves and cave systems appears to notably decrease. Cenotes are still abundant but they are not known to intersect extensive cave systems. Though these observations may be due in part to a combination of difficult land access and cave exploration bias, there may be differences in the geological and hydrological boundary conditions for karst development between the high cave density block and adjacent areas.

Within the high cave density block, sinkholes (both water filled (cenotes) and dry) are portals into various cave systems and are evident in great number across the area. The flank margin caves located in the coastal eolianites are indicators of maximum late Pleistocene marine transgression in the region and show that different hydrological conditions can co-exist within the same geographic area (Kelly et al. 2006). Coastal discharge from the conduit systems combined with mixing zone dissolution and littoral processes shape the morphology of the coastline forming caletas and crescent-shaped beaches (Back et al. 1979). The combination of extensive inland recharge, diagenetically immature carbonates, intersection of regional fault and lineament trends, and mixing zone dynamics, all within a glacioeustatic-driven fresh/saline water regime have formed a shallow, density-stratified coastal karst aquifer drained by an extensive network of conduits. Sea-level changes have stranded some of the conduit networks and the flank margin caves in the vadose zone. The conduits are indicative of turbulent flow and flank margin caves of laminar flow and their coexistence indicates cave development in different hydrologic regimes. Surface and littoral erosion facilitate the continued karstification of the inland and coastal areas.

16.2 Climate

The climate of the Yucatan Peninsula is classified as "Aw" based on the modified Köppen-Geiger climate classification and is considered to be tropical with distinct wet and dry seasons (Kottek et al. 2006). The average annual temperature is 26 °C, with a range in monthly averages between 23 and 29 °C (Beddows 2004). May to September is the hot, rainy season and October to April is the relatively cooler dry season. There is a significant east–west precipitation gradient across the peninsula (Neuman and Rahbek 2007). The Caribbean coast is the wettest side with >1,500 mm of precipitation per year (Gonzalez-Herrera et al. 2002).

Regional-scale evapotranspiration (ET) for the peninsula was initially estimated at 14 % by Lesser (1976) using simple water-balance equations. A similar value was calculated by Back (1995). However, Beddows (2003, 2004) contended that ET could be between 30 and 70 % based on her studies of coastal groundwater outflow over an 80-km stretch of coastline in Quintana Roo. Interestingly, Jones and Banner (2003) suggest that the Caribbean Yucatan displays rapid and efficient meteoric infiltration through a thick vadose zone that decouples most of the vegetative uptake from the water table, with only a few phreatophytes present to allow delayed groundwater withdrawal. Gondwe et al. (2010) determined ET for the entire peninsula from remote sensing data to be 17 % of the mean annual precipitation, which agrees with Lesser's (1976) and Back's (1995) estimates, and on field measurements by Thomas (1999). According to Bauer-Gottwein et al. (2011) actual ET is spatially variable across the peninsula with higher ET along the coasts and lower ET in the less densely vegetated and much drier northwest part of the peninsula. They attribute the discrepancy in ET values to be a function of groundwater flow from distant parts of the peninsula to the coastal areas. All current researchers agree that further investigations across a range of scales is required in order to determine a more accurate water budget for the Yucatan Peninsula.

16.3 Eogenetic Karst

The eogenetic karst model (Chap. 4) as formulated by Vacher and Mylroie (2002) consists of a dual porosity system, where aquifer permeability increases by the formation of small-scale secondary porosity features of touching vug channels and preferred passageways lacing through a high porosity intergranular matrix. Increase in permeability occurs from redistribution of small-scale porosity within the matrix due to freshwater flow towards the coast. This results in a network of very small horizontally oriented tubes with diameters of up to ~10 mm that effectively increase regional hydraulic conductivity over time, all while total porosity may remain near constant and high (Fig. 16.2). Horizontal orientation of the small phreatic tubes imparts significant anisotropy to the coastal aquifer. Vacher and Mylroie (2002) further posit that island size

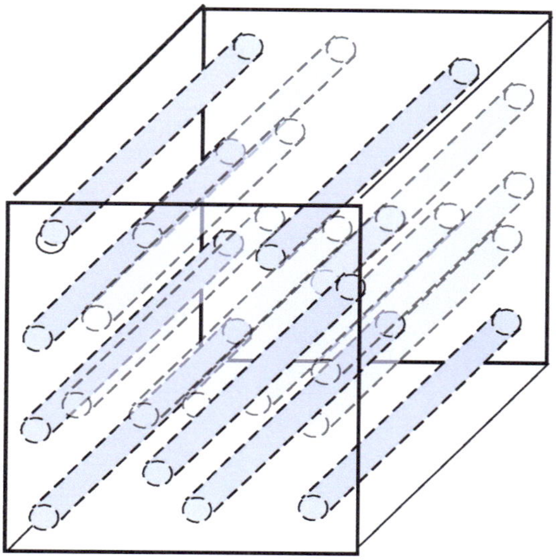

Fig. 16.2 Flow in eogenetic karst as modeled by Vacher and Mylroie (2002) as a series of small tubules

and thus catchment size control the nature of cave development on small carbonate islands. Small islands have a very large perimeter to area ratio, and meteoric catchment can easily be discharged to the sea as laminar flow, creating classic flank margin cave conditions. However, as islands (and carbonate platforms) grow larger via coral reef accretion and/or drops in sea level, island area increases by the square, with only a linear increase in perimeter such that meteoric catchment increases faster than the available discharge perimeter, and laminar flow paths become longer. Under these conditions, Vacher and Mylroie (2002) hypothesized conduit flow becomes sustainable and cave development makes a switch to an integrated turbulent flow system.

16.4 Lithologic and Structural Controls

The lithology of the Yucatan Peninsula is comprised of limestones, dolomites and evaporites overlying a basement of igneous and metamorphic rocks (Weidie 1985). The platform interior is composed of Eocene-Paleocene rocks surrounded by Miocene-Pliocene deposits, grading to and underlying Quaternary-age strata at the coasts (Ward 1997, 2003) (Fig. 16.3). The older carbonates in the northwest part of the peninsula are thickly bedded and highly permeable with a vertical extent of 1,000 m near Merida and at least 150 m at 50 km inland from the Caribbean coast (Neuman and Rahbek 2007). The eastern peninsular coastline consists of a 10-km band of off-lapping carbonates that were deposited during interglacial sea-level highstands in the Pleistocene (Ward 1985).

The eastern part of Quintana Roo was initially mapped as the Carrillo Puerto Formation, which is composed of undifferentiated, diagenetically immature carbonates of Miocene to Holocene age (Bonet and Butterlin 1962; Lopez-Ramos 1975; Ward 1985). More recent studies have divided the coastal carbonates into Upper, Middle, and Lower Pleistocene units comprised of marine and non-marine sequences that accumulated in shelf margin, reef, and back reef facies during interglacial highstands (Ward 1985), and separated by unconformities indicative of exposure and erosion of the platform surface on marine retreat (Lauderdale et al. 1979; Rodriguez 1982; Perry et al. 1995) (Fig. 16.4). Marine sequences include beach, nearshore and lagoonal strata, and coral-reef limestone; non-marine rocks consist of eolianites, freshwater lacustrine carbonate mudstone, and caliche (Ward 2003) (Fig. 16.4). Underlying the Pleistocene strata are Miocene-Pliocene carbonate rocks (Richards and Richards 2007).

Though absolute ages of the Middle and Lower Pleistocene units are not well known, Ward (2003) suggested that the Lower unit could have an age between 200,000 and 800,000 ybp based on dating of slightly altered coral; the Middle unit may have been deposited during the penultimate Pleistocene interglacial period based on dating of a cave deposit at 174,000 ybp (Ford, 1998, personal communication reported in Beddows 2003). The Upper Pleistocene carbonates have been dated to 125,000 ybp (Back et al. 1979) and correlate with the 6-m high stand above current sea level dated at 125,000 years ago (MIS 5e). A comparison of

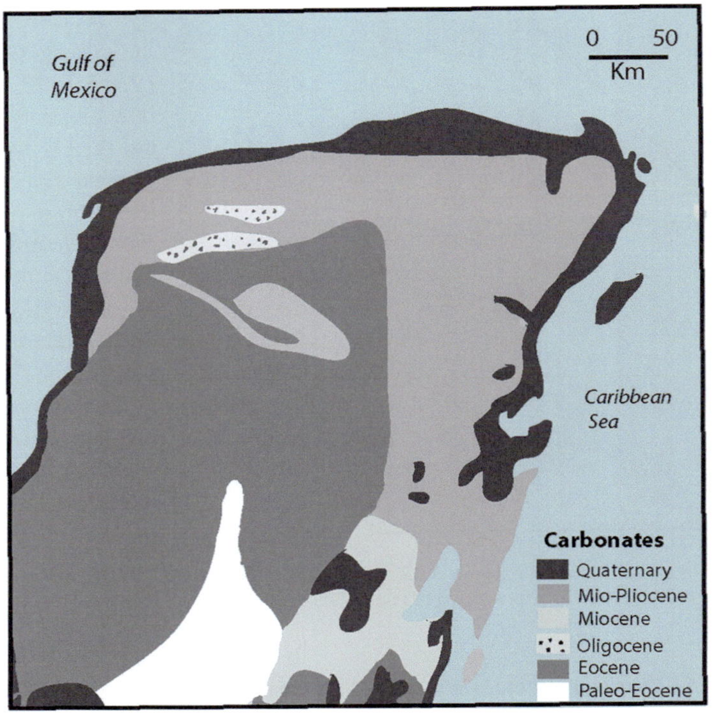

Fig. 16.3 Stratigraphy of the Yucatan Peninsula (Modified from Ward 1997)

elevation and age of the Upper Pleistocene unit on Yucatan's northeast coast to other similar-age areas in the Caribbean indicate that there has been very little vertical displacement of this unit since the last Pleistocene high stand of 125,000 years ago (Szabo et al. 1978). Kelly et al. (2006) demonstrated that the Pleistocene coastal eolianite sequences follow a depositional model associated with glacioeustasy that had been established for the extensive eolianite suites in the Bahamas.

The surface sediments in Quintana Roo are patchy and thin with terra rosa and organic soils present at the land surface (Gmitro 1986). Consequently there is rapid infiltration of meteoric recharge into a highly fractured surface that drains to the shallow, density-stratified coastal karst aquifer (Isphording 1974).

A narrow ridge and swale plain (Fig. 16.5) of Upper Pleistocene limestone is located 5–10 m above present sea level, characterizes the northeast coast of Quintana Roo between Cancun and Tulum (Ward and Brady 1979). Ridge crests are 1–5 m above the swales, and are spaced 50–200 m apart, paralleling the modern coastline (Ward 2003). There are as many as 20 ridges at the widest part of the plain but they all coalesce south of Akumal (Beddows 2003). The Middle Pleistocene unit, which is 150 km long, up to 4 km wide and 3–10 m thick, underlies the beach-ridge plain and is exposed at the surface as a low-relief karst plain due west of it (Ward 2003). East of the beach-ridge plain and still buried by it in many places are Upper Pleistocene barrier-reef limestones.

Lithologic changes in reef facies reflect the different environments of deposition of a reef structure and these differences are typically related to distance from the coastline that was active at the time of deposition. The reef facies that comprise the Pleistocene units of the Yucatan Caribbean vary depending on distance from the Pleistocene coastline, (Ward and Brady 1979; Ward 1985) which in most places is near-parallel and relatively close to the modern coast. Cave passage morphology and density in the region also vary as a function of distance from the coast so it is plausible that cave passage morphology

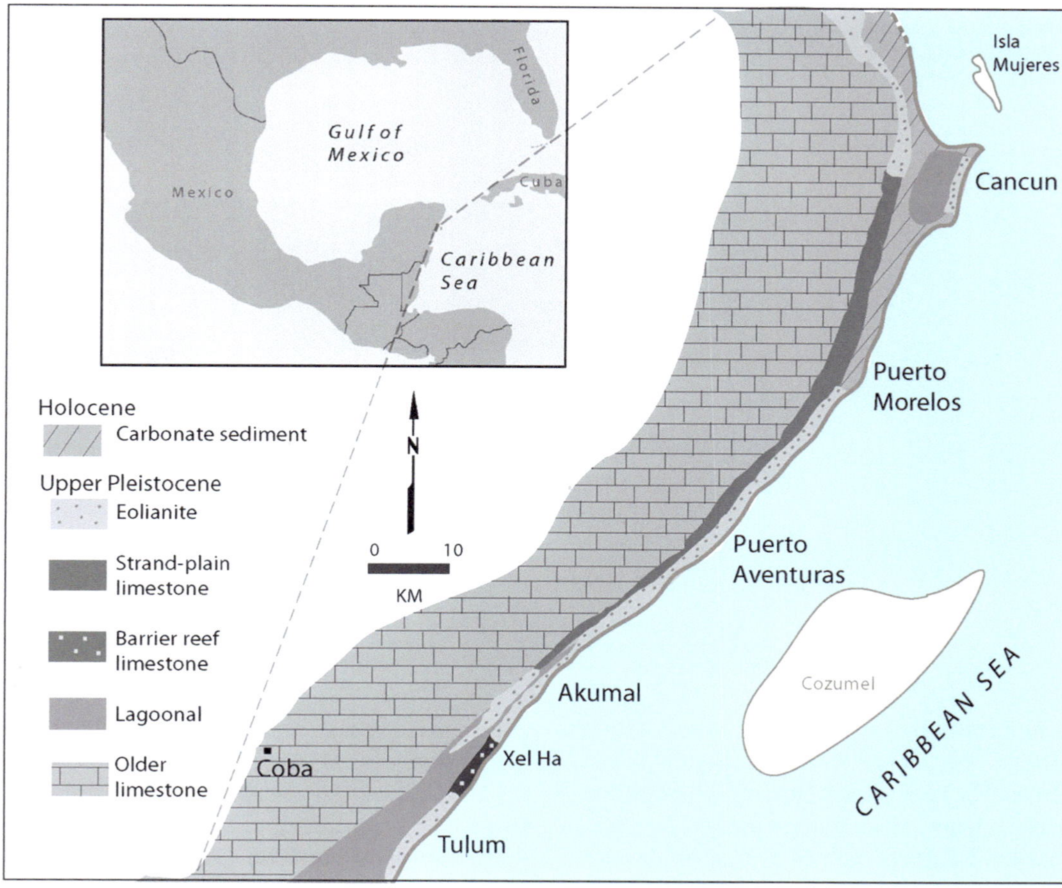

Fig. 16.4 Stratigraphy of northeast coast of Quintana Roo (Modified from Ward 1985)

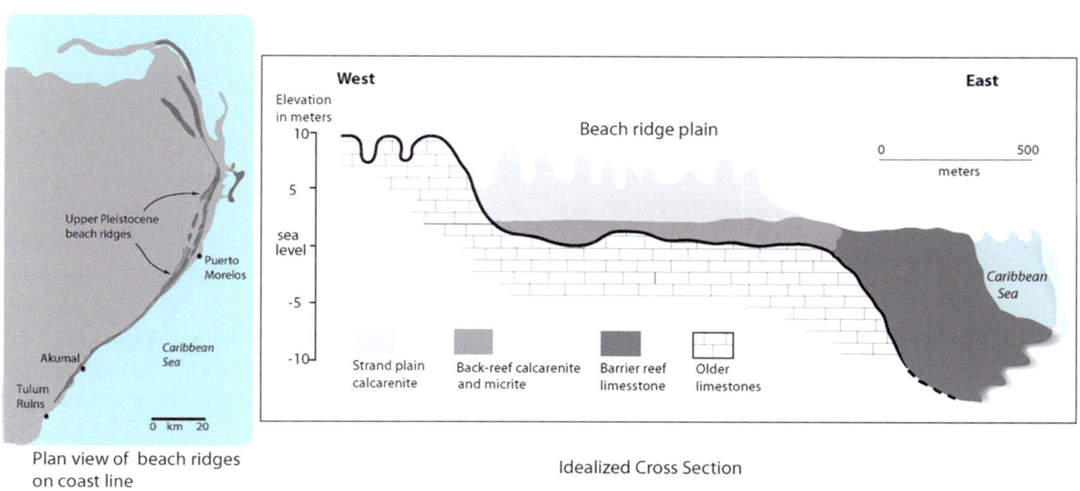

Fig. 16.5 Ridge and swale plain of the northeast coast of Quintana Roo (Modified from Ward 1985)

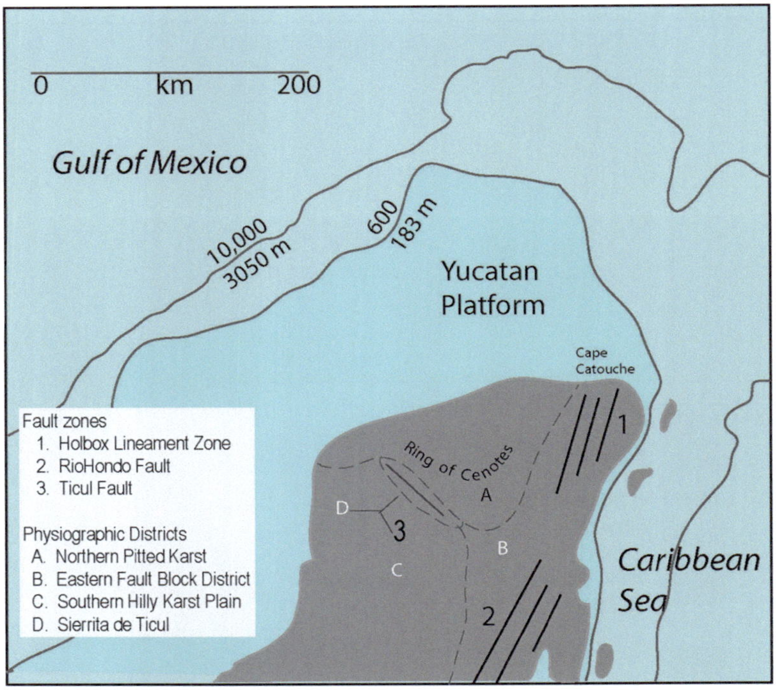

Fig. 16.6 Physiographic districts of the Yucatan Peninsula and fault and lineament zones of Quintana Roo (Modified from Beddows 2003)

is influenced by lithologies associated with reef facies (Neuman and Rahbek 2007). Its been observed that caves greater than 4 km from the coast contain large linear trunk passages vs. coastal-zone caves that are typified by smaller passage dimensions that form complex mazes (Coke 2009). An analog to this would be the Migjorn karst on the island of Mallorca, specifically Vallgornera Cave (see Chap. 11). Though that area is more structurally complex than the Yucatan peninsula, bedrock lithology strongly influences cave passage density and morphology in the caves of that region (Ginés et al. 2009a, b). A notable exception to the coastal location rule is seen in Sistema Ox Bel Ha, whose passages in the modern coastal zone share the same morphology as inland caves (Coke 2009). This area may be a geological transition zone between caves and karst in the high density cave block vs. adjacent areas to the southwest (Neuman and Rahbek 2007) and may also be where the location of the Pleistocene coastline makes a significant departure from the modern coastline (Beddows, 2012, personal communication). These possible changes in boundary conditions may account for the unusual passage morphology of Ox Bel Ha in the modern coastal zone.

The state of Quintana Roo is within the Eastern Block-Fault district that extends from Cape Catouche on the northeast coast, to the Yucatan's border with Belize (Fig. 16.6). It is one of five physiographic regions of the Yucatan Peninsula which are defined by the influence of prominent fracture or lineament systems (Isphording 1975). The two main faults/lineaments in Quintana Roo are the Holbox Lineament Zone and the Rio Hondo Fault Zone.

The Holbox Lineament Zone (HLZ), originates at the northeastern coast of the peninsula south to within 10 km of the coast inland from Tulum and trends N 5°E to N 10E (Bauer-Gottwein et al. 2011). The HLZ is expressed on the surface by the alignment of polje-like depressions that seasonally fill with water making narrow, aligned swamps (Weidie 1978). Remote sensing data indicate that development of regional dissolution features were strongly influenced by the lineament zone and result in high permeability and groundwater drainage (Southworth 1985; Tułaczyk et al. 1993). In a more recent study, higher subsurface electrical conductivity values relative to surrounding areas were detected in the vicinity of Tulum and were interpreted to indicate increased porosity and permeability (B.R.N. Gondwe, 2010, Technical University of Denmark, unpublished data, 2010 reported in Bauer-Gottwein et al. 2011).

The Rio Hondo fault zone (RHFZ) consists of a series of northeast trending (N30-32E) normal faults and has been identified as the on-shore continuation of an extensive horst and graben fault block system located off the southern Caribbean coast of Quintana Roo (Weidie 1985). This is supported by seismic data that confirms the fault system aligns sub-parallel to the southern Caribbean coast (Bauer-Gottwein et al. 2011). Surface expression of the RHFZ is seen in the alignment of shallow lakes, coastal bays, and the orientation of Cozumel that is identified as a horst block (Lesser and Weidie 1988). Gondwe et al. (2010) proposed, based on their interpretation of synthetic aperture radar (SAR) remote-sensing images that the Rio Hondo fault system extends northwards and intersects with the Holbox fracture zone in the vicinity of Tulum.

Weidie (1978) studied the distribution and orientation of fracture sets along the coast between Akumal and Tulum. He identified a well-defined fracture trend (N50-60 W) that occurs along the entire coast and indicates that fractures control the inland development and extent of coastal caletas (lagoons) and crescent-shaped beaches. He also describes a second set of fractures with a trend of N30-40E that parallels the coast and influence the lateral extent of caletas. Weidie (1978) notes that the fracture sets may form an orthogonal system that is genetically related to the RHFZ. He observed changes in fracture trend along the Caribbean coast and speculated the existence of a conjugate fracture system.

The location of coastal discharge features, caletas, and crescent-shape beaches correlate with areas of maximal fracturing. The existence of extensive underwater conduits with northwest and southeast trends support the idea that linear dissolution corridors are developed along the extensive fracture and lineament zones that occur in northeast Quintana Roo (Tułaczyk et al. 1993).

16.5 Hydrogeology

With the exception of the southern end of Quintana Roo where it joins to Belize and Guatemala, there is no surface drainage on the Yucatan Peninsula, save for surface bodies of water that exist where the ground surface intersects the water table (Beddows 2003; Coke 2009). Surface waters include aquadas, lakes, coastal mangrove swamps, and cenotes. Aguadas are small bodies of water that can be up to tens of meters in diameter and typically occur at or perched on clay layers just above the water table (Beddows 2004). Lakes can be hundreds of meters in diameter and are either large aquadas or low elevation areas that receive overland flow during large storm events (Beddows 2004). Cenotes result from dissolution of bedrock and/or conduit ceiling collapse. There are literally thousands of cenotes on the Yucatan peninsula. Many of those in Quintana Roo typically lead to extensive cave systems.

Because the surface soils in Quintana Roo are patchy and thin (Gmitro 1986) there is rapid infiltration of meteoric recharge into the fractured bedrock surface that drains to the shallow, density-stratified coastal karst aquifer.

Though precise vertical control is lacking on the peninsula overall, the elevation in Quintana Roo is about 30 m or less above sea level with local relief of 5 m but rarely exceeding 10 m. Lack of surface drainage features coupled with minimal vertical control make it impossible to identify individual drainages basins (Beddows 2004). Instead, sub-regional distinctions have been made based on fault zones and subtle variations in surface topography (Lesser and Weidie 1988).

The coastal karst aquifer of Quintana Roo is unconfined and recharged by precipitation from extensive, higher elevation areas near Muyil for the underwater caves in the vicinity Tulum and from higher elevations to the north in the cave systems north of Akumal (QRSS 2013). Neuman and Rahbek (2007) note that the aquifer responds to short terms conditions such as heavy rains, barometric pressure, tides, and ocean density but not to long-term changes in recharge, which

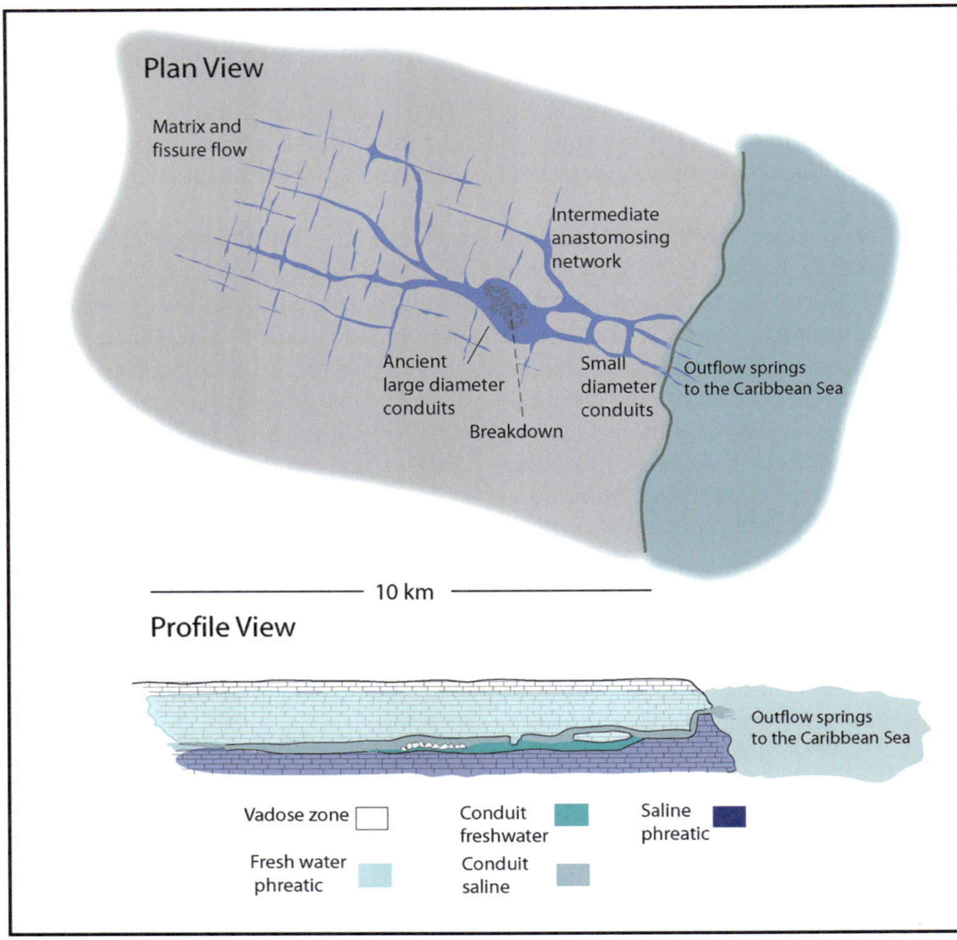

Fig. 16.7 Schematic of a stylized conduit system in the Caribbean Yucatan. Older conduits at lower depth or located inland omitted for simplicity (Modified from Beddows 2004)

supports the idea that base flow originates from the interior of the peninsula, a condition hypothesized by Neuman and Rahbek (2007).

Structural heterogeneities within the aquifer include bedding planes, fissures and fractures, some of which are dissolutionally enlarged to conduits via mixing zone corrosion. The Pleistocene strata have a primary matrix porosity of 14–23 % (Harris 1984). The structural heterogeneities, high matrix porosity, and the existence of conduits make for a triple porosity aquifer. The matrix acts as storage for 97 % of the aquifer but contributes very little to the flow; flooded conduit systems with very little storage capacity (3 %), link inland recharge to springs on and just off the Caribbean coast accounting for at least 99 % of the fresh water flux to the sea (Worthington 2000; Beddows 2004). The conduits increase aquifer transmissivity as evidenced by hydrodynamic response to changing hydrological boundary conditions with 84 % of the 30 cm amplitude semi-diurnal tidal signal transmitted to free water surfaces in cenotes at 1 km inland, and 39 % at cenotes 6 km inland (Beddows 2003).

Aquifer discharge of groundwater to the Caribbean Sea is via a network mesh of conduits of varying size from tens of millimeters in width to humanly enterable passages that can up to 80 m in width (QRSS 2013) (Fig. 16.7). Beddows (2004) has documented many large springs and numerous small outflows and seeps along the coast and has estimated that discharge of the Caribbean coast of the Yucatan may be 2.3×10^7 m^3/year per kilometer of coastline.

The aquifer is shallow in depth with a vadose zone that is up to 2 m thick in the coastal area and at maximum 20 m in thickness at inland areas (QRSS 2013). Though hydraulic gradient data are sparse on the peninsula, they have been measured to range from 7 mm/km (Marin 1990, cited in Neuman and Rahbek 2007) on the northwest side of the peninsula to 58–130 mm/km inland of the eastern coast and near the coast south of Playa del Carmen respectively (Moore et al. 1992; Beddows 2004). On a global scale, such nearly flat gradients are the lowest known in comparison to other karst areas (Ford and Williams 2007). The extremely low gradient and its value ranges are attributed to local and regional depressions in the water table caused by numerous conduits that locally attract groundwater flow (Ford and Williams 2007). Conduit density in the Tulum to Xel Ha area has been calculated to be >2 km/km^2 (Beddows 2003).

16.6 Aquifer Hydrodynamics and Hydrogeochemistry

The east coast of Quintana Roo receives approximately 2.5×10^6 m^3/year of marine inflow from the Caribbean Sea resulting in a density-stratified aquifer where a thin meteoric-derived fresh-water lens floats on the denser saline water. The transition zone between fresh water to marine water salinity is called the mixing zone or halocline. This zone is also a thermocline as evidenced by cooler temperatures in the freshwater lens and warmer in the underlying saline water (Beddows 2004). In the Quintana Roo region the mixing zone is thickest near the coast and decreases inland to a sharply defined boundary (Beddows 2003, 2004). The mixing zone responds to several factors including conduit cross section, turbulence from conduit discharge and to tides though the effects of the latter diminish inland (Beddows 2004).

Hydrological field research by Beddows (2004) has documented two types of saline flow on the Caribbean coast; a shallow two-way flow that corresponds to tidal frequency (up to >9 km inland), and a continuous incursion at a range of 5–45 m in depth. An interesting aspect of her studies is the opposing direction of saline flow between the shallow saline shuttle and the deeper unidirectional saline flows, and the decoupling of shallow saline water from the fresh-water lens. Though she notes that the saline inflows are tidally modulated, the deep saline flows occur continuously regardless of mean sea level and tidal change. Beddows (2004) field measurements showed deep saline inflows over a wide range of Caribbean Sea levels representing 63 % of the total annual range indicative that deep saline flows are not principally driven by head variations of the adjacent Caribbean Sea. She further notes that continuous inflow of the deep saline water along the Caribbean Yucatan coast must either be stored in the aquifer for later discharge, or must discharge continuously from an opposing margin. As with shallow saline shuttle waters, deep saline inflow water cannot be accommodated by the upwards displacement of the fresh water lens as the mixing zone depth was observed to be relatively constant. Furthermore it is hydrodynamically improbable that the even-deeper saline water is expelled from the platform, as the density gradients in the adjacent seas or within the aquifer would inhibit downward flows. Based on her field data, calculations of head difference between the Caribbean and Gulf of Mexico, and analogous studies in the Bahamas by Whitaker (1992), Beddows proposes that the continuous saline inflows into the Caribbean side of the Yucatan Peninsula may represent the inflows of a cross-platform saline circulation with saline water discharge occurring on the Gulf of Mexico coast.

Beddows (2004) shows that at 0–0.4 km from the coast the halocline gradient is steep and is accounted for by the low hydraulic conductivity of the area due to restricted size of conduits. For the zone >0.4–10 km inland, the depth to the halocline is lower than predicted by the

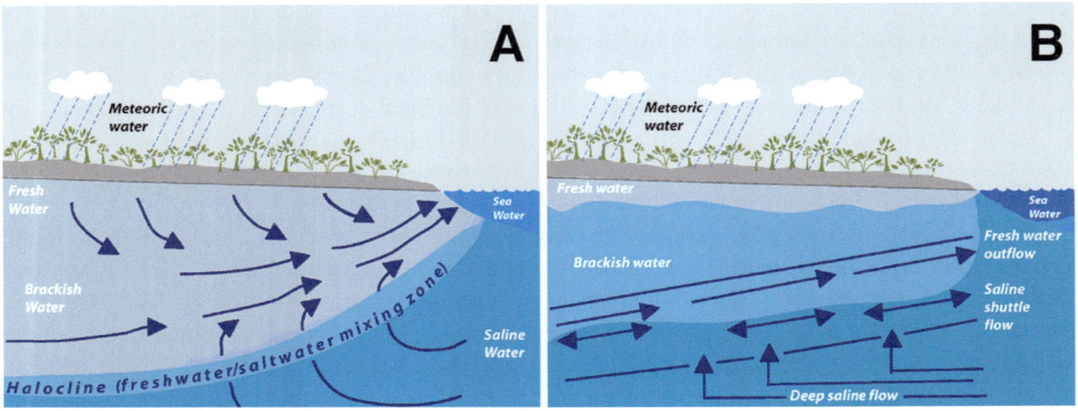

Fig. 16.8 Coastal hydrology with and without conduits. (**a**) Prior to the development of conduits, fresh water flows seaward as distributed diffuse flow. Mixing between fresh and saline water forms the halocline that also results in compensating saline inflow. (**b**) After conduit development, fresh water is discharged through cave systems to coastal springs and as a result the freshwater lens flattens and follows the level of the caves. Deep inflow may occur through paleo cave systems with a "shuttle" flow of saline water into and out of caves with tidal changes in sea level and other factors, and extending inland for several kilometers (Beddows 2004; Beddows et al. 2007a) (Modified from Richards and Richards 2007)

Ghyben-Herzberg principle (GHP) especially in areas of high conduit density. In those zones the high permeability of the conduits truncate the depth to the mixing zone because fresh water is quickly removed causing an upward flow of saline water from below the fresh-water lens. These observations do not fit the GHP model for a fresh-water lens but Beddows (2004) explains this by noting that her research area does not meet the classic porous media requirements for GHP and contains numerous conduits which affect the location of the halocline. For distances greater than 10 km from the coast, the depth to the halocline does seem to follow the Ghyben-Herzberg model (Neuman and Rahbek 2007). In summary, the flow of fresh and saline waters are decoupled within the aquifer with fresh water flowing toward the coast, tidally modulated two-way flow of shallow saline water, and deeper saline incursion inland (Beddows 2004; Beddows et al. 2007a). Figure 16.8a illustrates the coastal hydrology without conduits and Fig. 16.8b with conduits.

Extensive field research on the hydrology and hydrodynamics of the aquifer was conducted by Beddows (1999, 2003, 2004) in the area between Puerto Morelos and Tulum. Her data, in collaboration with observations from cave divers, showed the location of the halocline to be constant through time but varied in depth as a function of distance from the coast. Because the halocline maintains a near constant elevation within the aquifer, and cave passages increase in depth with distance from the coast, conduits can be exposed to all three zones of the aquifer that include fresh water, mixing zone, and saline water. The fresh-water lens is typically supersaturated with respect to carbonate and as a consequence is not active dissolutionally. Those areas within the mixing zone show the effects of chemical erosion in the form of pitted and fretted wall rock, and partial dissolution of speleothems (Beddows 2004). Vertical shafts provide access from one level to another (Fig. 16.9).

16.7 Cenotes and Dry Sinkholes

Cenotes are near-circular, water-filled sinkholes that intersect the groundwater table. There are two types of cenotes; pit cenotes and collapse cenotes (Neuman and Rahbek 2007). Pit cenotes are common to most all areas of the peninsula except in Campeche State and in the central

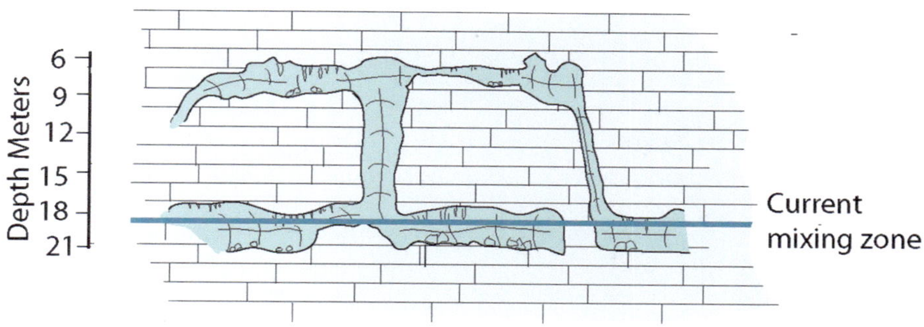

Fig. 16.9 Vertical connections between conduit levels (Modified from Beddows 2004)

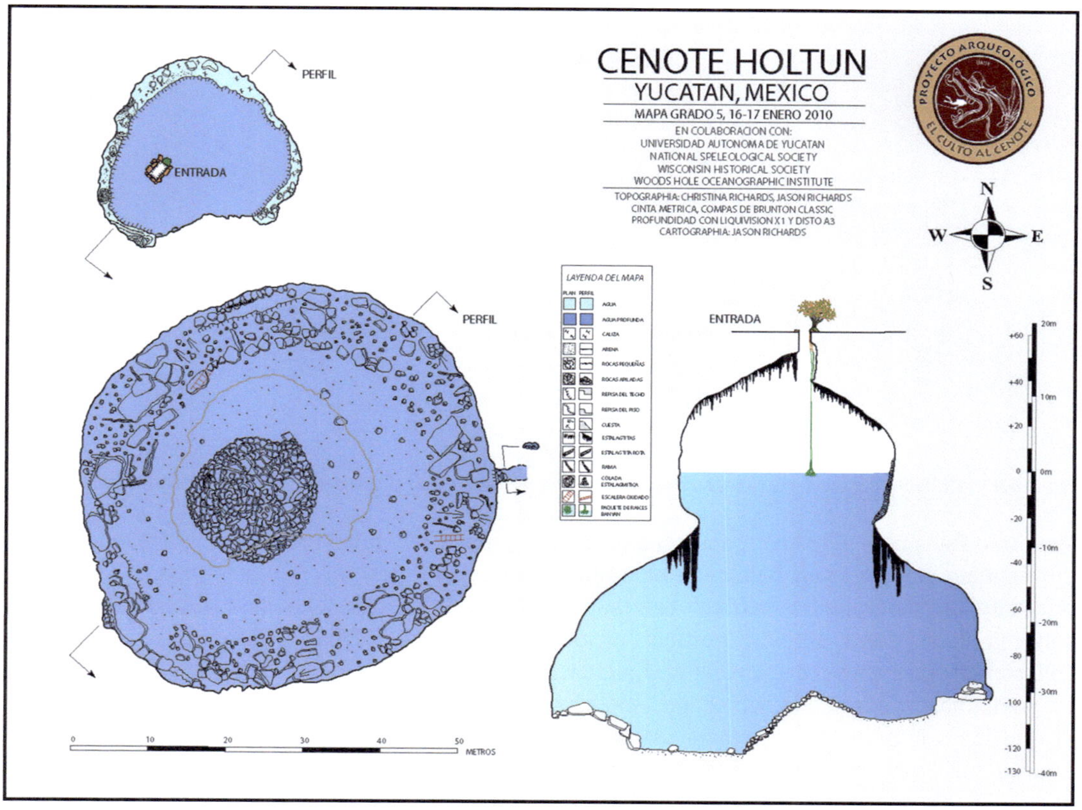

Fig. 16.10 Typical pit cenote (Cenote Holtun) (Cartography: Jason Richards 2010)

Serrita de Ticul (Coke 2009). Profile morphology of pit cenotes shows bell-shaped vertical shafts that may extend 100 m below the water table (Beddows et al. 2007a) (Fig. 16.10). They formed by dissolution of deep strata that collapse to the surface from past sea-level fluctuations. A small number of pit cenotes have been found to intersect horizontal conduits, but in all cases these conduits are not very extensive (Beddows 2004). In the northwest Yucatan, the Ring of Cenotes consists of pit cenotes in dense concentration forming a circular pattern associated with the Chicxulub impact crater (Perry et al. 1995). Mylroie et al. (1995) considered cenotes to be

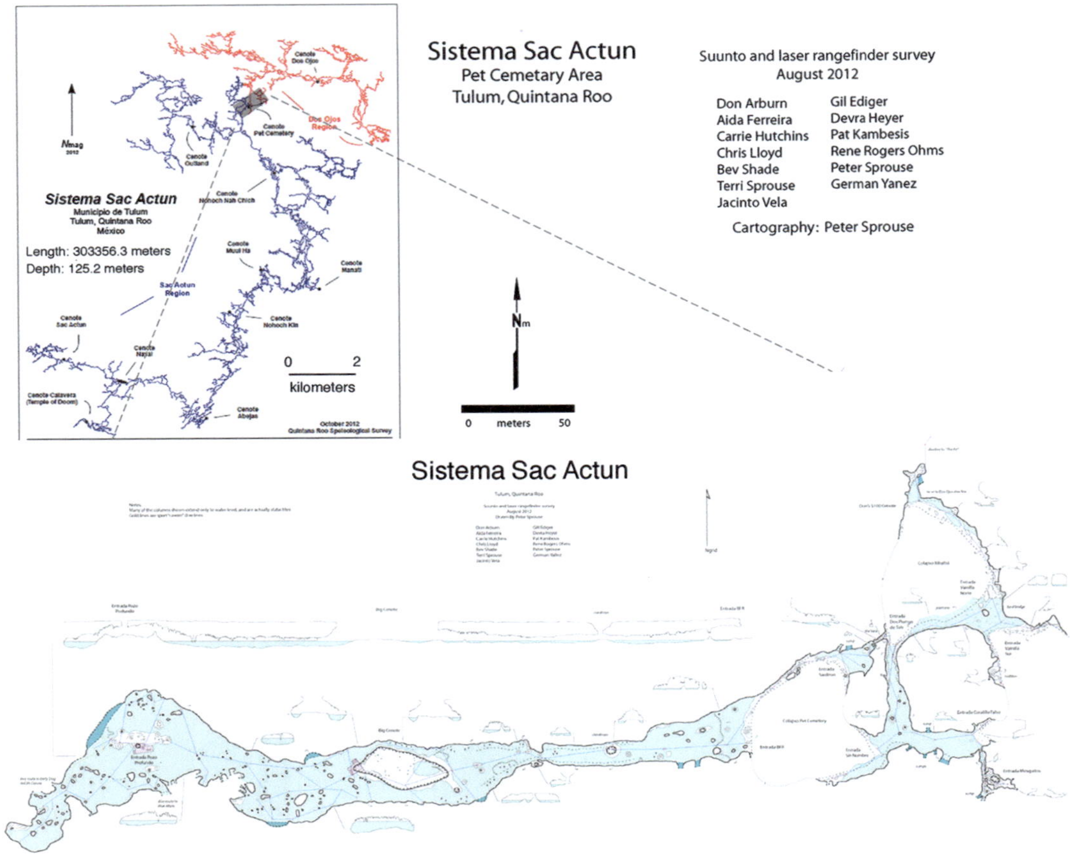

Fig. 16.11 Collapse cenotes, Pet Cemetary section of Sistema Sac Actun (Cartography: Peter Sprouse 2012)

similar to Bahamian blue holes, and attempted to differentiate cenotes from Bahamian blue holes primarily by the ratio of vertical distribution between vadose and phreatic portions of the collapse. In other words, Bahamian blue holes are progradational collapses in which the majority of the depth is under water, where as Yucatan cenotes are progradational collapses where commonly the majority of the depth is above the water surface.

Collapse cenotes are extremely common in Quintana Roo and are distinctly different from the typical pit cenote in that they form by mechanical collapse of the ceilings of shallow phreatic cave systems when buoyant support is lost as a result of low sea levels (Beddows et al. 2007b) (Fig. 16.11). Cenote density above the extensive underwater cave systems of Quintana Roo is about one cenote for every 300 m of cave passage (personal communications with cave divers reported in Neuman and Rahbet 2007).

The morphological difference between pit cenotes and collapse cenotes is a function of their hydrology. Groundwater circulation through pit cenotes is via matrix porosity and small, non-penetrable turbulent flow conduits. Collapse cenotes are associated with turbulent conduit flow within horizontally extensive cave systems (Beddows 2004).

Dry sinkholes are extremely common in the high density cave block between Cancun and Tulum. This feature has the same origin as the collapse cenotes that serve as portals to underwater cave systems. The dry sinkholes access cave passages within the vadose zone.

16.8 Caletas and Crescent-Shaped Beaches

Significant groundwater discharge to the Caribbean Sea occurs via springs and seeps associated with phreatic conduits. A cursory inventory of coastal discharge features documented large coastal springs and numerous small outflows and seeps estimated to discharge at least 2.3×10^7 m^3/year per kilometer of coastline (Beddows 2004). Coastal inlets (or lagoons) called caletas are narrow inlets that extend inland for several hundred meters and are associated with larger coastal springs (Back et al. 1979) (Fig. 16.12). Caletas form where discharging freshwater conduits mix with saltwater at their seaward margins causing an increase in local dissolution and inducing conduit collapse that migrates inland to form a cove (Beddows 2004). As dissolution continues to act on the caleta limestone, it becomes riddled and weakened by solution channels and becomes more vulnerable to the mechanical erosion by wave action (Back et al. 1979). As the inlet opening widens, waves have greater access to the caleta walls which eventually erode to form a crescent shaped beach (Back et al. 1979) (Fig. 16.13). Caletas are an example of coastal reentrants produced by dissolution that are common on carbonate coasts (e.g. Stafford et al. 2004; Kambesis et al. 2012).

16.9 Caves of Quintana Roo

Since the 1980's, the Caribbean coast of Quintana Roo has been the focus of intense underwater cave exploration. Cave divers have documented

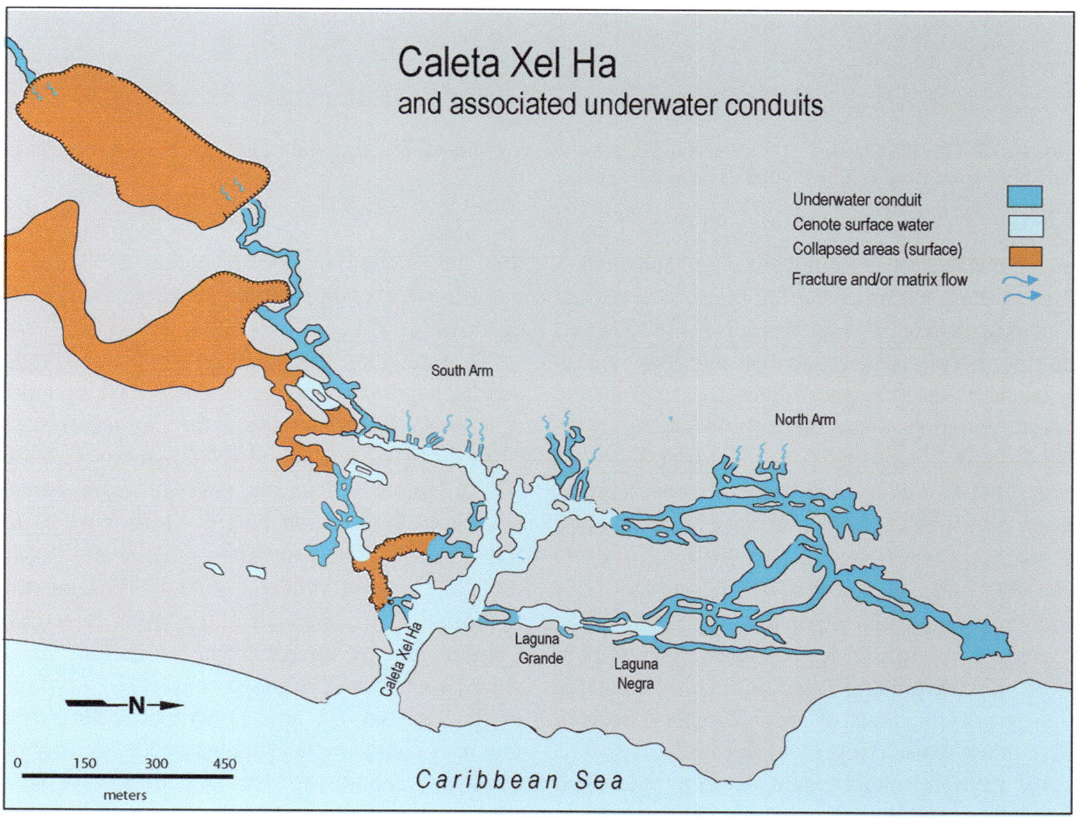

Fig. 16.12 Caleta at Xel Ha and associated conduits, Quintana Roo (Modified from Back et al. 1979 and Thomas 2005)

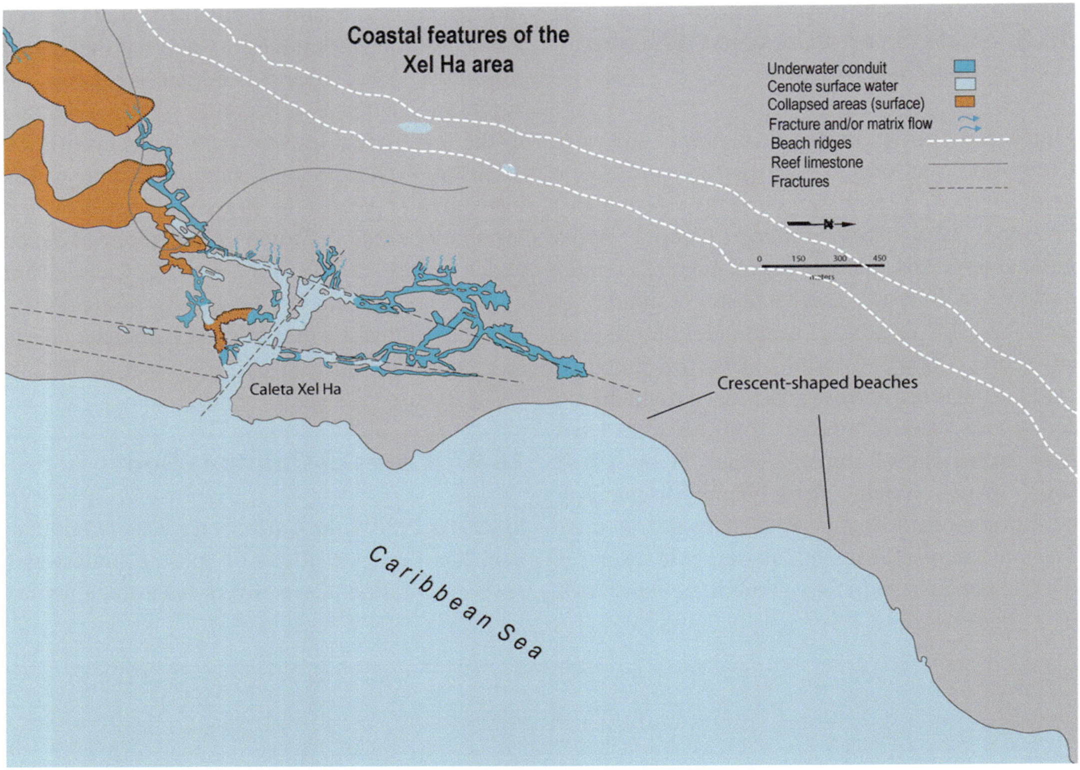

Fig. 16.13 Coastal features of Caleta Xel Ha area, crescent-shaped beaches, reef front, beach ridges, fractures (Modified from Back et al. 1979 and Thomas 2005)

an extensive series of interlinked and anastomosing conduits within an 80 km block of coastline that extends from Puerto Morelos south to Muyil on the northern boundary of the Sian Ka'an Biosphere Reserve, and inland 10 km from the coast, which, is near the eastern boundary of the Holbox Lineament Zone (Smart et al. 2006) (Fig. 16.14). The overall trend of cave passages is northwest to southeast and perpendicular to the Caribbean coast (QRSS 2013). Sistemas Ox Bel Hal (242 km), Sac Actun (220 km) and Dos Ojos (82) are respectively, the three longest underwater caves in the world. A dry passage connection between Sac Actun and Dos Ojos accords Sistema Sac Actun the status of second longest mapped cave in the world (Heyer and Sprouse 2012). The caves have an inland extent of up to 9 km and continuous passages have been documented to connect to coastal discharge points confirming that the conduit systems function as drainage pathways for a large interior recharge area (Beddows 2004).

To date over 1,000 km of underwater cave passages have been mapped in 223 cave systems (QRSS 2013) that range from the Caribbean coast to as far as 12 km inland (Neuman and Rahbek 2007). Humanly enterable passage widths range from 1 to 100 m and heights from 1 to 10 m (QRSS 2013). There are many smaller sized conduits that are not humanly enterable but are still hydrologically connected to the larger passages. Conduit density ranges from 1.4 km/km^2 in a zone 10 km inland from the coast to 2.2 km/km^2 in the Ox Bel Ha area (Beddows 2004). The region is characterized by high cenote density of about one cenote for every 300 m of cave passage (personal communications with cave divers reported in Neuman and Rahbek 2007).

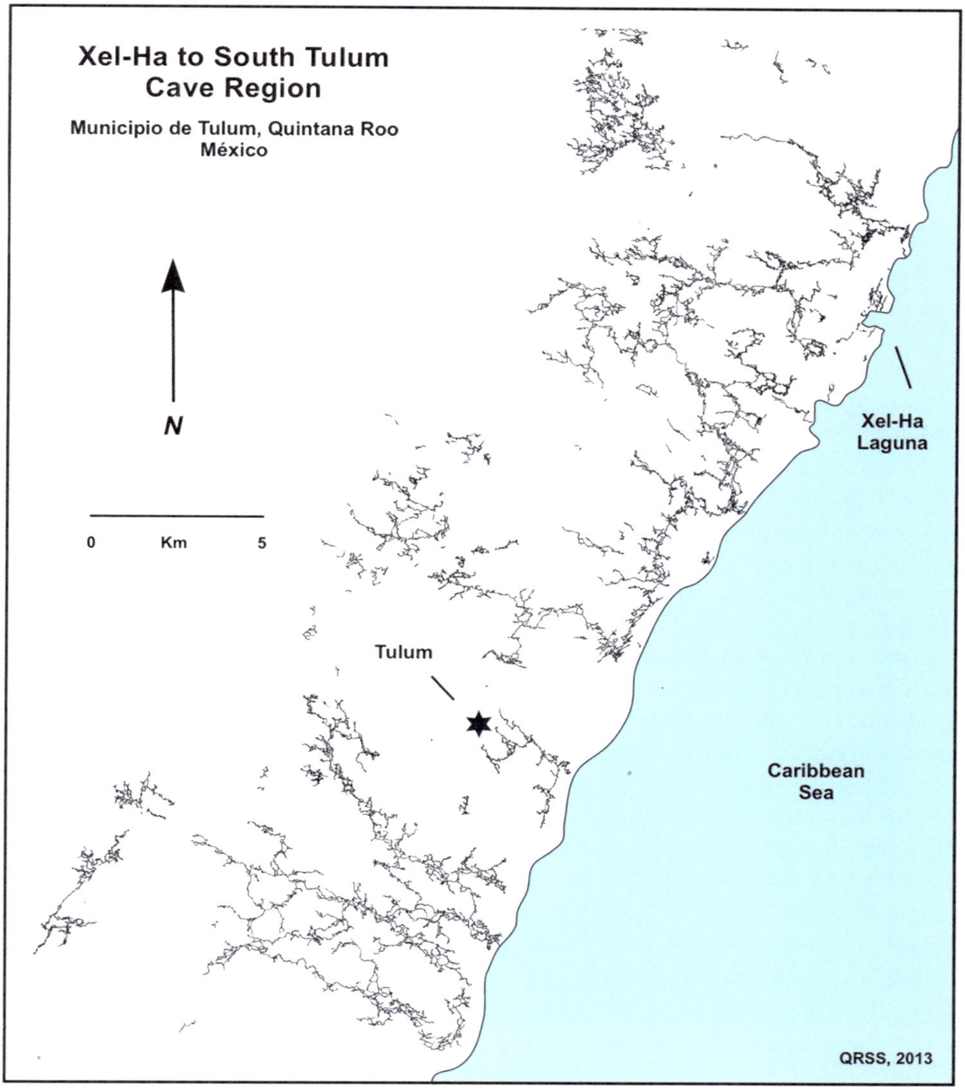

Fig. 16.14 Underwater caves in Tulum area, Quintana Roo (From QRSS 2013)

Caves currently within the vadose zone have been documented with the greatest concentration in the area between Akumal and Puerto Aventuras and extending from the coast and up to 7 km inland. Over 100 km of "dry" passages have been surveyed in 97 cave systems in the region. Of those, Sistema Pool Tunich is the most extensive at 30 km, followed by Sistema Sac Muul (11 km) and Sistema Dos Arboles (7 km). As exploration continues and more connections are made between the various caves within the vadose zone, the number and surveyed length of documented extensive "dry" cave systems will increase.

The location of caves within the vadose zone does not define their origin as vadose caves. Rather, their origin is likely the same as that of underwater caves as they share many of the same features and morphology, and some of the caves have underwater extensions. The 6-m drop in sea level since the last interglacial stranded previously submerged cave systems and segments of systems in the vadose zone. Small flank margin caves have also been documented in the eolianites exposed on the Caribbean coast (Kelly et al. 2006) and they too formed during the last sea level high stand.

16.10 Cave Development

The caves of Quintana Roo include linear phreatic cave systems and flank margin caves both formed by mixing zone corrosion within the density stratified aquifer. However, the flow regimes resulting in the two types of caves differ. Linear phreatic cave systems result from turbulent flow in the aquifer (Smart et al. 2006) vs. flank margin caves that form by laminar flow (Mylroie and Mylroie 2007).

Carbonate dissolution within the coastal carbonate aquifer is a function of the rock/water interaction between meteoric and saline groundwater. As meteoric water infiltrates into the bedrock it reaches chemical equilibrium with respect to carbonate and looses much of its dissolutional capability in the process. Saline groundwater is typically supersaturated with respect to carbonate and it has very little dissolutional capacity. When the two end members mix at the halocline they form a brackish water zone that is undersaturated with respect to carbonate (Wigley and Plummer 1976). These conditions result in distributed dissolution where meteoric water infiltrates, and enhanced dissolution at the mixing zone along the path of discharge of the undersaturated brackish water (Beddows 2003). While classic fresh-water dissolution has played a role in Yucatan cave conduit development, as is the case in continental interiors, mixing zone corrosion is a crucial mechanism for conduit development in the Caribbean Yucatan aquifer. In studies conducted by Back and Hanshaw (1980), it was determined that the groundwater of the Caribbean Yucatan arrived at the coast with a carbonate concentration of approximately 2.5 mmol $CaCO_3$ per liter. An additional half of that amount was added in the last kilometer before discharge as the water became increasingly brackish due to increased rates of fresh-saline mixing where the aquifer is strongly pumped by tides. The rate of fresh water-saline water mixing increases where the flow becomes turbulent and mixing-zone corrosion focuses dissolution creating a conduit (Beddows 2004). As the conduit enlarges, velocity and discharge increase resulting in a positive feedback loop that focuses dissolution at the expense of further permeability increases in the matrix or fractures and results in efficient conduit system development (Smart et al. 2006; Ford and Williams 2007).

Because the mixing zone maintains a near constant elevation within the aquifer, and cave passages increase in depth with distance from the coast, conduits can be exposed to all three zones of the aquifer that include fresh water, mixing zone, and saline water. These observations indicate that the conduits are old, and relate to several different past sea-level positions that helped generate the current conduit configuration and pattern. Those areas within the mixing zone show the effects of chemical erosion in the form of pitted and fretted wall rock, and partial dissolution of speleothems and breakdown blocks. Vertical breakdown shafts provide access from one level to another and indicate that the levels may have initially formed independently. The intersection of the mixing zone and the principal depth of cave development strongly supports that mixing corrosion is the major cave development process in the Yucatan.

Smart et al. (2006) identified four processes that influence continued development of linear phreatic cave systems in the Yucatan:

1. Mantling of the conduit floor and other near horizontal surfaces by limestone sediment and small pieces of wall rock that inhibits dissolution and results in the focus of dissolution on exposed surfaces. This may result in significant corrosion to or complete removal of existing speleothems.
2. During high sea stands, accretion of newly formed carbonates at the coast will modify the recharge, infiltration patterns, and the distribution of dissolution and deposition within the aquifer. The hydrodynamic equilibrium of the existing conduit system is affected by impedance of discharge through

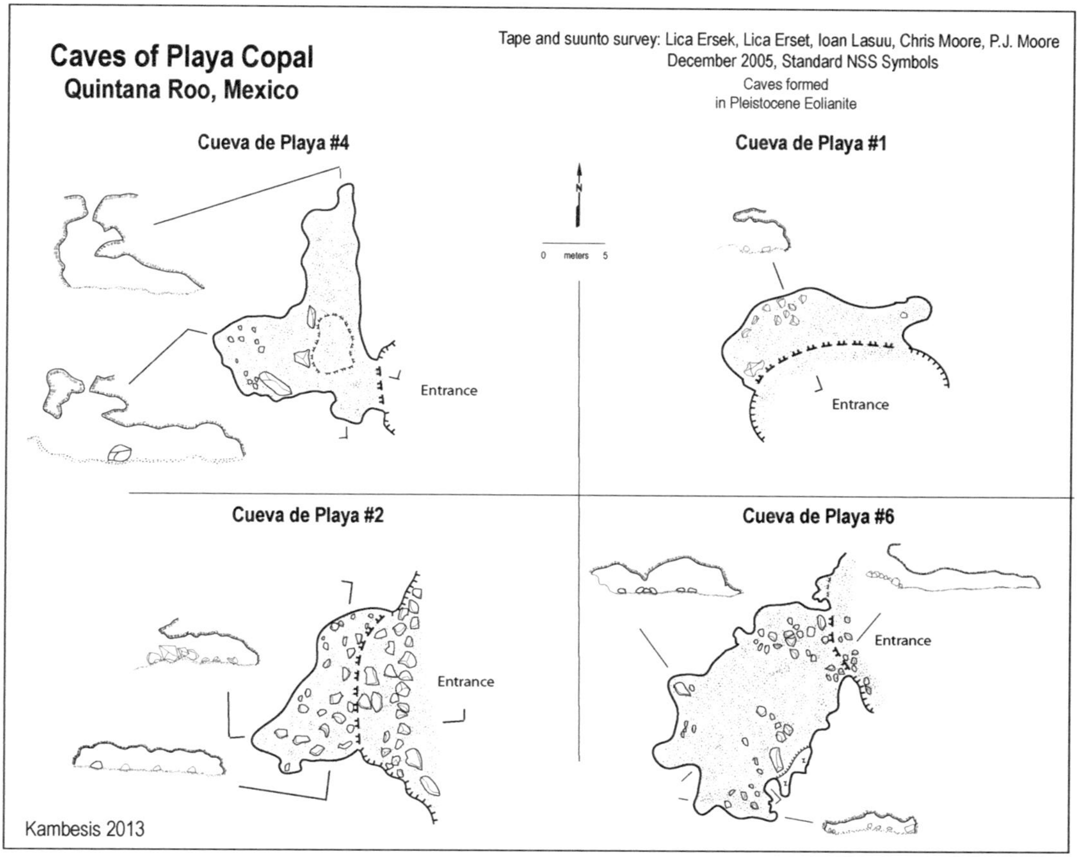

Fig. 16.15 Flank margin caves of Playa Copal (Cartography: Patricia Kambesis 2012)

newly accreted sediments. The amount of accretion combined with any subsidence will determine the space available to accommodate new conduit development during the next highstand.

3. Passage breakdown (collapse) results in lateral enlargement of cave passages within the mixing zone due to the increased, unsupported roof span. Conduit flow is forced around the breakdown/sediment obstacles resulting in an expanding series of anastomosing dividing and tributary flow paths, extending outwards from the original passage.
4. Extensive sediments that include breakdown, surface-derived clays, speleothems, and caclite rafts (formed by CO_2 degassing) can physically isolate the freshwater from the underlying saline water zone thus limiting mixing corrosion and resulting in the cessation of conduit enlargement [the authors note that this statement would require the sediments to be at the halocline and not at the water surface].

Flank margin caves also form from mixing zone corrosion though the hydrodynamics differ from those that form the linear phreatic conduits. According to the Carbonate Island Karst Model (Mylroie and Mylroie 2007) which describes the formation of flank margin caves, the caves are mixing chambers that form in the distal margin of a fresh-water lens, under the flank of an enclosing land mass. The greatest amount of mixing takes place near the margin of the land mass that is a shoreline when sea levels were elevated. The chambers are ramiform and cross-linked, which, is indicative of the inland migration of the dissolutional front (Fig. 16.15). The large width to low height ratio of the chambers take the form of the distal margin of the lens. Flank margin caves form without entrances, are drained when sea-level drops, and exposed to the surface by erosion and coastline retreat.

The significance of the flank margin caves in the Quintana Roo coastal region is that they must have formed totally independent from the extensive underwater caves indicating that fresh-water lenses in the coastal eolianites may be decoupled from the regional flow system and act with a separate flow identity (Kelly et al. 2006). This scenario would be possible during the last interglacial high-stand where sea level was 6 m higher. Marine inundation of the platform would have isolated some of the higher elevation beach ridges and eolianites such that they would have been small islands. Meteoric recharge may have formed small fresh-water lenses in the islands and would have subjected them to laminar flow mixing zone corrosion and consequently, flank margin cave development. As sea levels dropped, the independent fresh-water lenses of the small islands would have assimilated into the regional fresh-water lens and conduit flow would prevail. The eolianite dunes may not have been hydrologically linked to the conduit flow aquifer so when sea levels dropped, flank margin caves drained and were eventually exposed by erosion.

In the Caribbean Yucatan, glacioeustatic changes in sea level have resulted in the development of polygenetic caves and are reflected in the alternation between phreatic enlargement, and vadose incision and sedimentation within individual cave passages. Smart et al. (2006) determined that cave passage segments containing the current mixing zone and connected to underlying cave systems are subject to the inflow of saline water that maintains the rate of mixing zone dissolution, and removes low-stand fills. In the absence of those conditions, mixing zone dissolution ceases unless the flow of fresh water can be maintained by infeeders with high enough velocities to prevent the accumulation of sediments or to remove unconsolidated sediments from the passage. If the infeeders are blocked and freshwater flows are limited or absent, the passage segment will become filled with sediment and breakdown from roof collapse.

The rate of cave development in the Caribbean Yucatan is inferred to be rapid as more than 100 km of linear phreatic cave has been documented in the vadose zone even though sea level has been above current levels for only 5 % of the last 240,000 years (Smart et al. 2006). This is supported by the calculations of Hanshaw and Back (1980) that the Xel Ha caleta may have incised in less than 3,000 years assuming that all the additional dissolved calcite in the last kilometer of flow towards the coast occurs within the caleta.

16.11 Cave Morphology

Cave maps are available for a small portion of the underwater cave systems (Fig. 16.16) and for many of the caves in the vadose zone (Fig. 16.17). Inspection of maps reveals that passage dimensions for the caves are frequently >10 m wide and 1–10 m in height. Parallel flow paths occur (or have occurred) in most systems and speleothems and false calcite floors are evident. In the underwater caves this indicates periods of sub-aerial exposure when sea level was lower. Breakdown is common, especially associated with cenotes and can affect the local hydrology by causing the development of diversion passages (Beddows 2004; Smart et al. 2006).

There are at least four tiers of cave development in the region (Fig. 16.18). The depth below current sea level of submerged cave passages increases with distance inland. Passages up to 1 km from the coast occur at a depth of 10 m; and at a depth of 20 m in distances up to 10 km inland. However, several caves systems with sections at much greater depths (90–112 m) have been documented indicating that deeper tiers of conduit development likely exist (Coke, 2012, personal communication) but the explorational bias of deep, mixed gas diving limits the data base. The caves within the vadose zone are the highest elevation tier of cave development.

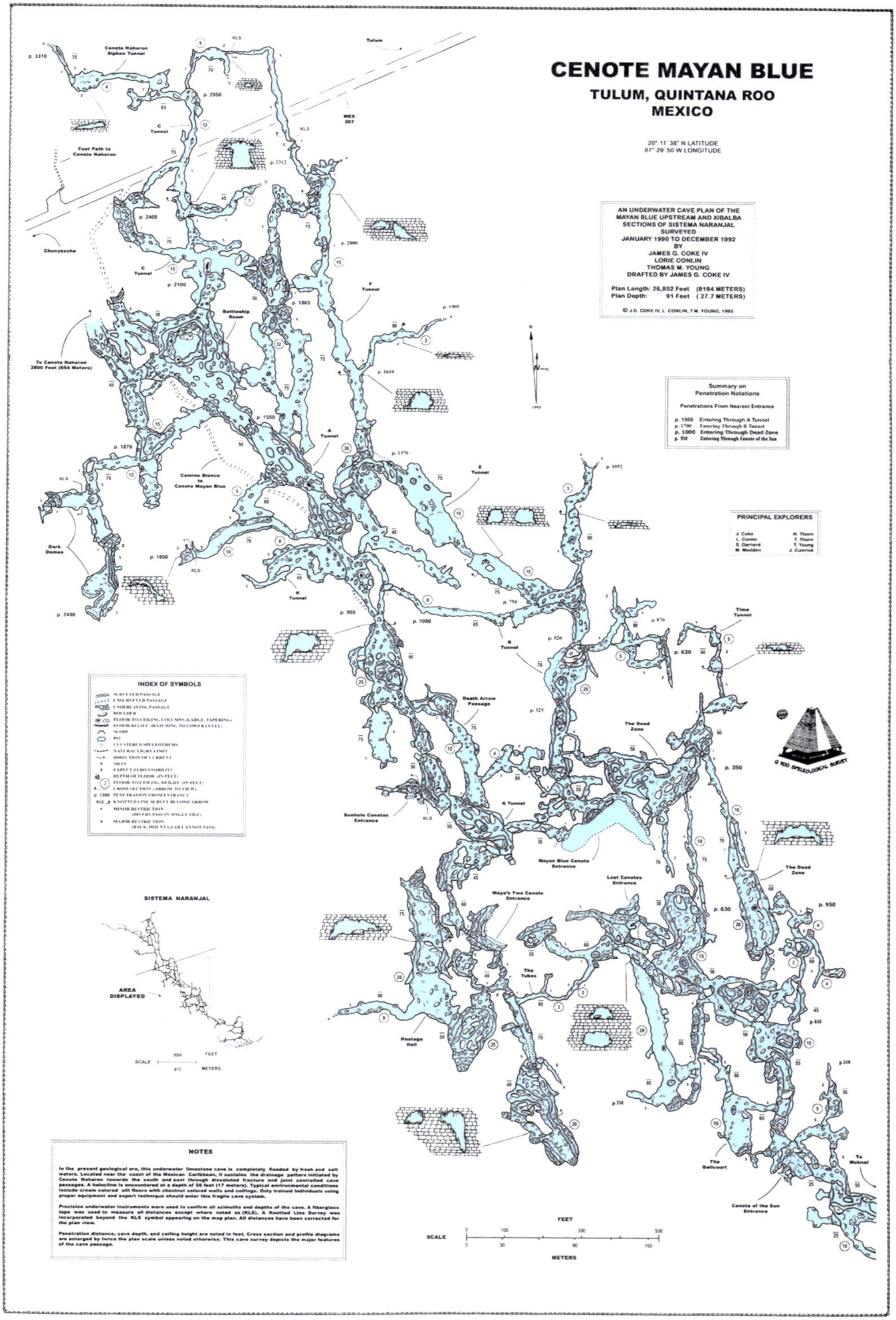

Fig. 16.16 Map of an underwater linear phreatic cave, Cenote Mayan Blue (Cartography Jim Coke 1993)

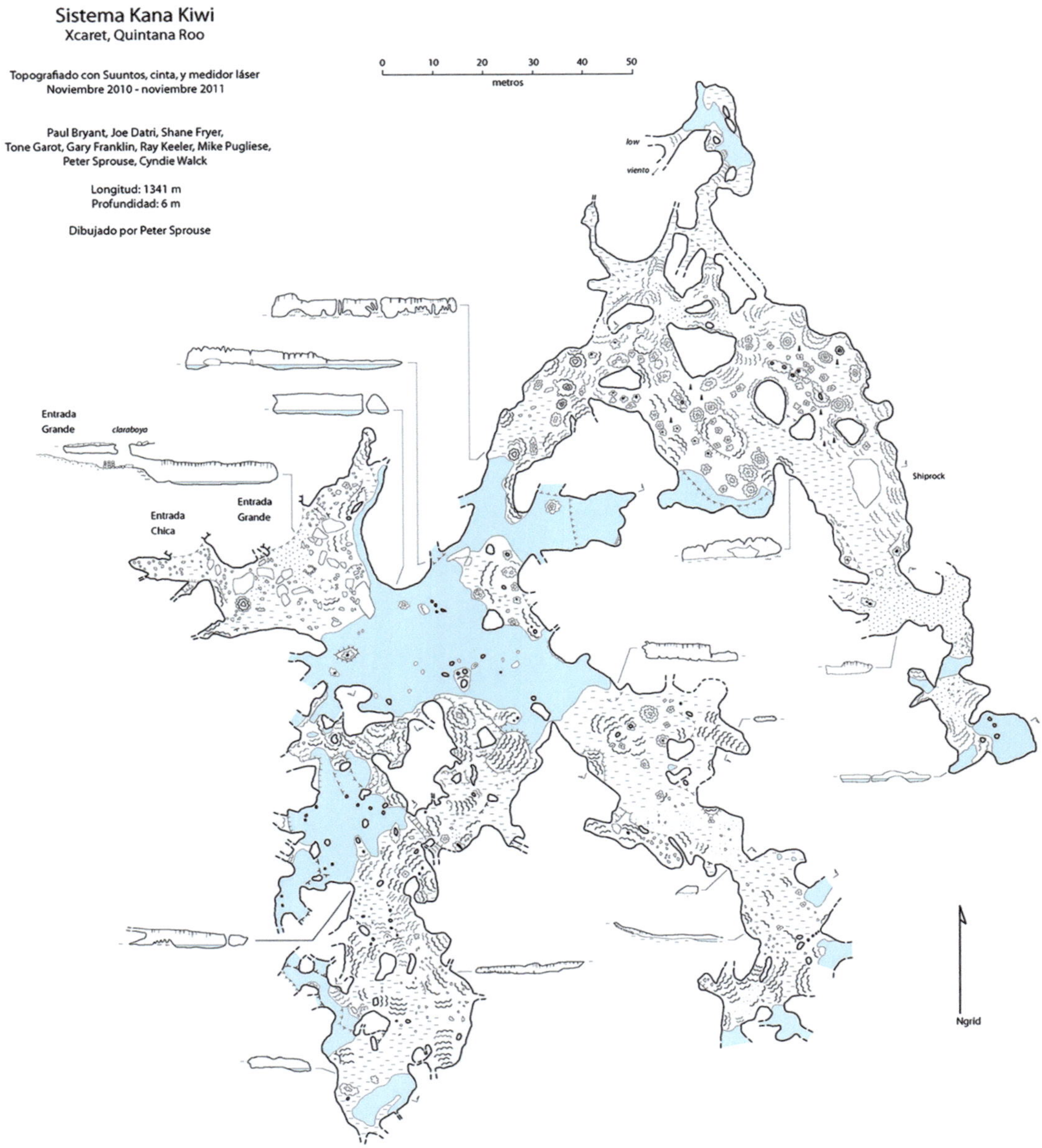

Fig. 16.17 Map of linear phreatic conduit in the vadose zone (Cartography: Peter Sprouse 2012)

A diversity of passage dimensions and cross sectional morphologies are exhibited among the caves and there appear to be two dominant passage morphologies: elliptical tube and fissure passage (Smart et al. 2006). Elliptical tubes tend to be laterally continuous over distances of up to a kilometer and have dimensions wider than high. Fissure passages are narrower in width but greater in height and they have a much more limited lateral extent (Smart et al. 2006).

Cave morphologies vary between and within cave systems. Near-coast caves and cave sections display small, fissure-controlled passages that form complex, laterally extensive mazes, some that are parallel with the coast, and their flow directions are tidally influenced (Beddows 2004; Neuman and Rahbek 2007). The caves are located within the mixing zone and as a consequence are actively enlarging (Beddows et al. 2007a). Their presence in the youngest

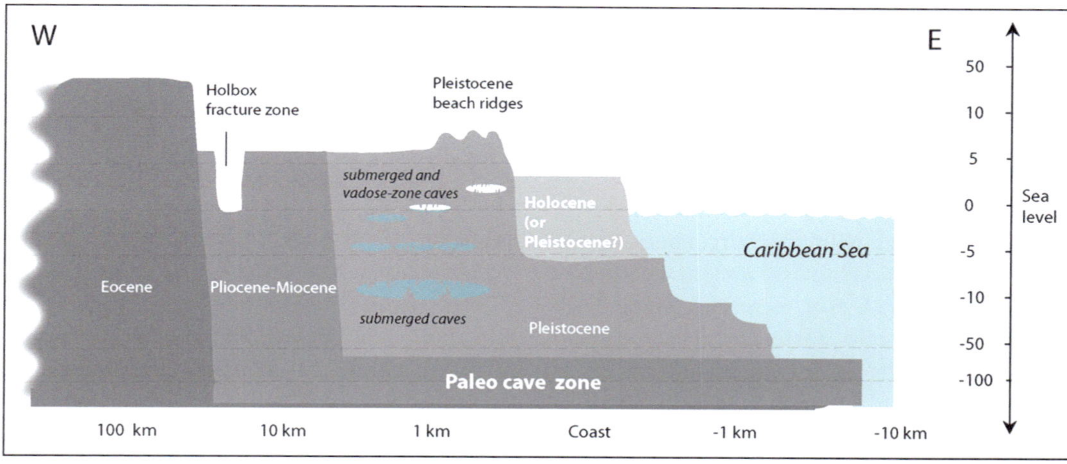

Fig. 16.18 Schematic of stratigraphy and tiers of cave development in the karst aquifer of Caribbean Yucatan (Modified from Richards and Richards 2007)

Pleistocene limestones, with the least overprinting by other processes, may also explain the passage morphology. Speleothems are less common or absent and passage incision is evidence of mixing zone corrosion (Smart et al. 2006).

Caves, and sections of cave, that occur greater than a kilometer from the coast and up to 12 km inland trend perpendicular to the coast and are characteristically elliptical in shape, linearly anastomotic, and large in size (Beddows 2004; Smart et al. 2006). Breakdown areas, sediment infills, and speleothems are common throughout and suggest polygenetic development (Smart et al. 2006). As noted above, Sistem Ox Bel Ha, though located in the near coastal zone, displays passage characteristics of an inland cave.

Smart et al. (2006) attribute the anastomosing nature of Quintana Roo caves to be caused by breakdown separation of penetrable passage, development of separate sub-parallel passages as diversions, and superimposition of passages, which may be from different phases of development (Smart et al. 2006). These conditions require pre-existing passages so the initial conduits that formed may have been fracture-controlled networks since these have been shown to be the precursors to the dissolutional conduits. Consequently, the anastomosing pattern that is common in the inland caves of Quintana Roo may be an overprint on incipient fracture-controlled passage morphology.

16.12 Environmental Concerns

The aquifer system of the Yucatan peninsula is one of the most extensive and significant karst aquifers in the world holding vast amounts of groundwater resources and hosting diverse, groundwater-dependent ecosystems including wetlands, tropical forests, the world's largest coral reef systems and the world's longest and most extensive underwater cave systems. In the Caribbean Yucatan, the natural hydrodynamics of the aquifer subject it to extensive saline intrusion that restricts the thickness of the freshwater lens. The high karstic permeability of the aquifer makes it vulnerable to the anthropogenic impacts of agriculture and human development.

Since Mayan times (2000 BC to AD 900) (Back 1995), water has been the limiting factor of human development on the peninsula. The Mayans understood the vulnerability of their water source and water played a major role in Maya religion and myths (Bauer-Gottwein et al. 2011). Cenotes influenced the location of most Mayan cities and Mayans were adept at the logistics of bringing water to their settlements for agricultural production. Paleoclimate studies indicate that the abrupt decline of the classic Maya civilization between AD 750 and 900 was caused by hydrologic extremes in the region (Webster et al. 2007).

Modern population centers are not exempt from the limitations and vulnerabilities of a karst aquifer. The Riviera Maya, the coastal district in Quintana Roo which is also the focus of this chapter, is undergoing large-scale development for tourism that poses significant threats to the groundwater resources. Extreme population growth associated with the burgeoning tourist economy and its tourist development are impacting the aquifer with ever increasing amounts of wastewater and solid waste as are the growing number of landfills (Bauer-Gottwein et al. 2011) Water moves rapidly through the karst conduits draining the aquifer, at times as much as kilometers a day (Beddows 2004). The thin soil cover typical of the peninsula, and fast groundwater flow do not provide the natural filtration and holding time necessary for the natural breakdown of pollutants.

The current practice of injecting effluents into deep saline water zones near the coast and pumping fresh water from wells 10 km inland may not be environmentally sustainable due to the hydrodynamics of the density-stratified, decoupled fresh and saline waters of the aquifer. The fresh/saline waters are dynamic in their flow, with fresh water moving coastward and saline water moving both coastward and inland (Beddows 2003, 2004). Since effluent is similar in density to fresh water, it tends to rise up either into the fresh water lens or to the fresh water-saline water mixing zone (Richards and Richards 2007). The shuttle flow dynamics of the saline water causes connected flow between cenotes and beaches, and the coastward flow of freshwater can bring effluent to beaches and reefs (Beddows 2004).

The construction of waterfronts and marinas for hotels, resorts, and housing developments can result in significant changes to the flow dynamics of the fresh water lens. Trenches cut inland from the coast and the removal of shallow bedrock allows freshwater to move much faster and easier to the coast and potentially causing the ultimate loss of the freshwater lens (Richards and Richards 2007).

Cenote/sinkhole density of the region poses serious potential threats to the construction of tourist facilities and housing developments. The extent and density of significant cave systems in the vadose zone is not well known, however in the past 4 years, 100+ km of dry cave passages have been documented in areas north of Akumal. Ongoing speleological fieldwork will continue to increase the number and known extent of vadose zone caves.

Preservation and protection of the groundwater resources of the Yucatan Caribbean require comprehensive regulation of wastewater treatment, continued scientific research of the aquifer and its related features, spacio-temporally resolved hydrological monitoring programs and efficient regional policy making (Bauer-Gottwein et al. 2011). Protecting and preservation of the coastal karst aquifer extends to the health of all of the features that make northeast Quintana Roo a popular tourist destination: first-rate beaches, Mayan cultural heritage, the over 600 km-long coral reef, outstanding nature preserves, and wonderful caves and cenotes.

16.13 Summary

The northeast coast of Quintana Roo displays the complex hydrogeology associated with a density-stratified coastal karst aquifer. Over a thousand kilometers of linear phreatic cave passages, submerged and within the vadose zone, have been documented in the region. Other significant karst features of note include cenotes that are portals to the linear phreatic cave systems, flank margin caves found in the coastal eolianites, large coastal springs and numerous seeps that discharge groundwater to the Caribbean Sea, and caletas and crescent-shaped beaches that are associated with zones of coastal discharge. The hydrological and geological controls on the development of eogenetic karst in the Yucatan Caribbean are as follows:

1. The coastal aquifer is recharged with precipitation from extensive, higher elevation areas inland in Quintana Roo.
2. The aquifer is triple porosity in nature with the combination of a high porosity/permeability matrix, extensive fracturing, and conduits.

3. Mixing zone corrosion is a significant mechanism for conduit development in the Caribbean Yucatan aquifer as evidenced by the coincidence of the mixing zone and the principal depth of cave development.
4. The intersection of regional fractures and lineaments serve as zones of linear dissolution along which have developed an extensive network of conduits.
5. Cave passage morphology and density vary as a function of distance from the coast as do depositional facies within a reefal environment. Consequently, lithology may influence conduit morphologies.
6. Fluctuations in sea level fundamentally control elevation (or depth) of cave development, and overprinting processes that have resulted in polygenetic cave systems.

Cave and karst development in the Yucatan Caribbean is a function of regional hydrology. The resultant cave types vary with rock/water interaction and hydrodynamics. Cave passage distribution, morphology, and density are controlled by lithology and structure. The hydrologic regime is affected by eustatic sea-level fluctuations, resulting in extensive polygenetic caves developed in different tiers and overprinted with features associated with turbulent and/or diffuse flow. The dissolutional processes that form the coastal caves of Quintana Roo also impact the modern coastline, especially when coupled with mechanical wave action.

Acknowledgements The authors would like to thank the Quintana Roo Speleological Survey for providing information about the underwater systems of the study area, to cartographers Peter Sprouse and Jason Richards for providing cave maps, and to John Mylroie for maps of flank margin caves.

References

Back W (1995) Water management by early people in the Yucatan, Mexico. Environ Geol 25(4):239–242

Back W, Hanshaw BB, Pyle TE, Plummer LN, Weidie AE (1979) Geochemical significance of groundwater discharge and carbonate solution to the formation of Caleta Xel Ha, Quintana Roo, Mexico. Water Resour Res 15:1521–1535

Bauer-Gottwein P, Gondwe BNR, Charvet G, Marin LE, Robellodo-Vieyra M, Meresiz-Alonso G (2011) Review: the Yucatan Peninsula karst aquifer, Mexico. Hydrol J 19:507–524

Beddows PA (1999) Conduit hydrogeology of a tropical coastal carbonate aquifer: Caribbean coast of the Yucatan Peninsula. M.Sc. Thesis, McMaster University, 162 p

Beddows PA (2003) Cave hydrology of the Caribbean Yucatan coast, Bulletin 11. Association of Mexican Cave Studies, Houston, 96 p

Beddows PA (2004) Groundwater hydrology of a coastal Conduit carbonate aquifer: Caribbean Coast of the Yucatán Peninsula, México. Ph.D. Thesis, University of Bristol

Beddows PA, Smart PL, Whitaker FL, Smith SL (2007a) Decoupled fresh–saline groundwater circulation of a coastal carbonate aquifer: spatial patterns of temperature and specific electrical conductivity. J Hydrol 346:28–42

Beddows PA, Hendrickson MR, Webster KH, Kras SM (2007b) Mapping flooded caves from above: Surface karst inventory or the Yucatan Peninsula. 2007 National Cave and Karst Management symposium proceedings, St. Louis, pp 193–197

Bonet F, Butterlin J (1962) Stratigraphy of the northern part of the Yucatan peninsula in Yucatan field trip guidebook. New Orleans Geological Society, New Orleans, pp 52–57

Coke JG IV (2009) Yucatan Peninsula (Campeche, Yucatan, Quintana Roo). In: Palmer AN, Palmer MV (eds) Caves and Karst of the America. National Speleological Society, Huntsville, pp 388–390

Ford D, Williams P (2007) Karst geomorphology and hydrology. Chapman & Hall, London

Ginés A, Ginés J, Gràcia F, Fornós JJ (2009a) Cave surveying and cave patterns in the southeastern coastal karst of Mallorca Island (Spain). In: White WB (ed) Proceedings of 15th international congress speleology, Kerrville, vol 1, pp 488–495.

Ginés J, Ginés A, Fornós JJ, Merino A, Gràcia F (2009b) About the genesis of an exceptional coastal cave from Mallorca Island (Western Mediterranean). The lithological control over the pattern and morphology of Cova des Pas de Vallgornera. In: White WB (ed) Proceedings of the 15th international congress speleology. Kerrville, USA, vol 1, pp 481–487.

Gmitro DA (1986) The interactions of waters with carbonate rock in Yucatan, Mexico. MSc thesis, Department of Geology, Northern Illinois University, 111 p

Gondwe BRN, Lerer S, Stisen S, Marín L, Rebolledo-Vieyra M, Merediz-Alonso G, Bauer-Gottwein P (2010) Hydrogeology of the south-eastern Yucatan Peninsula: new insights from water level measurements, geochemistry, geophysics and remote sensing. J Hydrol. doi:10.1016/j.jhydrol.2010.04.044

Gonzalez-Herrera R, Sanchez-y-Pinto I, Gamboa-Vargas J (2002) Groundwater-flow modeling in the Yucatan karstic aquifer Mexico. Hydrogeol J 10(5):539–552

Hanshaw BB, Back W (1980) Chemical mass-wasting of the northern Yucatan Peninsula by groundwater dissolution. Geology 8(5):222–224

Harris NJ (1984) Diagenesis of upper Pleistocene strandplain limestones, northeastern Yucatan Peninsula, Mexico. MSc Thesis, University of New Orleans, 130 p

Heyer D, Sprouse P (2012) The Sac Actun connection. Natl Speleological Soc News 70(12):12–17

Isphording WC (1974) Weathering of Yucatan Limestones; the genesis of terra rosas. In: Weidie AE (ed) Yucatan guidebook: New Orleans. New Orleans Geological Society, Louisiana, pp 78–93

Isphording WC (1975) The physical geology of Yucatan. Trans Gulf Coast Assoc Geol Soc 25:231–262

Jones IC, Banner JL (2003) Estimating recharge thresholds in tropical karst island aquifers: Barbados, Puerto Rico and Guam. J Hydrol 278:131–143

Kambesis PN, Mylroie JR, Mylroie JE, Larson EB, Owen-Nagel AM, Sumrall JB, Lace MJ (2012) Influence of karst denudation on the northwest coast of Curacao. In: 16th symposium on the geology of the Bahamas and other carbonate regions program, Gerace Research Centre, San Salvador Island, Bahamas, (abstract) p 34

Kelly K, Mylroie JE, Mylroie JR, Moore CM, Collins LR, Lica E, Lascl I, Roth MJ, Moore PJ, Passion R, Shaw C (2006) Eolianites and karst development in the Mayan Riviera, Mexico. In: Davis RL, Gamble DW (eds) Proceedings of the 12th symposium on the geology of the Bahamas and other carbonate regions, pp 88–99

Kottek M, Grieser J, Beck C, Rudolf B, Rubel F (2006) World map of the Koopen-Geiger climate classification updated. Meteorol Z 15:259–263

Lauderdale RW, Ward WC, Weidie AE (1979) Carrillo Puerto formation of northeastern Quintana Roo, Mexico. Gulf Coast Assoc Geol Soc Trans 29:275–280

Lesser JM (1976) Resumendel studio geohidrologicoe hidrogeoquimicode la peninsulade Yucatan. Boletin de Duvulgacin Tecnica 10:1–11

Lesser JM, Weidie AE (1988) Region 25, Yucatan Peninsula. In: Back W, Rosenhein JS, Seaber PR (eds) Hydrogeology. Geological Society of America, Boulder, pp 237–242

Lopez-Ramos E (1975) Geological summary of the Yucatán Peninsula. In: Nairn AEM, Stehli FG (eds) The ocean basins and margins, vol 3, The Gulf of Mexico and the Caribbean. Plenum, New York, pp 257–282

Marin LE (1990) Field investigations and numerical simulation of ground-water flow in the karstic aquifer of northwestern Yucatan, Mexico. PhD thesis, Northern Illinois University, DeKalb, 183 p

Moore YH, Stoessel RK, Easley DH (1992) Freshwater-sea-water relationship within a groundwater-water flow system, northeastern coast of the Yucatan Peninsula. Groundwater 30(3):343–350

Mylroie JR, Mylroie JE (2007) Development of the carbonate island karst model. J Cave Karst Stud 69(1): 59–75

Mylroie JE, Carew JL, Moore AI (1995) Blue holes: definition and genesis. Carbonate Evaporite 10(2): 225–233

Neuman BR, Rahbek ML (2007) Modeling the groundwater catchment of the Sian Ka'an reserve, Quintana Roo, AMCS bulletin no. 18. Association for Mexican Cave Studies, Austin

Perry E, Marin L, McClain J, Velazquez G (1995) Ring of cenotes (sinkholes), Northwest Yucatan Peninsula, Mexico: its hydrogeologic characteristics and possible association with the Chixculub impact crater. Geology 23(1):17–20

Plummer LN (1975) Mixing of seawater with calcium carbonate groundwater. Geolog Soc Am Mem 142:219–236

Quintana Roo Speleological Survey. Website: www.caves.org/project/qrss.htm. Accessed Jan 2013

Richards D, Richards S (2007) Overview of the geology and hydrology of coastal Quintana Roo. AMCS Activ Newsl 30:104–109

Rodriguez CJ (1982) Petrology and diagenesis of Pleistocene limestones, northeastern Yucatan Peninsula, Mexico. MS Thesis, University of New Orleans, 80 p

Smart PL, Beddows PA, Coke J, Doerr S, Smith A, Whitaker F (2006) Cave development on the Caribbean coast of the Yucatan Peninsula, Quintana Roo, Mexico. In: Harmon RS, Wicks C (eds) Perspectives on karst geomorphology, hydrology, and geochemistry – a tribute volume to Derek C. Ford and William B. Whiter: Geological Society of America special paper 404. Geological Society of America, Boulder, CO, pp 105–128

Southworth CS (1985) Application of remote-sensing data, Eastern Yucatan. In: Ward WC, Weidie AE, Back W (eds) Geology and hydrogeology of the Yucatan and Quaternary geology of Northeastern Yucatan Peninsula. New Orleans Geological Society Publication, New Orleans, pp 12–19

Stafford KW, Mylroie JE, Mylroie JR, Jenson JW, Taboroši D (2004) Coves and pocket beaches: littoral erosion and flank margin caves. Geol Soc Am Abstr Prog 36(5):284 (abstract)

Szabo BJ, Ward WC, Weidie AE, Brady MJ (1978) Age and magnititude of the late Pleistocene sea-level rise on the eastern Yucatan Peninsula. Geology 6:513–515

Thomas C (1999) Aspects hydrogéologiques du Yucatan (Mexique) [hydrogeological aspects of the Yucatan, (Mexico)]. Karstologia 34(2):9–22

Thomas C (2005) Dix années dans les grottes mayas 100km de première en siphon (Mexique). Spelunca 98:25–44

Tułaczyk SM, Perry EC, Duller CE, Villasuso M (1993) Influence of the Holbox fracture zone on the karst geomorphology and hydrogeology of northern Quintana Roo, Yucatan Peninsula, Mexico. In: Beck B (ed) Applied karst geology. Balkema, Rotterdam

Vacher HL, Mylroie JE (2002) Eogenetic karst from the perspective of an equivalent porous medium. Carbonate Evaporite 17(2):182–196

Ward WC (1985) Quaternary geology of northeastern Yucatan Peninsula, part 2. In: Ward WC, Weidie AE, Back W (eds) Geology and hydrogeology of the Yucatan and Quaternary geology of northeastern Yucatan Peninsula. New Orleans Geological Society, New Orleans, pp 23–53

Ward WC (1997) Geology of coastal islands, northeastern Yucatan Peninsula. In: Vacher HL, Quinn TM (eds) Geology and hydrogeology of carbonate islands. Elsevier Science Publishers, Amsterdam/New York, pp 275–298

Ward WC (2003) Salt water intrusion & coastal aquifer conference (SWICA) field trip to the Caribbean Coast of the Yucatan Peninsula (April), pp 13–22

Ward WC, Brady M (1979) Strandline sedimentation of carbonate grainstones, Upper Pleistocene, Yucatan Peninsula, Mexico. AAOG Bull 63:362–369

Webster JW, Brook GA, Railsback LB, Cheng H, Edwards RL, Alexander C, Reeder PP (2007)Stalagmite evidence from Belize indicating significant droughts at the time of Preclassic Abandonment, the Maya Hiatus, and the Classic Maya collapse. Paleogeogr Paleoclimatol Paleoecol 250(1–4):1–17

Weidie AE (1978) Lineaments of the Yucatan Peninsula and fractures of the Central Quintana Roo Coast. In: Field trip no. 10 – Yucatan, Road Log and supplement to 1978 Guidebook, 1982 GSA annual meeting, New Orleans, Louisiana, pp 21–25

Weidie AE (1985) Geology of the Yucatan Platform, part 1. In: Ward WC, Weidie AE, Back W (eds) Geology and hydrogeology of the Yucatan and Quaternary geology of northeastern Yucatan Peninsula. New Orleans Geological Society, New Orleans, pp 1–19

Whitaker FF (1992) Hydrology geochemistrv and diagenesis of modern carbonate platforms in the Bahamas. Ph.D. Thesis, University of Bristol, 347 p

Wigley TML, Plummer LN (1976) Mixing of carbonate waters. Geochim Cosmochim Acta 40(9):989–995

Telogenetic Limestones and Island Karst

17

John E. Mylroie and Joan R. Mylroie

Abstract

The development of flank margin caves in telogenetic carbonates is in most cases restricted to flow and mixing along fractures, joints and bedding planes, and the caves in this setting tend to have planar, two dimensional configuration. Under special circumstances, as when the telogenetic carbonates are highly fractured, as in New Zealand, or the telogenetic carbonates have been altered into a breccia faces, as in paleotalus in Croatia, a three dimensional series of phreatic chambers can develop. In both cases of fracturing and paleotalus, the three dimensional configuration of the void space and potential flow paths mimic eogenetic carbonates, and flank margin caves similar to those found in eogenetic carbonates have developed. These examples demonstrate that no single characteristic determines the morphological nature of dissolution in the coastal environment. While the coastal geochemical environment of mixing is unique, rock diagenetic maturity and structure also play an important role.

17.1 Introduction

As noted in Chap. 4, the flank margin cave model, and the subsequent Carbonate Island Karst Model (CIKM) were initially developed on tropical islands where the limestones are young and diagenetically immature, or eogenetic. While the vast majority of carbonate coastlines are eogenetic, being formed in the locale of calcium carbonate deposition, there are significant areas where diagenetically mature or telogenetic limestones are in coastal positions. The first work done in such a setting was that by Proctor (1988), in the Devonian limestones in coastal settings in Devon, UK. In contrast to the eogenetic caves and karst reported from the Bahamas at that time (e.g. Mylroie and Carew 1987), cave development in the Devonian limestones followed joints and bedding planes, as the matrix porosity was minimal (Proctor 1988). Little further work was done in telogenetic limestones, although limestone coasts from Gotland in the Baltic and the Greek islands in the Aegean were used as examples of telogenetic island karst (Vacher and Mylroie 2002), but

J.E. Mylroie (✉) • J.R. Mylroie
Department of Geosciences, Mississippi State University, Mississippi State, MS 39762, USA
e-mail: mylroie@geosci.msstate.edu

M.J. Lace and J.E. Mylroie (eds.), *Coastal Karst Landforms*, Coastal Research Library 5,
DOI 10.1007/978-94-007-5016-6_17,

without actual analysis. With increasing diagenetic maturity, even in environments where deep burial has not occurred, limestones can become sufficiently altered that they begin to behave more like telogenetic rocks. On Guam, Taboroši et al. (2004) demonstrated how with advancing diagenetic maturity in eogenetic rocks, surficial karren more and more displayed characteristics similar to that seen in telogenetic limestones. On the coastal cliffs of Guam, cementation and recystallization in the coastal environment diverted matrix flow to fracture sets (Jenson et al. 2006) in a manner similar to that reported by Proctor (1988) for teleogenetic limestones. In this chapter, we examine two disparate telogenetic island locations, New Zealand in the South Pacific, and Cres Island, Croatia, in the Adriatic, to provide examples of how the flank margin model works in dense, hard telogenetic rocks that have been modified by post depositional processes to mimic eogenetic rocks.

17.2 Cres Island

Cres Island is one of the northern Adriatic islands in Croatia (Fig. 17.1). For a more in-depth presentation of the material reviewed here, see Otoničar et al. (2010). At 66 km length and 405.7 km^2 area, Cres Island is the largest and longest Adriatic island. Tectonically, Cres Island belongs to the External Dinarides, the most external thrust unit of the Dinaric fold and thrust belt (Korbar 2009). Paleogeographically, the area corresponds to the northern part of the Cretaceous Adriatic Carbonate Platform (sensu Vlahović et al. 2005) and in the minor extent to the Eocene synorogenic carbonate platform (sensu Otoničar 2007). A regional unconformity denoted by paleokarstic surface separates these two carbonate systems (Otoničar 2007).

As reviewed by Otoničar et al. (2010), the tectonic environment is highly variable, with some areas subsiding and some undergoing uplift, including adjacent sections of island coastline separated by faults. The study area on northern Cres Island consists primarily of Cretaceous limestones and dolomites with occasional outcrops of Paleogene limestones. The rocks are folded and dip steeply, cut by reverse faults striking NNW-SSE (Otoničar et al. 2010), and with caves and karren typical of diagenetically mature carbonate rocks (i.e. karst on islands). The water budget near a lake in the island center is approximately balanced, over the time range 1927–1961, with average precipitation of 1,064 mm/a and evaporation of 1,161 mm/a (Rubinić and Ožanić 1992).

Flank margin caves are not common in coastal Croatia (Otoničar et al. 2010), as the limestone, while folded and faulted, is massive and provides only widely spaced fracture flow paths for ground water escape to the sea. However, on the west coast of Cres Island is a small embayment in the rocky, cliffed coastline (Figs. 17.1 and 17.2) which contains a series of small flank margin caves. The embayment is surrounded by steep hills and cliffs (Fig. 17.2), with a 100 m long and 35 m wide carbonate pebble beach which has infilled the back of the embayment. The cliffs around the embayment can be up to 10 m high. The limestone is Upper Cretaceous. The caves are separated into a series of independent chambers along the coastline, and continue inland where the beach meets the cliffs (Fig. 17.2). Almost all the caves are very small, only a few meters in dimension (e.g. Fig. 17.3), but one is more substantial, Plava Grota (Fig. 17.4). The cave, which has large interior chambers, speleothems, and phreatic morphology (Fig. 17.5), has a map footprint consistent with that of a flank margin cave (Fig 17.6). Fresh-water discharge enters the cave from the inland cave walls at several locations (Fig. 17.6). These characteristics indicate a dissolutional, and not wave erosion, mechanism for speleogenesis. The cave configuration and morphology are consistent with that of flank margin caves found in eogenetic carbonates of tropical island coastlines. The single most important aspect of these caves on Cres is that they are all developed in a breccia facies (Figs. 17.3 and 17.5) that has been identified as a paleotalus because the breccia material has a sharp contact with the underlying and adjacent

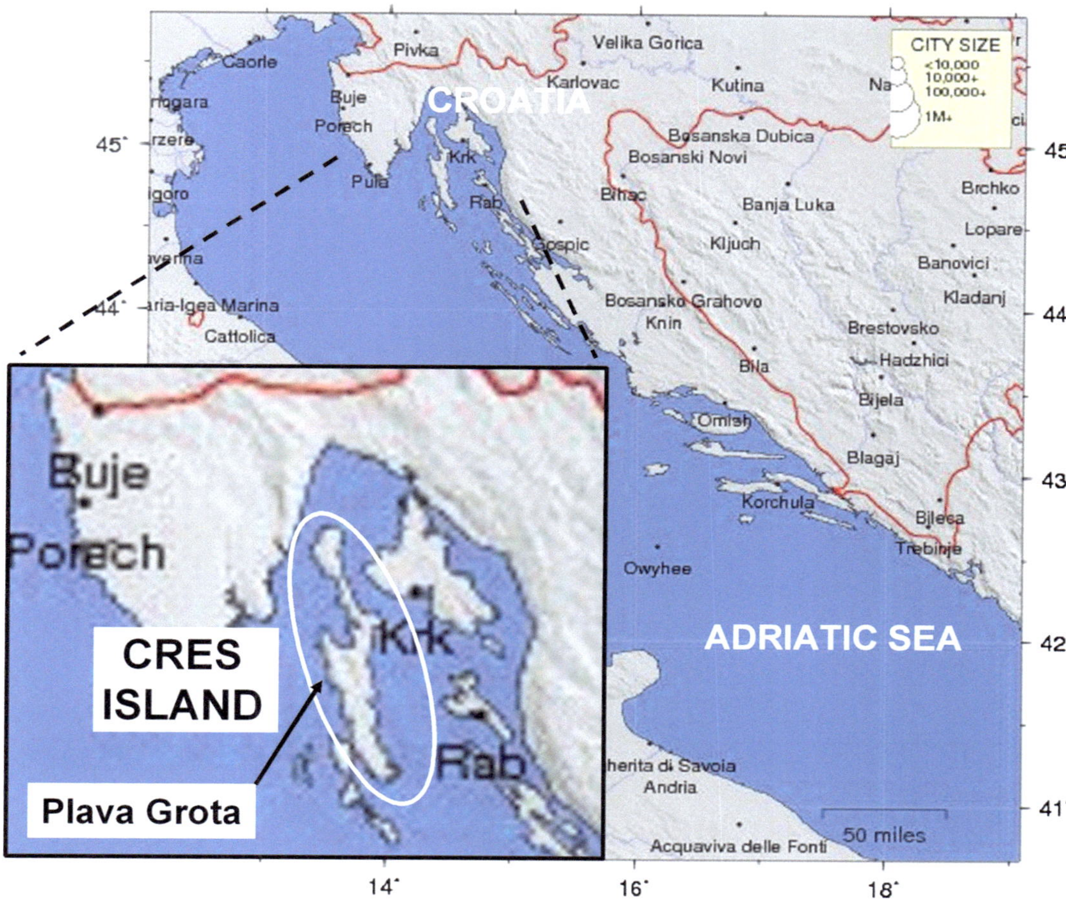

Fig. 17.1 Map of the Adriatic coast of Croatia, showing the location of Cres Island (*white oval*), and within the inset, the site of Plava Grota

Fig. 17.2 Overview of the Plava Grota site. (**a**) Looking down on the embayment, Plava Grota entrance is shown above the *white arrow*. (**b**) The limestone pebble beach; small flank margin caves are found both on the beach and seaward portions of the cliff wall

Fig. 17.3 Small flank margin cave at Plava Grota, flashlight in *white oval* is 12 cm long for scale. Note how the rock is made up of a variety of limestone clasts

limestone, and is made up entirely of carbonate clasts that are petrographically identical to the surrounding limestone, and there is an absence of tectonic influences within the breccia (Otoničar et al. 2010).

As presented in Chap. 4, when telogenetic limestones can be modified to allow fresh-water flow and mixing within a volume, as opposed to along joint, fault and bedding planes, volumetric dissolution occurs. As a result flank margin caves can develop with a morphology similar to that found in eogenetic limestones. Within the paleotalus outcrops on Cres Island, there is a high degree of porosity and permeability in the jumbled mixture of clasts. Not only is the permeability high, but there is also a high degree of matrix storage. In the adjacent host limestone, flow is restricted to planar features, as is mixing, and the dissolution is essentially two-dimensional as compared to three-dimensional in the paleotalus. In addition, there is low storage capability within the limestone. The Cres Island example is instructive because it allows direct comparison of telogenetic limestones with adjacent modified telogenetic limestones as a paleotalus that provide an entirely different fresh-water lens environment. The location of flank margin cave development faithfully follows the location of enhanced three-dimensional porosity and permeability of the paleotalus.

Fig. 17.4 The main flank margin cave, Plava Grota. The two main entrances are in the cliff face, seen above the bow of the boat and left of the boat bow. Immediately above the cabin of the boat, going right, the rock changes from paleotalus facies to the host limestone, with no flank margin caves

17.3 New Zealand

New Zealand has a complex and diverse geology, including limestones of Cenozoic age in coastal locations. A detailed study of these coastal limestones and their caves can be found in Mylroie et al. 2008. A synopsis of the more illustrative locations from that paper is presented here. New Zealand, on North and South Island, contain a variety of limestones in coastal locations in a variety of conditions. These rocks are geologically young, and are part of a very active tectonic environment and have been subjected to a variety of diagenetic conditions, leading to a variety of diagenetic maturities between eogenetic and telogenetic (Nelson 1978; Nelson et al. 1988; Caron et al. 2006). The caves described in this section are entirely from telogenetic carbonate rocks that are relatively young in age. The active tectonic setting also means that both glacioeustasy and tectonic movement have controlled the placement of sea level with respect to the coastal carbonate outcrops. As a result, the duration of a stable sea-level position, which in turn controls the time stability of the fresh-water lens, was not long. Data from the Bahamas (Mylroie and Mylroie 2007) and from Isla de Mona (Frank et al. 1998) indicate that a constantly moving sea level, and a constantly moving fresh-water lens, allows too little time for dissolution to create flank margin caves. On the other hand, the Bahamian data (Chap. 7) also show that even limited lens-stability time can create observable flank margin caves. The tectonic setting of New Zealand offers an opportunity to determine what minimum sea-level stability time would be necessary for a fresh-water lens to produce a macroscopic flank margin cave by mixing-zone dissolution. In addition, the New Zealand setting allows display of the controls of telogenetic limestones on subsequent flank margin cave development. Mylroie et al. (2008) examined three settings on North Island, and five settings on South Island (Fig. 17.7). For this review, one North Island setting and two South Island settings will be examined.

Fig. 17.5 Plava Grota. (**a**) Entrances shown in Fig. 17.4a, note the paleotalus facies that host the cave (portion of rubber boat at left for scale). (**b**) Large interior chamber, with low arch in wall leading to more passages, note paleotalus facies. (**c**) Looking out the left entrance of (**a**), with paloetalus facies clearly displayed. (**d**) Looking out right entrance of (**a**), swimmer's head in *white circle* for scale. Bedrock column in the entrance is unlikely the result of wave action, but more likely of dissolution origin

17.3.1 North Island, Napier Region

North of Napier (Fig. 17.7), at the southern end of Hawke Bay, are coastal outcrops of coarse-grained bioclastic Pliocene limestone assigned to various formations (Nelson et al. 2003) stretching from Waipatiki Beach south to Bay View, just north of Napier. Coastal reconnaissance failed to find any dissolutional caves except at a small coastal outcrop of Waipatiki Limestone to the north of the Whirinaki community and just south of Te Ngaru Stream. In this small, low hill, the seaward face has been cliffed by wave action, and a series of small caves revealed (Fig. 17.8). The caves are a collection of small tubes, small chambers, and vertical fissures (Fig. 17.8). The top of the outcrop is capped by a siliciclastic mudstone unit, which isolates the carbonate unit below from any epikarst development and overprint. The caves all show phreatic morphology, and are somewhat regularly spaced along the cliffed hillside (Fig. 17.9). The limited extent of the hill containing the caves makes it unlikely the caves are the result of conduit flow from catchment of precipitation on the hill surface. They are distributed over a vertical range of 4 m. The Napier area is famous for tectonic activity, and in 1931, a magnitude 7.8 earthquake resulted in coastal uplift in the vicinity of Napier of up to 2.4 m (Hull 1990). It is quite likely that the flank margin caves seen at Whirinaki are Holocene in age, and owe their current exposure to recent tectonics. The vertical, fissure-like appearance of the caves is a result of two effects. First, their

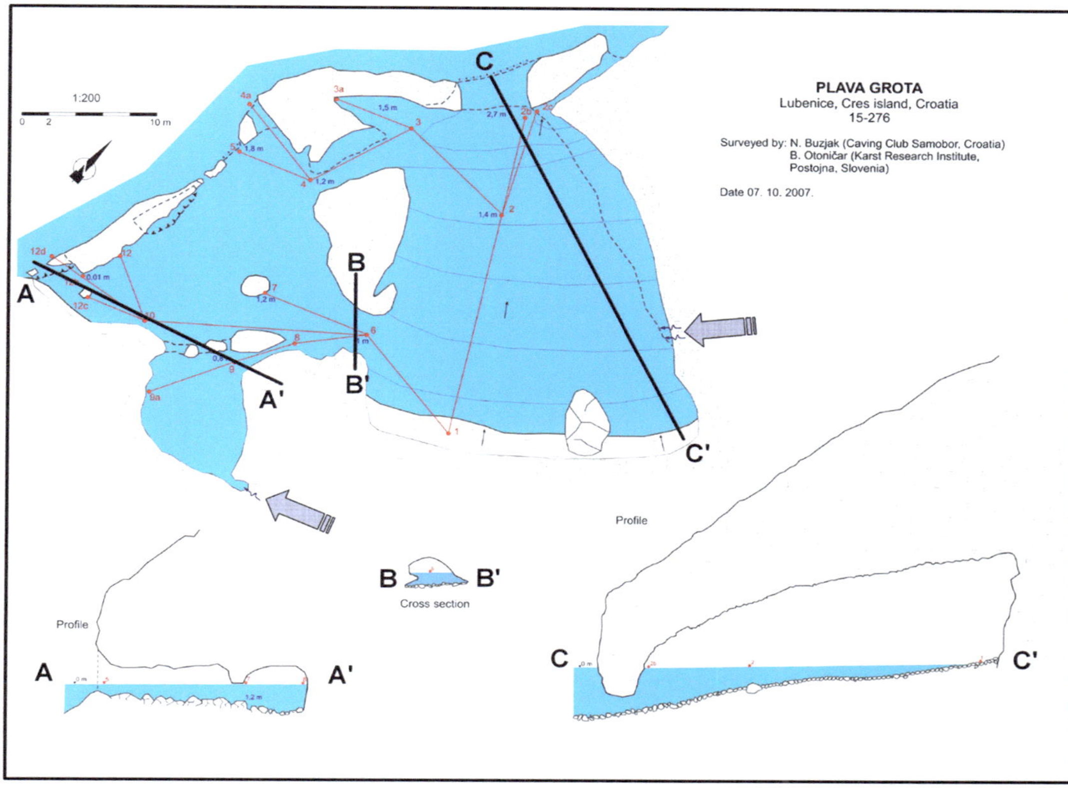

Fig. 17.6 Map of Plava Grota. In the plan view, *the two bold grey arrows* on the map identify fresh-water inputs. Line **B**–**B**′, and cross section **B** to **B**′ indicate the low arch seen in Fig. 17.5b (Adapted from Otoničar et al. 2010)

development in diagenetically altered limestones reduced matrix porosity and permeability, such that widely spaced joints were the major outward ground-water flow path and the site of mixing dissolution. Second, tectonic uplift moved the rock upward through the fresh-water lens, such that the dissolution of the carbonate rock continually moved downward, emphasizing the vertical signature of the dissolution voids. Small phreatic tubes and pockets represent stillstands in the tectonic uplift.

17.3.2 South Island, Kaikoura Region

Telogenetic Paleocene Amuri Limestone and overlying Oligocene Grey Marl crop out in a coastal setting on the Kaikoura Peninsula in northeastern South Island (Fig. 17.7) Compared to the medium to very coarse-grained bioclastic carbonates characterizing all the other New Zealand limestones studied, the Amuri Limestone is an indurated very fine-grained limestone, or micrite (Mylroie et al. 2008). The Kaikoura area is geologically complex, with folding and faulting leading to a complex outcrop pattern. The tectonics has created a very fractured and jointed limestone (Fig. 17.10).

On the southwest side of the Kaikoura Peninsula, at East Head, behind trees at the base of the cliff, is Kaikoura Point Sea Cave (it is marked as a sea cave on a tourist map), a large dissolution cave (Figs. 17.11 and 17.12). The entrance is 5 m wide and 5 m high, with a triangular cross section. The passage leads northeast, as an oval tube 3–5 m wide, and 3 m high, with an additional dissolutional slot in the ceiling a meter wide (Fig. 17.11b). The passage continues

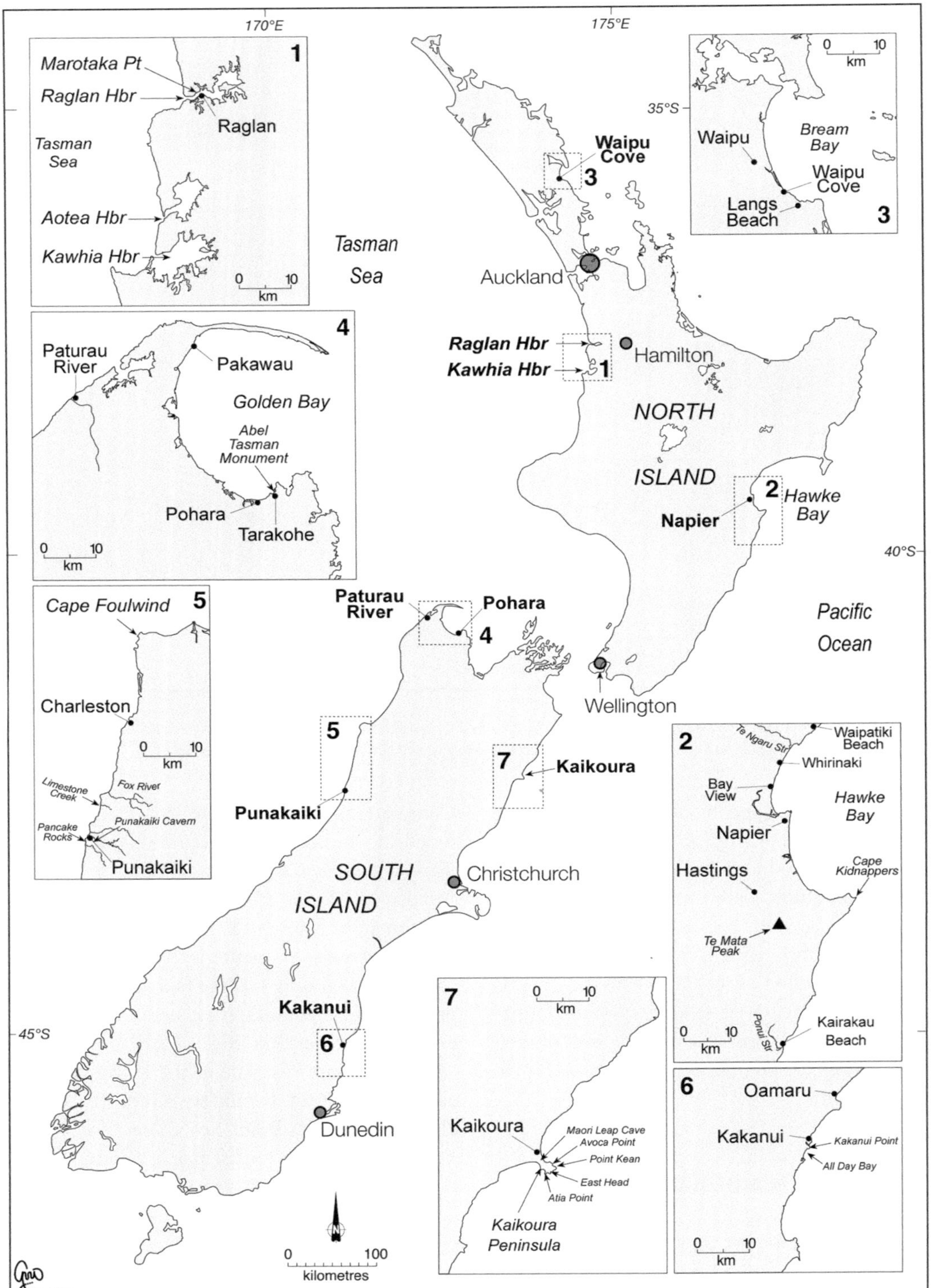

Fig. 17.7 Map of New Zealand, from Mylroie et al. (2008) showing all field locations from that study. The three sites discussed in this review are shown in inset maps 2, 5 and 7

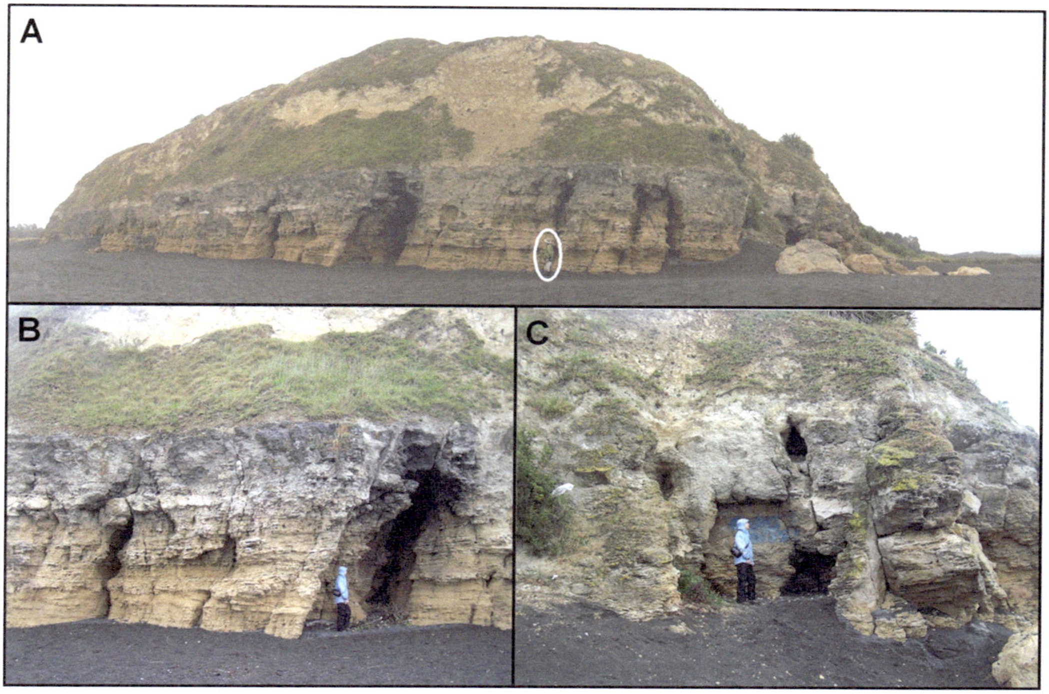

Fig. 17.8 Napier district (Late Pliocene Waipatiki Limestone). (**a**) The small beachside hill north of Whirinaki contains a series of short flank margin caves, which are exposed on the seaward side of the hill, where wave action has breached into them. Note person standing at cliff base in right center, *circled*, for scale. (**b**) Some of the caves at Whirinaki. Note that the limestone is covered by siliciclastic mudstone, so no epikarst development has occurred. (**c**) Isolated tubes and chambers at Whirinaki. The vertical extension of these caves, as seen in (**a**), and the separation into stacked tubes, as seen in (**b** and **c**), are evidence of rapid uplift

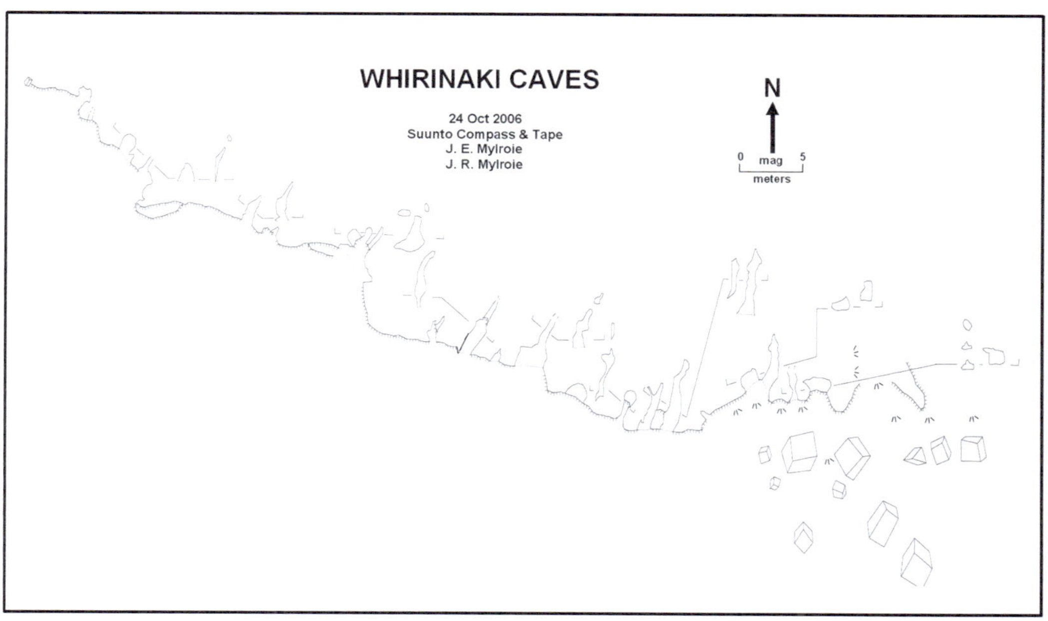

Fig. 17.9 Map of the Whirinaki Caves. The caves open southward on to beach seen in the foreground of Fig. 17.8

Fig. 17.10 Kaikoura (Paleocene Amuri Limestone) – evidence of tectonics. (**a**) Folded beds of the Amuri Limestone distorted to a vertical position, north side of Kaikoura Point. (**b**) Contact of the Amuri Limestone, *left*, with the Grey Marl, *right*. Note in foreground that faulting has displaced the contact

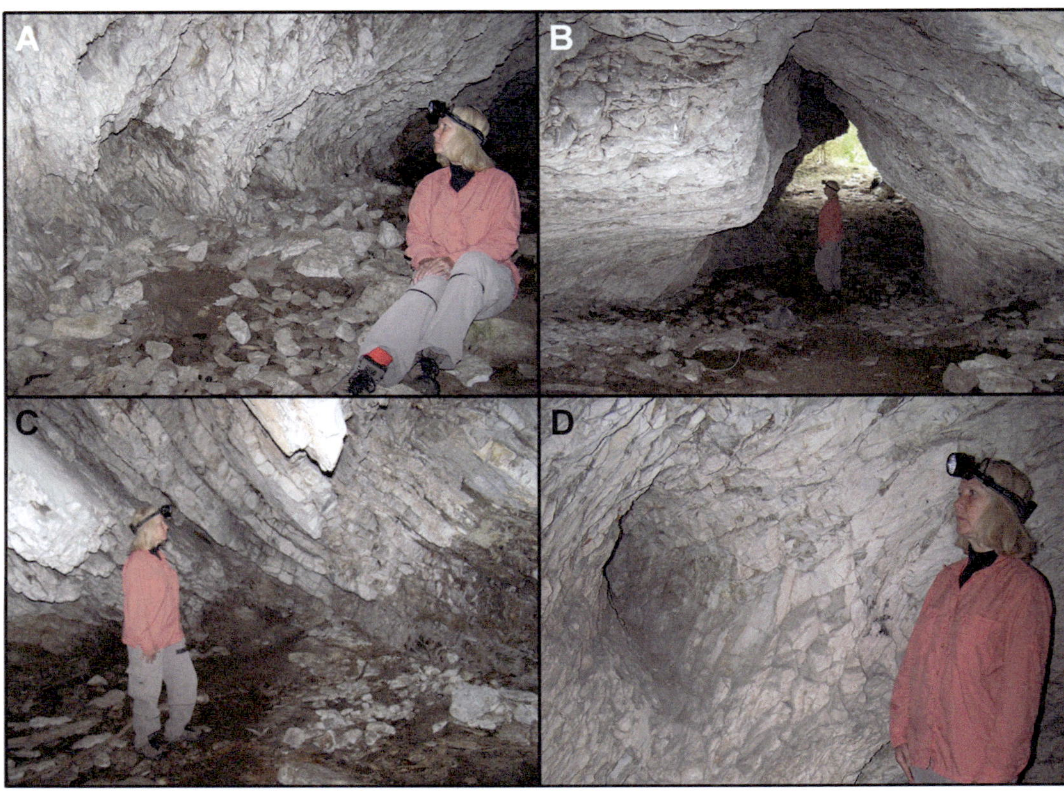

Fig. 17.11 Kaikoura Point Sea Cave. (**a**) West wall of the main passage, showing broken and shattered nature of the limestone (see map, Fig. 17.12). (**b**) Main passage in the cave. Note ceiling slot. Beds dip towards the camera. (**c**) Terminal room in the cave, looking west. Note steeply dipping (up to 55°) beds. (**d**) Phreatic pocket in west wall of the main passage of the cave. The pocket is directly behind the person in (**a**)

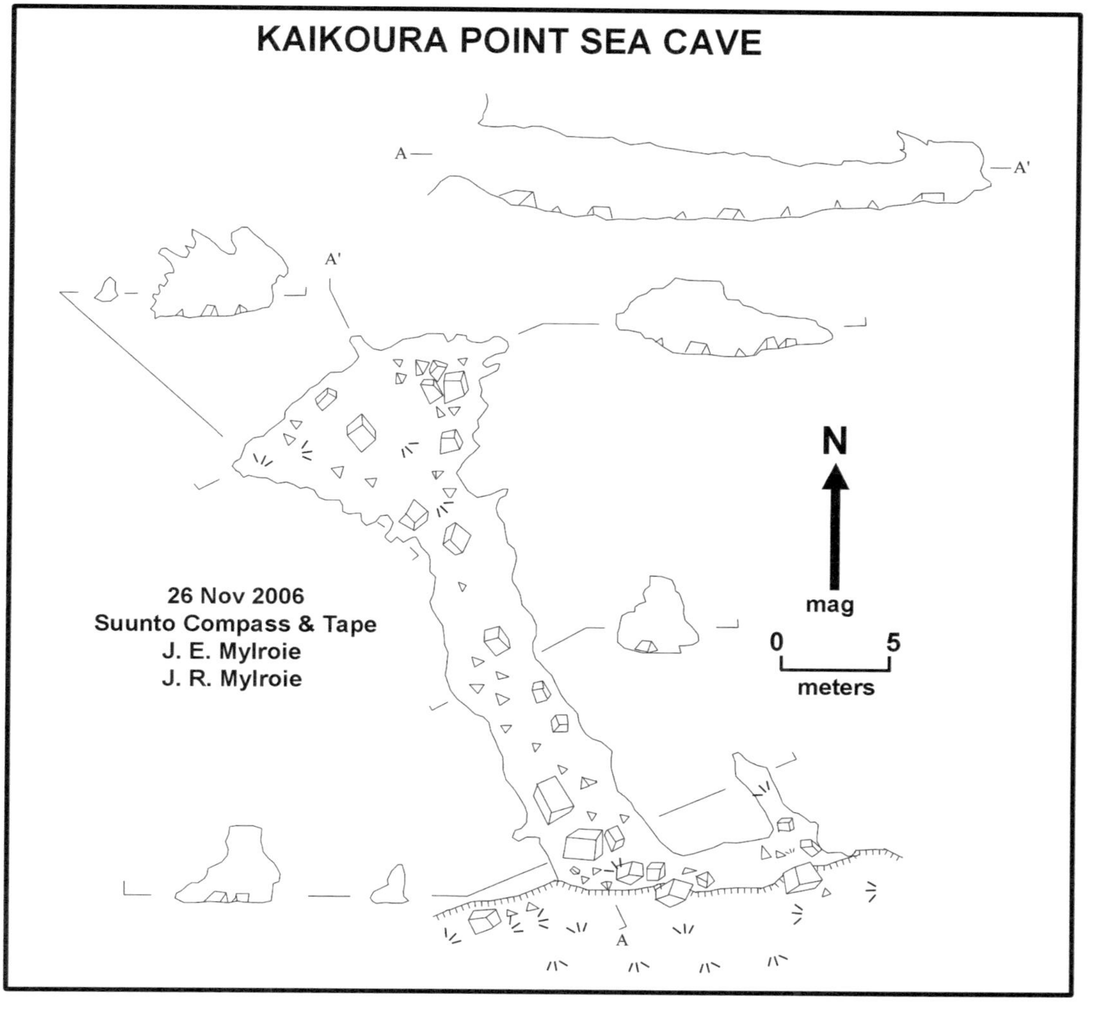

Fig. 17.12 Map of Kaikoura Point Sea Cave

for 30 m to a terminal chamber 14 m wide and 4 m high (Fig. 17.11c). The cave ends in bedrock walls. The cave walls contain numerous dissolution pockets, evident despite the shattered and irregular nature of the cave walls (Fig. 17.11d). A smaller passage exists 5 m to the right (east) in the main cave entrance alcove, and is 4 m long (Fig. 17.12). This cave contains the diagnostic phreatic sculpture and map footprint associated with flank margin cave development.

West and southwest from East Head, and approaching Atia Point, the limestone outcrop again ends, to be replaced by the Gray Marl. A few tens of meters east of this contact, a small limestone point contains two relatively large caves, collectively named Kaikoura Penguin Cave (map, Fig. 17.13). The caves run northeast to southwest, parallel to one another, one inland of the other, both with entrances on each side of the point. The more inland cave is entered by ascending and descending a large collapse pile, which leads to a voluminous entrance chamber (Fig. 17.14a). Trending northwest from the entrance chamber is a passage 16 m long, 5 m wide and 3 m high. This passage has well developed dissolution pockets and small chambers developed along its length, and it terminates in bedrock walls (Fig. 17.14b). Southwest from the

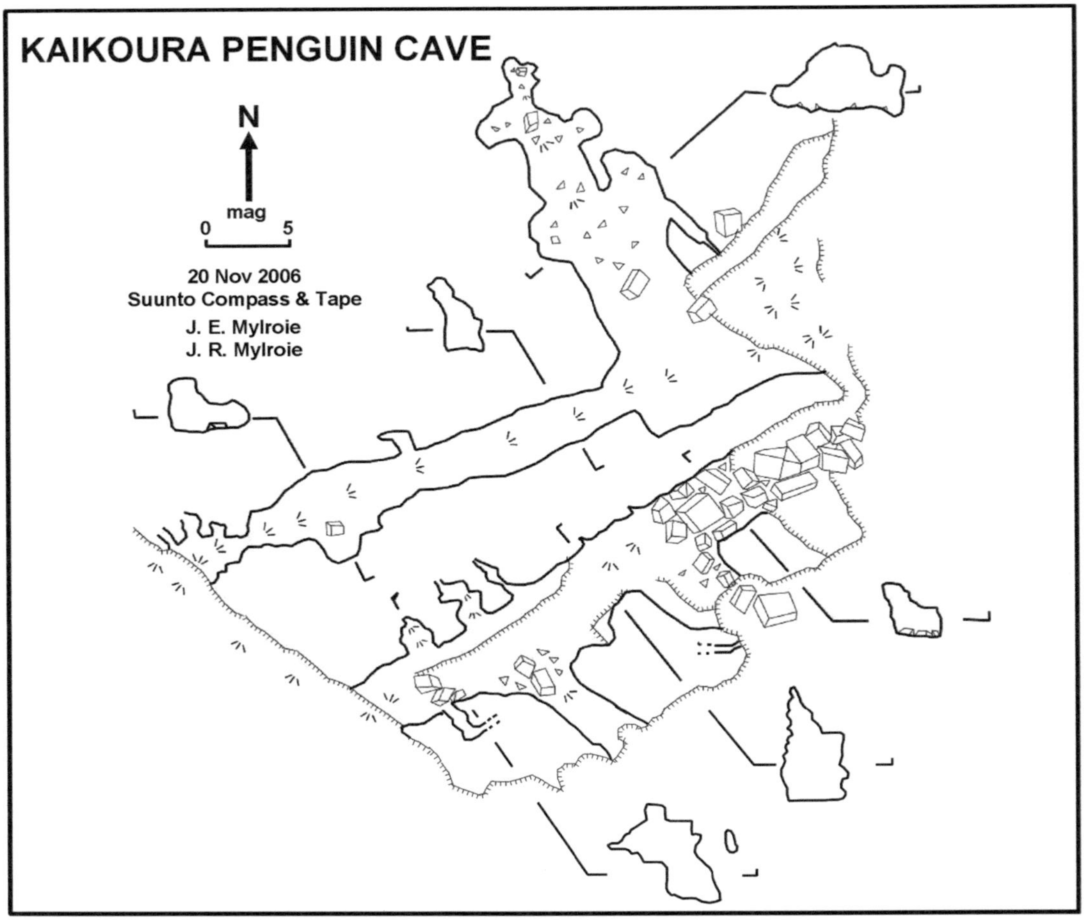

Fig. 17.13 Map of Kaikoura Penguin Cave, which is actually two separate caves

entrance chamber, a high, narrow passage heads 28 m directly through the limestone outcrop to a small entrance on the far side (Fig 17.14c). The passage starts out 10 m high and 3 m wide, sloping upward to the southwest so that at the far entrance, the cave is only 1 m wide and 0.5 m high. The cave floor is mud and sediment, and the nature and position of the bedrock floor cannot be determined. This passage has a number of dissolution pockets in the side walls and ceiling, consistent with a phreatic origin.

Seaward of the first cave is a second cave that also runs from the northeast to the southwest, through the limestone point (Figs. 17.13 and 17.15a). There are two additional entrances that open to the south on the face of the limestone point (Fig. 17.15a). The western of these two entrances is at an elevation low enough to have back-beach cobble rubble on the floor (Fig. 17.15b). The southwestern entrance on the far side of the limestone point has some collapse debris blockage. As with the inland cave, the northeastern entrance is dominated by a large pile of cliff collapse (Fig. 17.15d). The two entrances to the south are unobstructed. The main passage through the limestone point is 40 m long, and up to 5 m high and wide (Fig. 17.15c). Short side passages trend northwest towards the first cave but end in bedrock walls. The passages leading south to the southern entrances are 10 m long. The passage to the northeast is 5 m high and 3 m wide, while the passage to the southwest is 2 m high and wide.

Fig. 17.14 Kaikoura Penguin and Maori Leap Caves. (**a**) Looking south into the entrance chamber of the inland of the two Kaikoura Penguin Caves. (**b**) Side pocket in left wall of the passage shown in (**a**) displaying smooth, phreatic origin. (**c**) View northeast in the passage that cuts through the point in the inland cave. Note the smooth, curved ceiling of this passage, indicating a phreatic origin. (**d**) Folded and distorted wall rock in Maori Leap Cave

The two caves both show phreatic dissolutional surfaces and passage complexity, but are not currently connected. They pass through the limestone point containing them, and appear to have been opened by cliff retreat, demonstrated by the broad wave-cut bench in front of the point (Fig. 17.15a). The walls contain no turbulent flow markings. The caves fit the criteria for flank margin development.

A few kilometers south of Kaikoura Peninsula, just west of Highway 1, is Maori Leap Cave, a show or commercial cave. The cave was opened from above by limestone quarrying, but rubble infill was later removed to open the cave horizontally on the Amuri Limestone cliff face. The cave is a single passage, about 100 m long, up to 8 m high and 6 m wide, perpendicular to the cliff face. There is much collapse material in the cave, and all original dissolutional surfaces (if originally present) are gone. The cave is presented on the tour as a fossil sea cave, but there is no evidence for either that origin or a dissolutional speleogenesis. The cave is very long relative to its width, and propagation of wave energy to such a penetration seems unlikely. The walls of the cave display many folds and small faults, as well as chert, and the limestone is highly fractured (Fig. 17.14d). The collapse material is made up mostly of blocks less than 1 m in maximum dimension, and commonly only 20 cm or so in maximum dimension.

Fig. 17.15 Kaikoura Penguin Caves. (**a**) Limestone point containing the two Kaikoura Penguin Caves. Note that the caves have been breached by erosion and plantation of the bedrock outcrop in front of the point. (**b**) The two southern entrances into the seaward of the two Kaikoura Penguin Caves. Note the cobbles in the floor area of the western (*left*) entrance. (**c**) Main passage of the seaward cave. Penguin for scale is 1 m high. To the left is the southwestern entrance, passages ahead of the penguin end in bedrock walls before reaching the inland cave. (**d**) Looking northeast to the northeastern entrance of the seaward cave. The person is illuminated by light from the easterly (*right*) of the two entrances seen in (**b**). The wall behind the person shows dissolutional smooth, curvilinear morphology

17.3.3 South Island, Punakaiki Region

The telogenetic Oligocene Waitakere Limestone trends in a band parallel to the west coast of South Island from Charleston, 20 km south of Cape Foulwind, south to Punakaiki (Fig. 17.7). The outcrop is mostly inland, but reaches the coast south of Fox Creek at an area called Limestone Creek, and again at the classic Pancake Rocks outcrop at Punakaiki. At Limestone Creek, a few kilometers north of Pancake Rocks, coastal outcrops are accessible from the beach. A number of small caves exist at the south end of the beach next to Limestone Creek. Proceeding north, two very important outcrops are reached. Limestone Creek Through Cave is a phreatic passage developed along joints that passes for 30 m through a small limestone headland (Fig. 17.16), similar to the Kaikoura Penguin Cave. The passage averages several meters high and wide, but becomes smaller at each end (Fig. 17.17). The cave has no indications of high-velocity, turbulent conduit flow, such as scalloping of passage walls. The main cave passage splits, and a large side passage trends into the seaward side of the headland and ends in a blank bedrock wall, another hallmark of flank margin cave dissolution (Fig. 17.17d).

Farther north along the beach, is an area named Tube City. In this outcrop, numerous

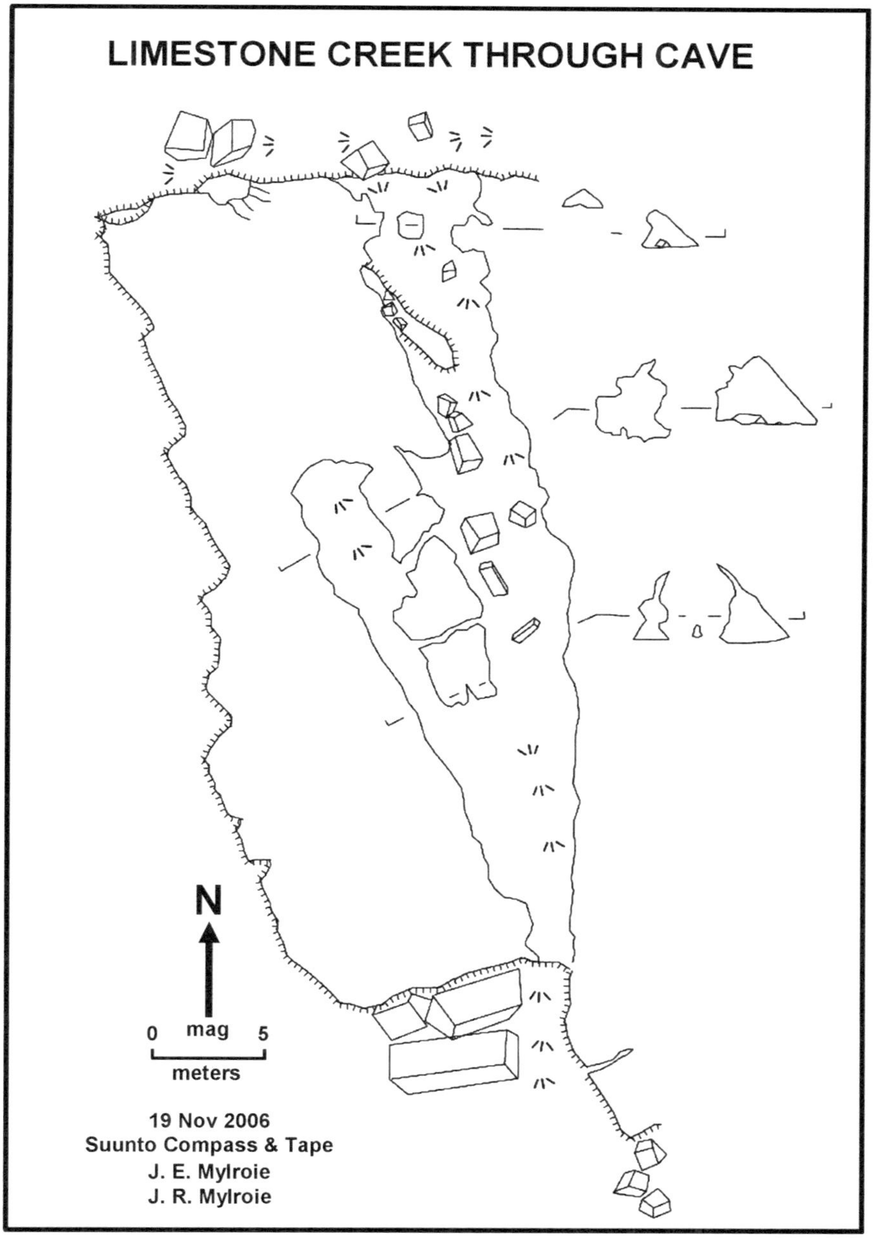

Fig. 17.16 Map of Limestone Creek Through Cave

phreatic tubes exist on a horizontal datum that cuts across the bedding and jointing (Fig. 17.18). The dip of the bedding in this outcrop is 20° to the SSE (Fig. 17.18a). The jointing is closely spaced, dipping at 65° to the NNE. The caves in the outcrop face, as shown by the wave-cut limestone bench in front of them (Fig. 17.18b), were once enclosed within the headland. The phreatic tubes trend inland up to 5 m, before closing down to a single joint (Fig. 17.18c). The tubes contain layers of flowstone on their walls, indicating they were once closed chambers and

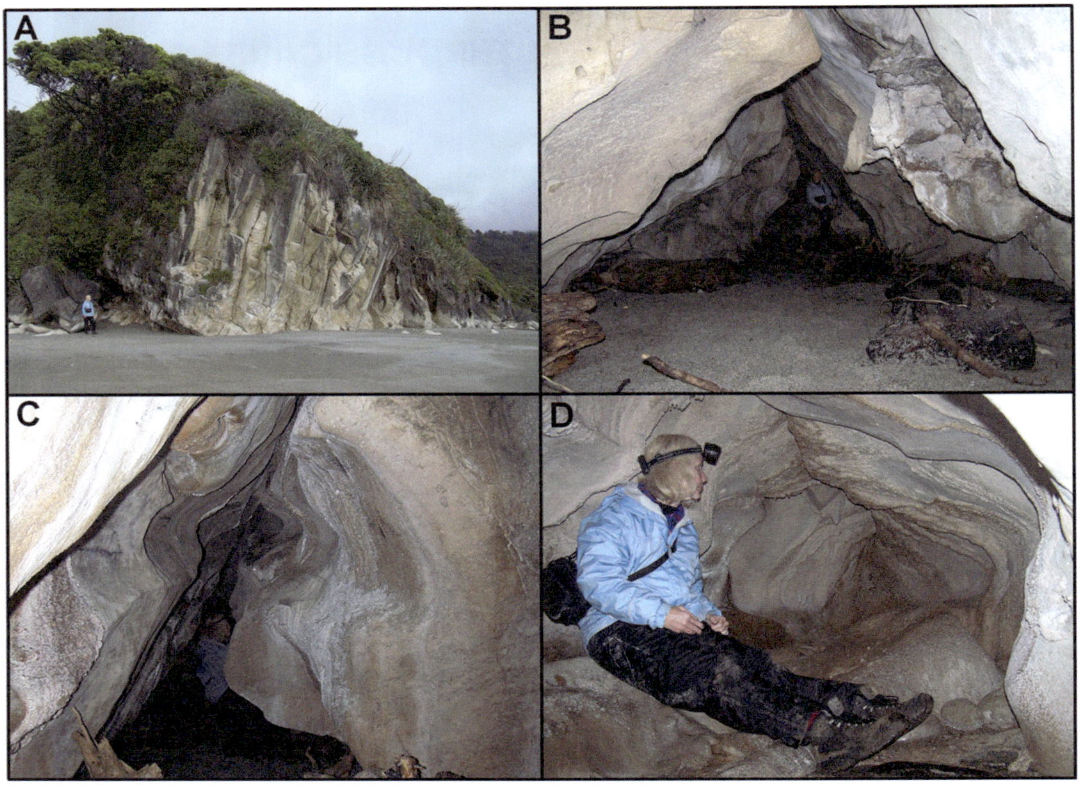

Fig. 17.17 (**a**) Point containing Limestone Creek Through Cave. Note that the bedding dips from left (north) to right (south), jointing from upper right to lower left. Person on left for scale. (**b**) The main passage in Limestone Creek Through Cave, formed by dissolution at the intersection of a bedding plane and joint. Person sitting at end of the passage for scale. (**c**) Passage developed along a joint, with a dissolutional widening that is horizontal despite steep dip of the bedding. This passage leads to the chamber seen in (**d**). Person sitting low in the distance for scale. (**d**) Terminal chamber, with passage pinching down along the guiding joint seen in (**c**)

have been breached by modern wave activity (Fig. 17.18d). A total of seven phreatic tubes are found in an outcrop span of 44 m, all tube floors being between 3 and 5 m above sea level (Fig. 17.19). Dead-end phreatic tubes without turbulent flow markings, such as scallops, at a constant elevation cutting across structure, are some of the indicators of a flank margin origin.

17.4 Summary

The flank margin caves at Kaikoura and Limestone Creek, South Island, New Zealand demonstrate how closely spaced joints and fractures in a telogenetic limestone can allow groundwater flow to occupy three-dimensional space, such that mixing dissolution can occur over a volume. The flank margin caves at Plava Grota on Cres Island, Croatia similarly achieved groundwater flow and mixing over a volume as a result of flow through a paleotalus breccia. At Whirinaki, North Island, New Zealand, the fracture pattern is more widely spaced, and the flank margin caves are isolated to those fracture locations. Whirinaki also highlights how active tectonism can move the carbonate rock relative to the fresh-water lens, migrating the site through the mixing zone dissolution horizon. All the New Zealand examples also emphasize the rapidity of

Fig. 17.18 (**a**) Tube City outcrop on the Limestone Creek coast. Seven individual tubes, all within 1 or 2 m of the same elevation, are in the outcrop. The bedding dips from left (north) to right (south), and the jointing is very closely spaced from upper right to lower left (as at Limestone Creek Through Cave). Person at left for scale. (**b**) Tube City outcrop from a distance, showing the wave-cut platform in front of the caves, indicating significant shoreline retreat in this location, exposing the caves. (**c**) View inward in the large phreatic tube seen on the right in (**a**). Note how the bedrock walls pinch down to the guiding joint, and the horizontal flowstone horizons on the right wall, and in the back. Flashlight is 15 cm long for scale. (**d**) Looking out the same tube as in (**c**), across the wave-cut platform

the dissolutional processes in the distal margin of the fresh-water lens, as the consistent tectonic activity allows even less time for a fresh-water lens to be in a single position than that which existed for glacioeustasy effects in the Bahamas (Chap. 7). Also instructive is the age of the New Zealand limestones. They range from Oligocene to Pliocene, relatively young geologically. However, the active tectonism of New Zealand has sent those rocks deep to be diagenetically matured, then brought them back to the surface as telogenetic rocks. In comparison, the Oligocene limestones of Florida (Chap. 15) have not been subjected to tectonics or deep burial, and are essentially still eogenetic. While rock age can be a good predictor of diagenetic stage (there are no eogenetic Paleozoic limestones), New Zealand shows that one must be careful of making diagenetic assumptions when dealing with Cenozoic carbonates. Maori Leap Cave does not display sufficient diagnostic indicators to allow its classification as a flank margin cave, a sea cave, or a traditional stream conduit cave, illustrating the sometimes frustrating field situations that can occur when attempting to classify caves in coastal locations.

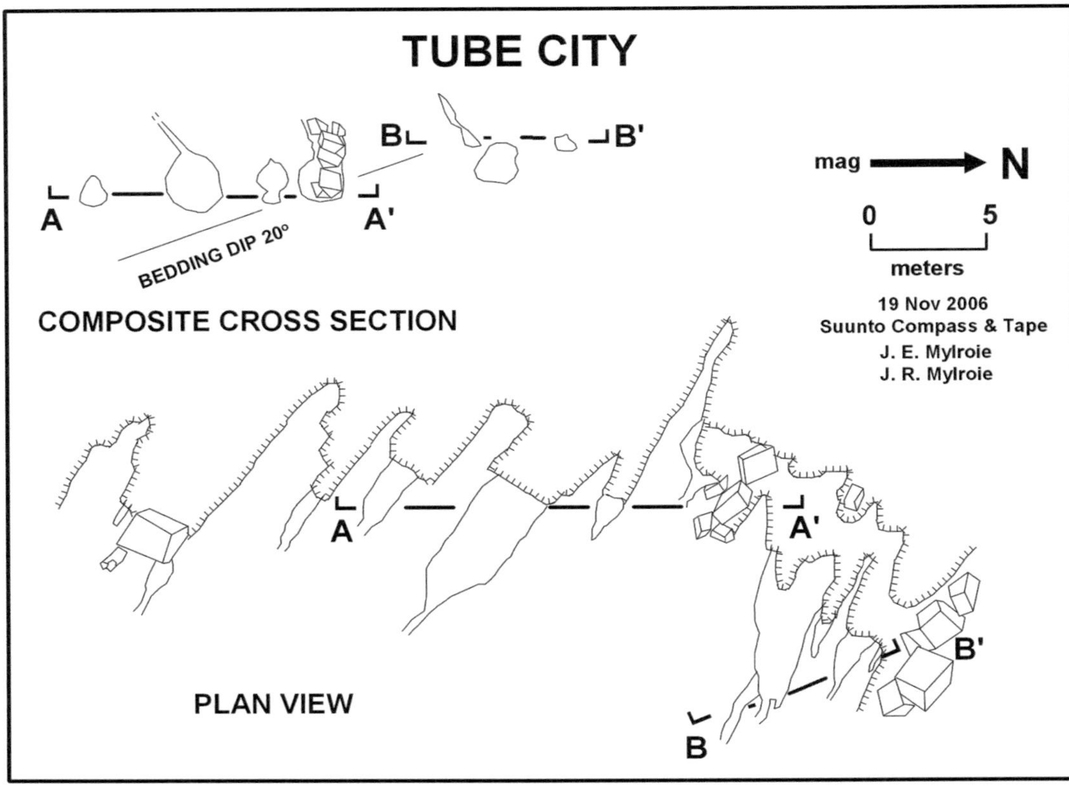

Fig. 17.19 Map of the Tube City outcrop, plan view and composite section. Compare to the wider cave spacing in Fig. 17.9 (Whirinaki Caves) where the joints are more widely spaced

References

Caron V, Nelson CS, Kamp PJJ (2006) Microstratigraphy of calcite cements in Pliocene cool-water limestones, New Zealand: relationship to sea-level, burial and exhumation events. In: Pedley HM, Carannante G (eds) Cool-water carbonates: depositional systems and palaeoenvironmental controls, vol 255, Geological Society special publications. Geological Society, London, pp 337–365

Frank EF, Mylroie JE, Troester J, Alexander EC, Carew JL (1998) Karst development and speleologenesis, Isla de Mona, Puerto Rico. J Cave Karst Stud 60:73–83

Hull AG (1990) Tectonics of the 1931 Hawkes Bay earthquake. New Zeal J Geol Geophys 33:309–322

Jenson JW, Keel TM, Mylroie JR, Mylroie JE, Stafford KW, Taborosi D, Wexel C (2006) Karst of the Mariana Islands: the interaction of tectonics, glacioeustasy and fresh-water/sea-water mixing in island carbonates. Geol Soc Am Spec Pap 404:129–138

Korbar T (2009) Orogenic evolution of the External Dinarides in the NE Adriatic region: a model constrained by tectonostratigraphy of Upper Cretaceous to Paleocene carbonates. Earth-Sci Rev 96:296–312

Mylroie JE, Carew JL (1987) Field evidence of the minimum duration for speleogenesis. Natl Speleological Soc Bull 49:67–72

Mylroie JE, Mylroie JR (2007) Development of the carbonate island karst model. J Cave Karst Stud 69:59–75

Mylroie JE, Mylroie JR, Nelson CN (2008) Flank margin cave development in telogenetic limestones of New Zealand. Acta Carsologica 37:15–40

Nelson CS (1978) Temperate shelf carbonate sediments in the Cenozoic of New Zealand. Sedimentology 25: 737–771

Nelson CS, Harris GJ, Young HR (1988) Burial-dominated cementation in non-tropical carbonates of the Oligocene Te Kuiti Group, New Zealand. Sediment Geol 60:233–250

Nelson CS, Winefield PR, Hood SD, Caron V, Pallentin A, Kamp PJJ (2003) Pliocene Te Aute limestones, New Zealand: expanding concepts for cool-water shelf carbonates. New Zeal J Geol Geophys 46: 407–424

Otoničar B (2007) Upper Cretaceous to Paleogene forbulge unconformity associated with foreland basin evolution (Kras, Matarsko Podolje and Istria; SW Slovenia and NW Croatia). Acta Carsologica 36: 101–120

Otoničar B, Buzijak N, Mylroie JE, Mylroie JR (2010) Flank margin cave development in carbonate talus breccia facies: an example from Cres Island, Croatia. Acta Carsologica 39:79–91

Proctor CJ (1988) Sea level related caves on Berry Head, South Devon. Cave Sci 15:39–50

Rubinić J, Ožanić N (1992) Hidrološke karakteristike Vranskog jezera na otoku Cresu (hydrological characteristic of the Vrana Lake on the Island of Cres). Građevinar 44:521–530

Taboroši D, Jenson JW, Mylroie JE (2004) Karren features in island karst: Guam, Mariana Islands. Zeitschrifft fur Geomorphologie 48:369–389

Vacher HL, Mylroie JE (2002) Eogenetic karst from the perspective of an equivalent porous medium. Carbonates Evaporites 17:182–196

Vlahović I, Tišljar J, Velić I, Matičec D (2005) Evolution of the Adriatic carbonate platform: palaeogeography, main events and depositional dynamics. Palaeogeogr Palaeocl 220:333–360

Australian Examples of Coastal Caves 18

John E. Mylroie and Joan R. Mylroie

Abstract
Rottnest Island and Kangaroo Island, eolianite-containing islands off Australia's west and southern coasts, respectively, display extensive coastal caves and karst that contain valuable geologic information. On Rottnest Island, Late Quaternary eolianites contain flank margin caves formed during MIS 5e, and may contain flank margin caves developed during the mid-Holocene sea-level highstand. Pit cave development, with subsequent cementation of infilling deposits, results in inversion of topography following surficial denudation. Pseudokarst forms such as sea caves and tafoni, and surficial polygonal structures, complicate karst interpretations. On Kangaroo Island, flank margin caves in Quaternary eolianites are preferentially preserved in protected embayments. Flank margin caves at ~30 m elevation indicate much greater uplift rates and magnitudes than previously believed. As with Rottnest Island, sea caves and tafoni complicate karst interpretations and provide a cautionary note to those working on coastal caves and karst.

18.1 Introduction

Australia has a long and extensive history of cave and karst science (e.g. Jennings 1972, 1984), but specific activity as regards application of the Carbonate Island Karst Model (CIKM) to its coastal regions has occurred only recently. Caves in coastal locations that were once identified as sea caves are now recognized to be flank margin caves (White et al. 2007; Mylroie and Mylroie 2009). A review of karst development in coastal eolianites on the Australian mainland can be found in White (2000). In this chapter, we discuss two specific locations, Kangaroo Island and Rottnest Island, which demonstrate flank margin cave development in Quaternary eolianites in coastal settings. The significance of the caves is the information they reveal regarding related geological processes in their local area. For Kangaroo Island, this information helps address a long standing controversy regarding tectonic uplift, for Rottnest Island, the caves help narrow the opportunities for Late Quaternary eolian deposition.

J.E. Mylroie (✉) • J.R. Mylroie
Department of Geosciences, Mississippi State University, Mississippi State, MS 39762, USA
e-mail: mylroie@geosci.msstate.edu

M.J. Lace and J.E. Mylroie (eds.), *Coastal Karst Landforms*, Coastal Research Library 5,
DOI 10.1007/978-94-007-5016-6_18, © Springer Science+Business Media Dordrecht 2013

18.2 Rottnest Island

Rottnest Island is located in the Indian Ocean on the continental shelf, 18 km west of the coastal city of Perth in Western Australia (Fig. 18.1). The island is approximately 11 km long, east to west, and 4.5 km wide, with 19 km^2 of area (Fig. 18.2). Rottnest is entirely Quaternary carbonate rock and sediments; in that regard it is extremely similar to eolianite islands of the Bahamian Archipelago. For a comparison of Bahamian and Rottnest stratigraphy, see Mylroie and Mylroie (2010). The standard geologic reference for Rottnest Island is the review paper by Playford (1997), who explains the island as a three-fold rock stratigraphy, with one Holocene subtidal unit (Herschel Limestone), one Pleistocene subtidal unit (Rottnest Limestone), and an eolian unit that spans the Pleistocene to early Holocene (Tamal Limestone). The Rottnest Limestone, which contains a fossil reef facies U/Th dated to MIS 5e (Szabo 1979), is limited to a single outcrop at Fairbridge Bluff (Fig. 18.2). The Tamala Limestone is entirely eolianite, and is named for, and correlated with, eolianites on the Western Australia mainland. Playford (1997) believed eolianite deposition was tied to sea-level lowstands, not highstands as is now understood to be the case (Carew and Mylroie 1995, 1997), which explains his ending of Tamala eolian deposition in the early Holocene, prior to the Holocene transgression (Fig. 18.2). Murray-Wallace et al. (1989, in Hearty 2003), made modifications to an earlier Playford stratigraphy as part of a review of the Perth area, but generally kept the three-fold stratigraphy. Hearty (2003) recognized that the Rottnest eolianites are tied to sea-level highstands. Hearty (2003), using amino acid racemization (AAR), also noted the basic three-fold stratigraphy, but used AAR to subdivide the Tamala's Pleistocene units, adding a MIS 5a-c eolian unit to the Rottnest Limestone in addition to the standard MIS 5e eolian units. Hearty (2003) was able to differentiate older Pleistocene units on the Australian mainland, but not on Rottnest, into MIS 7–9, 9–11, and >11 age units. His results, in contrast to Playford (1997), indicate that the majority of Rottnest Island is MIS 5 in age, and older Tamala units are

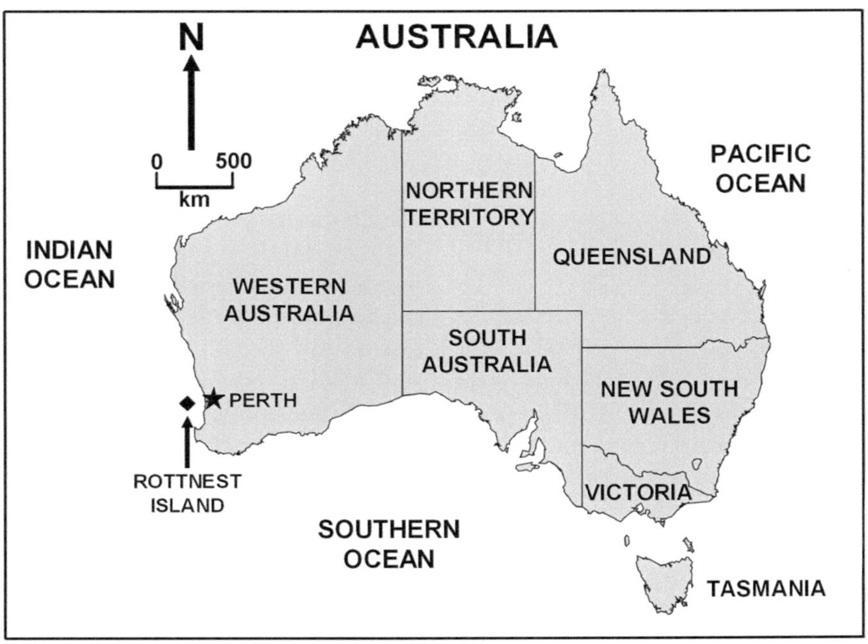

Fig. 18.1 Map of Australia showing the location of Rottnest Island

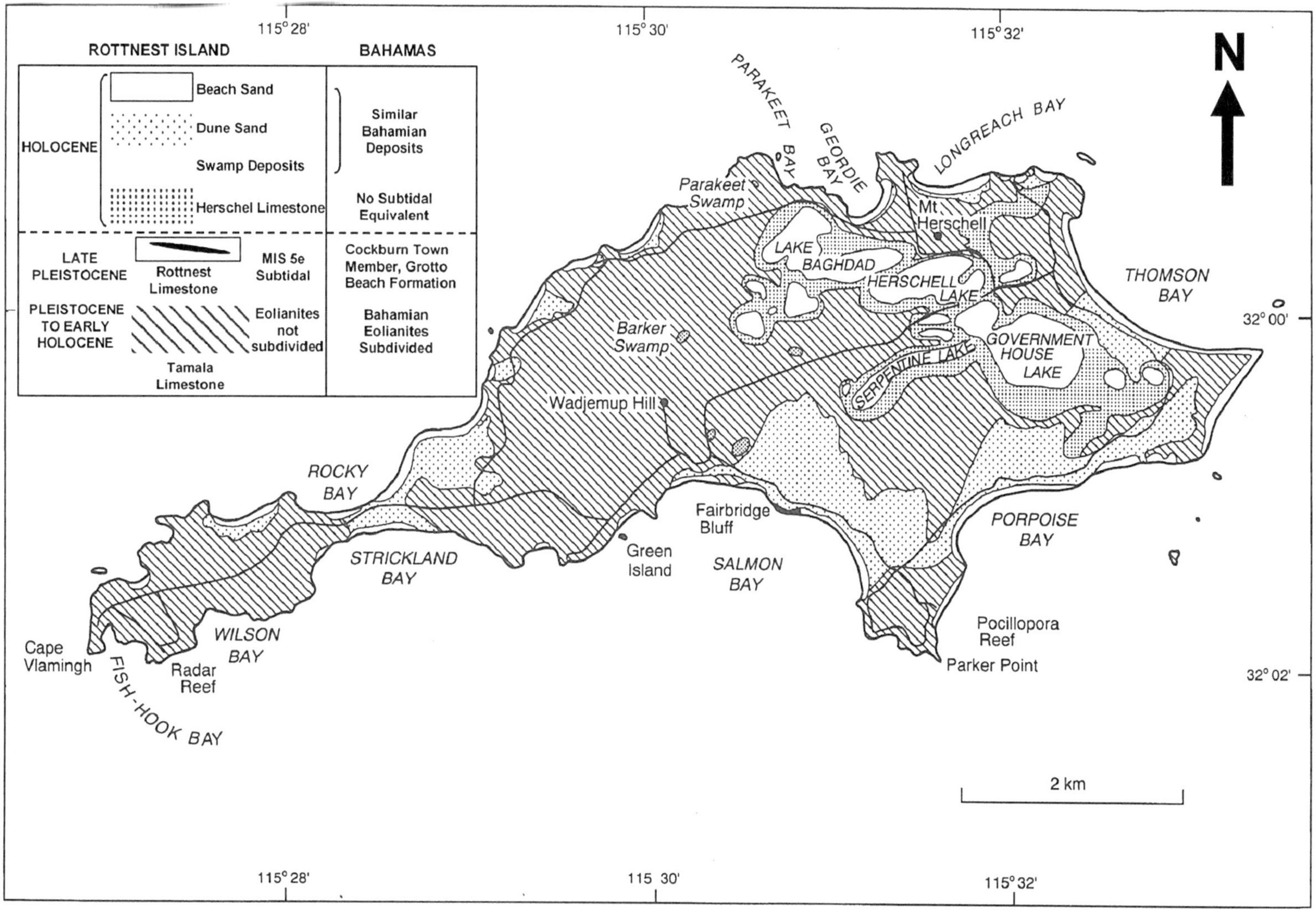

Fig. 18.2 Geologic map of Rottnest Island, showing comparison to Bahamian Stratigraphy (Modified from Playford 1997)

Fig. 18.3 (**a**) Outcrop of Herschel Limestone below Bathurst Lighthouse north of Thompson Bay. The upper eolianite lacks a terra rossa paleosol, indicating that it is Holocene in age. The person is standing on a terra rossa paleosol, indicating the lower unit is Pleistocene. The base of the Bathurst lighthouse is visible in the *upper right*; below the lighthouse, the apparent cave opening is a tafone (wind erosion pocket), not a flank margin cave. (**b**) Close up of the terra rossa paleosol seen in (**a**)

not common. These results also demonstrate the imprecision of AAR dating; however, there are few alternative options for dating eolianites.

18.2.1 Rottnest Island Stratigraphy and Karst Relationships

18.2.1.1 Bathurst Lighthouse

The Holocene/Pleistocene boundary is obvious at Bathurst Lighthouse on the eastern end of Rottnest Island. As seen in Fig. 18.3, the upper unit would be the Holocene Herschel Limestone, as it lacks a terra rossa paleosol. Playford (1997) defined the Herschel Limestone as the Holocene lagoonal facies surrounding the interior water bodies in the eastern interior of Rottnest Island. That name was expanded (Mylroie and Mylroie 2010) to label the Holocene eolianites reported here. Playford (1997) apparently did not recognize any lithified Holocene eolianites; Hearty (2003) found evidence of some Holocene eolianite production. The observations reported here indicate that Holocene eolianite production, as was the case in the Bahamas, is more voluminous than earlier workers believed. The lower eolian unit in the outcrop is under a terra rossa paleosol (Fig. 18.3) so it is Pleistocene.

Figure 18.3a also shows a small cave in the Herschel Limestone. This void is not a flank margin cave, which would be impossible at that elevation in a Holocene eolianite, but a tafone (plural, "tafoni"), a weathering pocket formed by wind action and wetting and drying (Owen 2007; see also Chap. 8, this volume). The void is a small, simple chamber without any secondary calcite deposits.

South from the Bathurst Lighthouse, a series of cave openings are found along the coastal cliffs (Fig. 18.4). These caves appear to be flank margin caves, but are currently being subjected to wave attack with consequent cliff retreat, so they are quite open to the elements in places and are being over-printed by marine processes. The host eolianite has a rich collection of plant trace fossils, or vegemorphs, which have been used in the Bahamas to identify regressive-phase eolianites (Carew and Mylroie 1995, 1997; Birmingham et al. 2008). MIS 5e regressive-phase eolianites could not host a fresh water lens above modern sea level, but these rocks appear to contain flank margin caves. If these eolianites were part of

Fig. 18.4 (**a**) Cave opening in the sea cliffs south of Bathurst Lighthouse. The cave is being attacked by modern wave action. (**b**) Entrance into a flank margin cave adjacent to the cave pictured in (**a**). Note the vegemorphs hanging from the cave roof. (**c**) Inside the cave shown in (**b**). Note the complex vegemorphs forming the cave roof and part of the pillar on the right side. (**d**) Back wall of a breached flank margin cave adjacent to the cave in (**b**). Note that the vegemorphs have significant penetration into the dune mass, a good indication of regressive-phase eolianite origin

a prograding strandplain, then they could have hosted a fresh-water lens during MIS 5e, as has been demonstrated for banana hole formation in the Bahamas (Mylroie et al. 2008a; Infante et al. 2011; see also Chap. 4). If the eolianites are truly regressive then they are older, pre-MIS 5e Pleistocene units not currently recognized on Rottnest Island. Such rocks would have been already in place to host a MIS 5e fresh-water lens. The rock cannot be MIS 5e transgressive-phase eolianites (French Bay Member equivalent of the Bahamas) as the abundant vegemorphs preclude a trangressive depositional environment (Birmingham, et al. 2008). Because these eolianites contain flank margin caves, they cannot be MIS 5a or MIS 5c in age.

18.2.1.2 Salmon Point

The sea cliffs developed in eolianites at Salmon Point (east side of Salmon Bay in Fig. 18.2) have a well-developed terra rossa paleosol with abundant vegemorphs (Fig. 18.5). The terra rossa paleosol indicates a Pleistocene age for the unit. The outcrop also contains numerous infilled dissolution pits, which being more lithified than the surrounding eolianite, have weathered out in positive relief in this coastal setting. The pits, when forming, conducted water from the epikarst into the eolianite. That fluid transfer, and subsequent infill, resulted in micritization of the pit walls and the infilling material, most likely at the time when the terra rossa paleosol covering the outcrop became micritized. As a

Fig. 18.5 Infilled dissolution pits at Salmon Point (next point west of Parker Point on the east side of Salmon Bay in Fig. 18.2). (**a**) Series of pits descending from a terra rossa paleosol surface. The pit infills are lithified to a greater extent than the surrounding eolianite, and weather out in positive relief in this coastal setting. (**b**) Close up of the largest infilled pit in (**a**). Note the numerous vegemorphs in this locality

result the pit and its infill are very resistant to erosion. Open pit caves were not observed during our reconnaissance, and infilled pits were rarely more than a few meters in depth. The resistance to erosion of the terra rossa paleosol mechanically supports the overhanging nature of the cliff, and allows weakly-cemented eolianite grains to weather from around the resistant infilled dissolution pits. The rock here fits the criteria for a regressive-phase eolianite.

18.2.1.3 Fairbridge Bluff

West of Salmon Point, at the midpoint of Salmon Bay, is Fairbridge Bluff with a classic MIS 5e fossil reef (Playford 1997). The reef outcrops from sea level to a little over 3 m elevation (Fig. 18.6). The corals are predominantly various *Acropora* sp and *Goniastrea* sp (Playford 1997), with molluscan-rich debris and coral fragments filling in the space between coral heads (Fig. 18.6). Szabo (1979) dated the fossil reef to 132 +/− 5 ka, indicating a MIS 5e age, which places it in the Rottnest Limestone. A terra rossa paleosol caps the fossil reef (Fig. 18.6), further proof of its Pleistocene age, and indicating that the reef was exposed to subareal weathering processes at the end of MIS 5e without any intervening eolian deposition. Infilled dissolution pits are present in the reef facies as well. Unlike those infilled pits seen at Salmon Point, which hang from an overlying terra rossa paleosol (Fig. 18.5), these are exposed on a wave-swept horizontal bench, such that their greater mechanical strength makes them protrude more than 10 cm from the bench surface (Fig. 18.6b).

18.2.1.4 Western Salmon Bay

The western side of Salmon Bay (Fig. 18.2) was mapped by Playford (1997) as Tamala Limestone. The cliffs are eolianites (Fig. 18.7) that have a terra rossa paleosol, but lack significant vegemorph development, and so are interpreted to be Pleistocene transgressive-phase deposits. These cliffs host a series of small to medium-sized voids that are unequivocally flank margin caves (Figs. 18.8 and 18.9). The voids are chambers entered by openings formed by slope retreat of the enclosing eolianite. The caves are made of globular chambers (Fig. 18.8c, d), and have entrances smaller than the chambers inside. The walls and ceiling show classic phreatic dissolutional morphology of cupolas and dissolution tubes (Fig. 18.10). Secondary calcite formations, such as flowstone (Figs. 18.10b and 18.11) are found. This accumulated evidence indicates a flank margin cave origin for these voids. Figure 18.11 is instructive, for flowstone found inside notches on cliffs in the Bahamas have been interpreted to be terra rossa paleosols *within* the

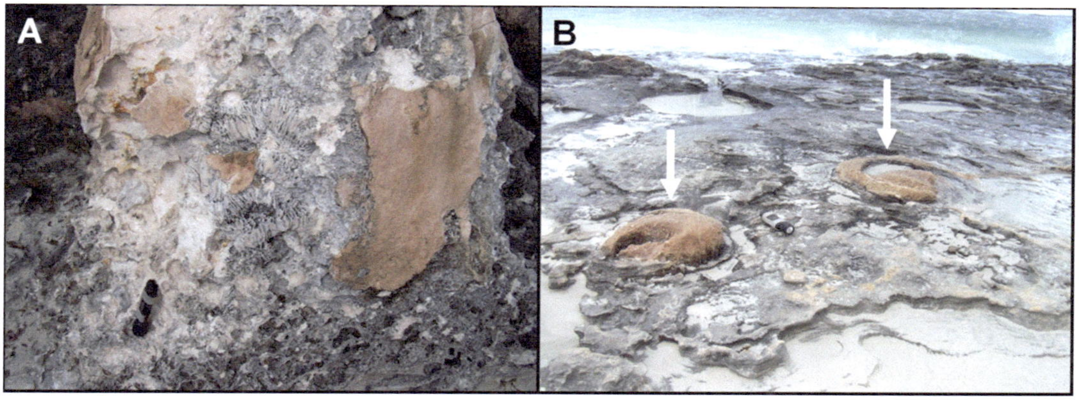

Fig. 18.6 Fairbridge Bluff (flashlight 15 cm for scale in both cases). (**a**) Coral head with patches of a red micritc crust from a terra rossa paleosol overlaying it. (**b**) Dissolution pits (*arrows*), infilled with resistant terra rossa material (*red* in image), only partially planed off by wave erosion. In color, these two mounds are bright red

Fig. 18.7 Eolianite outcrop on the west end of Salmon Bay, east of Green Island (Fig. 18.2). The outcrop is capped by a terra rossa paleosol, but lacks vegemorph development, indicating it is a transgressive-phase eolianite

rock, as opposed to flowstone *on* the rock, with misidentification of the stratigraphy as a result. Such notches have also been misidentified as fossil bioerosion notches instead of breached flank margin caves (Mylroie and Carew 1991). The rock here is either a transgressive-phase eolianite from MIS 5e or a transgressive–phase eolianite from an earlier highstand event.

18.2.1.5 Wilson Bay and Fish Hook Bay

Wilson Bay and Fish Hook Bay, at the extreme southwest end of Rottnest Island (Fig. 18.2), display outcrops that appear to be exact analogues for outcrops observed in the Bahamas. Sea arches are common along this coast (Fig. 18.12), a predictable outcome given the high eolianite cliffs and strong wave action. The problem in coastal

Fig. 18.8 Flank margin cave at the eastern end of the Tamala Limestone outcrop at the west side of Salmon Bay. (**a**) Eastern entrance to the cave. (**b**) Looking out the entrance shown in **a**. (**c**) Main chamber of the cave. (**d**) West entrance of the cave, at a lower elevation than the eastern entrance

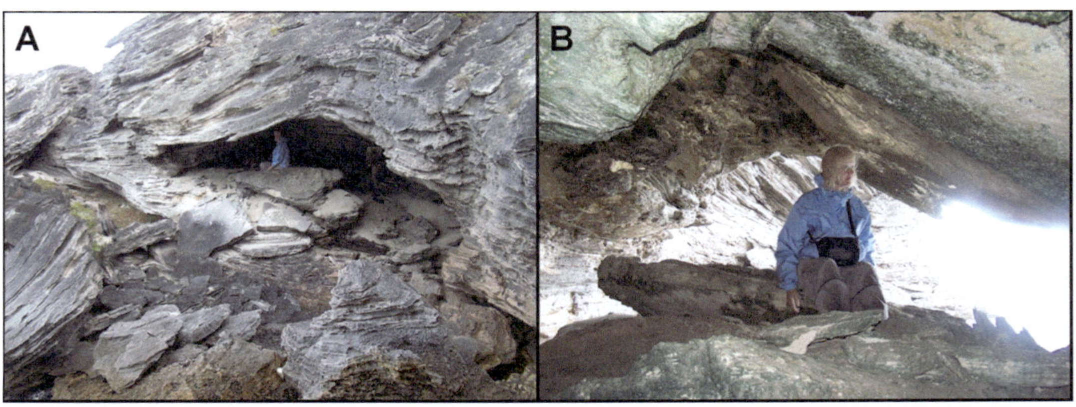

Fig. 18.9 Flank margin cave at the western end of the Tamala Limestone outcrop at the west side of Salmon Bay. (**a**) Main entrance to the cave. (**b**) Inside the cave looking out the main entrance

carbonates as discussed in Chap. 4, is successfully differentiating sea caves and arches produced by the mechanical action of waves, from flank margin caves that have been breached by coastline retreat and are now being over printed by wave action (Waterstrat et al. 2010). In the

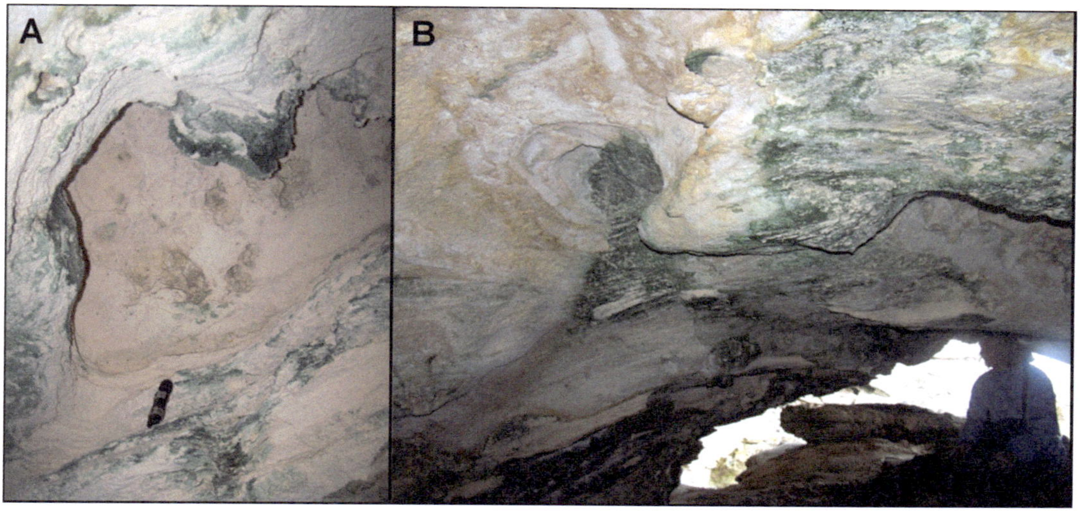

Fig. 18.10 Ceiling morphologies in the flank margin cave shown in Fig. 18.9. (**a**) Dissolution cupola in the cave ceiling (flashlight is 15 cm long for scale). (**b**) Dissolution tube encrusted in flowstone. Person seated in the lower right for scale

Fig. 18.11 Breached western extension of the cave shown in Fig. 18.9. Flashlight is 15 cm long for scale. The pale pink to red coating is a flowstone that developed on the cave floor before the cave was breached by surface erosion

two cases shown in Fig. 18.12, the cave walls lack calcite speleothems or phreatic dissolutional surfaces. The voids are at grade with current sea level (Fig. 18.12b), or have prograded upward from sea level by collapse (Fig.18.12a) whereas flank margin caves would reflect the higher sea-level origination position of MIS 5e. If the caves were from an earlier glacioeustatic

Fig. 18.12 Sea Arches along the coast at Fish Hook Bay. (**a**) Large arch, guardrail in upper left for scale. (**b**) Small arch in the active surf zone

sea-level highstand, it is unlikely they would have survived wave attack from both the MIS 5e and the Holocene sea-level highstands.

The surface landforms here are interesting. Terra rossa paleosol material has infilled dissolution pits, as at Salmon Point (Fig. 18.5), creating resistant features that weather out in positive relief, inverting the topography. Initially, the resistant infilled dissolution pits act as pillars supporting the terra rossa paleosol material as wind energy excavates the poorly-cemented sands beneath (Fig. 18.13a). Vegemorphs (plant trace fossils) are common (Fig. 18.13b). Continued weathering and erosion eventually strips the terra rossa paleosol and the underlying hillside landward, leaving the infilled dissolution pits as isolated pillars (Fig. 18.13c, d). The end result is a classic inversion of topography initiated by karst processes.

The eolian surface above Wilson Bay has weathered into a pattern of polygons (Fig. 18.14a, b). This phenomenon has also been observed in the Bahamas on Eleuthera Island (Panuska et al. 2002) and Cat Island (Mylroie et al. 2006); desiccation or salt wedging has been demonstrated as the cause (Glumac et al. 2011). Off shore, in the shallow lagoon, other polygonal features can be observed, at a larger scale (Fig. 18.14c, d). These are constructional features made by a kyphosid fish creating individual territories for algal "farming" (Playford 1997). While polygonal karst is well known from the karst literature (Ford and Williams 2007), the subaerial and subaqueous polygons found at Wilson Bay are not karst, and are a reminder that polygonal forms in carbonates can be polygenetic.

18.2.1.6 Herschel Lake

Herschel Lake is one of a series of inland water bodies located in the northeastern portion of Rottnest Island (Fig. 18.2). As noted earlier, this is the type locality for the Holocene Herschel Limestone and its lagoonal facies (Fig. 18.15). The Indo-Pacific basin and Rottnest Island show evidence of a mid-Holocene sea-level highstand of a couple of meters (Playford 1997), a feature that is not well documented in the Atlantic basin. The outcrops also show unusual features, such as seen in Fig.18.16, which may indicate the past presence of evaporites. The shoreline of Herschel Lake contains a number of Pleistocene eolianite outcrops with notches and voids in them. They were interpreted by Playford (1997) to be wave-cut or bioerosion notches, but in some cases they appear to be tafoni (Fig. 18.17) and in other cases breached flank margin caves (Fig. 18.18).

Fig. 18.13 Infilled dissolution pits and vegemorphs, Wilson Bay. (**a**) Outcrop forming a small cave by wind excavation of dune material from beneath a resistant terra rossa paleosol (tafone). Resistant infilled dissolution pits act as pillars to hold up the void roof. (**b**) View inside the cave in **a**, looking out a hole in the roof, with abundant and varied vegemorphs. Hole is 60 cm across. (**c**) Residual pillar formed from a resistant infilled dissolution pit. This situation represents a more advanced stage of (**a**), where the terra rossa paleosol and hillside have been stripped back, leaving the dissolution pit as an example of inverted topography. (**d**) Another example of the situation in (**c**), about 25 m away to the east

If they are flank margin caves, they could be relict from the MIS 5e highstand, or they could be Holocene in age. If Holocene in age, they would be one of the few flank margin caves known from the current sea-level highstand. Such mid-Holocene sea-level highstand caves have been recently documented from the Holocene Merizo Limestone on Guam (Miklavič et al. 2012).

18.2.2 Rottnest Summary

Rottnest Island demonstrates two main concepts. First, karst features can help constrain the age and subsequent history of the carbonate unit containing them, especially in Quaternary coastal environments where deposition of the units is tied to glacioeustasy. Second, karst features themselves must be properly interpreted to separate pseudokarst features such as sea caves from flank margin caves, or to separate surface karst from totally unrelated geometries such as polygons. Rottnest Island contains flank margin caves, dissolution pits, epikarst and fossil epikarst as diagnostic karst features, which not only constrain the age of the carbonate units containing them, but also subsequent rates of landscape denudation. The degree to which flank margin caves are exposed and over printed by coastal processes is a measure of coastline retreat, while the degree to which

Fig. 18.14 Polygonal forms, Wilson Bay. (**a**) Polygonal cracking of an eolianite surface. (**b**) Close up of (**a**), flashlight 15 cm long for scale. (**c**, **d**) Polygonal pattern expressed in the shallow off shore at Radar Reef, between Wilson Bay and Fish Hook Bay. This pattern is produced by territorial behavior of a kyphosid fish (Playford 1997)

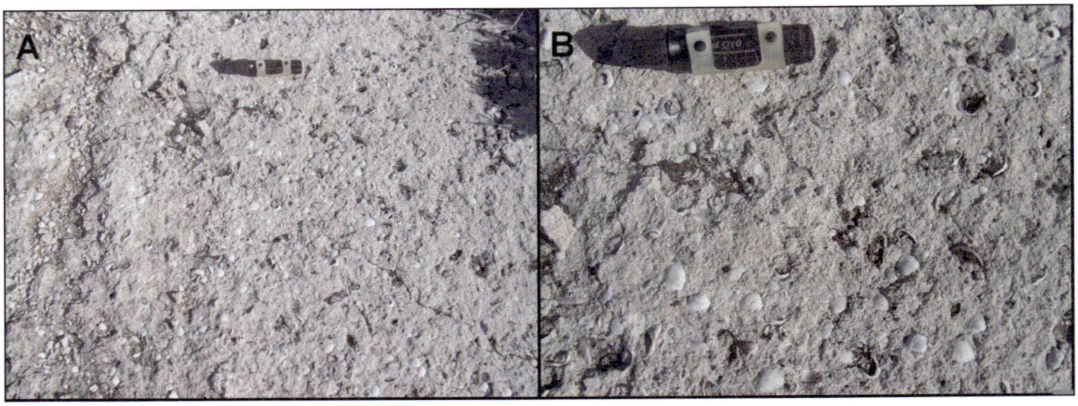

Fig. 18.15 Lake Herschel area, flashlight is 15 cm long for scale (note that the flashlight shadow in low-angle morning sun extends the length beyond 15 cm). (**a**) Lithified Herschel Limestone on the shore of Herschel Lake. This unit is a mollusk-rich lagoonal facies. (**b**) Close up of the same outcrop

Fig. 18.16 Flat lithified outcrop on the shoreline of Herschel Lake, showing a vertical bladed pattern that may relate to evaporite mineral activity. Flashlight 15 cm long for scale

Fig. 18.17 Eolianite ridge bordering Herschel Lake. (**a**) Probable tafoni, recognizable by the white interior and the layer-parallel configuration. (**b**) Another probable tafone

infilled dissolution pits are expressed as inverted topography is a measure of overall landscape denudation.

18.3 Kangaroo Island

Kangaroo Island is located 16 km southwest across the Backstairs Passage from Cape Jervis on the Fleurieu Peninsula, South Australia (Fig. 18.19). The island is a rectangle roughly 145 km east to west, and 60 km north to south with an area of 4,350 km^2 and a coastline length of 457 km (Short and Fotheringham 1986). Kangaroo Island is a geologically diverse environment with rocks present from the Proterozoic through the Paleozoic, Mesozoic, and Cenozoic (Belperio and Flint 1999). The island is a geologic extension of the Fleurieu Peninsula on the Australian mainland (Belperio and Flint

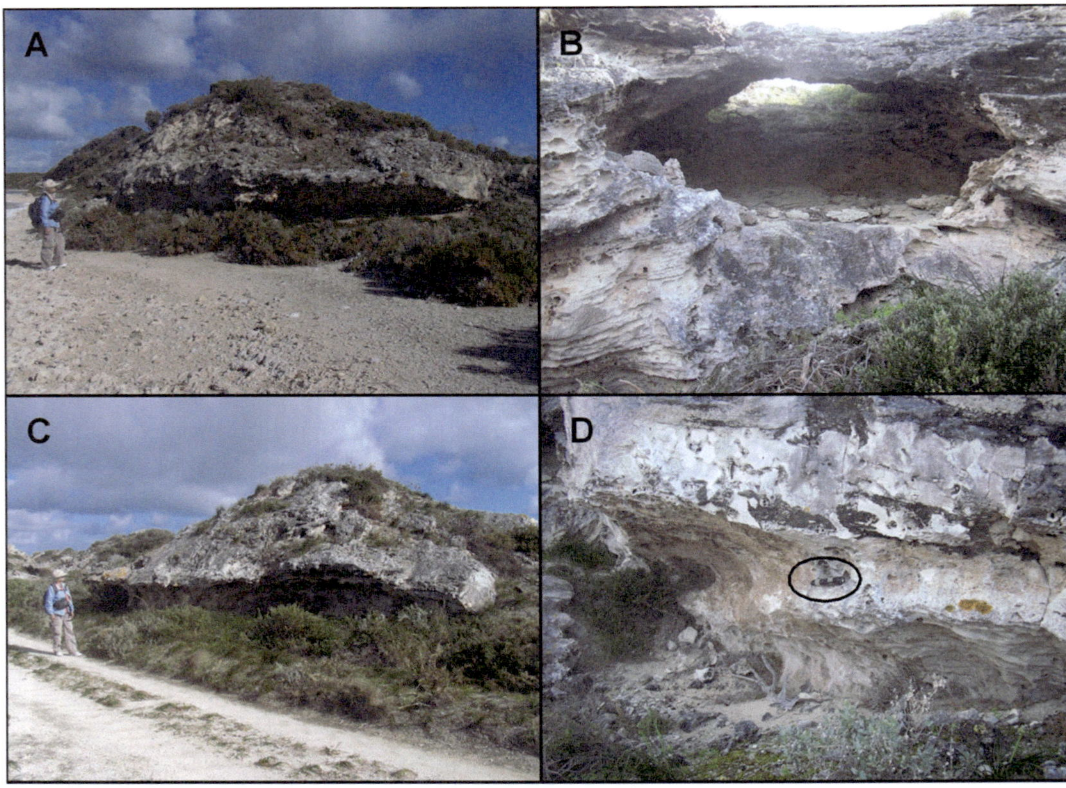

Fig. 18.18 Voids in eolianites along the shore of Herschel Lake. Described by Playford (1997) as wave or bioerosion notches from the mid-Holocene sea-level highstand, they appear to be flank margin caves. (**a**) Notch at the end of a dune; the notch does not have lateral continuity, but contains interior phreatic features. (**b**) Short tunnel with phreatic wall sculpturing. (**c**) Similar to (**a**), a discontinuous notch with phreatic wall sculpturing. (**d**) Close up of the back wall of one of the many notches along the shore of Herschel Lake. The notch contains curvilinear phreatic sculpture and a thin layer of red flowstone. Flashlight 15 cm long for scale (*in oval*)

1999; James and Clark 2002). Although isolated Paleozoic outcrops of carbonate rock exist on Kangaroo Island, the dominant carbonate units are Cenozoic (James and Clark 2002). Especially prevalent, primarily along the southern and western coasts, are eolian calcarenites of Late Pliocene through Holocene age (Ludbrook 1983; Short and Fotheringham 1986). The Cambrian Kanmantoo Group metasediments form the basement of the island (James and Clark 2002), and commonly the eolianites rest directly upon them along the southern and western shore of Kangaroo Island. Cambro-Ordovician granites also outcrop at the southwest end of Kangaroo Island (James and Clark 2002) and eolianites occasionally overlie those exposures.

The eolian calcarenites are assigned by most authors to the Bridgewater Formation, the name given to eolian calcarenites across South Australia and Victoria (Drexel and Preiss 1995). These eolian calcarenites represent depositional episodes associated with glacioeustatic Quaternary sea-level fluctuations and consist of at least 16 separate events on the Australian mainland (Drexel and Preiss 1995). On Kangaroo Island, the eolian depositional events are thought by some authors to represent only four glacial events because those authors recognize only the traditional, continental-based four Pleistocene glaciations (Twidale and Bourne 2002), while other authors indicate that more than 16 sea-level highstands occurred in the Quaternary

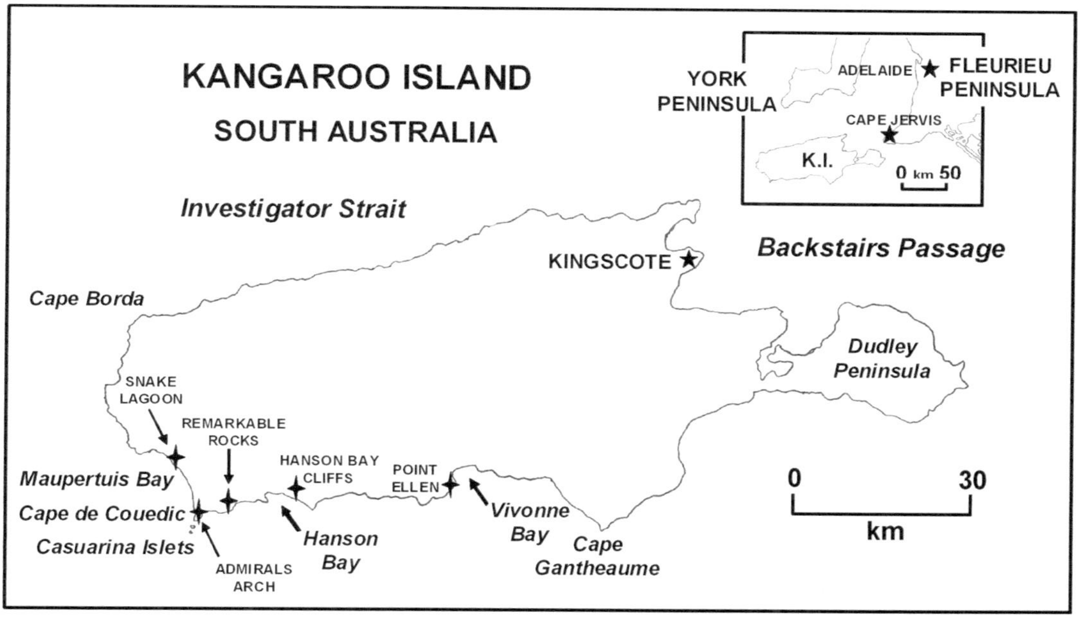

Fig. 18.19 Map of Kangaroo Island showing its position off the Australian coast, and locations discussed in the text

on Kangaroo Island (Short and Fotheringham 1986). Most authors (e.g. Drexel and Preiss 1995) consider the Bridgewater Formation to be Pleistocene in age. However, Ludbrook (1983) reports that eolian calcarenites of the Bridgewater Formation interfinger with Late Pliocene Point Ellen Formation subtidal facies at Point Reynolds, east of Point Ellen (Fig. 18.19). This observation raises the possibility that the initial deposition of the Bridgewater Formation eolian calcarenites predates the Pliocene-Pleistocene boundary on Kangaroo Island. Note that these uses of the Quaternary predate the recent shift in the Plio-Pleistocene boundary from 1.8 to 2.6 million years ago by the International Union of Geological Sciences in 2009.

Kangaroo Island is in an environment that has been uplifted during the Quaternary (Short and Fotheringham 1986; James and Clark 2002; Twidale and Bourne 2002); the debate has been over the amount of uplift. In high-relief coastal settings, such as on Kangaroo Island, cliff retreat under current sea-level conditions has removed many of the typical surface indicators of past, higher sea levels (Fig. 18.20), and uplift has displaced them from the elevation of their initial formation (Fig. 18.21). Numerous flank margin caves are found in the eolianite cliffs of the southern and western coasts of Kangaroo Island, and their location is a representation of a past fresh-water lens position, and hence, a past sea-level position. This discussion reports on how the use of caves, especially flank margin caves, can help refine our understanding of Quaternary processes on Kangaroo Island.

As with other Quaternary eolianite islands, such as Rottnest Island or the Bahamas, caves in coastal locations must be successfully differentiated from pseudokarst features such as tafoni and sea caves. This problem was addressed in Chap. 4, but Kangaroo Island will provide a new feature for consideration. The southern and western coasts of Kangaroo Island are areas of high wind and ocean energy (Short and Fotheringham 1986). The coastal outcrops of eolianite are cliffed and have retreated landward. An idea of how much retreat has occurred can be seen on the west coast of the island where the eolian calcarenites rest unconformably on the Kanmantoo Group basement rocks. Remnants of eolian calcarenite can be seen that are tens of meters away from the current eolian calcarenite scarp, indicating

Fig. 18.20 West coast of Kangaroo Island, north of Snake Lagoon and Rocky River. (**a**) Overall scene, with light-colored Bridgewater Formation eolianites overlying darker Kanmantoo Group rocks. The *white box* denotes the area presented in (**b**). (**b**) Isolated Bridgewater Formation eolian calcarenite outcrops (light-colored patches) on Kanmantoo Group rocks, demonstrating that a once-continuous eolian calcarenite unit has been stripped away by Holocene coastal erosion

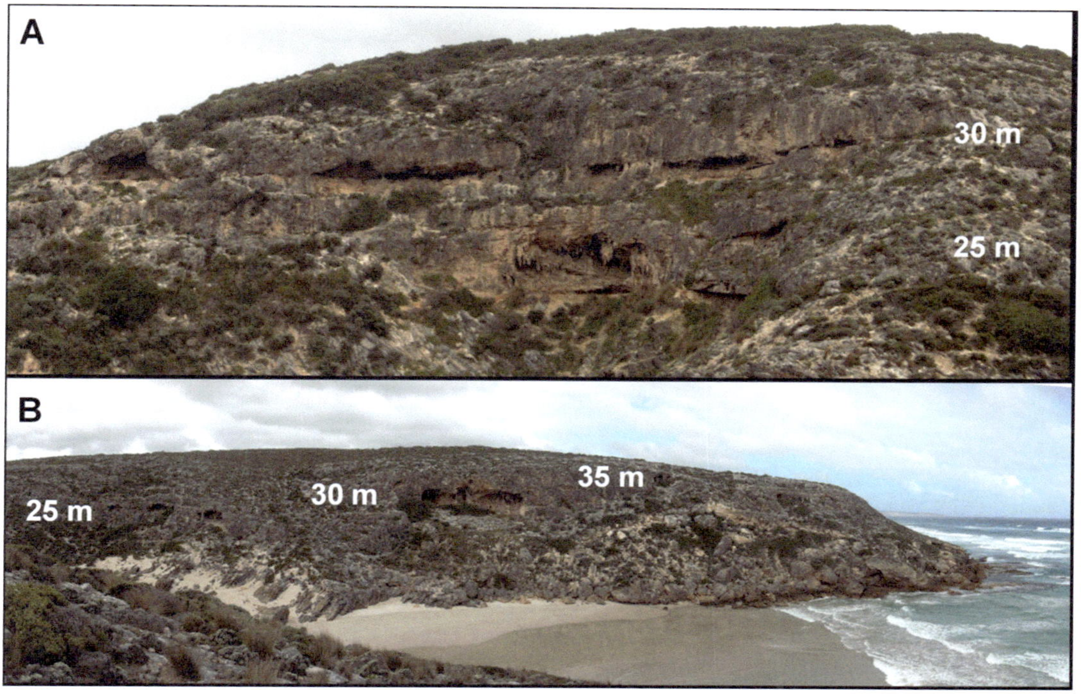

Fig. 18.21 Views of flank margin cave entrances in the lower Rocky River and Snake Lagoon area, developed in Bridgewater Formation eolianites. (**a**) Valley north wall, showing a well-developed band of caves at ~30-m-elevation, and other caves at ~25-m-elevation. (**b**) Valley south wall, showing three horizons of flank margin cave development, at ~25-, ~30- and ~35-m-elevation. As with the valley north wall, the best development is at ~30 m

significant Holocene erosional removal of eolian calcarenite material (Fig. 18.20). Such large-scale erosion not only strips off surface features such as intertidal deposits, but also removes sea caves, tafoni, and even flank margin caves, leaving no past sea-level record. To obtain a preserved eolian calcarenite section requires investigating embayments and stream valleys that are protected from direct marine assault, but which would have held a fresh-water lens in contact with sea water at a past, higher sea-level position. For these reasons, Rocky River, reaching the coast from Snake Lagoon to Maupertius Bay (Figs. 18.19, 18.21 and 18.22), was selected as the prime field investigation locality to search for flank margin caves and evidence of past sea-level highstands. Cape du Couedic was also investigated because offshore islands provided some wave protection (Fig. 18.23). A complete summary of the Kangaroo Island work can be found in Mylroie and Mylroie 2009.

18.3.1 Snake Lagoon

The Flinders Chase National Park trail along Rocky River from Snake Lagoon is a direct access route to the west coast (Fig. 18.19). At the location where the wooden footbridge crosses over Rocky River, Kanmantoo Group rocks are visible in the streambed and cave openings can be seen back to the east, on the north bank, high up on the eolianite valley wall. These caves are relict flank margin caves. Continuing west downstream, more cave openings are visible high on the north bank. As Rocky River and the trail approach the coast, numerous cave openings appear in the cliffs on both sides of the stream (Figs. 18.21 and 18.22).

The single most obvious aspect of the caves is their development as a series of chambers that commonly connect internally. This observation is known as *beads on a string* (Mylroie et al. 2001) and reflects the degree to which individual flank

Fig. 18.22 Flank margin caves at Snake Lagoon. (**a**) Cave entrance at ~25-m-elevation in the valley north wall, showing abundant flowstone, stalactite and stalagmite development. (**b**) Cave at ~25-m-elevation in the valley south wall, showing interconnecting chambers and stalactites. (**c**) Looking northwest from a flank margin cave in the valley south wall, showing the ocean and Kanmantoo Group basement rocks that underlie the Bridgewater Formation eolianites. (**d**) Series of eroded flank margin cave chambers on the valley south wall, showing smooth phreatic dissolutional surfaces and secondary vadose stalactites

margin cave chambers did or did not intersect as they grew by mixing dissolution in the distal margin of the fresh-water lens. The flank margin dissolutional origin of the caves is demonstrated by passage shape and configuration, abundant cave speleothems, and dissolutional wall morphologies (Fig. 18.22). The caves on the north side of the stream channel are found at two primary horizons, at approximately 25 and 30 m elevation (Fig. 18.21a). The hill on the south side is higher, and contains evidence of three cave horizons at approximately 25, 30, and 35 m (Fig. 18.21b). The two horizons on the north side appear to correlate with the two lower horizons (at 25 and 30 m) on the south side. At each cave horizon, the caves extend laterally over a distance of up to 100 m. The caves are not very deep, penetrating into the hillside generally less than 10 m. Because the caves have been breached by scarp retreat, the original voids had dimensions, perpendicular to the hillside, of over 10 m. The largest and most continuous band of caves, on each side of the valley, is the one at 30 m, indicating perhaps a longer sea-level stillstand at that horizon than occurred at the 25 or 35 m horizons.

The Snake Lagoon caves fit all the direct observational criteria that identify them as flank margin caves. As such, the caves represent past sea-level positions. All the caves described are above any past Quaternary glacioeustatic sea-

Fig. 18.23 Overview of Cape de Couedic, the Casaurina Islets, and Admirals Arch, Kangaroo Island. The two islands in the distance are the Casuarina Islets, also known as The Brothers. The east opening of Admirals Arch is labeled in the foreground. The *black vertical arrow* in the background points to the cave shown in Fig. 18.24a. Light colored Bridgewater Formation eolianites overlie dark Kanmantoo Group basement rocks

level highstand. Therefore, uplift of Kangaroo Island is required to have occurred to place the caves at their current position with respect to modern sea level. Records of sea-level highstands on Kangaroo Island above 10 m are regarded as equivocal (Twidale and Bourne 2002). However, Bauer (1961) indicated that a marine erosion terrace at 100–110 ft. (30.5–33.5 m) was the most significant of the five terraces he recognized at 20–25 ft. (6–7.6 m) and higher on Kangaroo Island. The cave observations presented here demonstrate a record of sea level well above 6 m, and at least three closely-spaced highstands are recorded. The duration of the highstands can be in part determined by how large the caves are. Cliff retreat since their formation has obviously decreased their size, but nonetheless, they are smaller than many flank margin caves in the Bahamas, which had 9,000 years to form (Chap. 7). The development of the largest and most continuous caves at ~30 m elevation agrees well with Bauer's (1961) best-developed terrace at that elevation. During glacioeustasy, the Bahamian record demonstrates that only when sea level is turning around from a lowstand or a highstand is it stable long enough to create large flank margin caves (Mylroie and Mylroie 2007). If uplift is also involved, then the time of sea-level stability will be even less. The New Zealand flank margin cave record, in a tectonically active environment, provides time limits on lens stability (Mylroie et al. 2008b, Chap. 17). The flank margin caves at Snake Lagoon indicate that uplift has definitely occurred, but it has been slow enough, and episodic enough, such that glacioeustatic stillstands can still leave a flank margin cave signature. Given that the last interglacial (oxygen isotope substage 5e) was 6 m higher than at present, such a eustatic sea-level elevation value should be considered when investigating the elevations of the caves found at Snake Lagoon today. In other words, the cave horizon at ~30-m-elevation could indicate an uplift of only 24 m, but that uplift would have had to occur in the last 120 ka, and other evidence at Point Ellen (Mylroie and Mylroie 2009), suggests stability for the last 120 ka. The cluster of caves between ~25 and ~35 m may represent two uplift events on a single glacioeustatic sea-level highstand, or no uplift episodes while three different glacioeustatic sea-level highstands occurred, or a combination of the two.

18.3.2 Cape du Couedic

This location (Fig. 18.19) is famous for Admirals Arch, a large void penetrating a rocky point composed of Bridgewater Formation eolian calcarenites overlying Kanmantoo Group basement rocks (Fig. 18.23). The steep eolianite cliffs to the east contain a number of breached flank margin caves (Fig. 18.24). Visual examination demonstrates that these features have phreatic dissolutional wall morphologies, calcite speleoethems, and align horizontally with other caves on the same cliff face.

Admirals Arch has clear views of the cave interior (Fig. 18.25). The public display on the tour path presents Admirals Arch as a product solely of wave erosion. However, visual examination of the north wall of the arch reveals that it has a series of phreatic dissolution pockets (Fig. 18.25c) and that the original floor of the arch was horizontal and developed in limestone above the dipping contact with the underlying Kanmantoo Group basement (Fig. 18.25a, d). The cave has abundant calcite speleothems.

Admirals Arch appears to be a breached flank margin cave, which has been modified by wave action on the current (and perhaps last interglacial) sea-level highstand(s). The phreatic dissolution surfaces and abundant calcite speleothems indicate a cave that formed by phreatic, mixed-water dissolution, then was drained such that vadose speleothems could develop in a sealed cave chamber. The remnant flat limestone floor on the north side of the cave is another indication of flank margin cave development. That floor has been mostly stripped away by modern wave action ramping up the sloping Kanmantoo Group basement rocks.

Observation from a distance of the nearer of the two Casuarina Islets (The Brothers) revealed two caves in the eolian calcarenites on the cliff facing Admirals Arch (Figs. 18.23 and 18.24a). Remnant speleothems can be seen, but little of interior configuration is observable. The caves are quite close to the Kanmantoo Group basement contact. They appear to have phreatic morphologies, and as the islands are too small to support conduit flow, the most likely interpretation is that they are flank margin caves.

18.3.3 Remarkable Rocks

The tafoni developed in Cambro-Ordovician granite at Remarkable Rocks (Fig. 18.19) provide a cautionary tale. The interior of several of the larger tafoni have wall sculpture that is especially cuspate and dimpled and bear a striking resemblance to dissolutional wall morphology as found in flank margin caves (Figs. 18.26c versus 18.26d). They also have been misidentified. Twidale and Bourne (2002, their Figure 14c) presented a photograph of the interior of a tafoni at Remarkable Rocks, calling the wall surface mammilated. This error is a result of printing the picture upside down, such that lighting and shadows invert the apparent relief in the picture. The features are clearly cuspate and not mammilary, as seen in Fig. 18.26b, c. Because such cuspate features are part of the visual inventory used to define a cave in limestone as phreatic in origin (Fig. 18.26d), the Remarkable Rocks example indicates that multiple lines of evidence should be used to identify a cave's

Fig. 18.24 Eroded flank margin caves in the Cape de Couedic area. (**a**) Cave entrance at the eolianite-basement contact, showing stalactites and stalagmites, on the landward of the two Casaurina Islets (Fig. 18.23). The prominent stalagmite in the cave entrance is ~1-m-high. (**b**) Back wall of a flank margin cave intersected by cliff retreat, in a small point just east of Admirals Arch. People are standing at top for scale. Note the curvilinear shape of the cave wall, and the stalactites and flowstone. (**c**) Large breached flank margin cave at the major headland between Admirals Arch and Remarkable Rocks. Note the Bridgewater Formation contact with the Kanmantoo Group basement rocks near sea level, and the many stalactites and stalagmites present. (**d**) Small flank margin cave, and associated phreatic pockets in cliff wall, between Admirals Arch and the breached cave seen in (**b**). The cave entrance is ~3-m-high

origin. A review of tafoni, their mechanisms of formation, and the techniques utilized to differentiate them from dissolutional caves can be found in Owen (2007) and in Chap. 8.

18.3.4 Hanson Bay

The east side of Hanson Bay (Fig. 18.19) begins as a stretch of beach and gradually trending southeastward becomes a high eolian ridge with sea cliffs down to the ocean below. High up on these cliffs are a series of planated notches (Fig. 18.27a) that could easily represent small wave-eroded platforms. The platforms have a rubble deposit of rounded clasts in a grey matrix (Fig. 18.27b). These are clearly not a paleosol layer, in which the clasts would be more angular, and the matrix would carry the red color of a terra rossa paleosol. Such deposits, when found in the Bahamas, indicate a back beach or rock platform rubble facies (Florea et al. 2001). It is therefore likely that these notches indicate a sea-level highstand approximately 30–35 m above modern sea level, which would require tectonic uplift. Such a sea-level interpretation supports

Fig. 18.25 Images from Admirals Arch. (**a**) Looking northwest through the Arch. Note the steep dip of the Kanmantoo Group basement rocks, and stalactites in upper foreground. Reclining seal, 1.5 m long, in the center of the image for scale (*white arrow*; same seal as in (**c**) and (**d**). (**b**) Looking west through the Arch, showing numerous stalactites silhouetted by the western entrance. The twisted and gnarly appearance of the stalactites is an outcome of modification by both evaporation and algal growth. (**c**) Phreatic pockets along the north wall of the Arch, formed in the Bridgewater Formation eolianites along a horizontal datum, just above the sloping Kanmantoo Group basement rocks. Seal, ~1.5 m long, in lower left foreground for scale (*white arrow*). (**d**) Surviving section of the original horizontal floor of the Arch. The phreatic pockets of Fig. 18.25c are in shadow ahead and to the right in the image. Seal lying in background for scale (*white arrow*)

the observations of flank margin caves at similar elevations at Snake Lagoon and Maupertius Bay to the west.

18.3.5 Summary

The observations made on Kangaroo Island are simple visual descriptions at the macroscopic scale, without detailed site survey or rock-sample analysis. On the other hand, the simple observations allow new interpretations to be offered that may help illuminate geologic processes on the island. None of the previous workers who interpreted the Cenozoic geology of Kangaroo Island utilized the potential data stored in caves on the island.

It is clear that flank margin caves are present on Kangaroo Island. The position of the flank margin caves at Snake Lagoon and Maupertius Bay reveal sea-level highstands of at least three elevations: ~25, ~30, and ~35 m, substantiating early claims by Bauer (1961) that Twidale and Bourne (2002) later called into question. The absence of flank margin caves from many high-energy coasts underlain by Kanmantoo Group

Fig. 18.26 Remarkable Rocks. (**a**) The Remarkable Rocks, where tafoni have developed in granitic rocks. (**b**) Classic cavernous weathering to produce a complicated tafoni. (**c**) Inside one of the larger tafoni, with pockets or cusps eroded into the ceiling. (**d**) Chamber in Salt Pond Cave, Long Island, Bahamas, in eolian calcarenites, showing the pockets and cusps considered as one of the diagnostic indicators of cave formation by phreatic dissolution. Two people in background, *left* and *right*, for scale. Compare with (**c**)

rocks is consistent with the inherent vulnerability of flank margin caves to destruction by powerful wave-generated slope-retreat processes. In such locales, flank margin caves are preserved in embayments and surface water course incisions, as at Snake Lagoon, or by offshore barriers, as at Cape du Couedic. Observations from Hanson Bay show marine erosion features consistent with development during one or more sea-level highstands at approximately 30–35 m, concurring with observations at Snake Lagoon and Maupertius Bay, which would require tectonic uplift.

At Cape du Couedic, an arch in Bridgewater Formation eolianites resting on Kanmantoo Group rocks is presented to the public as being the result of wave erosion with wave energy being focused on the point by the presence of the islands offshore. The evidence from the eolianite portion of the arch suggests that the original void formed as a flank margin cave and was subsequently breached by wave erosion. The offshore islands not only acted as a focusing mechanism for wave energy, but also provided a barrier function that has prevented the entire eolianite section at Admirals Arch from being removed by wave erosion.

Remarkable Rocks demonstrate how non-dissolutional erosive forces can produce surfaces in tafoni that mimic one of the classic indicators of flank margin cave development, and as such are a warning about using single lines of evidence to make important cave origin interpretations.

Fig. 18.27 Hanson Bay. (**a**) Long, linear and level notches cut into the eolianites of the Bridgewater Formation. *Vertical white bar* 1 m long for scale. (**b**) Rubble facies found on the floor of the notches shown in (**a**). The matrix is sandy and *white* or *gray*, not *red*, and the clasts are more rounded than normally seen in a paleosol, and are interpreted as a back-beach rubble facies. Pencil 15 cm long for scale (*black arrow*, same length as the pencil)

References

Bauer FH (1961) Chronic problems of terrace study, South Australia. Zeitschrift für Geomorphologie 3:57–72

Belperio AP, Flint RB (1999) Kangaroo Island biological survey: geomorphology and geology. In: Robinson AC, Armstrong DM (eds) A biological survey of Kangaroo Island, South Australia. Department of Environment, Heritage, and Aboriginal Affairs, Adelaide, pp 19–31

Birmingham AN, Carew JL, Mylroie JE (2008) The use of plant trace fossils to differentiate transgressive-phase from regressive-phase Quaternary eolianites, San Salvador Island, Bahamas. Geol Soc Am, Abstracts with Programs 40:230

Carew JL, Mylroie JE (1995) A stratigraphic and depositional model for the Bahama Islands. In: Curran HA, White B (eds) Terrestrial and shallow marine geology of the Bahamas and Bermuda, Geological Society of America special paper 300. pp 5–31

Carew JL, Mylroie JE (1997) Geology of the Bahamas. In: Vacher HL, Quinn TM (eds) Geology and hydrogeology of carbonate islands. Elsevier Science Publishers, Amsterdam, pp 91–139

Drexel JF, Preiss WV (1995) The geology of South Australia, vol 2, The Phanerozoic: Geological Survey of South Australia, Bulletin 54. p 347

Florea L, Mylroie JE, Carew JL (2001) Karst genetic model for the French Bay Breccia deposits, San Salvador, Bahamas. Theor Appl Karstology 13–14:57–65

Ford DC, Williams PW (2007) Karst hydrology and geomorphology. John Wiley & Sons, West Sussex

Glumac G, Curran HA, Motti SA, Weigner MM, Pruss SB (2011) Polygonal sandcracks: unique sedimentary desiccation structures in Bahamian ooid grainstone. Geology 39:615–618

Hearty PJ (2003) Stratigraphy and timing of eolianite deposition on Rottnest Island, Western Australia. Quatern Res 60:211–222

Infante LR, Mylroie JE, Kambesis PN, Leist JW (2011) Carbonate dissolution in a dynamic fresh-water lens: lens margin migration and the production of banana holes as syngenetic caves, San Salvador Island, Bahamas. Geol Soc Am, Abstracts with Programs 43:341

James PR, Clark IF (2002) Geology. In: Davis M, Twidale CR, Tyler MJ (eds) Natural history of Kangaroo Island. Royal Society of South Australia, Richmond, pp 1–22

Jennings JN (1972) Karst. MIT Press, Cambridge, MA

Jennings JN (1984) Karst geomorphology. Basil Blackwell, Oxford

Ludbrook NH (1983) Molluscan faunas of the early Pleistocene point Ellen formation and Burnham limestone, South Australia. T Roy Soc South Aust 107:37–49

Miklavič B, Mylroie JE, Jenson JW, Randall RH, Banner JL, Partin JW (2012) Evidence of the sea-level change since MSI 5e on Guam, tropical west Pacific: Studia Universitatis Babeşs-Bolyai. Geology Special issue 2012:30–32

Murray-Wallace CV, Kimber RWL (1989) Quaternary marine aminostratigraphy: Perth Basin, Western Australia. Aust J Earth Sci 36:553–568

Mylroie JE, Carew JL (1991) Erosional notches in Bahamian carbonates: bioerosion or groundwater dissolution? In: Bain RJ (ed) Proceedings of the 5th symposium on the geology of the Bahamas. Bahamian Field Station, Port Charlotte, pp 185–191

Mylroie JE, Mylroie JR (2007) Development of the carbonate island karst model. J Cave Karst Stud 69:59–75

Mylroie JE, Mylroie JR (2009) Caves as geologic indicators, Kangaroo Island, Australia. J Cave Karst Stud 71:32–47

Mylroie JE, Mylroie JR (2010) A rapid reconnaissance of a Quaternary eolianite island of Australia: Rottnest Island, with comparisons to the Bahamas. In: Martin JB, Siewers FD (eds) Proceedings of the 14th symposium on the geology of the Bahamas and other carbonate regions. Gerace Research Centre, San Salvador, pp 163–185

Mylroie JE, Jenson JW, Taborosi D, Jocson JMU, Vann DT, Wexel C (2001) Karst features of Guam in terms of a general model of carbonate island karst. J Cave Karst Stud 63:9–22

Mylroie JE, Carew JL, Curran HA, Freile D, Sealey NE, Voegeli VJ (2006) Geology of Cat Island, Bahamas: a field trip guide. Gerace Research Centre, San Salvador Island, p 43

Mylroie JE, Mylroie JR, Owen AM, Waterstrat MJ (2008a) Cave and karst inventory of the Primeval Forest, New Providence Island, Bahamas: unexpected discoveries. In: Freile D, Park L (eds) Proceedings of the 13th symposium on the geology of the Bahamas and other carbonate regions. Gerace Research Centre, San Salvador, pp 107–118

Mylroie JE, Mylroie JR, Nelson CN (2008b) Flank margin cave development in telogenetic limestones of New Zealand. Acta Carsologica 37:15–40

Owen AM (2007) Tafoni caves in Quaternary carbonate eolianites: examples from the Bahamas. Masters thesis, Mississippi State University. http://sun.library.msstate.edu/ETD-rdb/theses/available/etd-05142007-143443/

Panuska BC, Boardman MR, Carew JL, Mylroie JE, Sealey NE, Voegeli VJ (2002) Eleuthera Island field trip guide, 11th symposium on the geology of the Bahamas and other carbonate regions. Gerace Research Centre, San Salvador Island

Playford PE (1997) Geology and hydrogeology of Rottnest Island, Western Australia. In: Vacher HL, Quinn TM (eds) Geology and hydrogeology of carbonate islands. Elsevier Science Publishers, Amsterdam, pp 783–810

Short AD, Fotheringham DG (1986) Coastal morphodynamics and Holocene evolution of the Kangaroo Island coast, South Australia, Tech report no. 86/1. University of Sydney, Sydney, Coastal Studies Unit, 92

Szabo BJ (1979) Uranium-series age of coral reef growth on Rottnest Island, Western Australia. Mar Geol 29:M11–M15

Twidale CR, Bourne JA (2002) The land surface. In: Davis M, Twidale CR, Tyler MJ (eds) Natural history of Kangaroo Island. Royal Society of South Australia, Richmond, pp 23–35

Waterstrat WJ, Mylroie JE, Owen AM, Mylroie JR (2010) Coastal caves in Bahamian eolian calcarenites: differentiating between sea caves and flank margin caves using quantitative morphology. J Cave Karst Stud 72:61–74

White SQ (2000) Syngenetic karst in coastal dune limestone: a review. In: Klimchouk AB, Ford DC, Palmer AN, Dreybrotdt W (eds) Speleogenesis: evolution of karst aquifers. National Speleological Society, Huntsville, pp 234–237

White SQ, Mylroie JE, Mylroie JR (2007) Sea caves or flank margin caves: do we understand cave formation at Tantanoola and Taragal? Australian speleological meeting

Glossary

The following collection of terms is by no means complete but provides representative terminology associated with coastal karst geomorphology, and related karst features and processes, derived from references cited within this volume. Readers are also directed to the AGI (American Geosciences Institute) Glossary of Geology for a broader range of geological definitions and terms.

aeolites (eolites) mineral formation (e.g. flowstone) structurally elongated by air flow during deposition.

allogenic recharge water collecting on non-soluble rock or sediments and entering adjacent karst structures (see also *allochthonous*).

amblipygyd pseudo-scorpion species typically found in Caribbean caves and other tropical climates.

anchialine refers to a structure flooded by sea water or brackish water.

aquifer geologic structure in which water collects and flows.

atoll a raised coral reef or collection of coral islands enclosing a lagoon.

allochthonous formed or produced elsewhere than its present position (e.g. sediments).

angle of repose The maximum slope (vertical profile angle) at which soils or other materials within an embankment remain structurally stable.

authigenic or autogenic recharge karst waters derived from infiltration of water through the overlying soluble rock surface.

autochthonous formed or produced *in situ* (see also *authigenic* and *autogenic*).

backreef the inland perimeter of a tidal lagoon/reef structure.

banana hole Originally a Bahamian term for a shallow cave collapse that results in a sinkhole in which specialty crops, such as bananas, are grown.

bank margin failure landmass collapse associated with the steep coastal edge of a platform.

barrier island an elongate coastal island of sediment situated above high tide and parallel to an adjacent coast.

basalt extrusive black igneous rock formed by rapid cooling and solidification of lava flow (see *lava tube*).

beachrock sedimentary rock composed of carbonate sand and particulates deposited and cemented with $CaCO_3$ in an active tidal setting.

bell hole a vertical cylindrical cavity formed in a cave ceiling that is deeper than it is wide, usually less than a meter wide but sometimes several meters in height.

bioerosion removal of material by living organisms (e.g. littoral bioerosion by marine life).

bioherm reef-like mass of sedimentary rock formed by marine organisms, such as corals, and surrounded by rock of a different character.

blue hole A subsurface void formed in carbonate banks or islands in inland or oceanic settings. by a variety of mechanisms containing tidally influenced freshwater, seawater or a mixture.

boxwork a distinctive perpendicular array of blade-like mineral deposits exposed by dissolution or weathering of the surrounding rock.

breccia sedimentary rock composed of recemented angular-shaped rock fragments.

M.J. Lace and J.E. Mylroie (eds.), *Coastal Karst Landforms*, Coastal Research Library 5,
DOI 10.1007/978-94-007-5016-6, © Springer Science+Business Media Dordrecht 2013

calcarenite a consolidated carbonate sand; a limestone consisting of more than 50 % sand-sized grains.

cenote from the Mayan word for "covered well"; a vertical collapse of a cave roof revealing a chamber partially or completely filled with water.

chenier coastal beach or dune ridge parallel to a prograding shoreline.

coastal karst landscapes formed by the dissolution of soluble rock in a coastal setting.

conduit cave a cave that delivers (or once delivered) water by turbulent flow from sinking meteoric water to a spring.

cone karst conical-shaped hills formed by karst processes, preferentially in tropical climates.

constructional cave cave formed as a result of accretion of material as opposed to dissolution of rock (e.g. a tufa cave or coralloid cave).

continental karst landscapes formed by the dissolution of soluble rock in an inland or continental setting (as opposed to *coastal karst* within shoreline settings).

continental island an island formed of rock on a continental shelf (Ireland) or as a plate tectonic fragment of continental material (Madagascar).

diagenesis refers to change in chemical composition and/or structure of sedimentary rock over time, commonly associated with burial conditions.

doline European term for surface depression often resulting from dissolution of the surface or collapse of an underlying void. American version is *sinkhole*.

dolomite/dolostone carbonate rock containing magnesium carbonate as $MgCa[CO_3]_2$, in contrast to limestone which contains primarily calcium carbonate ($CaCO_3$).

dripline change in ceiling associated with a cave entrance, the limit of vertical rainfall into a cave.

El Nino Southern Oscillation (ENSO) the three to eight year cycle of oscillation of sea-surface temperature, wind, and atmospheric pressure in the tropical Pacific Ocean (thought to have a global impact on climate).

eogenetic sedimentary rock still on or near the earth's surface that has not undergone burial diagenesis.

eogenetic karren The term now used to explain the delicate and pervasive etching found on coastal carbonates in the spray zone, as a result of algal growth, invertebrate grazing, inorganic dissolution, and importantly, the presence of primary depositional features of the host rock (see *phytokarst*).

eogenetic karst karst in eogenetic rock that has not experienced processes associated with deep burial and is still within the range of meteoric processes.

eolianite sedimentary rock consisting of sand-sized material deposited by wind (e.g. dune ridges).

eolian calcarenite an eolian deposit made of carbonate sand grains.

epigenic cave a dissolutional cave formed as part of the surface hydrologic network.

epikarst the uppermost surface of a karst landscape composed of cavities and fissures which collect and channel water underground.

evaporite sedimentary rock composed of minerals precipitated from a solution as a result of evaporation (e.g. gypsum or halite).

fissure cave cave formed within a vertical fracture in bedrock by non-dissolutional means.

flank margin cave cavity formed within an enclosing landmass by dissolution processes associated with the margin of a fresh-water lens.

flowstone subaerial precipitated mass of mineral deposits, usually calcite, having the appearance of flowing down a wall or slope.

forereef seaward flank of a tidal lagoon/reef structure.

fresh-water lens layer of fresh water of varying thickness overlying a body of water with a greater density produced by dissolved salts (e.g. sea water).

fringing reef a coral reef attached directly to an insular or continental shore, often forming a shallow coastal lagoon.

glacioeustatic feature or process associated with past oscillating sea levels as a function of glaciation-deglaciation cycles.

gours (*see rimstone dams*).

guano sediment consisting of partially or completely decomposed and/or chemically altered animal fecal material (e.g. derived from bat or bird colonies).

gypsum mineral deposits composed primarily of hydrated calcium sulfate ($CaSO_4 \cdot 2H_2O$); unhydrated $CaSO_4$ is *anhydrite*.

halite mineral deposit of sodium chloride (NaCl).

halocline the contact zone between waters of different chemistry, such as the fresh-water lens and underlying saltwater; implies a sharp contact.

helictite a contorted speleothem which twists in any direction, seemingly in defiance of gravity, formed as mineral rich water is forced through a very tiny central tube and subjected to. hydrostatic pressure and evaporation.

hybrid cave a cave overprinted by a second speleogenetic event, such as a flank margin cave breached by the sea and modified by wave erosion.

hypogenic cave a dissolutional cave formed decoupled from direct surface hydrology.

intertidal the marine zone encompassed by low and high tides.

insular shelf marine zone surrounding an island or continent and extending from shoreline, through depths of up to 100 fathoms (183 m), to a transition area to greater depths.

karren dissolutional sculpting of exposed carbonate rock, generally on the centimeter to meter scale.

karst a landscape that is typically formed in soluble rock types, such as limestone or gypsum, characterized by internal drainage and a lack of surface water courses. Karren, sinkholes, sinking streams, springs and caves are typical.

laminar flow non-turbulent water movement (e.g. flow within a fresh-water lens).

lava tube conduit-like cave structure formed by subterranean lava flow.

limestone a sedimentary rock containing at least 50 % calcium carbonate.

littoral feature or process associated with a shoreline (e.g. sea caves).

littoral cell a coastal landscape separated from adjacent coastal landscapes by its sedimentology and geomorphology (see *barrier island*).

littoral sink vertically collapsed void within rock in a littoral setting (sometimes called a "punchbowl").

marine terrace (i.e. wave-cut bench or shore terrace) a terraced structure made by wave action or shore currents, exposed as a land surface by sea level change and/or uplift.

meseta provisional description of the plateau of a rock escarpment.

mesogenetic karst dissolution in the deep subsurface in diagenetically altered rocks to produce hypogenic caves.

mixing zone dissolution dissolution produced by mixing fresh and salt waters, which even if individually saturated with respect to $CaCO_3$, become undersaturated upon mixing.

mixing zone fracture cave a type of cave formed by mechanical processes that acts as preferential groundwater flow paths along fractures within the mixing zone (e.g. caves in the Marianas).

neritic refers to sublittoral zone above 200 m depth.

oceanic island an island not formed on a continental shelf (see *continental island*).

paleoclimate past environmental/climate settings.

paleokarst a karst feature formed by past dissolutional processes and preserved by burial in rock or sediment or by interruption of karstification.

paleosol fossil layer of deposited soil.

pendant an elongated, curved rock protrusion extending from a cave ceiling or wall, considered a *speleogen*.

petroglyph image incised into a rock surface.

phreatic associated with water beneath the water table.

phreatic lift tube a vertical or sub-vertical passage that carried water upwards under pressure in the phreatic zone.

phytokarst an older term used to explain exposed carbonate rock exhibiting a weathered appearance as due to bioerosion caused by endolithic boring algae. Replaced by the term *eogenetic karren*, which takes into account inorganic dissolution and the immature nature of the host rock.

karren channels, pockets, pinnacles or grooves formed in bedrock surfaces as a result of dissolution.

physiographic related to the evolution and genesis of landforms.

pictograph image applied to a rock surface by various techniques.

polje a karst feature defined as a large steep-sloped, sediment-floored closed depression that may be associated with disappearing streams.

pseudokarst karst-like landform or feature formed by processes other than traditional karst dissolution (e.g. sea caves or tafoni).

ramiform cuspate, branching cave morphology of a maze type typically associated with interconnected chambers and passages formed by mixing-zone dissolution.

recharge cave fluvial structure primarily associated with waters entering a karst drainage basin and sinking into a cave.

reentrant an angular indentation of a landform; on a vertical slope or cliff forms a horizontal or subhorizontal notch, in the horizontal plan view forms invagination of the coastline trend.

regressive associated with retreating sea levels.

rimout a cuspate-shaped pocket in a cave chamber formed by dissolution in slow moving phreatic waters (such as found in mixing zones).

rimstone dam (or gours) a vertical accumulation of mineral deposit (e.g. calcite) built up along the edge of a pool due to precipitation from a thin film of water and forming a dam wall allowing catchment of water. Flow over any low point has more turbulence, degasses, and has increased calcite precipitation, maintaining the level dam edge, which can grow several meters high.

saltwater intrusion encroachment of saltwater into a freshwater coastal aquifer (e.g. by overpumping fresh water from coastal wells).

sea arch A natural bridge or opening through a coastal headland formed by wave action; relict flank margin caves can assist this process in carbonate rocks.

sea stack column of coastal rock separated from adjacent shoreline cliff exposure by wave erosion.

seamount a peaked elevation rising 1,000 m or more above the ocean floor.

seiche free or standing wave oscillation of surface water of variable duration in an enclosed or partially enclosed environment, commonly caused by persistent wind action.

shore platform a flat surface at the base of a shoreline formed by littoral erosion processes.

skylight a hole in the roof of a cave passage that reaches the ground surface. It may be an inlet shaft, a section of collapse, or a breach due to surface lowering resulting from natural karst development processes.

soda straw a hollow tube composed of $CaCO_3$ deposited by water flowing through its interior and degassing CO_2 to force calcite precipitation in the cave environment.

solutional pan shallow, flat-bottomed depression in shoreline rock.

speleogen a structural cave feature composed of bedrock (e.g. a bedrock column) resulting from dissolutional processes.

speleothem cave formation composed of deposited mineral or sediments (note: a speleogen is a primary remnant, a speleothem is a secondary deposit).

spongework complex array of irregular interconnecting cavities formed within sedimentary rock.

stegamite a distinctive broad, fin-shaped projection rising from a cave floor (mechanism unknown).

stratigraphy refers to the ordered sequence of geologic formations deposited as a function of time.

subaerial located above the water surface.

subaqueous located below the water surface.

syndepositional refers to a structure or feature formed contemporaneously with sedimentation of its associated landform.

tafoni a pseudokarst cavity formed in a variety of lithologies and coastal settings by a variety of non-karst weathering processes.

talus displaced segments of rock associated with cliffs, some forming cavities beneath them (e.g. talus caves).

tectonic associated with mechanical stresses, such as tectonic plate movement.

telogenetic karst karst produced on diagenetically mature rocks after exposure following deep burial.

tidal pumping the transfer of the tidal force into an aquifer as a result of a kinematic wave, as opposed to a actual transfer of water.

transgressive associated with advancing or increasing sea levels.

troglobite an organism that lives its lifespan completely underground (obligatory cave organism).

troglophile an organism that completes its life cycle in a cave but also exists on the surface (cave lover).

trogloxene an organism that lives its lifespan on the surface but periodically enters a cave environment (cave guest).

tufa a variety of rapidly deposited spongy or porous travertine (see also *bioherm*).

upconing vertical intrusion of saltwater into a freshwater lens.

vadose associated with water movement below the land surface but above the water table.

vug a small cavity in sedimentary rock that may be lined with crystals.

vugular porosity pore spaces in soluble rock composed of small cavities or *vugs*.

unconformity a surface that represents a break in the depositional geologic record as a function of time, deposition and/or structure.

Index

M.J. Lace and J.E. Mylroie (eds.), *Coastal Karst Landforms*, Coastal Research Library 5,
DOI 10.1007/978-94-007-5016-6, © Springer Science+Business Media Dordrecht 2013

MIX
Papier aus verantwortungsvollen Quellen
Paper from responsible sources
FSC® C105338

If you have any concerns about our products,
you can contact us on
ProductSafety@springernature.com

In case Publisher is established outside the EU,
the EU authorized representative is:
Springer Nature Customer Service Center GmbH
Europaplatz 3, 69115 Heidelberg, Germany

Printed by Libri Plureos GmbH
in Hamburg, Germany